The Complementary Nature of Reality

The Complementary Nature of Reality

Peter Barab

Open Way Press

Portland, Oregon

Printed and bound in the United States of America

ISBN 978-0-9822637-0-9

Cover image courtesy of Tom Beshara, portfolio at www.turnyourhead.com

Thanks to Bryan Hull for all his wisdom and support.

Library of Congress Control Number: 2008943405

Library of Congress Subject Headings:

Science -- Philosophy.
Philosophy of nature.
Complementarity (Physics).
East and West.
Reality.
Knowledge, Theory of.
Cognition.
Dualism.

12 11 10 C 9 8 7 6 5 4 3 2

Open Way Press

www.openwaypress.com

Preface

It would be foolish to pretend that the established theories of science, and in particular physics, do not represent an enormously powerful framework for representing nature and reality. Yet it is also important to bear in mind that we are still dealing with *theories* about nature and reality, rather than with that underlying *essence* itself. For, incredibly useful as they are, these theories of physics nonetheless amount to a particular set of conceptual maps for depicting the world in which we live.

They may be the best maps that mankind has been able to come up with so far, but they are not good enough. For one thing, we are talking about map*s* in the plural: there is still no integrated map in the singular which can bring it all together — no unified model of nature — and the characters of the existing maps actually conflict with one another in crucial ways. And second, these maps defy comprehension, in terms of having any intelligible relationship with our usual meaning system and common experience of the world. This book argues that neither of these limitations is a necessary condition, and that a better map (singular) is within reach.

The map motif is a relevant one, in that it also describes how we got to where we are — not in the sense of "map" as noun, but rather in the sense of "map" as verb: physics has mapped out this or that realm of nature, as it has been encountered in our historical development, by means of the most convenient constructs and assumptions that were available at the time ... and by now in incredibly intricate detail. The main problem with this approach is that we end up with different theories that have been built upon a variety of haphazard and incompatible conceptual foundations which have no consistent theme, and which are characterized by arbitrary, one-sided representations of reality. The main solution is to be found in a two-sided premise which was put forth as a principle in physics early in the last century, by the great Quantum Theory pioneer Niels Bohr: it is called complementarity.

It is worth noting how the present project dovetails with Bohr's in developmental terms. This work began many years ago while I was a graduate student at Stanford University (bypassing a bachelor's degree) — both in the area of East Asian studies, when the ancient Chinese notion that we call complementarity became deeply embedded in my thinking; and also in psychology, when I was engaged in rigorous scientific research and analysis with one of the foremost theorists in the field. It was during my study of certain issues in cognitive psychology that it became obvious to me how naturally the concept of complementarity seemed to apply there, even though the prevailing models did not reflect this. I soon learned that, by contrast, this thesis played an important role in modern physics. I decided to explore that field further, with the hope that a better understanding of how the idea was implemented by the highly esteemed "mother science" would provide a powerful framework which could then be transported back to psychology with fruitful results.

That never happened — I never found what I was looking for. Of course, there are pieces of the puzzle in physics, Bohr's principle being the primary one; but it has never really borne full fruit. Indeed, far from being enthusiastically embraced and broadly integrated into its theories, the concept has been treated at arm's length by physics — essentially as something that was hard to avoid, but difficult to accept. In fact, it became apparent that complementarity was kept in a permanent state of virtual quarantine in the field — neglected whenever possible (and sometimes even rejected outright), but usually invoked in only a narrow and piecemeal fashion ... strictly for describing certain subtle patterns of microphenomena, most notably the wave-particle duality.

My disillusionment was substantial — not with the concept, but rather with the great theories of physics that we have all grown up to revere. For the conclusion became inescapable to me that physics itself had not correctly interpreted and fully applied the idea. That conviction was strengthened as I delved deeper into one area after another in physics where the core precept would seem to belong, but was nowhere to be found — and instead was usually superseded by other contrary themes. What began, therefore, as a brief detour ... turned into a lifelong pursuit: to develop the rudimentary conceptual framework for a new model of nature and science where complementarity reigned supreme.

For his part Bohr argued, especially in his later years, that complementarity was a fundamental theme of nature which went well beyond physics, and that over time our thinking would evolve to the point where we would be better able to understand this. In particular, he maintained that the fields of psychology and biology contained obvious manifestations of the principle. It is fitting, therefore, that at least in spirit this work now brings Bohr's quest full circle ... back to physics, and from psychology no less!

As to the "new model" reference just above, I am all too painfully aware of how utterly implausible such a proposition sounds — especially coming from someone without a background steeped in the physical sciences, who has also made notable contributions over the years. But this is precisely the point for explaining why so many practitioners in the field (Bohr included) did not themselves treat complementarity as primary, certainly not in the paramount terms being promoted in this book: there were too many other competing theoretical assumptions and conceptual loyalties — other established maps — for such a preeminent position to be adopted. Put in the simplest of terms, these dedicated workers were too fully engrossed inside the overall box of physics to gain the necessary distance. That is the case even though many amazing insights and crucial discoveries were being made along the way. Such extraordinary developments notwithstanding, the collective mind-set still did not allow for the necessary leap and reshuffling which would reveal how all these conceptual pieces could be fit together into a coherent whole ... with an expanded principle of complementarity at its core.

It was not genius that was required in order to transcend those limitations, for clearly that characteristic was in great abundance among the masters of modern physics. Rather, it was a completely fresh perspective that was called for, together with a willingness to question established concepts and theories which were incompatible with the central premise. That meant confronting one major topic area after another in physics, with the goal of showing that the underlying phenomenon had an inherent two-sided basis — in stark contrast to the one-sided manner in which it was conventionally depicted. Not only that, but even to pursue this agenda also required a reappraisal of the very rules which guide us in our theory-building endeavors: these *how* assumptions needed to be scrutinized just as much as the *what* conclusions of the models themselves. In particular, a singular criterion of prediction is not sufficient — for it has left us with an array of fragmented, narrow, arbitrary, and opaque ways of construing nature. Yet science also has a responsibility to provide a framework of understanding — one that is coherent, broad, consistent, and transparent.

All in all, my lack of inside-the-box imprinting or inculcation in the field was essential, in order to provide the necessary mind space for the novel ideas in this work to gestate and grow ... along with several decades of patient and persistent research, reflection, and analysis. That was by necessity a solitary affair, for there was no other way to fully explore and struggle with all the difficult issues in a wide-ranging and unfettered manner, without the burden of being constantly pulled back to the overpowering current of collective thought.

There is one distasteful aspect of the overpowering current which must be touched upon here — and that is the reflexive practice of treating any discussion of alternative interpretations which fall outside the mainstream as nonsense (if not heresy), while dismissing the proponents of all such views with disparaging labels like "cranks" and "crackpots". Surely there are preposterous ideas that are sometimes put forward, but that does not justify ridiculing every challenge to the prevailing doctrine. In an open society which values the search for truth, there must be place for respectful discussion of serious and thoughtful new ideas about nature and science. Needless to say, these are highly charged and salient issues, many of which will be discussed later in the book.

Putting that matter aside, let me close by saying I can only hope that the work presented here succeeds in demonstrating the viability of the underlying thesis — that the concept of complementarity is infinitely broader than is normally supposed, to the point of providing the central organizing principle for representing the patterns of nature. How well I have been able to implement that premise in all the many areas which are covered in this book remains to be seen — surely some interpretations are better than others, and I may have failed altogether in particular areas. What is important for the reader is to keep the larger picture in mind; to be open to new ways of thinking about these matters, even concerning topics that seem beyond question (if not sacrosanct); to realize that a transformative approach like this is an ongoing process as much as it is a finished product ... and that going further will require the input of others who are far more capable than I.

For now, however, I want to thank you for your company ... in whatever portions of the work you are able to share with me.

Oh, a minor point to note is the usage in this book of a modified British form of quotation, with non-intrinsic punctuation being placed outside the closing quotation mark, rather than inside as is the U.S. custom.

Contents

1. The Key to the Realm 1

A Complementary Vision 3 • Dimensions of Analysis 4 • Two Fundamental Forces 8 • Two Ways of Being 10 • Two Ways of Seeing 15 • Variations on a Theme of Symmetry 18 • Same and Different 21 • Multiple Meaning 24 • Is Complementarity a Sideshow? 27

2. The Muddle of Quantum Theory 31

The Discrete Quantum 32 • Waves and Particles 34 • The Principle of Complementarity 37 • The Uncertainty Principle 40 • The Correspondence Principle 45 • The Wave Function 46 • Pioneers in Conflict 54 • Physics Fumbles 59 • Complementarity Expanded 62 • The Quantum Paradox 63 • Making Reality 64 • Uncertainty Revisited 76 • Correspondence Revisited 81

3. Atom & Co. 83

The Conventional Approach 85 • A Complementary Approach 92

4. The Relationship Between Stasis and Motion . 116

The Basic Components 116 • The Principle of Relativity: Four Formulations 118 • Newton vs. Maxwell 123 • Revisiting the Principle of Relativity 125 • The Woes of the Inertial Frame 136 • The Dual Nature of Stasis and Motion 139

5. The Dubious Concept of Gravity 149

The Conventional Approach 149 • A Complementary Approach 157

6. The Great Light Hoax 181

A Prelude 182 • Confusion Between Two Light Postulates 184 • Two Relevant Experiments 188 • Measuring Light's Speed 189 • The Subjugation of Time 193 • The Non-Constant Speed of Light 196 • The Michelson-Morley Experiment 204 • The Necessary Function of the Medium 209 • A Complementary Model of Light 215 • Lifting the Veil of Confusion 221

7. Special Nonsense 226
Alterations of Space and Time 227 • The Triumph of Special Relativity 250

8. The Relationship Between Space and Time . 273
The Relativity World of Space-Time 273 • The Two-Force World of Space and Time 287 • Diverse Alterations and Transformations 291 • Complementary and Hierarchical Patterns 297

9. The Other Side of Nature 308
Disputing the Essence of Nature 313 • Complementary Connectedness 320 • Light Connects 331 • Physics Connects 335 • Order vs. Entropy 345 • Symmetry vs. Complementarity 355

10. The Establishment Rules 358
The Scientific Method 359 • Question Authority 380

11. The Myth of Modern Science 418
Understanding vs. Prediction 420 • The Concept of Compatibility 425 • Truth vs. Falseness Revisited 428 • Reconciling With Common Sense 435 • Slipping Backward from Truth 440 • The House Called Science 443 • The Ferromagic Physics Museum 445 • Science and Religion 447 • Science and Technology 455 • We Have Been Betrayed 461

12. The End of the Beginning 467
The End of Science 468 • Cosmology 478 • Biology 490 • Cognition 495 • The Beginning of Two-Force Theory 500

Bibliography 511

Indices 519
Index of Names 519 • Index of Subjects 522

Chapter 1

The Key to the Realm

This work is built upon a very basic premise: complementarity rules the world.

The pattern is all around us — night and day, light and dark, left and right, male and female, and so on. Yet we usually tend to think of each such paired relationship in its own right, if not merely in terms of one or the other side in each case. Rarely do we highlight the more general theme, which is part of why we do not recognize how pervasive the pattern is, instead treating complementarity as a somewhat quirky concept that can be casually applied to one or another particular situation.

The same is true in science, where the pattern of complementarity plays an equally minor role. This neglect also applies to the main discipline that has long been at the center of science, physics, which is the focus of the present work. The notable exception is the "wave-particle duality" of quantum physics — yet even there the complementarity interpretation has always been controversial ... and it has never been well understood, either in terms of its meaning or its scope. The upshot is that complementarity is usually treated by physics as a largely inscrutable phenomenon which can only be applied to a very narrow segment of nature — whereas so many other obvious instances of the pattern are ignored altogether.

By contrast, we maintain that complementarity represents the most general principle of nature that there is. Our mission in this work, then, is surprisingly simple: we take what is normally considered to be a curious but minor feature, which only applies to a very limited subset of natural phenomena ... and raise it up to the level of ultimate principle, which applies to all of nature. Our task is to elucidate this new principle and to propose a template for ordering the results. The goal is not just some superficial survey of fairly isolated and obvious examples of complementarity in this or that realm, but rather to demonstrate that a theory built upon its essence can provide a powerful and meaningful conceptual framework for understanding the nature of reality. Our goal is a unity that has long eluded the established paradigm — a complementary unity, to be sure, yet we maintain that this is the only kind of unity which nature knows.

Aside from presenting this new theory in its own right, we also propose it as an example of what a good theory of science should be striving for. We posit three criteria for such a theory — aesthetic criteria, if you will, that we will discuss more fully in Chapter 10. First, a good theory of science must be essentially *simple* in its basic formulation:

its central ideas should be straightforward, clear, and parsimonious — even if its application becomes complex and nuanced. Second, a good theory must be *general*, or comprehensive: it must use concepts which are universal, and which are not narrow or specific to particular regions or sets of phenomena. Third, such a theory must be *coherent*, in both senses of the word: its concepts must flow together smoothly, into a harmonious whole; and it must be comprehensible, i.e., have meaningful referents which can be understood. All of these criteria should apply both in terms of the way the theory's own conceptual parts are related to one another internally, as well as in terms of how the theory represents nature externally.

In some respects, the established discipline of science would also endorse theory-building goals like simplicity and generality, at least in principle. Yet in practice, we maintain that the conventional theories violate such premises — and they do not even accept the third criterion of coherence, let alone implement it. That missing coherence is evident in various respects — ranging from the haphazard conceptual foundations upon which these theories are built (and the lack of unity which results) to the acknowledged and even lauded inscrutability of the main theories themselves. These problems will become apparent in the pages ahead, as we apply the new model of complementarity to a wide range of important topics.

Our journey begins in the next chapter, where we examine some of the philosophical roots of quantum physics — including the established principles of complementarity, uncertainty, and correspondence ... as well as the role of the observer, and the meaning of the central concept of the quantum itself, along with that of the wave function. In Chapter 3 we look for complementarity in the micro world of the atom, and the forces and functions that apply there ... including the dual phenomena of electricity and magnetism, along with the notion of matter and antimatter. Chapter 4 deals with the complementary states of stasis and motion ... together with some of the basic constructs and theories that are built upon them. In the chapter which follows, we shall see how the conventional construct of gravity looks very different from a complementary perspective.

The topic of chapter 6 is the speed of light, where we shall find that a complementary approach makes the orthodox model seem very one-sided in its conclusion. Chapter 7 continues this discussion by applying the same ideas to the space and time transformations which are associated with high-speed motion. The following chapter looks more generally at the fundamental concepts of space and time — both in terms of the dimensions themselves, as well as in terms of alterations thereto. Chapter 9 examines a variety of crucial and long-standing disputes about connectivity and the essence of nature — like discrete vs. continuous changes and direct vs. indirect contact, as well as the second law of thermodynamics and the direction of time — and how these would be better understood in complementary terms.

With all of this theoretical reinterpretation in place, we can finally turn to a more substantive discussion in Chapters 10 and 11 about basic criteria for theory-building and theory-evaluating, such as the aesthetic factors we just mentioned. Finally, we close the book in the last chapter by contrasting different visions of a unified model of nature — as between the conventional approach and the complementary framework being pursued here.

The reader is welcome to jump ahead to any of those topics which immediately grab his or her attention — although the discussions there will probably be less meaningful without the more basic conceptual framework we provide in this introductory chapter. Here we will be presenting the main themes, and setting forth the central organizing constructs, for our new model of complementarity. But if reading these abstract and general "operating instructions" is too tedious, then just plunge right in and see how the model works more concretely in practice! One can always come back to this chapter later to look up the concepts on the run.

A Complementary Vision

The principle of complementarity which we are espousing is not intended as some kind of idle conjecture, to be occasionally invoked for describing certain commonplace relationships in everyday life — much less as a peripheral postulate which is conjured up as a last resort way of accounting for a particular set of peculiar patterns in science ... yet which otherwise has no intrinsic relationship with the larger model of nature to which it is attached. Instead, we are proposing a framework in which complementarity forms the kernel, and thereby represents the central postulate, around which the rest of the model is built ... from the ground up.

Indeed, this premise is so crucial to the present work that we take it to its ultimate conclusion: at the heart of the model is the radical notion of two fundamental forces of nature, as representing the essential basis of reality. In keeping with that core thesis, this model is called Two-Force Theory, which we will often abbreviate simply as 2FT. The resulting pattern is utterly simple and transparent: There are two fundamental force tendencies of nature, the interaction of which underpins the universe and comprises all aspects of the natural world.

We are putting forward the extreme thesis that everything which exists in nature is a manifestation of these two forces, in one form or another. This also means that the two interacting forces always come together, never just singly — they are complementary, and they are always there. This is true in the sense that there is no region or aspect of nature which is devoid of either of these fundamental forces; and it is also true by virtue of each force being the opposite of the other, and thereby entailing some form of the other. Plus, while their relative preponderance will vary from one situation to another, they are never perfectly balanced: one or the other force will always predominate in any given situation, in some form or other and to a greater or lesser degree.

And what is true of the two forces themselves is also true in whatever manner they manifest themselves: complementary partners do not exist apart from one another. Regardless of the situation, there is never the case that one side or aspect of nature exists solely alone, without the other side also existing in some or another different form and degree. This simple premise has profound significance for the way we understand the world, as will become clear throughout the book.

We need to have some way of identifying each of the two forces, for they serve different relative functions which we must discuss. We call one the *X*-Force of Structure (or simply the *X*-Force), and we call the other the *P*-Force of Process (or simply the *P*-Force). Of course, any such name tags are somewhat arbitrary, as we shall discuss below — the crucial point being not the labels, but the core complementary patterns which those labels represent.

The next most important parameter of our model incorporates the concept of hierarchy. Indeed, we often speak about complementarity and hierarchy as representing the main themes of this work — even though we still see the latter as also entailing a form of complementarity. To understand why, one need only reflect upon the fact that any hierarchy has a topmost pole and a bottommost pole, if only in principle, and that the very notion of hierarchy would thereby represent a form of complementarity between these two poles. The most obvious way of conceptualizing the hierarchy would be in terms of size — resulting in a range from the *smallest and most specific* levels at the bottom, to the *largest and most general* levels at the top.

One can also think about this pattern in more dynamic and bi-directional terms. For example, we can say that the upward direction of the hierarchy represents *growth*, while the downward direction represents *disintegration*. By the same token, and introducing another now verb-like meaning for our central concept, we can say that lower level parts combine together or "complement" upward into higher level wholes; whereas higher level wholes split apart or "uncomplement" downward into lower level parts. We can also talk about these two tendencies in Hegelian dialectical terms of higher level *synthesis*, vs. lower level *thesis-antithesis* — which can also be labeled as *generalization* and *convergence* in the upward direction, as opposed to *discrimination* and *divergence* going downward.

Here is a simple summary of the hierarchical characteristics we have been describing:

top pole
higher
↑
larger
synthesis
convergence / generalization / growth upward

↕

divergence / discrimination / disintegration downward
thesis-antithesis
smaller
↓
lower
bottom pole

Dimensions of Analysis

Next we want to talk about several very important characteristics of nature which have a special place in this model. We refer to them as "dimensions", or "dimensions of analysis", and we contend that they describe all types of phenomena. In what follows, we will first introduce the different dimensions in very broad terms, and then add some more specific comments afterwards. Also note that while these ideas are presented here in only a vague and general way, it is in the chapters which follow when we can apply them in more specific ways that their function and utility will become more apparent.

Starting at the top, the *stable-changing dimension* differentiates a stable situation, or region of nature, from a changing one. We could also describe the stable region or state with labels like uniform or balanced; by contrast, a changing region or state could also be labeled as non-uniform or unbalanced. Next, the *inside-outside dimension* differentiates an inside or inward-oriented situation, or aspect of nature, from an outside or outward-oriented one. We also sometimes describe the inside category as closed, and the outside category as open. One way of describing the *matter-medium dimension* is in terms of a difference between substance and void. Relative to one another, we can say that the matter domain is primarily based upon material objects, and consists of all that is relatively solid in the world — things like mass and particles; by contrast, the medium domain is primarily based upon the emptiness which exists in between — things like energy and waves. The *stasis-motion dimension* differentiates a static situation, or relationship of nature, from a dynamic one — or simply, static and moving entities, respectively. Significantly, we also associate stasis with space and spatial relationships, relative to motion with time and temporal relationships. Finally, the *dominant-recessive dimension* refers to the component entities or groups at the end of this dimensional chain.

See the full page chart just below, for how these pairs of respective characteristics could be represented in relation to one another. (For ease of presentation, the stable and changing regions are not presented side by side in this chart — even though they are complementary with each other in the same way as are the other attributes.)

Let us now elaborate further on some of these qualities. Starting at the bottom, we shall frequently speak about the dominant and recessive partners which comprise either a static or a dynamic relationship. The most obvious example of the former would be two interacting entities that are static in relation to one another — and in some cases, we talk about these dominant and recessive partners in terms of sink and source, respectively, for whatever is moving between them. We call what is moving between them the dynamic relationship — the dominant and recessive components of which might be understood as forward vs. backward, leftward vs. rightward, inbound vs. outbound, input vs. output, buildup vs. discharge, upward vs. downward, or clockwise vs. counterclockwise.

Consider the different ways that static and dynamic relationships manifest themselves in matter and medium domain terms. We already mentioned the most obvious example of the former, in terms of static and moving objects. In the medium domain, by contrast, stasis and motion can be understood in terms of standing and traveling waves, respectively. Another convenient pair of constructs that we adopt from physics is to speak about matter domain *frames* vs. medium domain *fields*. Expressed in terms of the next higher inside-outside dimension, the inside realm consists of an inner frame and a closed field, while the outside realm consists of an outer frame and an open field. Clearly, we are utilizing particular meanings for these "frame" and "field" terms, which are not the same as their conventional usage in physics; and we are incorporating the "medium" term as an essential construct in this model — which of course is a radical departure from the usual approach, as will become clear in later chapters.

Nesting Dimensions of Analysis

Changing

Non-uniform / heterogeneous / unbalanced / unbound

Inside

2D / flat / linear / river / horizontal / sideways

Matter

Inner Frame / mass / particle

- ***Stasis*** — Space — Static Relationship
  - ***Dominant*** / ***Recessive***
- ***Motion*** — Time — Dynamic Relationship
  - ***Dominant*** / ***Recessive***

Medium

Closed Field / energy / wave

- ***Stasis*** — Space — Standing wave
  - ***Dominant*** / ***Recessive***
- ***Motion*** — Time — Traveling wave
  - ***Dominant*** / ***Recessive***

Outside

3D / curved / circular / ocean / vertical / up-down

Matter

Outer Frame / mass / particle

- ***Stasis*** — Space — Static Relationship
  - ***Dominant*** / ***Recessive***
- ***Motion*** — Time — Dynamic Relationship
  - ***Dominant*** / ***Recessive***

Medium

Open Field / energy / wave

- ***Stasis*** — Space — Standing wave
  - ***Dominant*** / ***Recessive***
- ***Motion*** — Time — Traveling wave
  - ***Dominant*** / ***Recessive***

Stable

Uniform / homogeneous / balanced / bound

Inside

2D / flat / linear / river / horizontal / sideways

Matter

Inner Frame / mass / particle

- ***Stasis*** — Space — Static Relationship
  - ***Dominant*** / ***Recessive***
- ***Motion*** — Time — Dynamic Relationship
  - ***Dominant*** / ***Recessive***

Medium

Closed Field / energy / wave

- ***Stasis*** — Space — Standing wave
  - ***Dominant*** / ***Recessive***
- ***Motion*** — Time — Traveling wave
  - ***Dominant*** / ***Recessive***

Outside

3D / curved / circular / ocean / vertical / up-down

Matter

Outer Frame / mass / particle

- ***Stasis*** — Space — Static Relationship
  - ***Dominant*** / ***Recessive***
- ***Motion*** — Time — Dynamic Relationship
  - ***Dominant*** / ***Recessive***

Medium

Open Field / energy / wave

- ***Stasis*** — Space — Standing wave
  - ***Dominant*** / ***Recessive***
- ***Motion*** — Time — Traveling wave
  - ***Dominant*** / ***Recessive***

One prominent way that the difference between inside and outside realms can manifest itself is in terms of relative topology and orientation. In one situation or another, the inside realm might be variously described with labels like flat, linear, horizontal, sideways, and 2-dimensional (with parameters like length and width); by relative contrast, the outside realm might be described as curved, circular, vertical, up-and-down, and 3-dimensional (with parameters like length, width, and depth ... depth corresponding to height, altitude, or distance from center, e.g., of earth). We can talk about the inside realm in terms of characteristics like attraction and repulsion, in relation to the outside realm in terms of characteristics like contraction and expansion. One way of conceptualizing the difference is in terms of the contrast between plane geometry and solid geometry. Still another way would be with a body of water analogy — in which the inside realm is like a river, relative to the outside realm as like an ocean.

Let us go back to the basic point we made in the previous section, that everything is a manifestation of the two forces themselves — and now we want to give a couple of examples in terms of our dimensions of analysis. We can say that the *X*-Force manifests itself as stability, in relation to the *P*-Force which manifests itself as change; nested therein is an *X*-Force which manifests itself as inside, relative to a *P*-Force that manifests itself as outside; next comes an *X*-Force that corresponds with what we call the matter domain, relative to a *P*-Force that corresponds with what we call the medium domain; embedded next is an *X*-Force mode that we call stasis, relative to a *P*-Force mode that we call motion; and finally there is an *X*-Force representative that we can call a dominant partner, relative to a *P*-Force representative that we can call a recessive partner.

This also brings up something else that we mentioned in the previous section, regarding the two-ness of complementary partners — that they always come in pairs. Expressed here in terms of our dimensions of analysis, this means that change does not exist without stability, outside does not exist without inside, medium does not exist without matter, motion does not exist without stasis, and recessive does not exist without dominant.

Before we end this section, there are a couple of significant points to make. One is that we have not dealt explicitly with the crucial distinction between a discrete dichotomy and a gradual continuum. It is unclear whether this pair of characteristics warrants a separate dimension of analysis in their own right, or whether such a demarcation is inherent in the very distinction we make between complementarity and hierarchy ... which is how we generally approach the matter.

The other point is one that we already alluded to earlier — namely, that a good theory of nature should also describe itself. Put in other words, if a theory is putting forward viable patterns about (external) nature, then those same patterns should describe the (internal) theory itself and how it is constructed. Two-Force Theory is about complementarity and hierarchy in nature, and the dimensions of analysis which form the scaffolding of this model are a representation of those dual themes.

There is much more to say about the basic theoretical framework of the model, but at this point we need to step back and deal with one of the most obvious and important questions that has already arisen, regarding the fundamental force issue and the naming thereof. After that, we will return to elaborating the model itself.

Two Fundamental Forces

One of the most dramatic conceptual differences between the present model and that of the established paradigm is simply in terms of the notion of fundamental force — indeed, the contrast could hardly be more stark. As we shall see, this is true both in the contrasting character of the fundamental forces themselves, as well as in the way that each set of fundamental forces relates to their respective theoretical frameworks.

To start with, there is a crucial difference between the meaning of the "fundamental force" construct, as between the conventional model and the new approach being proffered here. The former is based on the proposition that all phenomena of nature can be explained in terms of a few specific categories of "fundamental force" interactions — whereas the latter is based on the proposition that every interaction, for all phenomena of nature, consists of two general "fundamental force" counterparts.

The conventional physics interpretation is that there are a handful of distinct fundamental force linchpins which provide the essential categories of interaction — and thus the crucial basis for all existence — in nature. That number has changed over time, and there are now supposed to be three or four such forces, with different modes of operation for each — yet which have no obvious inter-relationship with one another: These are the strong nuclear force, which mainly concerns interactions between nucleons; the electromagnetic force, which applies to interactions between atoms; the weak force, which deals with radioactivity; and the gravitational force, which concerns interactions that mostly relate to massive bodies. In practice, one might argue that there are actually more force factors than these, even though physics has perpetually talked about eventually unifying everything into one overall category.

One ramification of physics' fundamental force approach is that only the abovementioned levels can be said to truly exist as substantive phenomena — with the stunning consequence that all other levels, including most of our common experience and the macro objects of the world, are basically treated as illusory: they are said not to exist in any rudimentary sense, but only as mere compounds of more basic and real interactions. We maintain that this conventional approach is utterly arbitrary and misguided, and that it is based as much on historical happenstance as it is on anything exclusive about the force categories themselves. In Chapter 12, we will look more closely at this and other defects of the established fundamental force paradigm, but for the remainder of this section we will focus on the present model.

In contrast to what we have just described, the rudimentary nature of reality is very different in the present model. The *X*- and *P*-Force principles are primordial concepts which ultimately have no deeper explanation, beyond their essential complementarity. These two fundamental forces replicate themselves, in one guise or another, at every level of nature's hierarchy — and that number will never change, since "two" is precisely the essence of this entire approach … which, of course, is why it is called Two-Force Theory. Furthermore, these two forces have an intimate and evident relationship with one another — which is to say, again, they are complementary. We apply this simple theme ubiquitously, and the result is a pervasive and all-encompassing pattern of complementary interactions that manifests itself at every level of nature's hierarchy.

There are no preferred levels in the present model, to which our fundamental forces are attached ... and from which all the rest of the world derives. No particular levels take explanatory precedence, or are any more real or profound or pivotal for understanding nature, than any other level. Rather, all levels are equally worthy of being called real: every level of nature represents a legitimate layer of existence in its own right — and has its own *raison d'être*. Therefore, we ascribe a reality — a physical, lawful, independent, and even self-determining reality — to each level of the hierarchy. That is because this model is built upon the concept of interlocking steps of nature, whereby one level is enfolded into the next higher level — and so on, upward and downward, forming a whole hierarchy of nested realities. It is the interaction of the two complementary forces, at or between different levels of nature's nested domains, which is reality and all that exists ... in one form or another. *Nature molds itself*, complementarily, throughout its vast realm — from inside the nucleus to the far reaches of the cosmos.

One result is that for 2FT the specific is an instance of the general — but the specific does not supersede the general, as in conventional theory ... although, of course, there is ultimately no such general in conventional theory. Physics' arbitrary combination of level- or region-specific force explanations overlooks and obscures any general theme entirely that we are talking about — which is also why there is no unity in the established paradigm. But in our model, regardless of any particular name we ascribe to some unique existence — e.g., to any given set of "structural" entities or their "process" counterparts — we are always witnessing an interaction of the two fundamental forces of nature, in their ever-nesting forms. Every specific level and phenomenon can therefore be understood as a microcosm (speaking metaphorically, and not of micromatter!) of this fundamental two-force theme.

A very useful way of describing the overall pattern is to say that the *X*-Force binds or pulls (attracts/contracts) entities together, at all levels of nature; while the *P*-Force unbinds or pushes (repels/expands) entities apart, at all levels of nature. Therefore, our general analysis is broadly the same for understanding binding or bonding at the level of the nucleus as it is for binding or bonding at the level of the molecule — or "bonding" between the earth and some ordinary object. At whatever level we are dealing with, a given entity will be associated with one fundamental force tendency, relative to its counterpart being associated with the other fundamental force tendency.

If the interaction between the two entities is a complementary one (in some manner or form) then in hierarchical terms the two will bind together into a higher level unity — whereas if they are not complementary, then in hierarchical terms the two will not bind together (or they will unbind) thereby representing only lower level entities. Another way of putting this is in terms of the famous physics adage that "opposites attract and sames repel": To say that opposites attract would be another way of saying that two complementary entities bind together — whereas to say that sames repel would be another way of saying that uncomplementary entities unbind from one another.

Note that there is some precedent for the idea of attraction/contraction and repulsion/expansion as providing the basis for broad physical theory. For example, Jammer makes reference to the seven spirits of Böhme, one of which is "the principle of inward attraction or contraction", and another "in the opposite direction, a tendency to expansion" (Jammer, 1957, pp. 142-143). Moreover, "Kant attempts to account for the difference of individual existences by the combinations of two primary forces, attraction and repulsion.... [T]he very existence of matter requires a force of attraction ... Otherwise, owing to the repulsive force, if acting without opposition, matter would dissipate itself to infinity ..." (Jammer, 1957, pp. 180-181).

Before ending this section let us return to the point mentioned earlier regarding how our two fundamental forces of nature are named. While we will not get sidetracked here into a lengthy digression about the best approach to such terminology issues, let us just say that our strategy is to use simple and straightforward labels whenever possible, rather than inventing new words or creating cryptic usages; the tiny handful of novel exceptions will be very evident in the pages ahead. We adopt this position, in favor of existing terminology, even though there will not be a perfect correspondence between the common meaning of the old nomenclature and their altered application in our new model: they may carry some connotations which we do not intend, and they may not convey certain characteristics which we do.

One example should be evident from the discussion just above, where we associated the *X*-Force of Structure with attraction/contraction and the *P*-Force of Process with repulsion/expansion: obviously, the *X*-Force of Structure should not be construed solely in terms of *things*, and the *P*-Force of Process in terms of *actions*.

But the main point to be made here is that the names are not the point: how one chooses to label the two fundamental forces is fairly immaterial, for there will always be limitations with any naming scheme. That does not in any way alter the thrust of this model — which is that there are two complementary force principles of nature, with each representing a persistent tendency or character or goal. They need to be called something, in general terms ... just as they may have more specific names, at particular levels.

We will return to the terminology issue later in the chapter, but for now let us continue with our elaboration of the new approach in the form of what we call two different ways of being.

Two Ways of Being

One of the features of the new model of complementarity is that there are two qualitatively distinct ways that something can exist in the universe — two different ways of being in the natural world. In contrast to the established orthodoxy, we maintain that all of nature, and not just microphenomena, has these dual aspects ... such as are encapsulated in our dimensions of analysis.

There are many ways to describe the complementary states which result, but in all cases, one can differentiate what might be called a primary existence from a secondary existence. We use various general terms to label these two modes — such as focus, figure, or *X*-state for the primary ... and background, ground, or *P*-state for the secondary. But however they are labeled in any given instance, we maintain that the underlying pattern is pivotal — as fundamental as any there is. Put in a nutshell, *all of nature can be broadly conceived as a figure-ground complementarity*. This feature is a prime manifestation of the two-force basis of reality — and once again, it is so obvious a characteristic of nature that we rarely take notice.

Let us illustrate the figure-ground relationship, by means of a classic visual image that is typically credited to Rubens, and which is often discussed in the study of perception and, e.g., gestalt psychology.

The image is unusual in that it is almost perfectly balanced, in terms of its main visual components. The image can be just as readily perceived with the vase as figure, and the faces as ground; or vice versa, with the faces as figure, and the vase as ground. Either aspect can take on either role, because the complementary aspects are so well-balanced here that they are readily reversible.

But the more important point is that only one aspect can be predominant at any given time: while one can swap back and forth between them, it is impossible for both portions to manifest themselves as figure at the same time, in relation to one another. In that respect, this image serves as an excellent illustration of the principle which underlies all complementarities in nature. This is true even though many complementary relationships are more complex and less well-balanced than this. In all cases, however, the pattern is that either portion can serve as figure, in relation to the opposite status of the other as ground. Clearly, one of the main qualities which makes something a figure is that it can be contrasted with the background which it is not — and vice versa. Yet at the same time as figure and ground can each be construed as separate phenomena that exist independently from one another, they are also interlocked phenomena whose joint existence is dependent upon one another.

We must emphasize that the issue here is utterly fundamental; and while an image like faces-vase is a convenient visual illustration, this should not distract us from the fact that the underlying pattern is neither a spurious epiphenomenon, nor some peculiarity that is unique to sense perception (human or otherwise) ... or any limitation of its underlying mechanisms. On the contrary, as we shall discuss throughout this book, a figure-ground distinction and whole expresses the complementary nature of existence generally: it represents a crucial pattern which applies to all phenomena and all interactions in the universe, in one form or another.

One might think of the *X*-state of a given entity as active and excited, in relation to the *P*-state of that entity as passive and unexcited. An example for an animal species like ourselves would be the contrast between the waking organism and the sleeping organism: the former state is active and excited, in relation to the latter state as passive and unexcited. For the earth, one might think of daily and seasonal patterns like day vs.

night, or summer vs. winter. Every entity at every level in nature has two such complementary states of being. In this respect, the root cause of why we wake and why we sleep — or of why we live and why we die, and so on and on — is the complementary character of nature. And of course, the pattern we are describing is the same whether we are talking about the two different states of a single entity, or two different entities which are interacting together.

Now let us expand our vision broadly — very broadly, indeed — as we consider a whole host of theoretical constructs, phenomenological characteristics, and even belief systems ... all of which can be understood as examples of the two different ways of being that exist in relation to one another, expressed as a manifestation of the two underlying forces themselves. We will do so by means of a series of simple "complementary charts" — of the same sort that we will use throughout this book, as a way of illustrating the dual categories that are being discussed in each case.

Many of the following sets relate to concepts and phenomena that are dealt with by physics; some are far more general than that — but virtually all are accessible to most people, as part of their normal experience. Note that while there is a certain order to the way these pairs of constructs are grouped, there is nothing rigorous about it ... as we will explain further after presenting them. Also, we will not discuss any of the below-listed items here, because we are now looking for general themes; most of these qualities will come up during the course of the book, and their listing here is a foretaste of some of the discussions to come. Our immediate goal is simply to demonstrate the wide range of ideas and phenomena which are readily incorporated into this model — and to tickle the complementarity template inside our minds!

We start by restating the broadest and most abstract characteristics, which are represented by our dimensions of analysis:

X-Force	*P-Force*
stable	changing
inside	outside
matter / particle	medium / wave
stasis	motion
dominant	Recessive

Now consider a loose cluster of general attributes that relate to two basic modalities of being, some of which we have already mentioned:

X-Force	*P-Force*
discrete / sudden / collapse	continuous / gradual / evolution
focus / figure / central / precise / sharp / narrow / specific	background / ground / peripheral / vague / blurred / broad / general
primary / superior	secondary / inferior
dualistic / digital	pluralistic / analogue
black-and-white	shades (e.g., color)

This next cluster deals with related offshoots of different and same, negative and positive, and the Hegelian dialectic:

X-Force	*P-Force*
parts	whole
different / contrast	same / compare
divergence / discrimination	convergence / generalization
dichotomy	continuity
figure vs. ground	figure and ground
thesis vs. antithesis	synthesis
either-or	both-and
no / negation / negative / minus	yes / affirmation / positive / plus

Following is a grouping based on strength-weakness and absolute-relative, etc.:

X-Force	*P-Force*
strong / hard	weak / soft
solid / full / tight	fluid / empty / loose
concrete	abstract
absolute	relative
real	apparent

These deal with spatial, distance, and mathematical concepts:

X-Force	*P-Force*
local / central	distal / peripheral
short-range / inverse	long-range / inverse square
arithmetic progression	geometric progression
finite	infinite
inside	outside
square	round
flat	deep
plane geometry	solid geometry
linear	circular

This set deals with temporal, causal, and logical factors:

X-Force	*P-Force*
successive	simultaneous
sequential	synchronous
moment / points	duration / period
dependent	independent
causal	acausal
certain / definite	uncertain / indefinite
perfect / distinct / predictable	imperfect / indistinct / random
deductive	inductive

Finally, here are a melange of various attributes, including spiritual, emotional, and existential ones:

X-Force	*P-Force*
bright / hot / dry / land	dark / cold / wet / sea
alive / awake / active	dead / asleep / passive
male / yang / day	female / yin / night
devil / bad	god / good
hate / war	love / peace
existence / being / identity	non-existence / non-being / non-identity
appearance / somethingness / named	disappearance / nothingness / nameless
visible / finite / known / measurable	invisible / infinite / unknown / unmeasurable
unity / order / information	entropy / chaos / energy
form / manifest / actual	spirit / latent / potential
discipline	freedom

As we indicated above, there is nothing definitive about these lists, and many additions and reorderings can be made. Some of the categorizations might be done differently — both in terms of which fundamental force either characteristic is associated with, as well as in terms of how these sets of paired characteristics are grouped with one another. Moreover, a quality which is associated with one force, in relation to one counterpart, may be associated with the other force in relation to another counterpart. And while in most cases there is overlap in the underlying meaning between various characteristics within the same column, it should be apparent that this is not always the case. We emphasize that point in regard to the last set of characteristics — for we are certainly not suggesting any kind of ludicrous association between male and bad, or between female and good! The primary message here is the overall complementary pattern, rather than how particular qualities are allocated in any given situation.

We also want to make it clear that values, morality, and religion can be understood in complementary and hierarchical ways just as much as any other aspect of nature. That is why such concepts are not out of place in this analysis — nor do they merely represent arbitrary, human-developed philosophical or "metaphysical" interpretations about the world ... as opposed to the "real" world itself. Products of the human mind are not distinct from the brain of which they are part — nor from the larger world from which they were derived, and in which they are meshed. We will consider some of these issues near the end of the book, when we examine biological and cognitive manifestations of complementarity.

Before concluding this section, let us pick up on a point we just made, about complementary characteristics not being fixed in the way they are allocated between the two forces in different situations. There are other ways of making this same point — such that a given entity can take on either set of characteristics ... at the same time in relation to different partners, or at different times in relation to the same partner. Regarding the latter point, therefore, paired characteristics need not represent fixed states of the two partners involved — and they may change over time, with the two partners or components exchanging their roles. There is thus a relativeness in the way paired attributes manifest themselves.

Such inversions between partners could be either occasional affairs, or they could be regular — and we can even talk about a pattern of oscillating dominance. In fact, one might also think about the figure-ground disparity between the two states as being a property that can be passed back and forth between the two interacting entities — and this disparity might even represent the "glue" or binding which holds the two complementary partners or aspects together, as well as being swappable between them.

Two Ways of Seeing

We next turn to the obverse of the issue we have been discussing. For we now ask: whatever the different states of being which exist in nature, how do we detect them — how do we identify that which is out there, whatever its form?

One simple way of speaking about the mechanism by which we apprehend the world is in terms of seeing: we see that which is. So there is a complementarity between *being* and *seeing*. On the one hand are those phenomena which exist in nature, and on the other hand are the ways by which those phenomena are ascertained ... and we shall see that the two aspects are not divorced from one another. The upshot is that we are not merely dealing with some kind of "objective" reality which is independent of some element of "subjective" perception.

In the most general of terms, we maintain that there can be no credible theory of being without an equivalent theory of seeing to go with it — so a good model of nature must encompass both. That is because there can be no description of reality at any given level which exists apart from the way we view, interpret, and define that reality. Of course, the present work is not alone in highlighting the crucial role which observation plays in our interactions with nature. In particular, we shall see in Chapter 2 that one of the features which made Quantum Theory so revolutionary was in its assertion that the observer and the observed must be treated as a *single system*, in any examination of nature — as opposed to the classical assumption that there could be a neutral observer describing an objective reality.

So is there anything distinct about our approach to observation, given the crucial similarity we just mentioned? One of the most significant differences is simply that we approach observation in the same way as we approach everything else in nature — from a complementary perspective. Therefore, we start with the basic complementarity we just mentioned, which is that observation can be understood as the seeing part of a being-seeing relationship. Yet within that overall complementarity between being and seeing, there is also a nested one — between two complementary modes of being on the one hand, and two complementary modes of seeing on the other. We can talk about the former in terms of X- and P-modes of being (e.g., the observer and the observed); and we can talk about the latter in terms of X- and P-modes of seeing. The previous section was mostly concerned with the being pair, and now our topic is the seeing pair.

The two different modes of seeing are something that we each experience, all the time in our daily lives. That is evident in the fact that we can see something in the world either as a sharp figure, or we can see it as (part of) the more diffuse ground — or in terms of the other interchangeable pair of name tags we frequently use, we can either see a given entity as focus, or we can see it as (part of) the larger background. Aside from these two common pairs of labels, we can also use terms that are more specific to the present model and issue. Therefore, just as in the last section we used terms like *X*-state and *P*-state for the being pair, so we attach similar names to the seeing pair: *X*-view and *P*-view. Here is a complementary chart:

X-View	*P-View*
figure	ground
focus	background
sharp / precise / narrow	diffuse / hazy / broad

Using the Rubens example, we might have an *X*-view of the vase, relative to a *P*-view of the faces; or vice versa. It should also be evident that we could just as well have discussed that graphic illustration here in the Seeing section, as there in the Being section — for we apply the figure-ground (or focus-background) distinction to both.

We mentioned above that the being and seeing aspects of nature are intertwined, and therefore that the mode of viewing which one adopts will partly determine the outcome which one finds — there will be a correspondence between what we see and how we see it. Because of this interaction, we will sometimes talk about the seeing choice in terms of the being result — an *X*-view yielding an *X*-state, and a *P*-view yielding a *P*-state.

This correspondence also applies to the various dimensions of analysis. Thus, one might adopt a stable *X*-view of some given phenomenon, just as one might adopt a changing *P*-view of that phenomenon. Similarly, one can adopt an *X*-view perspective which will reveal an inner frame, just as one can adopt a *P*-view perspective which will reveal an outer frame. And so on, for the other dimensions, as depicted in the following complementary chart. In all these cases, the outcome represents a difference of relationship and perspective as much as it does an objective, underlying reality.

X-view	*P-view*
stable	changing
inside	outside
matter / particle	medium / wave
stasis	motion
dominant	recessive

We reiterate again, as we emphasized in the prior section, that there is nothing absolute about any of these categorizations ... as to which aspect goes with which view. What remains constant and crucial is the principle that there are complementary perspectives or aspects, in every situation — irrespective of how they are apportioned or labeled in any given case. Indeed, that is one of the lessons we must try to remember — the ever-present need for complementary and relative perspectives in describing things, as opposed merely to the singular and absolute. Thus any conclusion we draw about a phenomenon would seem to oblige us to identify which perspective that conclusion is drawn from (or corresponds with) ... but also to include the opposite and complementary conclusion which the opposite perspective would yield. There must always be two complementary views of reality, and they must be essentially contradictory to one another —

contradictory, yet fitting together into a coherent whole. One can only bounce back and forth between the two visions, absorbing all that one can from each. Both perspectives are necessary to understand a given phenomenon or relationship, and neither alone is sufficient.

Now that we have discussed the two complementary ways of seeing which separate the present model from the conventional approach, we can also state another difference which is more sweeping. That is, while we have been talking about seeing or observation as though it were a crucial category in its own right, we actually do not subscribe to such a thesis — at least, not in the way it is usually treated by Quantum Theory. In that framework, observation or seeing is construed in relation to perception and consciousness — and we shall see that those mechanisms are deemed to be critical in the process by which reality becomes manifest.

By contrast, we contend that such an approach unduly inflates the role of one particular kind of interaction over all others — and especially by treating our own species' role in nature's vast array of processes as being somehow pivotal ... when there is no logical basis or experimental evidence to support such a far-fetched assumption. Instead, we maintain that observation is merely the name we attach to one particular subset of interaction phenomena, which overall category is much more general than that. So while the construct itself is meaningful, as we have been describing so far, we also use the "seeing" label both as a manner of speaking and as a shorthand way of representing a far broader array of phenomena — simply, that of dynamic interactions generally.

In this regard, let us now introduce a set of complementary characteristics which can be applied to dynamic relationships:

<u>*X-Force*</u>	<u>*P-Force*</u>
inbound / forward / potential energy	outbound / backward / kinetic energy
concentrate / buildup / bind / inhale / input	disperse / discharge / unbind / exhale / output
centripetal / approach / arrive / come	centrifugal / recede / depart / go
attract / contract / pull / absorb	repel / expand / push / emit
stop / decelerate	start / accelerate
climb / ascend / above / emerge	fall / descend / below / submerge

Now we want to highlight one particularly vivid and utterly pervasive interaction pattern just mentioned — the dual process called *breathing*. Every structure, at every level of nature's nested hierarchy, breathes. Indeed, every periodic process is nature's inhaling and exhaling, at some given level of the hierarchy. Because of its ubiquitous character and obvious familiarity, we shall often refer in the book to this particular set of activity as an illustrative motif to describe the larger category of dynamic relationships generally.

We would suggest that the two components of breathing can be readily understood in a similar manner as we have described for the two modes of seeing — in terms of a drawing in and narrowing inhalation or buildup and focus which corresponds with the X-view, as opposed to a giving out and broadening exhalation or discharge and background which corresponds with the P-view.

There is also a connection with the notion we mentioned at the end of the previous section, of an oscillation of the X- and P-states which may exist between complementary partners, as they exchange primary and secondary aspects back and forth between them. We might think of one partner as exhaling in relation to the other partner as inhaling, corresponding with their alternating states. By the way, one interesting example of such "breathing" is the reciprocal exchange of buildup and discharge energy back and forth between the two modes of a swinging pendulum — and another is the peak and trough of a wave. Indeed, any sort of *simple harmonic motion* (SHM) is an instance of such basic oscillating relationships. So is a playground seesaw!

We should be clear about one point, as we end this section. While on the one hand we reject the notion that consciousness or perception per se are required for the *manifestation* of reality (being), on the other hand it seems obvious that some such (seeing) mechanisms are essential for the *description* of reality — and the latter category would include all the concepts and theories with which we represent that reality ... which largely means the symbolic content of our cognition about nature.

Variations on a Theme of Symmetry

This book is mostly about two-ness, in particular as that quality is represented by the concept of complementarity. But now we want to do a short compare and contrast examination between the two-ness of complementarity and the two-ness of symmetry.

To start with, it is important to acknowledge that the concept of symmetry has a pedigree all its own, both in culture generally and in science specifically — and there can be a variety of meanings for the term, with a range of referents from the very narrow to the very broad. The more narrow meaning is also the more conventional meaning, as referring to situations whereby there are two component entities or parts which have a one-to-one correspondence such that they are essentially mirror opposites of one another. Broader meanings for the symmetry concept would include cases of rotation and where there are more than two components, yet each of which is virtually interchangeable with the others. Perhaps the most expansive usage would concern an overall aesthetic sense of balance and harmony.

In the science arena, the symmetry concept is widely used and highly esteemed in the field of physics, particularly particle physics. Indeed, symmetry is frequently applied there in quite abstract and conjectural ways, often in mathematical and geometrical realms. Symmetry is also treated as providing the basis for the various conservation laws and invariance principles, by virtue of some quality or quantity remaining constant while others change. Whether it is wise to use the symmetry label for representing such a generalized feature of sameness is not a matter we will address in this book — but suffice it to say that this kind of usage is far removed from how the concept is usually understood by most people. In this regard, note that physicist Richard Feynman once said, "It is easy to understand how an object can be symmetrical, but how can a physical law have a symmetry? Of course it cannot, but physicists delight themselves by using ordinary words for something else." (Feynman, 1965, p. 84) We adhere to the common meaning in this book, and in particular we approach the symmetry issue in terms of the basic characteristic of two-ness that we have been describing.

That said, we take exception to a central premise which is inherent in the symmetry concept — namely, the idea that nature manifests this characteristic in a perfectly equal form. Thus, while the notion of pure symmetry is a very alluring concept, we contend that it is an idealization and a construction which is more a product of our minds than it is an accurate portrayal of the patterns which exist in natural contextual conditions. In reality, what we find in such situations is not completely symmetrical ... but rather is almost symmetrical.

This brings up the question of terminology. Thus, we can say that there are some situations where symmetry is predominant, and there are other situations where non-symmetry is predominant. How should these two states be labeled? For the latter case where symmetry is lacking, there is an established and viable term: *asymmetry*. But the former case is not so straightforward: there is no analogous and appropriate established term. We suggest that it is misleading to apply the "symmetry" label here, because we are specifically rejecting what is perhaps the most crucial and obvious connotation (of perfect equality) which is inherent in the concept. So instead we take a fresh approach, and go out on a limb now. Because our language lacks the crucial distinction we are looking for, we take the rare step in this work of proposing a novel term here. We call this new construct *esymmetry*, as referring to *essential symmetry*. (By the way, the new word would be pronounced with a short vowel — "eh-symmetry" or "ess-symmetry" — so as to clearly differentiate it from modern computer long vowel usage ... as in e-commerce, e-mail, and so on.)

Although the term "esymmetry" is itself novel, the concept behind it is not. It corresponds closely with the notion of "broken symmetry" (Morrison, p. 55): "It is probable that all symmetries are a little bit broken ..." (Morrison, p. 65; cf. Calder, 1977, pp. 178 and 180). Then there is "near symmetry" (Feynman, 1965, p. 107); "symmetries are never perfect" (Harrison, in Whyte, p. 96); and "biased symmetry" — meaning a symmetry whose perfection and precise equality have been tipped somewhere, in one direction or the other. Similarly, note Brown's assertion that "a universe comes into being when a space is severed or taken apart." (Brown, p. v) In this regard, Brown refers to the act of "indication", as a way of breaking perfect symmetry ... a process which he interprets as being required for nothingness to become somethingness. Perhaps a tongue-in-cheek way of putting the matter is to say that esymmetry is symmetry with a wink (or a nod)!

So there may be good reason why idealized forms of symmetry do not exist in nature. It seems that real symmetry must break one way or the other, in line with Brown's act of "indication", as we just noted. There must be at least a slight tilting, as in the case of two component entities binding together — e.g., with the figure-ground disparity "glue" that we mentioned in the Being section. The pattern of oscillating dominance we spoke of could be thought of in terms of an esymmetrical ball or parcel of energy (or even matter) that is being passed back and forth — which itself represents the disparity there ... the break from symmetry, the slight difference between the partners or aspects. The relatively small non-symmetrical portion of energy would represent the separation into figure vs. ground in this esymmetrical relationship ... as well as provide the unity which joins them. Consequently, symmetries in nature are not perfect because they *must not* be perfect, contrary theoretical predilections notwithstanding — rather than because a mistake was somehow made, tantamount to a flaw of nature ... let alone that we are simply dealing with a limitation of perception.

We suggest that the failure to recognize the sometimes subtle distinction between perfect symmetry and something less so lies at the root of several major misconceptions about the nature of reality, as exhibited by the established paradigm. Put plainly, the basic symmetry concept is flawed, in that it is too symmetrical: it implies a kind of perfect symmetry which does not actually exist. Throughout the book, we shall see that there are massive theoretical consequences which emerge from the pure symmetry concept, and the way it has been slavishly applied. We shall refer to this as *false symmetry*, or as stemming from a "symmetry conviction".

There is also a very different and virtually opposite problem that we will make repeated reference to, and we call it *no symmetry*. In these cases physics overlooks, ignores, minimizes, or rejects a great many situations where a two-sided symmetry-like approach belongs ... but is nowhere to be found. We are talking specifically about a very unsymmetrical pattern of one-sidedness which pervades far too many concepts and theories, rather than a more symmetrical two-sidedness. There is thus a serious gap between the lofty pedestal upon which the discipline places the symmetry construct in principle, as opposed to the very unsymmetrical way that it approaches so many phenomena of nature in practice. Indeed, we maintain that a great many of the field's primary concepts and theories are completely devoid of any semblance of two-sided symmetry, in terms of their composition and orientation.

Taking these two problems together, it should not be surprising that our approach to the underlying issues is very different. In fact, the alternative thesis that we are working with can be understood in terms of two premises which directly counter the two problems we have just been discussing. First, the present model assumes that everything in nature can ultimately be understood as representing an interaction between the two forces of nature, in whatever form they manifest themselves for some given situation: there is no instance whereby only one — either one — of the two forces is represented singly. In this respect, what we describe as physics' "no symmetry" inclination is unacceptable because it violates this two-sided premise. Second, symmetrical interactions never consist of precise equals — there must be a difference: there is no case or region of nature where there is a perfect balance between the two forces and the way they manifest themselves. Physics' "false symmetry" concept is unacceptable because it violates this premise of unequal components.

It should also be clear where we are going, in a broader sense. That is, there is an equivalent concept which plays the same overall role in the present model as does the notion of two-sided symmetry. That central concept, of course, is called complementarity — and we maintain that it captures and fulfills all the essential functions of the basic symmetry construct, and much more ... yet in a manner which conforms to reality.

We should also note that complementarity actually plays a far more prominent role in the present model than symmetry does for physics. The difference is apparent in the fact that the fundamental force and symmetry concepts are largely distinct from one another in the conventional approach, whereas the fundamental force and complementarity concepts are completely intertwined here. In that latter case, the two fundamental forces form a conceptual nucleus which replicates as it expands and cascades outward to generate an entire model, called Two-Force Theory. The result is that while symmetry is *a* characteristic of nature for physics (albeit nominally an important one), complementarity is *the* characteristic of nature for the new approach we are presenting.

Same and Different

Now we want to try and better understand the relationship between any two complementary partners in terms of a fundamental set of concepts. Along the way, we will also be further developing the framework for why a two-sided approach to nature is necessary.

Here is a puzzle to consider at the outset. The notion of complementarity entails within itself a degree of paradox, which is tantamount to the very meaning of "different" and "same". For, each partner, within every complementary pair, can be understood as representing a *different* and separate entity — each in its own right, apart (part) from the other. Yet each partner can also be understood as not being separate from the other, but rather of only existing in the form of a shared relationship with its partner ... thereby comprising one and the *same* larger (whole) phenomenon. These two different vs. same options can be construed as two ways of being: *X*-state and *P*-state. But they can also be construed as two ways of seeing: a narrow *X*-view of the different parts, as opposed to a broad *P*-view of the same whole.

One way of bringing it all together is to say that, "Reality is a unity of opposites" (as Mazrui put it succinctly). What a radical conception as a principle to be applied to all of nature! Yet, of course, the idea is very old. The pattern is inherent in the ancient Chinese belief system which is commonly associated with Taoism — represented graphically by the yin-yang symbol ... in which a black kernel exists in the white portion, and vice versa. By the way, the great Quantum Theory pioneer, Niels Bohr — who fathered the notion of complementarity in physics — adopted this symbol as the centerpiece for his coat of arms.

There are also similarities with Greek philosophy: Jammer refers to "Heraclitus' doctrine of opposing tensions, according to which all things, although in appearance most stable, are but the battlefields of antagonistic forces." (Jammer, 1957, p. 25) According to this thesis, there is an "inherent antagonism and balanced conflict of opposites in the individual object, each item of existence being the battleground of opposing forces and tensions." In a similar vein, Whyte writes about "the complex structured universe, with its opposed tendencies towards disorder and towards order." (Whyte, p. 3) And, of course, in talking about the two forces earlier, we cited Kant's description of existence in terms of opposing tendencies of attraction and repulsion; and we have mentioned Hegel's oppositional dialectics. This theme is also apparent in biology, in the form of dominant and recessive genes.

In 2FT, each of the two general force principles (as well as all their specific manifestations) only exists in relation to its opposite. Likewise, each force — every complementary partner or attribute — also entails or contains within itself aspects or instances of the opposite characteristic: everything is partly composed of its opposite, and contains within itself the seed or germ of its opposite (cf. Harris). Taken together as a unity, the two forces embrace and support one another with the power of their reciprocal essence.

Now we want to look at this issue in hierarchical terms. Just as we have talked about higher level wholes in relation to lower level parts — and an upward direction towards generalization and synthesis, vs. a downward direction towards discrimination and thesis-antithesis — so we can now add the concepts of same and different to these clusters. The unity of the higher level whole corresponds with the *sameness* (convergence) which they share together, just as the separateness of the lower level parts corresponds with the *difference* (divergence) which distinguishes them. Therefore, as we move higher in the hierarchy, in terms of more general and widespread nesting factors, we are dealing more with sameness — whereas conversely, as we move lower in the hierarchy, in terms of more specific and narrow unnesting factors, we are dealing more with difference. We can also describe this bi-directional hierarchy factor in the verb-like meaning for complementarity that we mentioned earlier, whereby partners "complement" upwards and "uncomplement" downwards. Here is a chart summary:

Different	Same
downward in hierarchy	upward in hierarchy
uncomplementary / unnesting	complementary / nesting
divergence / discrimination	convergence / generalization
parts / opposites	whole / unity
either-or	both-and
thesis vs. antithesis	synthesis

But now let us return to the puzzle we posed above, which also raises a seminal question about reality: Does each complementary partner even exist by itself, apart from its counterpart — or do the two only exist in relation to one another? As we shall see, the issue has enormous significance, and touches upon a great many areas of inquiry.

Take the issue of one-sidedness, which we will hammer away at throughout this book for being far too prevalent in the established theories of physics: that issue needs to be understood in this context. We even coin a couple of terms, in this regard, in that our usage may differ from the way the words are conventionally understood. We shall talk repeatedly about a "singularity" or "unitarity" approach, as opposed to a "duality" or "complementarity" approach — and we obviously mean by the former "one-sided", as opposed to the latter being "two-sided" (We will work with this interpretation even though the term of "unitary" itself also has a meaning of whole, or whole unit.). But if either side of a duality can exist independently in its own right as a satisfactory description of nature, then are we warranted in challenging physics' various one-sided assertions and conclusions in this regard?

Our position is that any such one-sidedness must be understood within a broader framework of two-sidedness. The result is that nothing exists completely alone by itself, and "conjoint pairs" are the rule of nature. In form or function, *everything exists in relation to some complementary counterpart*. Again, we shall encounter many violations of that principle by the established paradigm throughout this work. That was one of the main points we emphasized in the previous section, where we talked about the unsymmetrical or "no symmetry" basis of so much of the conventional approach. In its tendency to highlight one aspect of a given phenomenon, while neglecting or denying the parallel aspect — or in only accepting the joint unity between them, and denying the separate validity of the component parts — physics is violating the two-sided essence of the basic symmetry concept ... which pattern, of course, we call by another name.

The breadth of that two-sided pattern is staggering — which it is the purpose of this book to help illuminate. Expressed here in terms of the same-different paradox, we are arguing that one can draw diametrically-opposed conclusions about every complementarity in nature. Does male, for example, have an independent existence without female? *Yes* — but only in relation to *no*. On the one hand, it is obvious that male is a separate and independent phenomenon from female, or else there would be no basis for differentiating the two. Yet on the other hand, that very differentiation (and differentiability) also means that the one part does not exist without the other. Moreover, male and female also share most characteristics in common, and together they comprise a common entity (e.g., animal) — and in this respect, as well, male is not independent from female. Finally, of course, male and female can be understood as lower level components which may join together into a higher level synthesis of offspring.

Likewise, we mentioned earlier that a given entity can be active (e.g., awake) or it can be passive (e.g., asleep). Such complementary states of being can be understood as another example of *different* lower level partners which exist separately from — but only in relation to — one another ... just as they also join together into the *same* higher level unity, which is the whole organism we typically think about.

The same principle applies to the complementary characteristics for each of our dimensions of analysis. Does stability exist without change, and the other way around? Can inside exist without outside, and vice versa? Matter without medium? Stasis without motion? Dominant without recessive? It is the same answer we just gave above: Yes, and no. Therefore, both answers are correct, even though they do not agree with one another — and both answers are necessary: we cannot do without the two of them ... in relation to one another.

What about other common complementary phenomena in nature? Does lightness exist without darkness? Or chicken without egg? Or — dare we raise the crucial question that we shall discuss at length in Chapter 8 — space without time? Yet why stop there, what about the equally-salient issue of certainty without uncertainty, which we shall examine in the next chapter? Do any of these phenomena exist without their partners? How can there be any other answer except both yes and no, in each case? Both the separate (different) and joint (same) alternatives must be invoked ... but always in relation to one another.

The complementarity approach we advocate encompasses both sides of each respective pair or aspect of nature. What will not work is the unitarity position for any given phenomenon. Therefore, our contention is that we fall into a trap if we think about any one-sided quality as existing alone in a pure and isolated form, because it never exists that way in the real world — and any such descriptions by themselves are unsatisfactory. No phenomenon exists alone, completely apart from a relationship whereby its opposite also exists. Nothing that we experience, nothing in nature, is pure or complete in and of itself. Nothing in the real world is only one thing, at the utter exclusion of its complementary opposite. It is always the case that while a phenomenon may be (more or less) predominantly of one quality, the obverse side is inevitably lurking in some form ... to a greater or lesser degree.

We already specified the most basic form of this principle, early in the chapter: One never finds any manifestation of the *X*-Force of Structure in utter isolation, nor any manifestation of the *P*-Force of Process in utter isolation. There is no region or phenomenon of nature where only one of the two forces manifests itself, in an exclusive sense, and with no trace of its complementary opposite. Neither of the two forces of nature ever exists in isolation ... but only in relative interaction with one another, and as part of one another. Nature is dual, and the two sides always go together, in one proportion and relationship or another.

It may be a useful, or perhaps inevitable, analytical exercise to talk in terms of one side of nature only, at whatever level we are concerned with for that instance. For whatever reason, our normal mode of thought — at least our normal Western mode of thought — seems unable to avoid these absolutist or shorthand concepts. And no doubt such focused and one-sided visions are functional, in certain respects — so long as we do not lose sight of the other side.

That is because there is no singular manifestation in nature, of any construct or phenomenon, without some relationship to duality. This is the case whether we are talking about complementary partners that only exist at a given level in relation to one another; or in hierarchical terms, whereby a seemingly singular higher level whole (comprised of lower level parts) becomes, in turn, a part that must join with its opposite at that higher level ... to merge into the next higher level still. We argue that the isolationist perspective is only valid within an interactionist framework — and that including both is not just a matter of taste, but of necessity, for understanding the complementary nature of reality.

Multiple Meaning

We have been talking about how characteristics like convergence, same, generalization, and whole increasingly apply as one moves higher in the hierarchy ... relative to characteristics like divergence, different, discrimination, and parts as one moves lower. But just as nature itself can be understood in these terms, so can the language and concepts we use to describe nature. The pattern which emerges can be both illuminating and confusing, as we shall discuss in this section.

To start with, of course, the idea that the same word can have different meanings is obvious in any dictionary — and there is likewise nothing new in the notion that different words can share a common meaning. The point we are making here is that these factors can be understood in complementary and hierarchical terms, and that they also apply to the terminology we use for describing nature. Indeed, we would argue that the idiosyncrasies of our variegated concept system sometimes masks a common essence, which may blind us to crucial underlying patterns.

Consider a salient example of the issue we are concerned with. Attraction and repulsion are ways of describing the two fundamental forces of nature, *X* and *P*, respectively. But of course, *X* and *P* do not merely represent a general force of attraction and a general force of repulsion, but of two very broadly-conceived force principles. Thus, the *X*-Force and attraction correspond with related concepts like inward, inhalation, contraction, concentration, buildup, binding, construction, unity, order, and local — just as the *P*-Force and repulsion correspond with oppositely-related concepts like outward, exhalation, expansion, dispersal, discharge, unbinding, destruction, entropy, chaos, and distal. Here is a listing of these and some related characteristics:

X-Force	*P-Force*
attraction	repulsion
inward	outward
inhalation	exhalation
contraction	expansion
concentration	dispersal
buildup	discharge
binding	unbinding
construction	destruction
unity	entropy
order	chaos
local	distal
stasis	motion
structure	process

We maintain that such a shared essence (or common kernel) principle becomes almost inexorable or irresistible in the way it cascades and feeds upon itself ... and broadens our understanding of nature. Thus, consider the mutual meaning (or association) elements in the following proposition, which bears some resemblance to the position cited from Kant earlier in the chapter: If attraction were not stronger locally than repulsion, then there would be no local — there would be no stasis, no buildup, no structure! Conversely, if repulsion were not stronger distally than attraction, then there would be no distal — there would be no motion, no discharge, no process! Therefore, to speak about local is to speak about attraction ... is to speak about stasis, is to speak about buildup, is to speak about structure — just as to speak about distal is to speak about repulsion ... is to speak about motion, is to speak about discharge, is to speak about process. These respective pairs of concepts are not identical with one another, but they have considerable overlap — because of the sameness qualities that they share higher in the hierarchy, along with the different qualities which distinguish them lower in the hierarchy.

One conclusion is that a given phenomenon or construct not only has meaning in terms of its opposite, but both complementary partners exist and have meaning in relation to other complementary pairs. Therefore, every set of complementary relationships exists not just in its own right — but also by virtue of being linked together in a larger hierarchy ... a shared essence web of complementary overlap, with many other similar corresponding patterns. In this way, individual sets of complementary constructs and phenomena also exist as part of larger complementary clusters, giving and gaining strength and meaning from one another.

Of course, there is always the question of whether the issue we are talking about is the result of a faulty language and construct system that does not take into account the kind of subtle nesting factors we have been discussing — resulting in what might be construed as too much generality on the one hand, and too much specificity on the other — and that these would be rectified by a more satisfactory conceptual framework that has been methodically constructed to offset both. But for one thing, it is not clear how such a balancing act between same vs. different terminology could be accomplished. Plus, this (introducing new constructs) is precisely a step that we are resistant to taking or advocating here, given the strong bias that we mentioned earlier in favor of easy and understandable scientific terminology — which also means minimizing the use of special constructs and arcane conceptual name tags that might itself introduce layers of conceptual obfuscation and apparent gimmickry into the model.

Having said that, we must also be explicit in recognizing the kind of terminological confusion which can result — and does describe the present model. In particular, we will encounter various instances in this work where the same term is applied in different ways — starting with the core construct of complementarity itself. Thus, we have said that while complementarity represents the most general overall principle of nature, there are also subsidiary meanings of complementarity which we contrast with other characteristics. The most prominent example of that would be the contrast between this subsidiary form of complementarity on the one hand, and hierarchy on the other. There are also subsidiary nested meanings of complementarity within the hierarchy mode. One is a verb-like attribute: things complement (or nest) upwards, into higher level wholes — and they uncomplement (or unnest) downwards, into lower level parts.

The important concepts of horizontal and vertical are also used in varying ways. For one thing, we frequently characterize complementarity as having a horizontal orientation, in relation to hierarchy as having a vertical orientation. Taken together, we can say that there are many different horizontal levels, each of which consists of complementary partners or portions ... all of which collectively comprise an overall vertical hierarchy of nature. Yet subsidiary to this very broad usage, we have also described our inside-outside dimension of analysis with these labels — the inside realm being characterized in horizontal terms, in relation to the outside realm being characterized in vertical terms. So even though we use the same "horizontal" and "vertical" terms, they have different (nested) meanings in different conceptual contexts.

Another notable example of multiple meaning in the pages ahead is our usage of the dominant and recessive terms — for they not only describe a particular dimension of analysis, but we shall also use the labels in a more generic way to describe the relationship between various sets of complementary partners, including the other dimensions of analysis. For instance, we might say that the inside realm is dominant in some circumstances, meaning that the outside realm would be recessive in that situation. Could we have used different terminology than dominant-recessive in these generic cases, reserving the labels themselves for dimension of analysis application? Yes, but the conceptual cost did not seem to make it worthwhile.

In closing this section, our position is that the problem we have been discussing is not primarily the result of vagaries in the present model or with deficiencies of our language: it is a limitation of our mode of thought. And the solution is to be more nimble in our thinking, requiring a new degree of relativism that may not come easily. We are trying to facilitate that outcome not with an arbitrary add-on or excuse here, but rather by providing a viable conceptual framework for understanding the issue. By helping us to embrace the complementary and nesting character of reality, we will thereby be more comfortable with the complementary and nesting character of our constructs about reality.

Is Complementarity a Sideshow?

Before concluding this opening chapter, it is vital that we deal with some obvious skepticism about our overall approach, regarding the importance of complementarity in nature.

To begin with, we should note that any contention about a significant role for complementarity at all would have been laughable a century ago — utterly lacking in credibility, and completely beyond the pale of reason — prior to the insights of modern science. Indeed, it is physics itself that is responsible for having legitimized the basic premise upon which we are building, thereby making our work far easier than it would otherwise be. And for that, we send our heartfelt thanks!

Such kudos would not be reciprocated, however. For while it would be generally acknowledged that complementarity has a useful role to play, it would also be argued that this function is properly fulfilled in the established paradigm — whereby complementarity is treated as a rather peculiar mechanism that is useful for describing a specific set of characteristics (notably that of the wave-particle duality), which applies only to a restricted range of natural phenomena (the micro world). This is a far cry from our contention that there are no limits to the role which complementarity plays in nature — including the most general notion that there are two fundamental forces, the interaction of which provides the ultimate basis for reality.

In this regard, the counter-argument would be that science must concern itself with actual forces of nature — specific and truly *fundamental* forces, whose interactions can be described and measured in concrete ways. After all, how can a discipline which has progressed so far and produced so many obvious results be expected to take seriously such simplistic notions as an *X*-Force of Structure and a *P*-Force of Process — when there already exists a solid and sophisticated theoretical foundation for describing nature ... in relation to which such vacuous constructs make little sense? These so-called "forces" would be treated as little more than metaphors — almost poetic or rhetorical flourishes, at that.

It should be clear that the issue is not whether the conventional interpretation, based upon primary interactions which take place at particular levels or regions, has immense predictive power — but rather whether there is a broader framework, according to which the "fundamental forces" of physics amount to specific and subsidiary instances of more general patterns ... thereby leading us to a more coherent conceptual framework for understanding the world.

Of course, there is one perplexing question which must be answered, in regards to all these issues: If complementarity is, indeed, such an all-encompassing characteristic of nature as we maintain, then why are we not more aware of it — both in our everyday lives, as well as in science generally? Let us address this issue now, in terms of three main explanations for why the central role of complementarity has not been recognized. One concerns the very pervasiveness of the phenomenon, a second concerns the nature of scientific theory, and a third concerns general cultural issues.

We touched upon the first point at the beginning of the chapter, for the pattern of complementarity itself is so omnipresent and commonplace — so deeply embedded in our beings and in the nature all around us — that we simply pay no heed to it. The pattern is everywhere: it is part of everything we are, do, and see — but like our own hearts beating and our own lungs and breathing, we rarely take notice. The result is that we have become accustomed to not seeing the dualistic pattern which lies right before our eyes (and is represented by our eyes)! At most, we are left with thinking of complementarity as some strange or exotic way of describing or experiencing reality, or as a convenient (if not fashionable) label for characterizing certain relationships — but having little to do with the essence of nature itself.

The result is that it is very easy to dismiss the basic pattern — nay, take no note of it at all — which is what established science has mostly done. And this is the second reason why we are not more aware of complementarity: the conventional scientific regimen has failed to deliver — it has overlooked the primary theme of nature, and so it has not integrated this most general principle into its theories.

Yet how could this be a credible explanation? Science's whole mission is to study the patterns of nature, and to formulate its theories accordingly. So if science is not detecting a widespread pattern of complementarity, then how likely is it that such a generalized principle actually exists? That is a very good question, which we shall address in various ways during the course of this book. But one overall way of answering the question here is simply to say that science is too rigid and too narrow in its thinking — in terms of its methodology, its goals, and its content. We shall discuss the methodology topic in Chapter 10, the goals topic in Chapter 11, and the content topic throughout this book. In brief, we would turn the tables and argue that it is the established paradigm which has been too distracted by its own sideshows — as a result of the theoretical ruts that it is stuck in ... and the way it builds its theories.

However, if we were to mention a single point of issue now, which we will revisit repeatedly in the pages ahead, it would be that scientific theory has developed in an arbitrary manner — in which theoretical constructs have an obvious *chronological* relationship, in historical terms ... but very little apparent relationship, in *logical* terms. One result is a big, gaping conceptual schism between the way science sees the world in its theories, and the way the world looks to most of us.

There is an even more serious consequence of science's haphazard mode of development — which is that the established paradigm has been unable to satisfy a crucial requirement for a good scientific theory, namely that of unity. And we maintain that there are two main reasons for why this unity has always eluded science. First, arbitrary parts are unlikely to fit together into a coherent whole; and second, science has been unable to identify the primary character of that coherent whole ... which we claim is nothing else but complementarity itself. Plus, these two points are related to one another — which is to say, complementarity is not more apparent to conventional science because complementarity is not compatible with the disparate constructs of conventional science ... and it has been eclipsed by them. Our task in this book is to identify as many of those dissonant constructs as we are able — and explain how the underlying phenomena would be better understood in terms of a broad (and unifying) principle of complementarity instead.

The third and final reason we shall mention as to why we are unaware of the ever-present pattern of complementarity can be traced to the cultural styles and intellectual traditions that have long been predominant in the Western world. The issues go well beyond the scope of this book — but suffice it to say that the concept of complementarity has a long and respected history in Asian thought, as we mentioned earlier in the chapter. But, alas, this Eastern mode of thinking has had very little impact on the powerful and technologically-driven culture of the West, from which the established paradigm of science derives.

In sum, our position is that science has it all wrong: as we shall discuss at considerable length in the next chapter, the crucial wave-particle duality which quantum physics unearthed or stumbled upon is no fluke. Wave-particle is not unique in representing the fundamental pattern of the universe — the same principle applies equally well to the rest of nature. For example, let us single out here the ubiquitous relationship between stasis and motion — usually expressed in the unfortunately one-sided notion that all motion is relative — and the universal acceptance of the premise that motion only exists in relation to stasis. This relativistic stasis-motion pattern is so simple and so prevalent in our experience — and in the statements of science — that who would even think that we are thereby talking about a particular instance of the general principle of complementarity itself? We will try to demonstrate in Chapter 4 that the stasis-motion relationship is as compelling a statement of complementarity as any could be — only, it has not typically been conceptualized in those terms.

More broadly, regardless of whether two interacting entities or phenomena may appear to be complementary or not on the surface, we maintain that the relationship can *always* be described in terms of several fundamental complementary characteristics of nature — stability vs. change, inside vs. outside, matter vs. medium, stasis vs. motion, and dominant vs. recessive. These paired patterns are hardly trifling abstractions: As we shall argue throughout this book, each of these compelling features of the natural world — which we call dimensions of analysis — represents an obvious and irrefutable demonstration of the most basic principle that there is.

We must open our eyes to a new way of thinking about these matters: complementarity is not a limited phenomenon or freak-like aberration, as conventional theory and thinking would have us believe. Moreover, complementarity is not only a pervasive pattern throughout nature, but it also provides an invaluable conceptual tool of analysis for comprehending nature, and for constructing our models about nature — not to mention providing a basis for thinking, in general.

As far as this model's ultimate complementary interpretation — of two fundamental forces — we maintain that such an answer is sitting there in the world, waiting to be plucked. That would be clear to modern physics, as well, if its thinking weren't so clouded by antiquated theoretical cobwebs. The two-force conclusion is an almost inevitable outcome of the rest of the conceptual picture we are dealing with — and, as we suggested earlier, probably represents the ultimate and unavoidable essence of any thoroughgoing principle of complementarity.

So the answer to the question is *no*, complementarity is not a sideshow — it is the main act on the big top! For nature roams in pairs, at and between each of its many nested levels, in whatever form the two forces manifest themselves. It is the interaction which we are, which we know, which we see: for the two forces drive the universe — nay, they are the universe. Neither ever exists in isolation, but only in tandem with its opposite counterpart — the two dancing together throughout the cosmos, thereby providing the very heartbeat of nature ... including inside ourselves.

Chapter 2

The Muddle of Quantum Theory

We turn our attention now to one of the twin pillars of modern physics, Quantum Theory.[1] Since its inception in the early part of the twentieth century, this model has had a profound impact on most of our concepts about the physical world. Many of these ideas have permeated the rest of science, as well as more commonplace discourse and analysis. The discrete quantum, the wave-particle duality, the Principle of Complementarity, and the Uncertainty Principle are four areas where Quantum Theory has offered powerful new ways of thinking, which have challenged many of our conventional theories and ideas about nature and reality.

And yet, where has this metamorphosis taken us — this "quantum leap forward" in our vision of nature — and how thoroughgoing have our transformations of thought been? Indeed, how soundly based are the principles which have been put forward by Quantum Theory? While few would doubt the radicalness of many of these propositions, there are serious questions about whether they lead us toward a more coherent and meaningful model of nature.

In the first part of the chapter, we will examine some of the fundamental ideas of Quantum Theory, and how they came about — which will necessitate considerable digression into historical background, as well as an examination of theoretical disputes from the past ... some of which continue today. We begin by looking at the central notion of the discrete quantum itself, and then turn to the famous wave-particle duality. Next

1 Or Quantum Physics, or Quantum Mechanics. Some people differentiate Quantum Theory from Quantum Mechanics, based on the degree of abstractness of the issues (e.g., Zukav). Others differentiate an older form of the Quantum model from the later approach put forward by Schrödinger and Heisenberg in 1925-1926, referring to the latter as Quantum Mechanics. Jammer differentiates Quantum Physics generally from its more theoretical subset of Quantum Theory, while the latter is said to comprise both Quantum Field Theory and Quantum Mechanics (Jammer, 1989, p. 149). We make none of these distinctions here.

we examine three pivotal principles of Quantum Theory — namely, those dealing with complementarity, uncertainty, and correspondence. Then we look at another core concept of the theory, called the wave function, including two significant interpretations of how this construct is to be understood in reality. After that, we consider two important sets of theoretical conflicts that existed between the theory's creators, which expose some serious weaknesses in the model's epistemological foundations.

These internal conflicts provide a segue to our own critical stance, which we turn to in the second part of the chapter. We will raise serious questions about some of the central assumptions of the theory, as well as identify significant contradictions therein. We start by re-examining the wave-particle duality and the Complementarity Principle which underlies it, and see how physics fundamentally mishandled this seminal thesis — as well as describing how complementarity conflicts with the theory's namesake notion of a discrete quantum. Then we discuss different aspects of the reality theme which lies at the theory's core, and see how these can be reformulated in more meaningful terms. After that, we look again at the Uncertainty Principle, and propose ways that it could be understood in terms of the central precepts of the present model. Finally, we come back to the Correspondence Principle, and argue that its underlying thesis has no place in the kind of consistent framework that a good theory of science should be providing.

Obviously, then, we are profoundly skeptical of the main themes of Quantum Theory, and we will challenge many of these assumptions here. Our goal, of course, is to reframe these matters and findings with an eye towards the criteria we mentioned in the first chapter — simplicity, generality, and coherence.

The Discrete Quantum

The single most crucial feature of that massive theoretical edifice known as Quantum Theory is, precisely, the construct of the discrete quantum. Indeed, Quantum pioneer Max Born described the conceptual pivot of discrete energy as being equivalent in importance for Quantum Theory as the constant speed of light is for the Theory of Special Relativity.

It is the emergence of the quantum concept that we want to consider in this section. Several key problems and phenomena provide background for this development. We will mention the blackbody problem, the photoelectric effect, the Compton effect, and Planck's constant.

The methodology involved in studying the blackbody phenomenon entails heating a given body, thereby altering its state and producing varying ranges of radiation at different temperatures. The problem at the dawn of the twentieth century was that classical physics was unable to adequately explain the detailed results of this process. This failure represented such a serious challenge to the conventional interpretation that it was called the "ultra-violet catastrophe" — a catastrophe in the sense that it represented a crucial breakdown in classical theories, based upon the example of ultra-violet light.

Max Planck brought a fresh perspective to the issue, when he was investigating the anomalies of blackbody radiation in 1900. One of his conclusions was that the traditional assumption concerning the continuous emission of blackbody radiation was erroneous, and needed to be cast aside: contrary to expectation, radiation appeared to be emitted (and absorbed) only in discrete, discontinuous chunks or packets. Planck's formula to account for this variability is considered the first great success of the new Quantum Theory.

Another crucial problem related to the photoelectric effect. The basic mechanism underlying this phenomenon is simple: when light (of certain frequencies) is absorbed by matter, electrons are emitted. The details are very complex — but, again, it was these particulars which the classical theory was unable to account for satisfactorily. One example is that the *intensity* of the incoming light determined the *number* of emitted electrons — rather than determining the *speed* of those emitted electrons, as classical theory would have expected. Instead, it was the *frequency* of the incoming light which determined the *speed* of the emitted electrons. Finally, incoming light (of the proper frequency) resulted in *instantaneous* emission of electrons — rather than being a gradual, cumulative process, as the orthodox wave theory of light would have predicted (Hecht, p. 258).

Oddly enough — given the fundamental philosophical disputes which later arose between Quantum Theory and Relativity Theory — it was Albert Einstein who proposed a satisfactory solution to this problem. Einstein's 1905 interpretation became another of the building blocks for Quantum Theory. It dealt with the above three problems in the following ways: first, the intensity of light would be explained in terms of the number of photons — a higher number of incoming photons resulted in a higher number of emitted electrons; second, the frequency of the incoming light would be explained in terms of its energy — higher energy (i.e., frequency) incoming light (photons) led to higher energy (i.e., speed) outgoing electrons; lastly, the instantaneous emission of electrons was explained simply in terms of the incoming photon particles hitting and immediately knocking out electron particles.

What was unique about the new interpretation was the notion of a discrete unit or packet of energy — which were called by Einstein *quanta*. Einstein "postulated that all electromagnetic radiation occurs in the form of quanta (or *photons*)." (Marion, pp. 418-419) "It is a central feature of the quantum theory that energy is transmitted in discrete lumps or quanta, which give the theory its name." (Davies, 1983, p. 148) "By showing that the energy of light was quantized, Einstein contributed directly toward the acceptance of Planck's original quantum theory" (Holton and Brush, p. 442). "Einstein's explanation of the photoelectric effect was so elegant, and fit the observations so well, that the quantum theory sprang suddenly into prominence." (Asimov, p. 135)

The more so several years later, when even more detailed support for the predictions was found, in the form of what is called "Compton scattering". The associated research represented another major triumph of the quantum approach — and failure of classical theory — and it came in a 1924 experiment by Arthur Compton. Conceptually, it was a very simple affair. Compton directed light (in the form of X rays) at electrons. In the resulting interaction, the X rays appeared to bounce off the electrons — just as though both entities were particles. This experiment "practically clinched the photon as a proven material particle." (Murchie, p. 457)

Once again, the essential thesis of Quantum Theory, which clearly differentiates that theory from the classical Newtonian approach, is that all energy changes occur in discrete "units of action" rather than by gradual change. The discrete unit of the quantum is defined in terms of *Planck's constant*, which is conventionally labeled *h*. And although it refers to an unimaginably small number,[2] the important thing to grasp conceptually is not the size of this "universal constant" — but its very existence.

Niels Bohr said that the "essence [of the quantum theory] may be expressed in the so-called quantum postulate, which attributes to any atomic process an essential discontinuity, or rather individuality, completely foreign to the classical theories and symbolized by Planck's quantum of action." (Bohr, quoted in Jammer, 1974, p. 90) Indeed, "To Bohr, discontinuity was a fundamental truth." (Wolf, p. 85)

Planck's constant came to be understood as representing a *fundamental and discrete unit of energy* throughout all of nature. "Planck had discovered that the basic structure of nature is granular, or, as physicists like to say, discontinuous." (Zukav, p. 48) This meant that, "[E]nergy in all its forms is 'quantized' — that is, can only be considered as behaving as though it were made up of indivisible quanta." (Asimov, p. 135) Indeed, these energy units have "been accepted as bona fide particles" (Capra, 1975, p. 56). Planck's constant, in turn, underlies *Planck's law*, and it is that law "which forms the basis of Quantum Theory" itself (Pitt, p. 285). From this idea that "energy is not continuous ... but ... exists in lumps ... evolved an entire theoretical formalism known as quantum mechanics" (Hecht, p. 205).

Note how the crucial interpretation actually unfolded here. Einstein's idea was "similar to what Planck proposed, but actually it is a leap beyond. Planck discovered that energy is absorbed and emitted in packets. He described the *processes* of energy absorption and emission. Einstein theorized that *energy itself* is quantized." (Zukav, p. 53)[3]

Waves and Particles

Pre-quantum physics assumed that every phenomenon had a fixed and absolute identity of either "wave" or "particle". But the notion of the quantum seemed to be telling us that phenomena, which had been previously thought of as continuous and wave-like, were deep down particle-like in their character. Thus light — which had been widely understood in terms of waves — was now said to be ultimately quanta-like in nature ... discrete and granular.

In fact, the absolute wave and particle categories of classical physics were crumbling — but not just in one direction, as the name of Quantum Theory suggests. In 1924, Louis de Broglie suggested that just as waves could behave like particles, so particles could behave like waves. In particular, he proposed that electrons, which had thereto-

2 Specifically, $6.62196 \, x \, 10^{-34}$ joules *x* seconds.

3 In particular, Jammer refers to "Planck's energy quanta, associated with the interaction between radiation and matter, and Einstein's light quanta, associated with radiation in transit." (Jammer, 1989, p. 36) But Popper seems to argue otherwise: "In Einstein's theory, which thus was the model of de Broglie's, light is emitted and absorbed in the form of 'particles' or 'light quanta' or 'photons'; and thus in the form of things which have a pretty sharp spatio-temporal location, at least while they *interact with matter* by being emitted or absorbed. Light is, however, *propagated* like waves." (Popper, 1982, p. 48)

fore been widely understood in terms of particles, could also display characteristics of waves. More generally, de Broglie indicated that any moving body could behave like waves: these became known as *matter waves*, a characteristic which would be associated with any moving particle — e.g., manifested in terms of wavelength (Trefil, p. 301). "De Broglie's wave was envisioned as accompanying any particle wherever the particle went. Like a shadow, the matter wave traveled alongside its particle." (Wolf, p. 92) De Broglie "argued for the reality of the matter wave.... a guiding wave determining the real trajectory of the particle in its journey through space and time." (Wolf, p. 118)[4]

The pattern which was emerging turned out to be a radically new concept in physics, whereby microphenomena in nature manifested themselves in both wave and particle forms: this pattern became known as the *wave-particle duality*. Eventually, de Broglie stated the specific parameters for such a wave-particle relationship — and voilà, a few years later, the wave properties of the electron were demonstrated experimentally (e.g., by Davisson, Germer, and Thompson).

Actually, the most well-known wave-particle demonstration is that of the *double-slit experiment*, first conducted by Young in 1803 with the intent of demonstrating that light was wave-like in character. In fact, this experiment is often presented as the quintessential illustration of the wave-particle duality. In some cases we are merely talking about thought experiments, although certain actual tests which are related in concept have also been conducted. They all serve to illustrate the same basic premise, which is that such microphenomena possess an inherently dual character — though not at the same time; and that whether one finds a wave-like or particle-like manifestation is largely dependent upon the nature of the experimental situation in which one chooses to make the observation ... even as part of the same experiment. As a result, the double-slit experiment has "become a standard paradigm for the illustration of the wave-particle dualism ..." (Jammer, 1974, p. 129) Indeed, Feynman says that this experiment "has in it the heart of quantum mechanics; in reality, it contains the *only* mystery" of the theory which cannot be explained in any classical way (Feynman, quoted in Jammer, 1974, p. 129).

Modern versions of the double-slit experiment consist of aiming either electrons or photons through openings in a wall, towards a screen beyond. Thus, for example, if coherent light is emitted from a source towards a barrier that has two side-by-side slits in it, and then the light moves through those slits to a detecting screen beyond, a clear pattern emerges which indicates that the light moves as waves: when both slits are open in the barrier, the light waves passing through one slit interact with the light waves passing through the other slit. Such a conclusion is drawn from the fact that the pattern of registration upon the final detecting screen can only be explained in terms of an interference pattern that is based on a combination of two different sets of waves, as each passes through each slit. The wave-like — as opposed to particle-like — character is clear in this phase of the experiment, because the joint wave pattern of the two interact-

4 In another description, Schrödinger referred to the "de Broglie-Einstein undulatory theory, according to which a moving particle is nothing but the foam on a wave radiation in the basic substratum of the universe." (Schrödinger, quoted in Jammer, 1989, p. 258)

ing waves is essentially opposite to that which is obtained if one or the other slit is closed off, and only a single wave passes through.

The result is paradox. As Richard Miller puts it, "[T]he two-hole experiment by itself ought to shock anyone who thinks that what the [source] filament transmits must either be corpuscles or waves." (R. Miller, pp. 523-524) Others maintain that the crucial question is not so much what leaves the source or what arrives at the sink (or end point), as what moves between them — which is a distinction we mentioned just above. Consistent with other wave-particle interpretations from Quantum Theory, like that of the photoelectric effect, the presumption would be that discrete particles are emitted from the source and arrive at the target — and indeed, it is particle-like impressions which register on the detecting screen. At the same time, it should be noted that no dynamic waving-like entity is manifest in this experiment: the only thing that waves is the static form of the particle-like pattern on the final screen.

Yet we are still left with the question of what is traveling in the realm between source and sink — particle or wave? On the one hand, it appears to have wave-like characteristics, as we have just described.[5] Then again, is the intervening entity a particle which is carried by waves, almost as though the particles have been guided into position by some wave-like force — to wit, de Broglie's matter waves (also called "pilot waves", as we shall discuss below) ... ? Or is it a particle which has been transformed into waves for the voyage — only to turn back again into a particle "when the clock strikes midnight", as it reaches its destination? Physics sometimes embraces this interpretation: "Light propagates from one place to another as if it were a wave ... and yet ... light interacts with matter in the processes of absorption and emission as if it were a stream of particles." (Hecht, p. 260)

Whatever specific analysis one came up with, the net result seemed clear — heretical as it was: what were understood to be waves appeared also to exist as particles, while what were understood as particles also appeared to exist as waves. Quantum Theory is said to have shown "particles and waves to be interchangeable manifestations of the same entities ..." (Harris, p. 159) Capra says that, "The subatomic units of matter are very abstract entities which have a dual aspect. Depending on how we look at them, they appear sometimes as particles, sometimes as waves." (Capra, 1975, p. 55) Redhead describes this wave-particle duality as "[t]he central philosophical puzzle posed by the quantum theory" (Redhead, p. 65) — and one which has caused physics no end of conceptual indigestion!

So here we are: welcome to the wave-particle dilemma — a world turned upside down ... and physics in a quandary. Suffice it to say that there has never been a satisfactory solution to the wave-particle enigma, and it remains an ongoing conceptual riddle in modern physics.

The quandary continues in the next section.

5 One can complicate the discussion by saying that each of the double slits serves as an intervening (and mutually-synchronized) target and thence secondary wave source. But that does not alter the fundamental issue here.

The Principle of Complementarity

What is generally understood as emerging from the various experiments is that one could set up conditions whereby a wave phenomenon would appear, *or* one could set up conditions whereby a particle phenomenon would appear. For the fact was that while it seemed that one could observe, e.g., a photon as a wave *or* as a particle, one could not observe a photon as both a wave *and* a particle in the same moment: "[N]o experiment has ever been designed which shows light behaving as both a wave and a particle simultaneously." (Asimov, p. 139) Likewise with the electron. And in general, "The wave and particle features cannot coexist in their pure form." (March, p. 237)

So in one sense the traditional wave-particle dichotomy remained sound: there were two distinct categories of phenomena. On the other hand, these valid classical categories no longer applied in an absolute sense: no given entity (e.g., photon or electron) was always a wave or particle in *every* situation. A note of relativism had been introduced into our classical models. Not only that but, "The use of two such mutually exclusive, and therefore contradictory, notions was rightfully regarded as a serious dilemma at the very foundations of physical science." (Jammer, 1989, p. 173)

In 1927, the Danish physicist Niels Bohr proposed what was to become — and still is — the most widely accepted interpretation of the wave-particle dilemma. It is the dualistic characteristic which Bohr's famous Principle of Complementarity describes. Here is one way in which this thesis can be phrased: "[I]f an experiment allows us to observe one aspect of a physical phenomenon, it simultaneously prevents us from observing a complementary aspect of the phenomenon." (Marion, p. 438) Bohr's own wording is similar, referring to "the mutual exclusion of any two experimental procedures, permitting the unambiguous definition of complementary physical qualities ..." (Bohr, quoted in Popper, 1982, p. 58)

Bohr "considered the particle picture and the wave picture as two complementary descriptions of the same reality, each of them being only partly correct and having a limited range of application. Each picture is needed to give a full description of the atomic reality, and both are to be applied ..." (Capra, 1975, p. 145) Oppenheimer: " ... [A]s we have learned to say in physics, complementary views, each supplementing the other, neither telling the whole story." (Oppenheimer, 1954, p. 69) Bohr's approach is therefore a combination of either/or and both/and: on the one hand, the underlying phenomenon will manifest itself *either* one way *or* the other way — but not fully together, at the same time. On the other hand, *both* the one way *and* the other way are necessary, in order to provide for a complete description of the phenomenon.

In practice, Bohr used the notion of complementarity in a variety of ways, and it is still a matter of some dispute as to how the concept was — and today, is — meant to be applied (Jammer, 1974, p. 87). Jammer says that "Bohr never gave a clear-cut explicit definition of the term, 'complementarity.' " (Jammer, 1974, p. 95) Nonetheless, "Bohr's discovery of complementarity, it is often said, constitutes his greatest contribution to the philosophy of modern science." (Jammer, 1974, p. 105)

Yet how widespread is this phenomenon? There are many who treat the principle of complementarity more broadly, and that it goes well beyond any set of microphenomena in its application. For example, C.F. von Weizsäcker felt that there was a fundamental choice to be made, and that, in Jammer's words, "[P]hysical reality is either pointlike concentrated or else spread out in space; the former is described by the model of a particle, the latter by that of a field or wave." (Jammer, 1974, p. 94) Likewise, Wolf says that:

> [T]he physical universe has a complementary nature. It is known as the *wave-particle duality*.... This wave-particle duality appears in all matter, including light. There are no exceptions.... For nature is dualistic ... The most general physical properties of any system must be expressed in terms of the complementary sets of that system. These sets are joined as complements of each other, and the more we determine or define a system in terms of one of these complements, the less we know about the other.... [T]he complementarity principle ... explained that the universe had to be composed of complementary and contradictory pictures of reality ... (Wolf, pp. 132, 139, and 153)

Unfortunately, there is nothing approaching unanimity, or perhaps even consensus, about the way Quantum Theory interprets complementarity and the wave-particle phenomenon. Trefil notes that, "[T]he question of whether elementary subatomic particles are 'particles' or 'waves' has never been resolved" (Trefil, p. 284). On the most fundamental level, for example, there is sharp disagreement as to whether a duality is even involved here. And there are even a significant number of people who reject the notion of complementarity outright, as either an affront to a kind of naive realism, or as an unnecessary layer of conceptual obfuscation.

For example, Karl Popper emphasizes the lack of any dynamic waves as accompanying the particles — which we noted in describing double-slit above — in order to criticize the very notion of a wave-particle duality, arguing instead for a pure particle interpretation:

> There is thus no symmetry or duality between particles and waves: the waves describe dispositional *properties* of the particles. As Alfred Landé puts it: "If this ... is the famous *duality*, then swans must be ... dualistic birds, since on the one hand they *are* individuals or particles; on the other hand, they *have* the quality of wavy necks; so that 'one will ever remain uncertain whether swans are really particles or waves'.".... [T]here really seems no basis for the assertion that quantum theory incorporates a dualism of particles and waves: in all these theories, the "waves" play the part merely of determining the probabilities for particles to take up certain states ... (Popper, 1982, pp. 127 and 196)

Indeed, Popper is utterly cynical about the whole complementarity approach — speaking of "the great quantum muddle" (Popper, 1982, p. 50), "which led to the emergence of the notorious 'wave-particle duality' and hence to the complementarity interpretation" (Jammer, 1974, p. 448). Popper goes on to attack the underlying notion: "[T]he principle of complementarity has ... remained completely sterile within physics. In twenty-seven years it has produced nothing except some philosophical discussions, and some arguments for the confounding of critics (especially Einstein)." (Popper, 1965, p. 101) "There is no symmetrical relationship, and thus no 'duality', between particles and waves, or between particles and their associated field." (Popper, 1982, p. 52) On the contrary, Popper says that, "[T]alking of 'dualism of particle and wave' has created much confusion ..." (Popper, 1982, p. 84)[6]

So one extreme approach is to deny that there is anything mysterious about the duality problem — or even that there is a duality issue — because we are not actually observing changes of character here. "The slogan here is that, despite appearances, a particle is a particle is a particle, and *mutatis mutandis* neither need we equivocate about waves." (Redhead, p. 65) Miller also disparages " 'the wave-particle duality' or a general 'principle of complementarity,' which many physicists came to regard as a metaphorical statement of a problem, not a solution." (R. Miller, p. 565) Others, too, have spoken equally unkindly about the "mumbo jumbo" of complementarity and wave-particle dualism, as opposed to "physically more reasonable ideas." (Whitten, quoted in Jammer, 1974, p. 463)

Some critical approaches are less explicit, concerning the character of the underlying phenomena. Among the alternatives put forward in this regard is the notion that the underlying reality is not wave or particle. Thus,

> [I]t is incorrect to say that the microphysical entities are both corpuscular and undulatory, since they are *neither*. It is more correct to say that by some of their properties they *remind us* of the behavior of particles, while other features *remind us* of the behavior of waves. (Capek, 1988, p. 99)

Max Born (who first proposed the probabilistic interpretation of quantum events, discussed below) said that it is "beyond our power to produce proof that it is actually particles or waves with which we are dealing" (Marion, p. 434). Trefil says that the phenomenon has "some properties in common with waves and other properties in common with particles. It is an entirely new kind of beast." (Trefil, p. 283). One solution is simply to merge the dual concepts together, into a single combined entity called the "wavicle" — and this composite entity may behave in a manner that we wrongly characterize as wave or particle. Similarly, Redhead says that, "Reality on this view is not a wave or a particle in complementary situations but rather transcends and unites both these concepts.... [R]eality is neither a wave nor a particle, but a quantised field." (Redhead, p. 66) For his part, "Heisenberg thought it irrelevant whether to use words like 'waves' or words like 'particles' for the description of what was going on" (Jammer, 1974, p. 68; cf. also Heisenberg in Zukav, pp. 113-114). We shall further discuss Heisenberg's important views on this subject later in the chapter.

Wolf also seems to reduce the issue down to a limitation of our concept system: "The wave-particle duality and the pictures associated with this duality ... were a result of the fundamental clash between two contrasting mental constructs of the appearance of reality." (Wolf, p. 123) Indeed, Wolf appears to explain the wave-particle duality in terms of the act of observation itself: "[P]hysicists discovered that their acts of observing the atomic world introduced a duality, a double or paradoxical way of seeing." (Wolf, p. 129)

6 Popper clearly detests the concept of duality as it is applied in Quantum Theory, because of what he sees as all the unfortunate misconceptions which go with it. Indeed, he endorses Landé's description of the "three articles of faith" — duality, uncertainty, and complementarity, together "with all those other gratuitous quantum mysteries and horrors such as subject-object confusion, the 'transitions from the possible to the actual' as a consequence of our measurements, and quantum jumps." (Popper, 1982, p. 133)

It is unclear whether Einstein was among these extreme cynics. On the one hand, for example, he was very explicit in asserting "the fact that the wave-particle duality is a reality which cannot be dodged by metaphysical tricks." (Letter from Einstein to Renninger, 1954, quoted in Jammer, 1974, p. 495) Even Popper acknowledged that Einstein accepted the wave-particle duality (Popper, 1982, p. 53). On the other hand, Einstein also made some quite caustic comments about wave-particle — and certainly the associated concept of duality — a point we shall return to in Chapter 9.

In sum, there have always been skeptical responses by physics to the wave-particle and complementarity dilemma. There are those who deny the very existence of any anomaly whatever, and who decry attempts to find deeper meaning. Others acknowledge the apparent *duality*, but explain it in terms of what we shall later describe as essentially an underlying *singularity*. We talk about the basis for the latter interpretation next.

The Uncertainty Principle

Bohr's Complementarity Principle tells us that we cannot have it both ways: we cannot fully identify both a particle and a wave at the same time. Yet how close can we come to that ideal goal? Heisenberg's famous Uncertainty Principle (also called the Indeterminacy Principle)[7] is a mathematical statement, based upon Planck's constant, which in effect describes how precise we can be in specifying both wave and particle outcomes *simultaneously*.[8] The Uncertainty Principle can thus be seen as providing the particular interchange or limitation parameter described generally by the Principle of Complementarity (e.g., Marion, p. 438), thereby quantizing the wave-particle relationship. We have thus gone from the quantum as a fixed and certain unit of minimal energy, to the notion of an unfixed and uncertain ultimate reality which exists only within the limitations of this minimal energy chunk.

One corollary of the principle that an entity cannot simultaneously manifest both its wave and particle attributes is that one cannot precisely measure a moving entity's *momentum* and *position* at the same time: if one wishes to determine a moving entity's position with absolute certainty, then one can only determine its momentum within a range of error; conversely if one wishes to determine a moving entity's momentum with absolute certainty, then one can only determine its position within a limited range of

7 One might use the term "uncertainty" when it refers to the state of our knowledge about some phenomenon in nature, and "indeterminacy" as regarding a more objective description of the state of the entity itself. Jammer, 1974, p. 61, discusses this question.

8 "Planck's constant, Bohr realized, establishes in these equations a relation between two descriptions of radiation that are mutually exclusive, but equally necessary.... [T]he indeterminacies, expressed by Heisenberg's relations, are precisely the price we have to pay if we nevertheless attempt to apply such mutually exclusive sets of classical conceptions simultaneously." (Jammer, 1989, p. 363; Jammer, 1974, p. 101) Born and Heisenberg phrase it differently: "[T]he real meaning of Planck's constant *h* is this: it constitutes a universal gauge of the indeterminism inherent in the laws of nature owing to the wave-particle duality." (Born and Heisenberg, quoted in Jammer, 1974, p. 114)

accuracy.[9] Again, one can specify both qualities simultaneously — if one does not demand absolute precision in both cases.

But the issue goes beyond both wave and particle, as well as momentum and position — or even another linked pair, that of time and energy. Rather, the Uncertainty Principle represents a general restriction upon our ability to measure all such "conjoint variables" in nature with absolute precision. Moreover, the issue also extends beyond the present: for example, since we are limited from specifying the exact position and momentum of an entity at any one point in present time, we are thereby precluded from extrapolating what the entity's precise position and momentum will be at some time in the future. As a result, the Uncertainty Principle means that we cannot predict future events with certainty, but only with probability.[10]

But of what use are predictions which are not precise? There are two caveats, in this regard. First, this limitation of prediction is said to apply only to microphenomena, as opposed to the kind of real-world events which we all experience; we discuss this issue in the next section. Second, while we are constrained in predicting the behavior of individual micro-entities, we have lost none of our ability to predict the behavior of large numbers of such entities, and thus the *likelihood* of particular event outcomes. "Quantum Theory cannot predict the result of any single event, but the *average* of a large number of events can be predicted with precision." (Marion, p. 437) In fact, "in Bohr's view, quantum mechanics can be consistently regarded as a computational device for obtaining the probability of every measurement ..." (Jammer, 1974, p. 198) Quantum Theory is thus a statistical, or probabilistic, model — very different from the sort of strictly mechanistic or deterministic theories which preceded it.

That classical approach was based upon a picture of the world as a grand machine, with spatially separated entities interacting over time in a perfectly lawful manner — such that if one were able to completely specify all the conditions affecting a system at any given time, then one would be able to use the laws of nature discovered by physics to specify the course of that system into the indefinite future. Newton's model was built upon this mechanistic approach, which derives from Descartes — who "gave scientific thought its general framework — the view of nature as a perfect machine, governed by exact mechanical laws." (Capra, 1983, p. 60) According to such a causal view of nature, every event has clear and ultimately identifiable antecedents: there is always a cause underlying every effect. "All that happened had a definite cause and gave rise to a definite effect, and the future of any part of the system could — in principle — be predicted

9 Schrödinger proposes a meaningful example with which to understand the scope of the constant which underlies the Uncertainty Principle. "For an electron, this constant happens to be about 1 if the units centimetre and second are used. Thus, if the velocity of an electron is considered sharp with a latitude of only 1 cm/sec, its location has to be considered as blurred within the latitude of 1 cm." (Schrödinger, 1950, p. 110)

10 "Quantum mechanics clearly shows that nothing can be predetermined, no matter how events may appear." (Wolf, p. 214) In fact, Heisenberg challenged the law of causality both in terms of an ability to determine the present state of an individual entity, as well as our independent inability to determine the future of state of an individual entity. See Jammer, 1974, pp. 75-76.

with absolute certainty if its state at any time was known in all details." (Capra, 1983, p. 66)

Yet Quantum Theory has found that subatomic phenomena do not operate in accord with such strict causal determinism, and that some irreducible level of uncertainty is inevitably involved in any such predictions. We have talked about this outcome in terms of the wave-particle pattern — and it is also apparent in another kind of important non-causal pattern that we shall discuss in Chapter 9, under the topic of the Einstein-Podolsky-Rosen experiment.

It is not just the classical approach which is contradicted by the indeterminism that is inherent in Quantum Theory: the Theory of Relativity is also challenged in the same way. For his part, Einstein rejected such indeterministic or probabilistic interpretations, encapsulated in the oft-cited phrase that is attributed to him: God does not play dice. More specifically, he wrote to Max Born that, "The quantum mechanics is very imposing. But an inner voice tells me that it is still not the true Jacob. The theory yields much, but it hardly brings us nearer to the secrets of the Old one. In any case I am convinced that he does not throw dice ..." (Letter from Einstein to Born, 1926, quoted in Jammer, 1974, p. 155)[11]

So what is going on? What is the source of the limitations we have been describing which are specified by the Uncertainty Principle? We shall consider three potential explanations. One interpretation is that we are dealing with a failure of theory: Quantum Theory is inadequate, because it is unable to provide full precision in these conjoint situations — which might, in turn, be due to the theory's being based upon unsatisfactory concepts to begin with. This first issue is sometimes posed in terms of *whether Quantum Theory is both a consistent and a complete theory*. A second idea is that there is nothing wrong with the theory or its concepts, but rather that the pattern of uncertainty is simply an artifact of the observation process — a *limitation of measurement* which is an inevitable result of examining nature. The third explanation is the most fundamental of all — that the factor of uncertainty is in the essence of nature itself, and there is no way to ever get around it, because it reflects a more profound *limitation of existence* ... that phenomena in nature only exist in a probabilistic, rather than definitive, way.

We shall deal with the first explanation presently — and especially in the context of a major dispute between Bohr and Heisenberg that we shall discuss extensively in a coming section. Aside from that, however, the choice between the remaining two options is significant — and stark. For example, if we are measuring a moving entity's momentum with absolute precision, does it continue to have a specific position — but we cannot simultaneously measure such position, e.g., because we are disturbing it by the very act of measuring the momentum? Or, do conditions which manifest the moving entity's momentum thereby preclude that entity from even having any distinct position at all, in the same moment? Thus, if our experiments reveal the electron to possess a wave-like character, does the specific particle-like character continue to exist at that moment ...

11 There seems to be some question about the specific date. Thus, Holton cites a 1965 work by Born as referring to a letter from Einstein to Born of 1944 which states that, "You believe in the dice-playing God, and I in the perfect rule of law in an objectively existing world which I try to capture in a wildly speculative way." (Holton, p. 276)

but we are unable to simultaneously measure it? Or is the issue more fundamental, and the wave-like electron simply has no precise particle-like attribute at that moment? Therefore, is the probability restriction, which the Uncertainty Principle specifies, only a limited probability of seeing — or a more profound probability of being?[12]

So which is the issue, measurement or existence? On the side of the former is the simpler explanation, which is consistent with " ... the naive realist assumptions: that a quantum system always has a dynamical state specifying a precise real value for every dynamical variable, and that this is the value which a successful measurement of that dynamical variable would reveal." (Healey, p. 180) According to Lockwood, for example, "Einstein and Schrödinger [assumed] that an electron invariably *has* some exact position and momentum ..." (Lockwood, p. 202)

If the source of the problem is not in the nature of existence itself, then one might conclude that the uncertainty is due to "the disturbance of an object by the act of observing it." (March, p. 224) As Miller paraphrases Bohr: "[M]easurement requires physical interaction." (R. Miller, p. 545) Or in Wolf's terms, "[T]o observe is to disturb." (Wolf, p. 117) Healey refers to " ... unexplored indeterministic processes ..." and he speaks of the underlying "indeterministic transitions among dynamical states, taking place throughout the [measurement] interaction." (Healey, p. 45) Zukav says that, "Whatever it is that we are observing *can* have a determinable momentum, and it *can* have a determinable position, but of these two properties, *we must choose*, for any given moment, which one we want to bring into focus." (Zukav, p. 114)[13]

For his part, Jammer argues that disturbance itself is not the issue, and he criticizes the "widespread view that it is the disturbance of the object by the observation that entails the principle of indeterminacy and thus eventually constitutes the foundation of the whole theory." (Jammer, 1974, p. 160) He goes on to state that such disturbances could, in principle, be quantified and integrated into the calculations. Yet, "The meaning of the principle of indeterminacy is precisely the statement that such a corrective theory is not possible." (Jammer, 1974, pp. 160-161) Wolf makes a similar point in arguing that, "[T]he uncertainty principle ... denies existence to objects having both spatial locations and well-defined paths of motion simultaneously." (Wolf, p. 200)

Obviously, these are very profound, and complex, questions. Nor, as we mentioned, did two of the main fathers of Quantum Theory agree as to the source of the problem — something we shall consider at greater length later in the chapter. On the one hand, the "reduction of definability to measurability was unacceptable to Bohr" (Jammer, 1974, p.

12 Here is another way of describing these issues: "The exact statement of the basic (Born) probability rules has also been a highly controversial matter: Do these specify probabilities that a quantity has a certain value, that it will or would acquire a certain value on measurement, or that a measuring apparatus will or would record a corresponding result if the quantity is measured?" (Healey, p. 6)

13 A related argument is that the two conjoint variables cannot be simultaneously measured to begin with. Thus, Jammer cites the work of Suppes (1961), in which he concludes that, "not only are position and momentum not precisely measurable simultaneously, they are not simultaneously measurable at all." (Suppes, quoted in Jammer, 1989, p. 355) The point is that what we take as an inability to simultaneously measure conjoint partners with precision actually results from the fact that we are only measuring them in succession.

69), who treated the uncertainty as due to "not just uncertainty of knowledge. Rather it is uncertainty of being." (Murchie, p. 493) Therefore, measurability was not the issue for Bohr, who saw the explanation in a much more fundamental pattern of nature, described by his Principle of Complementarity. On the other hand, for Heisenberg, it was meaningless to speak about a hypothetical reality which could not be observed or measured.[14]

Indeed, Heisenberg's position goes back to the first explanation we put forward above — although not so much in terms of an incomplete (Quantum) theory, but rather in terms of inherently inadequate concepts. That is, Heisenberg saw the limitation not in nature, but rather, "a limitation of the applicability of classical notions, like position or momentum, to microphysical phenomena ..." (Jammer, 1974, p. 65) Others echo this interpretation: "[O]ur everyday experience does not deal with everything that there is in the world.... It should not surprise us, therefore, that categories which are appropriate to the description of objects in the everyday world of normal-sized objects are *not* appropriate to the description of objects whose size is very small." (Trefil, p. 283) As d'Espagnat puts it, " ... it is impossible to extend indefinitely the linguistic framework of the world of things into the sphere of the very small." (d'Espagnat, p. 71)

Heisenberg's thinking about unsatisfactory concepts went further. Indeed, his goal here can be characterized as essentially finding a conceptual bridge between conjoint constructs like wave and particle — but which would not necessarily validate either concept in its own right. In this respect, he was looking for an interpretation which would perform a similar function with microphenomena as Einstein's redefinition of simultaneity did for the general concepts of time and space — which we shall discuss in Chapter 8:

> Deep in his heart Heisenberg cherished the hope that an operational analysis of the concepts of position and velocity, or rather their reinterpretation, would do for the mechanics of micro-objects just what Einstein's [re]analysis of the notion of simultaneity had done for the mechanics of high-speed phenomena. Just as it was meaningless to speak of the simultaneity of two distant events before the introduction of an appropriate synchronization of clocks, so it is "meaningless to speak of the place of a particle with a definite velocity," said Heisenberg. (Jammer, 1974, p. 58)

We shall return to this dispute later in the chapter, but by now it should be apparent that the question of what the Uncertainty Principle refers to is unclear ... on various levels: The principle may be due to an incomplete or inconsistent theory, perhaps by virtue of the inherent inadequacies of our conventional concepts; the principle may reflect more pragmatic problems which are an inevitable outcome of the process of measurement; or most profoundly of all, the principle may represent fundamental limitations of existence. These are grave issues, which deal with the very nature of reality itself. Yet Quantum Theory provides no clear answers to these questions, and they remain the subject of deep metaphysical debate.

14 "[P]hysics must confine itself to the description of the relationships between perceptions." (Heisenberg, quoted in Jammer, 1974, p. 76) Heisenberg's very approach was to "establish a basis for theoretical quantum mechanics founded exclusively upon relationships between quantities which in principle are observable." (Heisenberg, 1925, p. 261)

The Correspondence Principle

We now want to mention an important precept which Niels Bohr proposed, regarding the relationship between new and old theories. The most general form of the principle is a straightforward statement that any new theory must encompass an older theory as a special or limiting case, to the extent that the older theory is valid in particular circumstances. As Rogers puts it, the "[n]ew theory and old must overlap and agree in the region where the difference between their assumptions does not matter." (Rogers, p. 737) Such a principle would describe the relationship between General and Special Relativity, and that between Special Relativity and Newton's model. But the particular relationship which Bohr was concerned with was that between the newer Quantum Theory and the older classical model.

The laws of classical physics which describe the macrophenomena of our experience would be interpreted as a special or limiting case of the more general laws of quantum physics. As Holton puts it, "[C]lassical physics becomes the limiting case of the more complex quantum physics" (Holton, p. 114). Consequently, the principles of Quantum Theory, which are applicable to microphenomena, must also be applicable to macrophenomena — where, of course, they should yield the same results as do the classical theories (Pitt, p. 84). Needless to say, the reverse will not hold: classical physics theories, being a limiting case of general physics, do not apply to the realm of microphenomena.

But what would this mean, in terms of Quantum Theory's basic discoveries about nature? Given the assumption that the newer theory is supposed to encompass the older theory, then the central ideas and constructs of Quantum Theory should apply pervasively throughout nature. Yet in practice the central constructs of Quantum Theory are typically applied only to the micro world — whereas the classical constructs are usually thought to describe the everyday macro world of our normal experience. For example, Capra says "At the atomic level, matter has a dual aspect" (Capra, 1975, p. 136) ... but apparently not so at the classical level. So why should there be a wave-particle duality, or even a discrete quantum, in the realm of microphenomena — while in the realm of classical physics and ordinary life, there is no sign of them?

The conventional explanation for why we are not more aware of the discrete energy characteristic in our common experience is simply that on the "macroscopic level this aspect of nature is not evident" (Zukav, p. 52). "[I]n a flood of radio emissions, the granular nature of the radiation is generally obscured, and one detects only a smooth wavelike transfer of energy." (Hecht, p. 473) "[T]he mode of being of matter in the macroscopic domain depends on the cancellation of chance fluctuations arising in the microscopic domain." (Bohm, 1957, p. 24) Referring to Planck's constant, Pitt notes that, "The effects of the uncertainty principle are not apparent with large systems because of the small size of h." (Pitt, p. 396) Or Holton: "[T]he discreteness of individual events is dissolved, owing to the large number of events, in an experienced continuum." (Holton, p. 115)

As far as the wave-particle duality goes, we saw earlier in the chapter that some observers think the very idea is due to the false assumption that concepts which apply in the macro world are equally applicable in the micro world. The solution would presumably be that more sophisticated concepts would dispense with the duality altogether. Another idea is that the wave-particle duality straddles the two worlds. Thus, Holton writes that the Correspondence Principle served to join the discrete quantum postulate of Quantum Theory with the gradual continuum presumption of the classical model[15] — and as such served as a historic way station or forerunner of the Complementarity Principle itself, which Bohr put forward later (Holton, p. 114).

All in all, we are left wondering how broadly basic constructs like the discrete quantum, complementarity, uncertainty (along with probability) and the observer-observed whole (if not causality) are meant to apply. There is no clear answer, and there are some who think that the Correspondence Principle is not a very satisfactory resolution of the issues here: it seems ad hoc and "patched-up" (Murchie, p. 403) as it "attempted, not very convincingly, to reconcile the classic and quantum worlds" (Murchie, p. 468).

Sir William Bragg captured the confusion we are dealing with in his famous remark, "We use the classical theory on Mondays, Wednesdays, and Fridays and the quantum theory on Tuesdays, Thursdays, and Saturdays." (cf. Jammer, 1989, p. 173) For his part, Einstein wrote that, "I do not believe in micro- and macro laws, but only in (structural) laws of general validity. And I believe that these laws are logically simple, and that reliance on this logical simplicity is our best guide." (Letter from Einstein to Bohm, 1954, quoted in Jammer, 1974, p. 219)

The Wave Function

We now come to another very important concept in Quantum Theory — and one of the most difficult to understand.

When we think about the wave-particle duality, the particle side of the equation is fairly clear in terms of what it represents — some kind of obvious, concrete, physical entity. But the wave phenomenon is far more obscure. Are we talking about the sort of physical waves we see in everyday life, such as when we shake a rope? Or are we talking about some abstraction that can never be observed directly — not because its substance is less material or concrete, but because it does not correspond with anything physically "real" that we could ever conceivably see or touch?

We are here talking about the seminal Quantum Theory construct known as the wave function, often represented by the Greek letter ψ. d'Espagnat writes that in general, "[T]he wave function … is the essential element of the fundamental mathematical formalism underpinning the whole [quantum] theory …" (d'Espagnat, p. 99)[16]

Most interpretations of the wave function do not treat it as corresponding to anything concrete, but rather in terms of a wave of probability — the probability that a concrete particle is likely to be found at some location.

15 Holton talks about "the continuity-discontinuity (or wave-particle) duality" (Holton, p. 117).

16 Note that there are some who disregard the wave function altogether as "only a superfluous auxiliary construct." (Holton and Brush, p. 495)

> At the subatomic level, matter does not exist with certainty at definite places, but rather shows "tendencies to exist," and atomic events do not occur with certainty at definite times and in definite ways, but rather show "tendencies to occur." In the formalism of quantum theory, these tendencies are expressed as probabilities and are associated with mathematical quantities which take the form of waves. This is why particles can be waves at the same time.... They are "probability waves," abstract mathematical quantities with all the characteristic properties of waves which are related to the probabilities of finding the particles at particular points in space and at particular times. All the laws of physics are expressed in terms of these probabilities. (Capra, 1975, p. 56)

This is the predominant interpretation of what wave, as in wave-particle, represents: merely an abstract probability pattern — in contrast to some physically real wave, such as we might see at the ocean.[17] No actual physical wave manifests itself, in the case of the wave function, but instead there is the waxing and waning of the likelihood that a concrete physical particle will be found.[18] Zukav describes the wave function itself as "a dynamic description of possible occurrences ..." (Zukav, p. 80) Jammer paraphrases Fock as claiming "that the wave function denotes something physically possible and not something actually realized ..." (Jammer, 1974, p. 447) Jammer also quotes von Weizsäcker, who goes so far as to say that, "[T]he ψ-function expresses the probability of every possible result of every possible experiment." (von Weizsäcker, 1941, quoted in Jammer, 1974, p. 486) Finally, de Beauregard says that, "[W]ave and particle are the continuous and the discrete aspects inherent in the concept of probability ..." (de Beauregard, p. 119).

17 Cf. Baker, p. 216. Schrödinger states, with skepticism, that according to the orthodox position, "[A]ll waves, including light, ought rather to be looked upon as 'probability waves.' " (Schrödinger, 1950, p. 110) Schrödinger, along with Bohm and Popper, are three who do champion a reality-linked interpretation of the wave function. Bohm felt that, "[T]he wave function ψ likewise represents an objectively real field whose 'field equation' is the Schrödinger equation just as the Maxwell equations are the field equations of the electromagnetic field ..." (Bohm, quoted in Jammer, 1974, p. 282) Indeed, "Schrödinger ascribed to ψ an electromagnetic meaning, defining it in the early sections of his paper as a continuous distribution of electricity in actual space." (Jammer, 1989, p. 299) Popper interpreted the wave function in terms of a real physical property referred to as a "propensity"; Jammer mentions some of the historical antecedents for this view. (See Popper, 1982, p. 129; R. Miller, p. 548) Popper says that these functions are "*physical and real, and not merely a mathematical device*" (Popper, 1982, p. 84). It is also noteworthy that Heisenberg himself "thought it necessary, in view of the fact that these ψ-waves evolve in time and propagate in space in accordance with Schrödinger's equation, not to regard them as merely a mathematical fiction but to ascribe to them some kind of physical reality." (Jammer, 1974, p. 44)

18 One view, sometimes called the "Born hypothesis", says that Quantum Theory predicts the probability of the result of a measurement performed on a single entity; whereas the alternative "Einstein hypothesis" says that Quantum Theory is only dealing with the relative frequency of exact measurement results performed on an ensemble or collection of individual entities. "The ψ function does not in any way describe a condition which could be that of a single system: it relates rather to many systems, to 'an ensemble of systems' in the sense of statistical mechanics." (Einstein, quoted in Jammer, 1974, p. 440) In any case, Jammer notes that, "Many physicists deny the logical legitimacy of such a distinction." (Jammer, 1974, p. 441)

In terms of these ideas, then, the only sense in which the notion of a wave applies at all, with the wave function construct, is simply in terms of the wave's shape and properties — including the usual wave-like interference pattern. If one were to superimpose the wave upon a section of physical space, then regions which lay at the higher amplitudes would correspond to points where a physical particle was likely to be found, whereas regions which lay at low amplitudes would correspond to points where a physical particle was unlikely to be found. Similarly, a very steep and narrow portion of the wave would indicate a very high probability that the particle would be localized within that region, whereas a very low and broad wave would indicate a low probability of finding the particle in that area. Another way of describing the situation is to say that the square of the wave's amplitude at any given location corresponds to the probability of actually finding a particle there.[19]

Some of the genesis for the wave function comes from de Broglie, but ultimately it was Schrödinger who defined the construct in its own right. As far as a mechanism which might underlie or illustrate the wave function concept, consider a model from Renninger, which in turn is similar to one put forth by de Broglie. Renninger proposed an experiment in which the photon is a particle which traverses a continuous trajectory in space and time but also spreads out like a field or a wave.[20] It led Renninger to conclude that, "[E]very quantum is an energy-corpuscle which is 'carried' or 'guided' by an energy-free wave." (Renninger, quoted in Jammer, 1974, p. 494) In de Broglie's "theory of double solution", later truncated into his "pilot-wave theory", the energy corpuscle or particle is also "guided by an extended wave ψ ..." (Jammer, 1974, p. 49)[21]

In historical terms, it was Schrödinger who "generalized de Broglie's limited speculations into an all encompassing mathematical formulation.... [which] marked the

19 Zukav summarizes the distinction between the wave function itself and the square of its amplitude: "The wave function is calculated with the Schrödinger wave equation. It deals with possibilities. The probability function is based upon the wave function. It deals with probabilities." (Zukav, p. 73) By contrast, Popper refers to the basic wave function as a "probability distribution function", whereas he calls the square of its amplitude a "density distribution function" (Popper, 1982, p. 52). The difference between these two approaches is at least partly based on whether we are dealing with a single particle or an ensemble thereof, as mentioned in the previous footnote.

20 One of the objections to such pilot-wave approaches is in the assumption that any moving "wave packet" (associated with the particle) would eventually spread or dissipate to infinity. In the latter part of Chapter 9, we speculate about some of these issues, in terms of such spreading waves also being a search for a complement. We would suppose that it would normally not take until infinity for the spreading wave packet to reach or find such a complement.

21 "The wave-particle duality, in brief, has been reduced by de Broglie to a wave-particle synthesis: not wave *or* particle but wave *and* particle constitute physical reality!" (Jammer, 1974, p. 49) Popper claims that de Broglie's theory is itself based on a similar model of Einstein's. (Popper, 1982, p. 48) According to that approach, "Einstein regarded the wave field as a kind of 'phantom field' whose waves guide the particle-like photons on their path ..." (Jammer, 1974, p. 41) Wolf says that light "follows the least time paths along the shortest routes by going along all possible paths from start to finish. It does this by sending out little wavelets that bob and weave along their paths in all possible directions. But it is only the least time path that reveals itself; all other paths cancel out in confusion and noise, as the troughs and crests of the light waves slosh together." (Wolf, p. 196)

arrival of the modern conception of *quantum mechanics*, or *wave mechanics*." (Hecht, p. 270) In this process, Schrödinger developed a very important differential equation, since called the *Schrödinger wave equation*, which "is an equation of motion of de Broglie waves." (Hecht, p. 271) The solution of this equation is called the *wave function*, or ψ. Jammer says that according to de Broglie's model, Schrödinger's wave equation "admits two different kinds of solution: a continuous ψ function with statistical significance and a 'singularity solution' whose singularities constitute the physical particles under discussion." (Jammer, 1989, p. 309)

These two different modalities form the basis for two subsidiary concepts, which are crucial here: one is the *evolution of the wave function* itself — evolving in the sense that it is not static, but changing in a wavelike fashion (as described above), as it continually spreads out in different directions. By contrast there is the *collapse of the wave function*, which is generally correlated with the manifestation of an actual physical particle.[22] Zukav summarizes the difference in this manner:

> Physicists speak as if the wave function exhibits two very different modes of development. The first is a smooth and dynamic development, which we can predict because it follows the Schrödinger wave equation. The second is abrupt and discontinuous ... This mode of development is called the collapse of the wave function.... The transition from the first mode to the second mode is called a quantum jump. (Zukav, p. 75)

Physics has found a consistent correlation between the statistical aspects of the wave function and certain experimental results. But the meaning of these discoveries remains totally unclear. In fact, two of the main approaches regarding the interpretation of the wave function concept are so fundamental and so divergent that they amount to totally different schools of thought, within the Quantum Theory framework (Davies, 1980a, p. 145). The conventional view is part of the "Copenhagen Interpretation", which represents the consensus approach to Quantum Theory, generally; whereas the more extreme view is labeled the "Many Worlds Interpretation". As we shall see next, these two approaches differ radically, both in terms of the degree of reality which corresponds to the evolving wave function, as well as in terms of whether there is any collapse at all.

The Copenhagen Interpretation. The Copenhagen Interpretation of Quantum Theory was put forward in 1927, largely under the influence of Niels Bohr, and it remains to this day the orthodox position of modern physics. It is a very pragmatic philosophy that entails several components. One stream is to be found in Bohr's own Principle of Complementarity; a second is Heisenberg's Uncertainty Principle; and a third is Max Born's probabilistic interpretation of the wave function.[23] The result is that, "[T]he Copenhagen interpretation is not a single, clear-cut, unambiguously

22 One can distinguish between the upper case Ψ, which "stands for the complete wave function including time" (French and Taylor, p. 119), and the lower case ψ, which "stands for the spatial part" of the wave function (French and Taylor, p. 119).

23 Healey substitutes Pauli for Born in saying that "Bohr, Heisenberg, and Pauli each held significantly different views on how quantum mechanics should be understood." (Healey, p. 2)

defined set of ideas but rather a common denominator for a variety of related viewpoints." (Jammer, 1974, p. 87)

The Copenhagen Interpretation is also associated with a quantum postulate which treats that which is observed and the observational mechanism (including the observer) as an indivisible whole. Also inherent is the controversial notion that consciousness and the act of observation are necessary, in order for any aspect of reality to become actualized. As Wolf puts it, "Consciousness is the process wherein potential reality becomes actual reality.... It is the wave function collapsing." (Wolf, p. 240)[24] Thus, observation is involved not only in identifying the nature of reality, and not only in distorting the nature of reality, but in actually making manifest any reality at all:

> Is the momentum hidden when we measure the atom's position? Is the atom's location hidden when we measure the momentum? Not in any ordinary sense of the meaning. Both of these attributes, momentum and position, are *potentially* present in nature, but not *actually* present, until an attempt is made to measure these attributes. How we choose to compromise will determine whether the wavelength (momentum) side of reality or the particle (location) side of reality is manifested.... [W]e alter potential reality, making it actual. (Wolf, p. 140)

According to the Copenhagen Interpretation, therefore, the difference between an evolving wave function and a collapsed wave function (or reduced "wave packet") is precisely that between potentiality and actuality. And the sole catalyst which brings about this crucial transformation is the observer — indeed, the observer's consciousness. Thus, one new factor which the wave-particle duality introduces — with a bang — is to thrust the observer smack into the middle of how we define reality.[25] So we no longer have the classical situation of a neutral observer exploring, and e.g. measuring, an objective reality. The paradox of the wave-particle duality implies that the observer is part of the reality which (s)he is observing — observer and observed form a single system and are not separate, as had been previously assumed.

This new realization about the important role of the observer was one of the most significant impacts which the wave-particle duality had, and it was really a kind of shock to science, philosophically. Somehow man, or at least science, had lost its innocence or aloofness: now we were an inextricable part of the scene we were examining, and not just a neutral outside observer of it. As Wolf puts it, "[N]o clear dividing line exists between ourselves and the reality we observe to exist outside of ourselves. Instead, reality

24 There are some physicists who identify only the Principle of Complementarity with this assumption — including Wolf, who says for example that, "According to the tenets of the complementarity principle, there is no reality until that reality is perceived." (Wolf, p. 127) Again, "according to Bohr's complementarity principle ... reality does not exist until it is chosen." (Wolf, p. 130) Jammer equates the two positions in speaking about the " 'complementarity interpretation' or 'Copenhagen interpretation' of quantum mechanics." (Jammer, 1974, p. 87)

25 "It is well known that the reduction of wave-packets always appears in the Copenhagen theory when the transition is completed from the possible to the actual ... which is done by the 'observer' ..." (Heisenberg, 1955, p. 23)

depends upon our choice of what and how we choose to observe.... [R]eality does not exist until it is chosen." (Wolf, pp. 128 and 130)

This meant that the experimenter could determine, by virtue of the conditions which were established for the experiment, whether the phenomenon (s)he was observing would be a wave or a particle ... as we saw in the double-slit situation. "It is the way in which we make our measurements that determines whether we obtain a wave or particle result." (Marion, p. 434) Or in Heisenberg's words: "We have to remember that what we observe is not nature herself but nature exposed to our method of questioning." (Heisenberg, quoted in Murchie, p. 469)

The Copenhagen creed here is partly based upon von Neumann's dual-process model of measurement, which can also be understood in terms of the evolution vs. collapse distinction mentioned above.

> One process is that described by the time-dependent Schrödinger energy equation. It is continuous, i.e., values are reached via a continuous gradation of intermediate values. And it is deterministic in that an initial quantum state, generally described through a superposition of wave functions, dictates all subsequent quantum states. The second process consists of measurement interventions.... The episode is discontinuous, since the other components of the quantum state [or ψ] simply disappear. It is nondeterministic ... (R. Miller, p. 534)

As Cartwright puts it,

> Von Neumann said there were two kinds of evolutions. He wrote down two very different looking equations.... There are evolutions that are correctly described by the Schrödinger equation, and there are evolutions that are correctly described by something like von Neumann's projection postulate. (Cartwright, p. 198)[26]

The Schrödinger equation is therefore treated as describing the evolving wave function itself, while von Neumann's own projection postulate describes the wave function's collapse.[27] The two-stage aspect of this interpretation is crucial, as is the role of measurement and consciousness in moving from the one to the other. According to the Copenhagen Interpretation, we cannot escape "von Neumann's far-reaching conclusion

26 The transformation from evolution to collapse is sometimes colorfully referred to as a "von Neumann cut".

27 Cartwright goes on to say that the two kinds of evolution "are not different kinds in any physically relevant sense.... Quantum statistical mechanics has developed a more abstract formalism, better suited to these kinds of problems. This formalism writes down only one evolution equation.... Schrödinger's evolution and reduction of the wave packet appear as special cases of a single law of motion which governs all systems equally.... The two evolutions are not different in nature; their difference is an artifact of the conventional notation." (Cartwright, pp. 198, 199, and 164) Likewise, Lockwood argues that, "[T]he projection postulate is false.... The collapse theory does violence to quantum mechanics, effectively obliterating one of its most beautiful features. [i.e., time-symmetry]" (Lockwood, p. 210) Lockwood argues for a variant of the Many Worlds approach.

that it is not possible to formulate the laws of quantum mechanics in a complete and consistent way without a reference to human consciousness." (Jammer, 1989, p. 394)

Beyond this is the positivist or instrumentalist assumption that the results of experiments are all there is:

> [T]he physical world has just those properties that are revealed by experiments, including the wave and particle aspects, and ... the theory can only deal with the results of observations, not with a hypothetical "underlying reality" that may or may not lie beneath the appearances. Any attempt to go further than this, to specify more precisely the microscopic details of the structure and evolution of an atomic system, will inevitably encounter nothing but randomness and indeterminacy. (Holton and Brush, p. 500)

In fact, the extreme instrumentalist view, which the Copenhagen Interpretation represents, denies that the laws of physics actually describe phenomena of nature so much as our *knowledge* about those phenomena of nature: "[T]he wave function represents man's knowledge of the state rather than the state of the system itself ..." (Jammer, 1974, p. 445) Again, the idea of an objective reality, apart from the observer of that reality, completely disappears. "[O]bservations not only *disturb* what has to be measured; they *produce* it!" (Jammer, 1974, p. 161)

Therefore, it is not only how we perform our observation, as to whether we find a wave-like or particle-like result. For we must make a choice, one way or the other, in order for reality to even manifest itself. Yet this leads to another issue, regarding observation: what was the objective reality *before* we carried out our observation? Was the entity a wave, or was it a particle — or was it some kind of combination of these conventional concepts? Furthermore, how could we possibly know that prior state without performing a measurement — which would inevitably influence the result?!

But again, the issue is even more profound: since we can never know the "true" state of an entity prior to our examining it, then what basis do we have for even assuming there *is* a true state prior to the observation — *does the entity even exist*? And if the entity does exist prior to the observation, then in what form or sense does it do so ... without the triggering factor of observation? Needless to say, no matter how one decides such a question, there would be absolutely no way to prove it: the only acceptable proof is an observation — and since we know that this would transform the observed, then such a proof has no validity. Finally, rather than go endlessly round and round in unprovable and metaphysical arguments about the "real" state of unobserved reality — when there is good reason to think that *any* observed reality will (or at least, may) be different from its unobserved counterpart — are we not better off from a scientific vantage point to simply conclude what we know ... which is that the only reality worth talking about in science is an observed reality, and thus that reality only exists if it is observed or measured?

According to this perspective, then, it is meaningless to speak about an abstract reality apart from the means by which it has been apprehended. Indeed, we need not even burden ourselves with the assumption that an unobserved reality even exists, as reality — at all! Without the observer and his/her observation, then, there would be (in effect) no observed: reality would not exist unless someone is there to make it manifest. Clearly, this approach amounts to "an absolute renunciation of any realistic conception of nature." (Jammer, 1974, p. 162)[28]

The Many Worlds Interpretation. The Many Worlds hypothesis was primarily developed by Everett in 1957, and was first promoted by Wheeler, Graham, and DeWitt. It takes a very different perspective on how to interpret the wave function — very different, especially, because it completely rejects our common sense ideas of what reality consists of, to begin with. In this regard, the Copenhagen Interpretation is far more accessible, because it affirms the single reality we experience: the act of observation is said to create reality (by collapsing the wave function), so there is no contradiction: *experience is* — by virtue of *creating* — *reality.* But Many Worlds disputes this approach, and with a bizarre twist: it says that all the *potential* realities which correspond with the evolving wave function *actually* exist. So in effect, for Many Worlds every possible world is real and manifest ... somewhere, and sometime.

Essentially, Many Worlds eliminates the distinction between actual and potential worlds. Furthermore,"the role of consciousness, which was central to the Copenhagen Interpretation ... is incidental to the Many Worlds theory." (Zukav, p. 84) According to this approach, the wave function never collapses (Coveney and Highfield, p. 133; Jammer, 1974, p. 512; Lockwood, p. 226)[29]: it just gets split by the observational interaction "into a multitude of equally real 'worlds,' each of which corresponds to a definite component of the superposition ... [I]n each separate 'world' a measurement yields only one result, though this result differs, in general, from 'world' to 'world' ... [in a] continual splitting of the world into a stupendous number of branches." (Jammer, 1974, pp. 512-513)[30] Moreover, such observation does not take place from without, by an external observer interacting with an idiosyncratic and objective reality. Rather, the observer is part of the universe or world which is being observed — and every observation spins off an infinite array of new universes (and observers) based upon every possible different outcome of the situation. Thus, our own consciousness is continually splitting into different worlds: "[D]ifferent editions of us live in many worlds simultaneously, an uncountable number of them, and all of them are real." (Zukav, p. 87) As DeWitt says, "Here is schizophrenia with a vengeance ..." (DeWitt, quoted in Coveney and Highfield, p. 133)

Furthermore, the Many Worlds approach speaks about a "superspace" which is composed of all the usual spaces (or universes or worlds or realities) — ours, as well as those which are not ours and which we think do not exist. They do exist, Many Worlds says, all kinds of universes or realities — coexisting equally with our own. "Whenever a

28 Here is another way of describing the situation, as Jammer reflects upon it: "Since, in Bohr's view, quantum mechanics can be consistently regarded as a computational device for obtaining the probability of every measurement involving both the entity to be observed *and* the measuring experimental setup, no attributes of physical reality can be ascribed to the former alone." (Jammer, 1974, p. 198)

29 Everett felt it was "unreal that there should be a 'magic' process in which something quite drastic occurred (collapse of the wave function), while in all other times systems were assumed to obey perfectly natural continuous laws." (Everett, quoted in Jammer, 1974, p. 508)

30 Coveney and Highfield are critical of the theory in this regard, pointing out that the theory does not "explain what is so special about measurement processes that they should lead to splitting universes." (Coveney and Highfield, p. 134)

choice is made in the universe between one possible event and another, the universe splits into different branches" (Zukav, p. 304). And thus all possible events occur, in dichotomous but equivalent (and perhaps similar) worlds. The one we know about is simply the particular one which we happen to have selected. "Each edition of us is convinced that *our* branch of the universe is the entirety of reality." (Zukav, p. 304) Davies describes the Many Worlds hypothesis as being "not one space and time but perhaps a whole infinity of space and times, all existing alongside one another — very difficult to visualize, but nevertheless it makes sense mathematically" (Davies, 1981)[31]

Needless to say, the Many Worlds approach is full of possibilities and has a real science-fiction ring to it. Jammer describes it as "one of the most daring and most ambitious theories ever constructed in the history of science." (Jammer, 1974, p. 517) What it actually means, though, is harder to assess.

Pioneers in Conflict

The differences of interpretation regarding the wave function are not unique to that crucial construct. More generally, it is important to distinguish between the *formalism* of Quantum Theory, which is essentially the mathematical statements upon which it rests, and the *interpretations* which are placed upon those statements. Unfortunately, there is nothing approaching consistency, in terms of a coherent model for how the theory should be understood. There are various approaches, even within a given "school" or interpretation, and none is especially satisfactory. That is the case even though there is one very dominant reading of the model, which is used by almost everybody in the field — the Copenhagen Interpretation, which we discussed above — even though it has no clear meaning.

For his part, physicist F. David Peat said years ago that:

> Quantum Theory still isn't interpreted. We still don't understand what Quantum Theory means — there are many interpretations ... Although there may be equations that describe how certain things work, the interpretation of those equations still hasn't been achieved — if you get a group of scientists together, they're still arguing about what they mean. (Peat)

Morris put the problem like this:

> No one really understands the meaning of quantum mechanics.... The problem is that quantum mechanics can be interpreted in a number of different ways, and that no one is sure which interpretation is most likely to be correct. Nor is there any general agreement about the theory's implications. Quantum mechanics can be used to construct several different pictures of reality. But all of these pictures contain elements that appear to be a bit

31 Consider the description of space-time as put forth by Wheeler, trying to integrate the Theory of Relativity base with some Quantum Theory details: his space-time world is called "Jiffyland", and it includes bizarre "worm-holes" and "bridges". Of course, "There is no evidence whatever that such features really exist, but neither can they be ruled out." (Davies 1980a, p. 99) Enough said, for now!

> paradoxical. It appears that whichever interpretation one chooses, the implications of quantum mechanics are not easy to accept. (Morris, p. 216)

In the present section, we will examine two seminal conflicts in the development of that model. We first consider a dispute between Niels Bohr and Werner Heisenberg regarding the primacy between their two principles, of complementarity and uncertainty, both in historical and logical terms. Then we shift to a controversy between Erwin Schrödinger and Heisenberg, regarding the essential wave vs. particle character of physical reality. We shall see that very distinct visions emerge, with both these two splits — which are instructive in their own right, as far as delineating the issues ... and which are also readily amenable to interpretation in terms of the present model, as we shall see later.

Bohr vs. Heisenberg. The main features of Quantum Theory were developed in a remarkably short period of time, largely in the 1920's, and primarily by a very few highly creative and energetic individuals. We have already talked a fair bit about the views of two of the most important, Niels Bohr and Werner Heisenberg — creators of the Principle of Complementarity and the Uncertainty Principle, respectively. These two men worked together very closely during those early, formative years of the theory, and indeed their two crucial principles were presented within only a few months of one another. These developments were not smooth, however, and there were very bitter and intense clashes between Bohr and Heisenberg over some of the fundamental aspects of their respective approaches. (Jammer, 1974)

In a way, it was the primacy issue — which principle was more fundamental in nature — that Bohr and Heisenberg were struggling about. Bohr's position was that complementarity was a primary feature of nature, and that this was reflected in the Uncertainty Principle's limitations of measurement.[32] Heisenberg's view was that the Uncertainty Principle's mathematical formalism itself was primary — and that there was no reality to consider short of the given observation or measurement. The result was that, "Whereas for Heisenberg the reason of indeterminacy was [the fundamental quantum] discontinuity, for Bohr the reason was the wave-particle duality." (Jammer, 1974, p. 66)

It is partly an issue which we shall discuss in Chapter 10, as to the proper role of mathematics in science: Heisenberg claimed that mathematics should take precedence over everything else, and he rejected Bohr's complementarity proposition as speculative, metaphysical, and scientifically irrelevant. For his part, Bohr retorted that, "[M]athematical clarity had in itself no virtue" and that "a complete physical explanation should absolutely precede the mathematical formulation." (Bohr, quoted in Jammer, 1974, p. 66) "For Bohr, in contrast to Heisenberg, it was not the formalism as

32 For Bohr, "the limitation of measurability [viz, the Uncertainty Principle] confirms the limitation of definability [viz, the Complementarity Principle] but does not logically precede it.... [I]ndeterminism was a consequence of the wave-particle dualism and hence ultimately of the utilization of different pictures for the complementary modes of description ..." (Jammer, 1974, pp. 69 and 98) Not surprisingly, "Bohr's derivation of the indeterminacy relations ... differs fundamentally from Heisenberg's derivation." (Jammer, 1974, p. 93)

such but rather its underlying logic which should be given priority in the attempt to reach an appropriate interpretation of the theory." (Jammer, 1989, p. 362)

The contrast could not be clearer: Bohr felt that with phenomena like the wave-particle duality, we were dealing with a *fundamental new aspect of nature* which science had to accommodate. But Heisenberg maintained that we were merely confronting flawed remnants of an outmoded concept system — and that this explained the *apparent complementarity*, rather than any new underlying patterns of nature.

In fact, Bohr's position on complementarity was not something so simple that could be characterized in terms of mere "appearance". In the case of light, for example, he said that,

> The two views of the nature of light are to be considered as different attempts at an interpretation of experimental evidence in which the limitation of the classical concepts is expressed in complementary ways.... [This] forces us to adopt a new mode of description designated as *complementary* in the sense that any given application of classical concepts precludes the simultaneous use of other classical concepts which in a different connection are equally necessary for the elucidation of phenomena. (Bohr, quoted in Jammer, 1974, pp. 92 and 95)

But this was not meant to indicate that Bohr thought our classical concepts were lacking. Jammer describes the situation:

> Heisenberg agreed with Bohr that any interpretation should make use of the terminology of classical physics. But whereas Heisenberg was satisfied with the fact that either the particle language or the wave language — and either of them independently of the other — could be used for optimal description ... Bohr insisted on the necessity of using both. For Bohr the indeterminacy relations indicated that not the classical concepts but the classical conception of explanation has to be revised. "Thus, according to the view of the author," wrote Bohr in 1929, "it would be a misconception to believe that the difficulties of the atomic theory may be evaded by eventually replacing the concepts of classical physics by new conceptual forms." And in the same year he declared: "We must, in general, be prepared to accept the fact that a complete elucidation of one and the same object may require diverse points of view which defy a unique description." It was [thus] the breakdown of the classical ideal of explanation Bohr had in mind when, referring to the wave-particle duality of light, he rejected Heisenberg's approach.... In other words, in Bohr's view, the indeterminism of quantum mechanics has its origin in the unavoidable "rupture" of description; for what Heisenberg later called the "reduction of the wave packet" was for Bohr, so to say, the switching over from one mode of description to its complementary mode. (Jammer, 1974, pp. 97-98)

> [A]ccording to Bohr the conceptual framework of classical mechanics is indispensable for a description of quantum phenomena and hence logically prior to any application of quantum mechanics. (Jammer, 1974, p. 493)

In brief, Bohr felt that

> [T]he problem with which quantum physics found itself confronted could not be solved by merely modifying or reinterpreting traditional conceptions.

> What was needed, he concluded, was a new logical instrument. He called it "complementarity" ... (Jammer, 1989, p. 363)

It should be noted that Heisenberg also realized

> that it would be impossible to construct an independent appropriate conceptual apparatus (in the sense of a descriptive language) which would give an adequate intuitive interpretation of the abstract formalism, since all our conceptualizations are inseparably connected with spatio-temporal ideas, Heisenberg saw no other alternative than to retain the classical intuitive notions but to restrict their applicability. (Jammer, 1989, p. 345)

That restriction, of course, referred to the limitations which were specified by the Uncertainty Principle. "In fact, it is precisely this uncertainty ... which makes the use of classical notions admissible." (Jammer, 1989, p. 349) Thus for Heisenberg, the Uncertainty Principle established the limitations of classical concepts — without which principle those concepts simply could not fit and would presumably have to be abandoned.

Jammer points out that the Principles of Complementarity and Uncertainty are neither identical to one another, nor even equivalent to one another, as is sometimes claimed. (Jammer, 1974, p. 60) Nor did either principle lay the groundwork for the other's emergence: they appear to have been developed independently, over a limited period of time. In this respect, there is no real basis for giving one either logical or historical precedence over the other.[33]

We shall return to this dispute later in the chapter, when we try to reach a coherent interpretation of the Uncertainty Principle itself.

Schrödinger vs. Heisenberg. There is another major strand of thought, which stands at the apex and genesis of Quantum Theory, and it deals with the ultimate "stuff" of reality — and in particular whether it is particle-like or wave-like in essence. The debate here is not between Heisenberg and Bohr, but between Heisenberg and Schrödinger.

We have referred to Schrödinger previously. Erwin Schrödinger was another of the relative handful of physicists who fathered Quantum Theory, and we have already noted that he was the creator of the wave functionconcept. Schrödinger never embraced certain assumptions of the orthodox Quantum Theory. He was particularly offended by the notion of discontinuous "quantum jumps", as describing the shift from wave function evolution to collapse — which was a transition state that Schrödinger did not accept.

33 This would run counter to Jammer's conclusion elsewhere: "That complementarity and Heisenberg-indeterminacy are certainly not synonymous follows from the simple fact that the latter ... is an immediate mathematical consequence of the *formalism* of quantum mechanics..., whereas complementarity is an extraneous *interpretative* addition to it." (Jammer, 1974, pp. 60-61) This takes us back to the role of mathematics: What value does any formula have without meaningful concepts with which to understand — nay, to *interpret* — them? And this is as true with Heisenberg's indeterminacy principle as with anything else. Indeed, Jammer himself makes a similar point, in a totally different context: "[A] formalism, even if complete and logically consistent, is not yet a physical theory. To reach this status, some of its symbols have to be given an operationally meaningful interpretation ..." (Jammer, 1989, p. 343)

"Are there quantum jumps?" This was the title of a pair of articles by Schrödinger that were published in 1952 (Schrödinger, 1952a and 1952b). Schrödinger suggested that the wave function concept could be applied to states of the electron: "One may, of course, be tempted to associate the function ψ with a *vibrational process* in the atom, a process possibly more real than electronic orbits whose reality is being very much questioned nowadays." (Schrödinger, quoted in Jammer, 1989, p. 260) "It is hardly necessary to point out how much more gratifying it would be to conceive a quantum transition as an energy change from one vibrational mode to another than to regard it as a jumping of electrons. The variation of vibrational modes may be treated as a process continuous in space and time and enduring as long as the emission process persists." (Schrödinger, quoted in Jammer, 1989, p. 261) Schrödinger felt that such constructs were unnecessarily redundant ways of viewing what could just as well be understood in terms of a wave-oriented approach: "[P]hysical reality consisted of waves and waves only. He denied categorically the existence of discrete energy levels and quantum jumps ..." (Jammer, 1974, 27) Schrödinger believed that the "wave picture" accounted for all quantum phenomena, including "even the Compton effect, the paradigm of particle physics" (Jammer, 1974, p. 29). In brief, Schrödinger saw the field of the wave function "as the ultimate physical reality." (Jammer, 1974, p. 219)[34]

In 1925, Heisenberg developed some of the main foundation of modern quantum theory, in the form of what is referred to as *matrix mechanics*. Independently, Schrödinger was working on the model upon which the wave function rests: published a few months after Heisenberg's, it is called *wave mechanics*. The result is that the "Schrödinger-format statements and Heisenberg-format statements are different ways of saying the same thing ..." (R. Miller, p. 585) Jammer comments on the extraordinary juxtaposition of these two approaches: "It is hard to find in the history of physics two theories designed to cover the same range of experience, which differ more radically than these two." (Jammer, 1989, p. 270) Here is how Holton and Brush describe the respective Schrödinger and Heisenberg approaches:

> The former emphasized the *continuity* of physical processes and the wave-like behavior of the electron — almost visualizable, despite the obscure meaning of the wave function. The latter proceeded from the *discontinuity* of physical processes, suggested by the observed discreteness of spectral lines, and regarded the electron as a particle, though without assigning it any definite space-time description in the classical sense. Nevertheless it was quickly discovered that the two theories are mathematically equivalent ... The development of physics since 1926 has not undermined in any significant way the validity of the Heisenberg-Schrödinger theories, considered

34 Schrödinger interpreted the quantum postulate as a "resonance phenomena, analogous to acoustical beats." (Jammer, 1974, p. 28; we discuss resonance in the next chapter). And, "The energetic property of the individual partial oscillation is its *frequency*." (Schrödinger, quoted in Jammer, 1974, p. 29) The Schrödinger approach "eliminated from the Quantum Theory the even more vexing quantum 'jump' from one orbit to another. The transition from one state ... to another, as described by Schrödinger, was a quite orderly process.... One wave pattern gradually fades out, while the new state fades in." (March, p. 216)

now as two alternative formulations of a single theory, *quantum mechanics*. (Holton and Brush, p. 496)

Mathematically equivalent or not, the two pioneers viewed each other's work with dismay:

> ... Werner Heisenberg, whose physics was built on the theme of the discrete, atomistic, quantized, *discontinuous*, wrote about Erwin Schrödinger's physics: "The more I ponder about [his] theory, the more disgusting it appears to me." And Schrödinger, in turn, whose physics was based on the idea of *continuity*, wrote that Heisenberg's approach left him "discouraged if not repelled." (Holton and Brush, p. 217)

It should be clear that while each of these radically-divergent philosophical positions is one-sided, taken together they are two sides of the same coin! First, we have the concept of the particle-like quantum — which is quintessential Heisenberg, in the philosophical terms we have just described ... and leaving aside historic attribution. Then we have the concept of the wave — and in particular, Schrödinger's wave function, ψ. Here is a way that these two very different approaches would be understood as representing complementary *X*- and *P*-views, accordingly:

X-View	*P-View*
Heisenberg's Matrix Mechanics	Schrödinger's Wave Mechanics
matter domain vision	medium domain vision
discrete and particle-like	continuous and wave-like
quantum	wave function

One might argue that Heisenberg and Schrödinger are subsets of Bohr here, in that complementarity subsumes both!

Even so, the parallel is not complete. For one thing, the wave function itself can be understood in complementary terms, Schrödinger's views on this topic notwithstanding — and we adopt just such an interpretation later in the chapter. But it is difficult to find such a complementary interpretation of the discrete quantum concept, which would justify its central place in the theory which holds its name ... and upon which the theory is largely built.

Physics Fumbles

We now begin our reanalysis of the central constructs and premises of Quantum Theory. As we loop back through these ideas, we will also talk about how the underlying issues and phenomena can be better understood in terms of the present two-force model, and the fuller concept of complementarity that we are developing here.

To set the stage for what is to come in the rest of this chapter, let us return to the wave-particle phenomenon and the Principle of Complementarity — for here is where the nub of the issue lies. Indeed, the importance of these concepts cannot be overstated, and Quantum Theory's observations and insights in this regard represented an enormous step forward for science and mankind. The Principle of Complementarity amounted to a revelation — both in terms of providing a novel and ingenious conceptual framework for approaching a new and otherwise inscrutable set of phenomena ... as well as in terms of exposing a massive breach in both the conventional patterns of thinking about nature, along with the standard methodology for examining nature. This was truly an epiphany for science — which we will boldly describe as the most profound that it will ever have.

That may sound like hyperbole — but bear in mind that we are talking about the kernel of the present work ... which we claim is also the main theme of nature. Of course, it is hardly the main theme of the established paradigm — so if our position is correct, then the obvious question is why physics missed this sublime opportunity, when it was perfectly poised to take the leap. Bohr's great discovery could have become the centerpiece of the new physics, rather than being one among various principles — which some even doubted in its own right, thus leaving the nascent idea stillborn.

Let us mention some of the reasons for why the new concept was never pursued and developed to its full potential. To start with, complementarity was a completely alien concept for science — one which was not only very subtle and difficult to understand in its own right, but was probably also quite threatening because of its philosophical (if not metaphysical) origins and overtones. The idea surely went against the grain of a more absolutist or deterministic mode of thought — and even culture of thinking — which was (and is) so overwhelmingly predominant in Western science generally, and in physics particularly. Likewise, the reigning methodology and rules for theory-building, including a commitment to logical positivism and the central role of mathematics, were not especially amenable to complementarity.

Then there was the point we already alluded to, regarding other prominent principles that were equally (if not more) embraced as (than) was complementarity itself. The competition or conflict with all these other significant constructs and theories — and the divided loyalties and conceptual compromises — never allowed complementarity to come to the fore as both primary and pervasive. In any case, no one was advocating a full-fledged version of complementarity, as opposed to a niche concept that applied to microphenomena; and even if they had, it would have been a massive struggle for physics to completely reorganize itself around such a new core.

Nor are we even touching upon various psychological and sociological factors which must have also come into play — in terms of all the usual rigidities and intransigence which characterize any well-established discipline or organizational endeavor like physics, where individual and collective success and conformance to the dominant paradigm are primary motivating factors[35] ... even if they can be effectively challenged on rare occasions. All in all, it is rather amazing that Bohr even got as far as he did with the new thesis, so far outside the mainstream of physics was he reaching and so great was the resistance that he obviously encountered.

Alas, all such explanations aside, we are talking about the path not taken, the path which physics was unwilling or unable to pursue. On the contrary, far from representing the road to enlightenment, complementarity was treated by many as representing an unfortunate detour, which needed to be tolerated until a more satisfactory solution emerged. In this way, Bohr's thesis has never been fully welcomed, and its significance

35 See Chapter 10 for more discussion of these issues.

has always been narrowed as tightly as possible. The result has been to leave this principle on its own, with no apparent relationship to the rest of the theoretical framework.

Indeed, we would suggest that complementarity has left physics with a kind of identity crisis, if not schizophrenia: on the one hand, the established paradigm cannot really do without this principle[36] — and yet on the other hand, they do not really know what to do with it, either. It amounts to a "skeleton in the closet", as it were — which physics opens up and looks at sometimes, but then slams the door behind ... in exasperation and bewilderment, if not embarrassment. At best, complementarity has become the star of a puzzle show, which physics trots out to illustrate the mysteries of which it is master.

"Master" — fat chance! For in spite of the new way of thinking that was initially engendered, and the fresh enthusiasm that went with it, the complementary wave-particle interpretation turns out to have been a false start, which fizzled out all too quickly. By virtue of utterly failing to grasp the breathtaking magnitude of the breakthrough that had been attained, in terms of its very essence and scope, physics mostly missed its meaning: Wave-particle was only one manifestation of a far larger pattern, whose infinite range the established paradigm has never even begun to imagine. It has never been able to follow through with the idea, and take the principle to its proper and ultimate conclusion. So while Bohr's thesis about wave-particle had the potential to represent physics' coming of age, it instead represents precisely the place where physics lost its way.

Look at the paths we have been on instead, lo these many years. There was the path set forth by people like Heisenberg, who did his best to establish a one-sided premise about the particle-like nature of the physical world — and who essentially dismissed complementarity as an undesirable and misleading result of certain unfortunate word problems (if not word games), which did not accurately reflect the underlying reality. Then there was the path of Einstein, who was successful in getting his singular vision of relativity established — and who did his best to squelch and fend off the very notion of complementarity or duality ... as having no actual place in nature, or viable role in science. Both of these topics are dealt with at different points in this book, so we shan't get sidetracked by them here.

In a very real yet paradoxical way, the same wave-particle phenomenon which physics discovered many years ago is itself crushing testimony to the inadequacy of today's theories. That is because in many respects, here in a nutshell lies the very crux of physics' failure as a science — there could be no better example: with wave-particle, physics indicts itself. For this esteemed discipline had the ball of nature in its hand — complementarity — the new centerpiece of modern science. But then instead of running with it, physics dropped it: it was too hot to handle, and the players didn't know what to do with it. The result was that physics fumbled — badly.

36 "The notion of complementarity has become an essential part of the way physicists think about nature ... " (Capra, 1975, p. 145)

While wave-particle can be described as instigating "the biggest revolution in the whole history of physical theory" (French and Taylor, p. 85), that revolution has never succeeded — the reactionaries in physics have both suppressed and co-opted it. These forces of regression fought this central precept with all their might; and while they could never succeed in vanquishing it altogether, they did succeed in keeping it at bay. So the real revolution is still waiting to happen; and it will be a very different physics — a very different attitude towards nature — which will result.

Complementarity Expanded

It should be quite clear that the principle of complementarity that we are advocating is intended to be far more pivotal and expansive than that put forward by Niels Bohr, in whose trailblazing footsteps we follow. We make that assertion even though Bohr's eventual convictions on this count were boundless, a point we will return to in the final pages of this book.[37] In any event, it is worthwhile to try and summarize a number of ways in which the two approaches differ.

First, we apply the principle to the very essence of nature, by postulating two — and only two — fundamental forces, the interaction of which provides the dynamics and structure of the universe. Second, this principle manifests itself in both what we can describe as a horizontal and a vertical orientation — which we label in terms of complementarity (again) and hierarchy. Third, we apply the principle to several crucial, pervasive, yet deceptively ordinary characteristics of nature, in terms of our multi-leveled dimensions of analysis — whereas Bohr's is most commonly understood to apply primarily to only one ... which we call the matter-medium dimension. Fourth, the complementarity we are espousing applies everywhere in nature — to everything, and not just to a peculiar dualistic mode in which microphenomena manifest themselves; indeed, complementarity applies not only to all types and levels of existence, but to all the ways that we view or interact with that existence. Fifth, we presume that complementarity can be understood as manifesting itself in terms of a general, non-symmetrical figure-ground pattern. Lastly, as we shall discuss in the next section, the two-sided notion of complementarity that we are developing could not possibly coexist with the core, one-sided concept of a discrete quantum.

We contend that complementarity has been beleaguered in science for far too long, and now the time has come for it to have its day and flourish. Complementarity is neither a linguistic or conceptual artifact — nor is it merely a curious pattern, which can be applied to a particular subset of natural phenomena: it is *always* and *everywhere*, in one form or another. Moreover, complementarity is not only the most pervasive characteristic of nature, it is likewise the most powerful conceptual tool for viewing and analyzing nature. These are the themes we are trying to implement in this book.

37 There is also the obvious fact that it takes a good deal of time, as well as the right kind of circumstances, for significant and far-reaching new ideas to develop and mature to the point where their full potential is revealed.

The Quantum Paradox

Next we want to examine the relationship between two of the central precepts of Quantum Theory — which we maintain stand juxtaposed, in complete and irreconcilable opposition to one another. This profound conflict, in turn, makes a mockery of Quantum Theory itself.

On the one hand is the two-sided wave-particle duality, which depicts "the particle picture and the wave picture as two complementary descriptions of the same reality, each of them being only partly correct and having a limited range of application." (Capra, 1975, p. 145) On the other hand is the one-sided notion of the discrete quantum, and the fervent claim that nature is essentially particle or particle-like: "[T]he basic structure of nature is granular …" (Zukav, p. 48), and the discrete energy units known as quanta have "been accepted as bona fide particles" (Capra, 1975, p. 56).

Will the real Quantum Theory please stand up: Is the essential nature of reality one-sided or two-sided — quantum or complementary? There is no middle ground, and it is impossible to bring these two currents together, for they are manifestly incongruous with one another. Indeed, it is a remarkable paradox that a definitive conclusion about the discontinuous and particle-like nature of reality was developing at the same time — and within the same theory — as a view of nature that was dualistic and consisted of both particle-like and wave-like characteristics, in essentially equal measure. Yet such are the contradictions and philosophy of modern science, that these two totally incompatible visions of nature emerged — and continue to exist within the same theory — even today.

These two diametrically opposed approaches may continue to exist, but we contend that only one has been primary all these years — that of the discrete quantum and its particle picture … with complementarity being kept on the sidelines. As we noted in the discussion above, even that secondary role has been a begrudging concession to a mode of thinking, or conception of nature, which many scientists find distasteful at best. Again, that was certainly the case with Werner Heisenberg, who treated complementarity as a minor byproduct of a flawed concept system.

Recall Heisenberg's underlying premise — that a notion like complementarity was not only an unnecessary and superfluous distraction from the real business of science, in terms of what was strictly observable and measurable, but that it was actually a kind of artifact. That is, as we discussed earlier in the chapter, Heisenberg felt that the source of the supposedly-mysterious duality was not to be found in the underlying reality of nature so much as in the defective man-made concepts with which we describe nature. It is the inadequacy of our everyday concept system that is the problem, he said, and not a limitation of science's theoretical outlook — which latter interpretation represented precisely the perspective of Niels Bohr.

Actually, it is surprising that Bohr's position on the one-sided particle-based quantum postulate is not entirely clear. Jammer claims on the one hand that, "Bohr's point of departure in developing his complementarity interpretation was … the quantum postulate" itself (Jammer, 1974, p. 160). Yet elsewhere Jammer says that, "Bohr's conception of complementarity originated from his final acceptance of the wave-particle duality" (Jammer, 1989, p. 361), and that Bohr argued for "the necessity of two mutually exclusive descriptions of physical phenomena [as] the ultimate foundation of the whole [quantum] theory" (Jammer, 1974, p. 67). Again, Bohr "regarded the wave-particle duality as the ultimate point of departure for an interpretation of the theory. He disliked Heisenberg's idea that 'nature imitates a mathematical scheme' and his one-sided preference for the intuitive particle conception ..." (Jammer, 1989, p. 362) For his part, Jammer states "that discontinuous change is not necessarily an essential characteristic of quantum theory.... [whereas] the wave particle duality may be regarded, in our view, as the essential characteristic of quantum theory ..." (Jammer, 1989, p. 35)[38]

There can be no mistaking our position, as to the proper outcome: the notion of a discrete quantum makes no sense as a primary principle of nature, which is what the very name of Quantum Theory maintains. Such a thesis is utterly *anti*-complementary, and thereby amounts to a one-sided distortion of nature. As we shall discuss at greater length in Chapter 9, *discontinuousness* is the very essence of one side of nature, in the same way that *continuousness* is the very essence of the other side.[39]

In many respects, the notion of a discrete quantum unit seems like a regression to the kind of mechanistic outlook which was supposedly overcome by modern physics. Indeed, one might consider the notion of the discrete quantum as a conceptual throwback to, if not replacement for, the classical vision of the atom as representing the ultimate building block — both serving the role of an absolute and fundamental unit of nature. Popper makes a similar point, in arguing that Quantum Theory's "dualism is interpreted by [Max] Born in a way which may be described almost as a reversion to atomism in its original sense: what exists are corpuscles or particles ..." (Popper, 1982, p. 164) And Schrödinger made the telling comment that, "Atomism in its latest form is called quantum mechanics." (Schrödinger, 1950, p. 109)

So, the quantum is the culprit. Tiny as it is, the discrete quantum as it is presently conceived represents an enormous impediment on the path to understanding nature — being not only the ultimate *unit* of physical reality, but also representing the ultimate *essence* of physical reality. Physics will never understand nature, nor grasp the true meaning of the wave-particle duality — not to mention a more generalized form of complementarity — until its love affair with the quantum is ended.

Making Reality

We have been arguing that Quantum Theory overlooked the scope and salience of the complementarity theme, in general — as well as specifically contradicting that theme in establishing the discrete quantum as its namesake principle. Of course, there are count-

38 Holton notes that, "Bohr, longer than most major physicists, continued to be skeptical about the fundamentality of the light-quantum concept ..." (Holton, p. 185) which he found to be characterized by "an inherent irrationality" (Bohr, quoted in Holton, p. 185).

39 It is worth emphasizing that the basic notion of complementarity entails both dichotomous and continuous characteristics — with the result being that there are either-or and both-and varieties of complementarity, throughout all of nature. Thus wave and particle may be an absolute either-or, but other patterns are a relative more of one and less of the other ... such as position and momentum.

less other ways that physics' concepts and theories ignore or violate the crucial pattern of complementarity, which we will be discussing throughout this book. But now we shall examine another important instance, in terms of the very way that Quantum Theory identifies the content of reality itself, and the role that observation plays in identifying that unitary state. We discussed these issues in the first part of the chapter, and now we would like to critique them from our 2FT perspective.

We start by recalling the two main approaches to the wave function, in this regard. The Copenhagen Interpretation says that an entity's evolving wave function represents its potential existence, whereas the collapse of that wave function — brought about by the act of observation — corresponds to its actual existence or manifestation in reality: the entity does not truly exist until the wave function's collapse, instigated by the act of observation. This approach therefore presumes a nature which only exists if it is observed — an amazingly circumscribed view of the universe. Presumably such observation requires the consciousness of our own species ... or else, how broadly would observation be construed? Such an observation-as-reality approach also puts Quantum Theory's whole ontological status in doubt. As Healey puts it,

> [A]ccepting this interpretation means accepting that quantum mechanics is descriptively incomplete on a massive scale. It does not describe the structure of elementary particles, nuclei, atoms, metals, superfluids, superconductors, lasers, crystalline solids, and semiconductors: It merely describes what we find when we observe such things. Indeed, on this view, quantum mechanics would have nothing to say about a world in which no observations ever took place; and depending on exactly what is taken to be required for an observation, it may not even have anything to say about our world before there were entities such as humans capable of observing it. (Healey, pp. 17-18)

How, then, could anything have existed in the universe before our human consciousness developed to the point of being able to trigger or produce it — or did mankind somehow perform that reality-creating trick after the fact, retroactively?

Many Worlds comes along and — perhaps realizing how suspiciously narrow is the standard Copenhagen line — rejects this limited-world view, and substitutes instead an agglomerate all-worlds theme, whereby an external observer is not necessary to create reality. According to this interpretation, the evolving wave function represents an entity's actual existence or manifestation in reality: Many Worlds sees no such thing as a collapse of the wave function, but rather its continual evolution and replication (or "fissure") — into an endless multitude of permutations. The Many Worlds interpretation describes all potential worlds as being actual and real, without entailing any external catalyst such as observation to bring this about — but instead requiring an ad hoc explanation for deciding which reality is subjectively experienced or apparently measured.[40]

We argue that there are fundamental flaws with both of these models. Just as Copenhagen's observation-is-reality approach is so narrow and specific as to be implausible, Many Worlds' all-is-reality approach is so broad and indiscriminate as to be meaningless. We go from one extreme to another, with Copenhagen making too sharp a distinction between existence and non-existence, and Many Worlds seeming to make none at all. Both interpretations miss the mark as we see it — which concerns the dual nature of reality that exists pervasively throughout the universe, regardless of whether it is observed or not.

That said, most of our comments will focus on the Copenhagen Interpretation, which remains the dominant paradigm in Quantum Theory today — and is much closer to the model we are proposing here. By contrast, Many Worlds is more like scattershot in its all-encompassing approach, and the requisite differentiation between two modes does not even exist as part of that interpretation ... and so there is not much more to analyze, in the terms that we are concerned with here. But this is not the case with Copenhagen, and so we now look more closely at the paradoxical puzzles which emerge from that approach, regarding the relationship between observation and existence. We start by considering these issues in the form of a famous experiment.

Schrödinger's Cat. In 1935, Erwin Schrödinger put forth a fairly simple thought experiment — but one with rather audacious implications.

Schrödinger proposed the idea of putting a cat into a closed container — together with a cyanide capsule, and a mechanism for puncturing the capsule at some random subsequent time point ... thereby killing the cat. The precise details of the execution device which bring about the poison release can vary, and need not concern us here. But here is a step-by-step synopsis of what one quasi-automated scenario might look like: first there would be the random and unpredictable decay of an atom, followed by the release of radiation, which would be absorbed by a photoelectric cell, that would in turn trigger a mechanical device, which would release a poison capsule ... followed by the death of the cat. There is nothing mysterious or controversial about any of these processes, or the normal expectations regarding the results.

Except for one thing: according to the usual Copenhagen approach, there would be no result — until an external observer actually comes along to examine the state of the system. For, Copenhagen presumes that there will be no change or progression in the situation, such as in the sequence we just presented — indeed, there would be no situation at all — until and unless an external observer (presumably together with his or her *consciousness*) came along to make that reality manifest. Up until that point — and disregarding the usual expectations about the development of the cat's state over time, in

40 Healey criticizes Everett's Many Worlds interpretation on this point: "[W]e always observe a single, definite result of any properly conducted quantum measurement, even though in the many-worlds interpretation there are often supposed to be many different results: This is the empirical problem. Proponents of the interpretation have made strenuous and ingenious efforts to solve these problems, but in my opinion without success." (Healey, pp. 20-21) Furthermore, he says that, "A proponent of the Everett interpretations must try to show that his view is neither inconsistent nor empirically inadequate. The view seems inconsistent insofar as it claims that following a good observation an observer may be in incompatible states; and it seems empirically inadequate insofar as it fails to accommodate the fact that just one of these states is actualized." (Healey, pp. 210-211) Finally, he concludes that, "[T]he fantastic ontological cost of this view is not balanced by any corresponding explanatory benefit, as compared to a view according to which all but one of the worlds emerging from a quantum measurement are merely possible, whereas there is a single actual world in which the observer records a definite result." (Healey, p. 211)

relation to the randomly-timed poison release — the cat is said to exist in some kind of suspended animation, an equal mixture of live cat and dead cat.

Schrödinger put forth this experiment, which became known as the Paradox of Schrödinger's Cat, largely to demonstrate just how untenable (or "incomplete") the Copenhagen position was (Lockwood, p. 197).[41] Schrödinger said that it would be "naive" to presume that observation was necessary in order for reality to manifest itself here (Jammer, 1974, p. 217). Yet it is interesting to note that this fable is usually presented not as an illustration of the Copenhagen Interpretation's inherent untenability, but rather as a demonstration of its uncanny insight into the depths of reality!

We completely concur with Schrödinger. What brings about the eventual outcome in this situation need have no particular relationship to any observer, or act of observation. It is largely a matter of time and chance as to when the decay event will occur, and no external observation is necessary for a reality outcome to manifest itself at some point in time — be it that the atom has not decayed and the cat is alive, or that the atom has decayed and the cat is dead — regardless of whether Quantum Theory knows about it or not. As Healey puts it in describing one approach to the problem: "Schrödinger's cat is definitely alive or dead before it is observed, but quantum mechanics cannot say which." (Healey, p. 15)

There is thus no issue here, no paradox — beyond illuminating the limitations of the Copenhagen position. For there is a difference between *something which happens* (e.g., Schrödinger's cat dies or not) and *the knowledge that something happens* (e.g., knowing whether the cat is dead or not). The Copenhagen conclusion — that the cat lingers in a suspended state of life-death until an observation occurs — is utter nonsense, and rests on no factual grounds at all. It is not the case that there is a spatial-temporal vacuum in the universe unless there is a human being to observe it. Erwin Schrödinger's skeptical thinking was right on the mark here.

The Measurement Problem. The cause of these bizarre mental gyrations lies in what is known as the measurement problem. Healey describes the issue in the case of the cat paradox, and in relation to a measurement "pointer" as revealing the outcome of the experiment:

> [T]he key difficulty may be stated quite simply. It is that many initial states of an object system give rise to final compound object+apparatus quantum states which, in the present [Copenhagen] interpretation, imply that the apparatus fails to register any result at all (its pointer points nowhere at all)!.... [I]n the absence of any observation or other intervening interaction, at the end of the period the quantum state of the joint system nucleus+cat (+other coupled intermediate systems) will correspond to the cat's being

41 It is argued that the cat paradox experiment was also designed by Schrödinger as a way of challenging the notion of a micro-macro split — that " 'wave-particle dualism' and the whole formalism of quantum mechanics apply only to the microscopic quantities of matter, while the physical behavior in the macroscopic region must follow from the rules of classical physics." (Sachs, 1988, pp. 107-108)

> *neither* dead nor alive: though, of course, there would be a 50 percent chance of *observing* the cat to be dead. (Healey, p. 12)

The measurement problem can be understood as dealing with how one gets from the different possibilities — which the statistics of Schrödinger's wave function put forth — into a single manifestation, which is how concrete reality is said to exist when we examine it. Heisenberg talks about the collapse of the wave function when he says that, "The discontinuous 'reduction of wave-packets' ... cannot be derived from Schrödinger's equation" (Heisenberg, 1955, p. 27) — and that the observer is required in order to produce a result. "[T]he actual is selected from the possible ... by the 'observer' " (Heisenberg, 1955, p. 23).

Zukav elaborates the dilemma more generally:

> The unavoidable flaw in this [Copenhagen] interpretation is the Problem of Measurement. Some type of detection by an observing system is required to collapse the wave function of the observed system into a physical reality, otherwise the observed system does not physically exist except as an endlessly proliferating number of possibilities generated in accordance with the Schrödinger wave equation. (Zukav, p. 82)

Popper also cites the problem here as representing a fundamental weakness in the overall theory, pointing out that the necessity — for an external observer to perform a function, which is not explained within the model — is an obvious demonstration of the incompleteness of Quantum Theory. (Popper, 1982, p. 123) Lockwood, too, says, "Some extra *deus ex machina* is called for." (Lockwood, p. 207) For her part, Cartwright brings up the von Neumann interpretation which we mentioned earlier, by noting that, "Von Neumann postulated that measurements are special because they involve a conscious observer." (Cartwright, p. 163) Cartwright goes on to argue that, "It is clearly not a very satisfactory solution. The measurement problem is one of the long-standing philosophical difficulties that troubles the quantum theory." (Cartwright, pp. 164)[42]

The upshot is that Copenhagen has found itself in a trap: "[T]he measurement problem has proven intractable ..." (Healey, p. 12) There have been many attempts to get around this problem, but without success. Indeed, Jammer goes so far as to conclude that, "The immense diversity of opinions and the endless variety of theories concerning quantum measurements ... are but a reflection of the fundamental disagreement as to the interpretation of quantum mechanics as a whole." (Jammer, 1974, p. 521)

Rampant Subjectivism. With the Copenhagen Interpretation, the concept of the wave function essentially degenerates from being an objective representation of all phenomena of nature, into a mere subjective reflection of only that very narrow subset of relationships as between individual human consciousness and a given

42 We would also disagree with other aspects of von Neumann 's model — such as the nature of the determinism which is said to underlie each mode (e.g., Schrödinger's "causal" wave equation). Likewise we would question Cartwright's conclusion that, "Deterministic motions are continuous; indeterministic are discontinuous." (Cartwright, p. 201)

phenomenon of nature — a wave of knowledge in particular, not of nature in general. Our whole analysis becomes confused and distorted in the process, and we are left with a theory which conflates the scientist's internal state of knowledge with the universe's external reality. Thus, Jammer writes about "the spirit of Heisenberg's interpretation, according to which the wave function represents man's knowledge of the state rather than the state of the system itself ..." (Jammer, 1974, pp. 444-445) Lockwood offers the view that one may think of "the collapse of the wave function, not as any kind of physical event, but simply as the collapse of ignorance into knowledge ..." (Lockwood, p. 198)

Popper has written at length about the rampant subjectivism which results. He cites the outrageous example of Copenhagen advocates Pauli, von Neumann, and Born all arguing that the *entropy* in a physical system is negatively correlated with our *knowledge* of that system — rather than with the concrete physical conditions of the system itself:

> They interpret the tendency or propensity of the system to increase its *disorder* (or "*entropy*") ... as a tendency of our *ignorance* to increase. That is to say, instead of interpreting the "entropy" of a system as a measure of its objective state of disorder or randomness, they interpret it as a measure of our own subjective state of ignorance of the system. (Popper, 1982, p. 109)

In this regard, it is true that Bohr makes a valid point when he says that, "[I]t is clearly impossible to distinguish sharply between the phenomena themselves and their conscious perception ..." (Bohr, quoted in Jammer, 1989, p. 182) Yet this does not change the fact that phenomena may exist in the world regardless of whether we consciously perceive them — or whether we actively observe or measure them.

It is not only wrong to think that observation or measurement is the culprit in *causing the inherent uncertainty of nature*, as we shall discuss in the next section — even taking into account the inevitable disruption which the measurement action entails; it is also wrong to think that such observation is a prerequisite for *producing any reality at all*. Thus, the wave-particle duality is not the result of a measurement problem, nor is reality in general dependent upon such measurement — or upon any other specific type of interaction.

Popper likewise challenges such a human-centered interpretation:

> The central issue here is *realism*. That is to say, the reality of the physical world we live in: the fact that this world exists independently of ourselves; that it existed before life existed ... and that it will continue to exist, for all we know, long after we have all been swept away. (Popper, 1982, p. 2)
>
> [T]he transition from the possible to the actual takes place whenever a new state of the world emerges ... whether observed, or measured, or not.... As long as anything happens, as long as there is any change, it will always consist in the actualization of certain potentialities.... *The world changes without reference to us*.... [I]t looks to me very much like a symptom of either myopia or megalomania to allow one's view of the world, or of science, to be dominated, or even coloured, by the disturbances created by one's own experiments. Transitions from the possible to the actual and quantum interactions were going on before anybody ever interfered with anything, and they will continue to go on long after we have all left off interfering.... [T]he interference of the observer with the object may be visualized as *some-*

> *thing of no fundamental importance* — a *consequence* of the prevailing laws of interaction ... (Popper, 1982, pp. 185, 186, and 191; italics in originals)

Healey states:

> [Q]uantum mechanics is *not* fundamentally about observations. Terms such as "observation," "measurement," and "classical system" do not appear in a careful formulation of the basic principles of the theory. Rather, measurements are treated as elements of a class of physical interactions whose characteristic features may be specified quantum mechanically without any use of such terms. (Healey, p. 185)

> [I]n neither the naive realist interpretation nor the present interactive interpretation are the fundamental probabilities in quantum mechanics probabilities *of measurement results*. Rather, in both interpretations, measurement involves just a special kind of physical interaction, subject to basically the same physical laws as other interactions. (Healey, p. 181)

> Dynamical states do undergo indeterministic as well as deterministic changes, but these indeterministic transitions occur during *any* type of interaction, not just during measurement. (Healey, p. 43)

Likewise, Lockwood says that,

> When physicists speak of a *measurement interaction*, what they have in mind is merely a particular category of physical process; that a specific such process is regarded by us, in virtue of our interests and intentions in setting it up, as the *measuring* of something, is neither here nor there. (Lockwood, pp. 210-211)

Similarly, Cartwright raises a question ... and then answers it:

> How then do measurements differ from other interactions between systems? It turns out to be very difficult to find any difference that singles out measurements uniquely.... [T]he measurement problem is not a real problem.... Reductions of the wave packet go on all of the time, in a wide variety of circumstances. There is nothing peculiar about measurement, and there is no special role for consciousness in quantum mechanics. (Cartwright, pp. 163, 164, and 195)

We concur. Whereas orthodox Quantum Theory treats observation or measurement as some kind of sacrosanct form of interaction, it is in principle no different from any other.

At the same time, it seems vital to differentiate our internal, symbolic, cognitive existence from an external, physical, environmental existence — a distinction we will return to in Chapter 11.[43] Things may exist in our minds which do not correspond with something which exists in our environment, and likewise things may exist in our environment which do not exist in our minds. Because the perception of reality requires observation of the external environment does not mean that the environment does not exist without such observation and knowledge. Such a Copenhagen assertion is pure sophistry.[44]

Complementary States. We noted earlier that Quantum Theory in general works on the assumption that there is some sort of unitary and all-or-none reality alone, which is (or is not) made manifest — either singularly (e.g., Copenhagen), or multiply (e.g., Many Worlds). Thus, most Copenhagen theorists connect *only* the collapse of the wave function with actual physical reality, in a singular sense; and they see the evolving wave function largely as some kind of abstract probability pattern. On the other hand, Many Worlds theorists see *only* the evolving wave function, as being associated with an endlessly proliferating array of unitary realities.

Not surprisingly, we take a very different position on these matters, in keeping with the fundamental tenets of our model. So in this and the following subsection, we will attempt to reframe the issues in terms that are compatible with the main thesis which guides this work. First we will talk about complementary states which can manifest themselves in any situation, and then we will look at the before and after complementary stages in which these patterns occur.

One of the most useful general ways for describing the two states is in terms of our figure-ground motif. Cast in terms of the Rubens illustration that we described in the first chapter, some entity can manifest itself as figure or it can manifest itself as ground: when faces are figure, vase is ground, and v.v. What this also means is that any figure entity which exists does so only in relation to a given background of existence, and not merely in some abstract and isolated (not to mention unitary) sense. And not only does the figure exist in relation to the ground of existence — but vice versa, the background exists in relation to the figure's existence. This means, in turn, that it is folly to speak about existence or reality in merely a unitary sense, for reality is dual: there are two fundamental kinds of existence, and both must be incorporated into our theories.

The whole 2FT model can be understood as an attempt to describe these two kinds of reality, as they exist in so many different ways. We can use generic terms like figure vs. ground to describe them, or we can talk about our nesting dimensions of analysis, each of which represents a fundamental characteristic or quality of existence — with each complementary partner or state existing in relation to its counterpart: dominant vs. recessive, static vs. moving, matter vs. medium (e.g., particle vs. wave), inside vs. outside, and stable vs. changing. Thus:

43 Neither in this chapter, nor the later one, will we discuss (e.g., "postmodernist") schools of thought in philosophy which are consistent with the Quantum Theory position we are critiquing here — namely, that reality does not exist apart from our knowledge of that reality.

44 Jammer cites Höffding to the effect that, "Knowledge, however rich and powerful it may be, is after all only a *part* of Being ..." (Jammer, 1989, p. 180). It is interesting that Höffding was an important influence on Bohr — yet Bohr's attribution of reality only to the results of observation flies in the face of such a crucial differentiation between observation and knowledge, on the one hand — and the larger reality (or "Being") of which it is a mere subset.

X-State	*P-State*
stable	changing
inside	outside
matter (particle)	medium (wave)
static	moving
dominant	recessive

Existence is all of these nested and dual characteristics, and more — not some isolated and singular event, which is how it is typically depicted in Quantum Theory. Aside from a confused and limited nod to the matter-medium dimension (in terms of particle and wave), or the stasis-motion dimension (in terms of position and momentum, or in terms of the one-sided catechism that all motion is relative), the conventional model mostly deals only with the unitary and figure aspect of reality, thereby ignoring the nested layers of background in which the figure is embedded.

In more general terms, of course, when we speak about nested layers we are also talking about the hierarchy of factors in relation to which some phenomenon exists — which we can also describe in terms of context.[45] The important point is that context counts: Nothing is anything in relation to nothing; everything needs something else to make it what it is. Every phenomenon of nature manifests its reality only in relation to other entities. As Bohm says, "[E]very entity ... is dependent for its existence on the maintenance of appropriate conditions in its infinite background and substructure.... [A] thing does not and could not exist apart from the context from which it has been conceptually abstracted." (Bohm, 1957, pp. 144 and 146) Or as Smith puts it, "Nothing is a thing by itself; it takes meaning, indeed existence, only as it interacts with something else." (Smith, p. 23)

Put another way, any world which might exist, potentially ... does exist, really — from *some* point of view, even if it is only an imaginary one: we make this extreme assertion hesitantly, at this point, whereas we will explain its meaning more fully in Chapter 11. The only reason we raise the issue here is because it is in this narrow sense that the Many Worlds thesis is viable — but only barely. That is because in the present model, nature is a cumulative affair, with different branches amounting to different hierarchies ... all nesting together into a coherent overall framework. But Many Worlds does not elucidate these orderly and nesting dualistic patterns. That is the problem with the Many Worlds "superspace", as comprising everything in the universe in a kind of chaotic junk heap.[46]

45 There are obvious similarities between our concepts of background or context, and Popper's "propensity" forces and fields, which we noted in an earlier footnote is the way he describes the wave function. His model is based not on the duality of particle and wave as such, but on "the particle and its '*associated*' propensity fields ... [These] fields depend not only on the particle but also on other conditions" (Popper, 1982, p. 84). But the contextual background we are talking about does not merely carry a future potentiality (or propensity), as is associated with the wave function, but is also part and parcel of what defines the present in what are largely spatial terms.

Complementary Stages. Much of what we discussed in the previous subsection can be understood in terms of the dual way that reality manifests itself in quasi-spatial terms, as opposed to how such factors can be understood as progressing over time. Now we will look briefly at the latter, in terms of the complementary modes in which things develop — in a manner that is equivalent to the evolution vs. collapse of the wave function, but which we will talk about here in simple terms of pre-stage vs. post-stage.

To start with, we have rejected any contention that the pre-stage portion should be understood as something stagnant and corresponding with what we have called a period of suspended animation, whereby some phenomenon remains fixed in a virtually static mixture of both-and states — until some specific event occurs that breaks the logjam and thereby instigates the post-stage and thus the phenomenon's discrete unitary existence in one form or the other.

By contrast, we would describe the relationship between the two stages in terms of the interaction between the two complementary states — e.g., between two complementary partners regarding which takes on the role of figure and which of ground, or whether that status oscillates between them. Think of the faces-vase image, or even the child's seesaw. One might interpret such situations in terms of a gradual passage of energy from one side to the other — in buildup vs. discharge terms — until the tipping point is reached, and then the "balance of power" shifts from one side to the other.

Aside from the internal interaction between the two complementary states or partners themselves, another factor would obviously be any external intrusion or input that would alter the balance of energy in the situation. Again, the standard interpretation would be to describe this category in terms of observation — but as we keep saying, any appropriate catalyst from outside (be it observation or not) will do the trick in bringing about the change from one stage to the other.

Regardless of any distinct causative factors which can be identified, be they internal or external to the complementary relationship, we must emphasize the transitory or dynamic aspect of the evolution or pre-stage period: it is not merely something static, which is waiting to be triggered by one or another explicit event ... but an ongoing process which will culminate at some (e.g., uncertain) time point.

Here is a complementary chart for representing the two temporal stages of existence that we have been describing:

X-Stage	*P-Stage*
post-stage	pre-stage
collapse of the wave function	evolution of the wave function
discharge / disperse / exhale	buildup / concentrate / inhale
actual	potential
figure	ground

46 "The Many Worlds theory says that there is one universe and that its wave function represents all of the ways that it can be decomposed into different possible realities. We are all together here in a big box ..." (Zukav, p. 85)

Lab Experiments. Let us mention certain aspects of some recent physical research that has been conducted (e.g., Friedman, et al.; Leibfried, et al.; van der Wal, et al.) which relate to the issues we have been discussing.[47]

A typical procedure is to arrange the conditions of these experiments such that it is random as to which of two opposite states — e.g., clockwise motion vs. counterclockwise motion — will result, once an observation has been made … meaning that we are dealing with a 50-50 likelihood that one outcome will be produced, as opposed to the other outcome. Of course, these are the same odds as are said to apply to the critical atom in Schrödinger's thought experiment, where it is just as likely not to decay as to decay — the latter, in turn, triggering the release of the cyanide and the death of the cat.

What is new and unique about these actual experiments is that they are said to demonstrate the simultaneous occurrence of the two opposite states prior to the actual observation, after which only one (either one) of the two states is manifested. Leaving aside the details of how such a subtle and contradictory pre-stage pattern can be demonstrated — a quantitative measure of some subtle characteristic is said to be explicable only as a result of such joint behavior — the stated conclusion is that we are dealing with a situation which can be variously described as a "quantum superposition", a "cat state", or simply as a snapshot of the "smudge" which underlies reality.

The "cat state" label is paradoxical. On the one hand, the term is meant to describe the hypothetical situation with Schrödinger's cat — in the sense that the cat is said to remain in some kind of intermediate stage mixture of alive vs. dead, until the observation occurs and the cat is collapsed (randomly) into one or the other state. Yet on the other hand, it is commonly recognized that such a bizarre combination state of affairs is impossible to imagine for a cat — and therefore that we are dealing with a pattern which exists in the case of microphenomena (like in these physical experiments), but not in the case of macrophenomena (like with the hypothetical cat). After that, the reasoning goes, where is the micro-macro dividing line between these two worlds, such that the pattern exists in the former world but not in the latter?

Given the principles which guide our work, it should not be surprising that we do not embrace this analysis or conclusion. So how are we to interpret what is going on in these experiments? We will discuss the issues in terms of three points: complementarity, observation, and the micro-macro split.

Our comments about the first two points are fairly brief, given that we have discussed the issues previously. To begin with, we would describe the overall pattern which emerges from these experiments in terms of complementarity — although this is not at all how the phenomena are normally depicted. As we have explained in the prior two subsections, we are dealing here with complementary modes of being on the one hand, and complementary periods of time on the other. The former is represented by what are usually called opposite states, such as alive vs. dead cat or clockwise vs. counterclock-

47 There are other aspects of these experiments which are taken as demonstrating faster-than-light interactions, sometimes described in terms of "spooky action at a distance"; we discuss this issue in Chapter 9.

wise motion. The latter can be understood in terms of the two aspects of the wave function, evolution vs. collapse ... which we have been describing simply in terms of pre-stage vs. post-stage.

As far as the second point concerning observation goes, there is little more to say here. We have explained why we reject the proposition that observation has any intrinsically-crucial role to play in these matters. Observation, such as is involved in these experiments, is merely a subset of a far more general pattern of interaction.

Regarding the third point, our position is that Einstein was correct in his dismissal of a micro-macro split. We will consider the topic in more general terms shortly, but for the moment let us address the question in relation to the experiments we have been discussing. We contend that the essence of the problem revolves around the issue of probability or likelihood, specifically in terms of a subtle confusion between different ways for determining this factor — and that the issue has nothing to do with a micro-macro split. Let us explain.

To start with, consider the pattern which is described for these physical experiments. Conditions have been arranged such that the post-stage likelihood is 50-50 as between the two options (e.g., clockwise vs. counterclockwise motion); and this is reflected in what is presumed to be a corresponding composition of the pre-stage — in which both options appear to be equally represented (e.g., both clockwise and counterclockwise). It is as though the pre-stage provides a proportionate setting (e.g., ψ-evolution) for the post-stage outcome (e.g., ψ-collapse) which follows.

Again, the same pattern would apply to the critical atom in Schrödinger's thought experiment: circumstances are arranged such that there is a 50-50 chance that the atom will not decay as that it will decay. In that case, one would assume that there is a corresponding atom pre-stage which is equally composed of not decay vs. decay — just as in the experiments being discussed here.

But here is the crucial point: while such an analysis about likelihoods applies to the *atom* in Schrödinger's experiment, we argue that it does not apply to the *cat* there. This is why we dissent from the usual reasoning — that because there is a 50-50 chance that the critical atom will decay (thereby leading to the release of the cyanide and the subsequent death of the cat), that this also means there is a 50-50 chance that the cat will live or die ... in the sense that there should be a bizarre pre-stage mix of the (live-dead) cat which would reflect these post-stage odds (assuming, of course, that there was no micro-macro split).

We beg to differ, and on fairly simple grounds: We reject the presumed linkage — between the atom's pre-stage likelihood of not decaying vs. decaying, and the cat's pre-stage likelihood of living vs. dying. Such a pre-stage linkage is unwarranted, because while it may make analytical or logical sense, it is physically spurious.

That is because the relationship between the two systems, atom and cat, should be understood in terms of immediate physical factors. Unless there is an inherent physical linkage between the likelihood of one system developing in a certain way and the likelihood of the other system developing in a certain way, then the fact that we humans can propose or deduce — not to mention stage a series of events so as to produce — such a *logical* linkage is irrelevant to the actual *physical* situation, especially as represented by pre-stage likelihoods.

In the terms that we are describing, then, there is never a 50-50 pre-stage chance that Schrödinger's cat will die — even though there is a 50-50 pre-stage chance that the critical atom will decay. On the contrary, until the atom decays, there is a very small chance that the cat will die. Once the decay event happens, of course, a new set of physical circumstances will quickly intervene to alter the cat's prospects — changing them from very likely to continue living to very likely to suddenly die.

Again, the issue has nothing to do with micro-macro, for this interpretation (based on differentiating physical and logical linkages) would apply regardless of whether we are speaking about a macro cat or some analogous micro entity, presuming each of the two systems has a comparable physical likelihood of living/existing vs. dying/decaying. And the fact that our manipulation or logic might seem to make such a likelihood dependent upon some other system's developing in a certain way (e.g., an atom decays and cyanide is then released) does not justify such a linkage in physical terms — until the latter event occurs and thereby produces such a changed likelihood.

Here is another way of illustrating our point that the issue is not micro vs. macro, by way of an example we keep coming back to. Thus, in contrast to the macro case of the cat — where we argue that there *is not* a 50-50 likelihood of one outcome vs. the other — consider another macro case where there *is* a 50-50 likelihood as between the two outcomes: that of the faces-vase image. Given that this image is similarly designed as the actual experiments we have been discussing, in being equally likely to yield one post-stage outcome (e.g., faces as figure and vase as background) as the other (e.g., vase as figure and faces as background), then it is reasonable to suppose that there would also be a brief moment in time, when the viewer is first exposed to the image, that there is a pre-stage mixture which is equally composed of both possibilities — after which there is a post-stage outcome one way or the other (even if the partners can alternate in a subsequent time period).

We will return to the micro-macro issue shortly.

Uncertainty Revisited

Earlier in the chapter, we noted that the Uncertainty Principle is a very confusing concept, and that it is unclear precisely what it refers to. While the principle provides a useful computational tool, it lacks a sound epistemological foundation. Our goal in this section is to try and come up with some coherent options.

To start with, given all the issues we have discussed in the interim, we will make short shrift of some of the explanations we described earlier — and simply state our position that the source of the problem is not due to a shortcoming of Quantum Theory itself (in terms of whether it is consistent or complete), nor is it a byproduct of the observation or measurement process. As we have said regarding the latter, there is nothing unique about the underlying process of observation or measurement, as distinct from other types of interaction phenomena; and the issue certainly has nothing special to do with consciousness, human or otherwise. Our conclusion is that the Uncertainty Principle ultimately relates to the nature of existence, in some form or other. But how?

The most obvious explanation is that there is an inherent uncertainty in nature — and it seems apparent that at least some of the principle's roots lie there. For example, Peirce wrote in the late 1800's that, "[C]hance is a factor in the universe ..." (Peirce, quoted in Jammer, 1989, p. 175) Peirce "analyzed the process of experimental observation and concluded that absolute chance, and not an indeterminacy arising merely from our ignorance, is an irreducible factor in physical processes and hence an ultimate category for the existential universe." (Jammer, 1989, p. 176) Karl Popper also endorses the reality of such a characteristic of nature: "We take indeterminism as a cosmological fact which we do not attempt to explain." (Popper, 1982, p. 181) Indeed, Jammer notes that, "[T]he general idea of a fundamental uncertainty or indeterminacy in the physical world is as old as the idea of strict determinism." (Jammer, 1974, p. 73)

The notion of an inherent uncertainty in nature is integral to the present model — and is just as fundamental as that of certainty. Of course, we are also speaking about the two forces of nature, as they can be described (or manifest themselves) in these terms. The *X*-Force of Structure corresponds with certainty, just as the *P*-Force of Process corresponds with uncertainty. Clearly, then, the phenomenon of uncertainty is not some after-layer in this work, to be added on as a footnote or explanation for why perfect prediction and determinism do not reign in the universe. Rather, uncertainty, randomness, and chaos are an integral part of nature, and are just as salient as — indeed, are complementary to — certainty, pattern and order. Our position is that the two taken together represent the dual nature of reality.

One particular example of uncertainty can be found on the molecular level, in the form of "random microscopic motions", or "molecular chaos". For physics, this phenomenon poses "the intriguing problem to try and find a *cosmic* explanation of [this] molecular chaos." (Davies, 1977, p. 193) Of course, the "cosmic" explanation of the particular pattern is inherent in the present model's general *P*-Force of Process. Whether we call it random behavior, chaos, or uncertainty, such variability merely represents another (level-specific) manifestation of the same underlying force principle.

While it is evident that in the present work there is nothing mysterious about the existence of a phenomenon of uncertainty, and that it can be understood to correspond with one of the two force tendencies of nature, the same clarity is not apparent when we speak about the Uncertainty Principle. That is partly for semantic reasons: while the phenomenon of uncertainty and the Uncertainty Principle need not be the same, their common morphology makes it difficult to differentiate them. In some ways, these two constructs overlap — but in other ways they diverge, especially in the current model. That is evident from the description we just put forth — that in 2FT, there is a basic phenomenon of uncertainty, which exists in relation to a basic phenomenon of certainty; and this duality is an obvious instance of the complementary pattern of nature. But this same two-sided clarity is not apparent when we speak about the Uncertainty Principle. So what can that principle be representing?

Our interpretation is that there are some respects in which the Uncertainty Principle may represent the phenomenon of uncertainty itself, as opposed to certainty, in the manner we just described. But in other respects, the Uncertainty Principle may represent some mixture of both certainty and uncertainty — which combination obviously represents a very unfortunate clash of terms. That is a tough pitfall to contend with, but we shall do our best to clarify these two alternatives in the rest of this section.

The most obvious approach is in keeping with the first description, and would simply say that the Uncertainty Principle codifies the influence of its uncertainty namesake as a factor in the universe. Using the template of the wave-particle duality, the idea would be that the Uncertainty Principle is essentially based on the wave factor, and its associated characteristics — and not on the particle factor, and its associated characteristics. Properly speaking, one might even argue that such process-like wave phenomena *cannot* be observed or measured finitely, because we are talking about the uncertain side of nature — which, again, we identify with the *P*-Force — in one or another manifestation. So by definition, this category does not even exist in distinct and granular ways that can be clearly isolated and quantified. By contrast, structure-like phenomena such as particles can be isolated and measured finitely: that is their nature. One way of putting the matter is to say that structure is a finite quantum — but by contrast, process is a non-quantum continuum. Put another way, Wolf says that while our concept of a particle "is localized.... Waves are different; they are not localized." (Wolf, p. 97) Or as Marion states it, "Waves cannot be localized in space and so any measurement of the position of a wavelike object must be subject to uncertainty. The Heisenberg principle gives a quantitative description of this uncertainty." (Marion, p. 437)

We would argue that if one attempts to identify the uncertain side of nature, then one must accept its hazy nature: any attempt to sharply delineate the parameters of uncertainty will only distort it, into what it is not. This could be described as attempting to transform inherent uncertainty into reluctant certainty! That may be what is happening here, in that while the Uncertainty Principle acknowledges the limits of absolute certainty, in the form of perfect predictability — even though it substitutes probability for such absolute predictability — it does so only from compulsion, and without coming to terms with the essential duality of nature which underlies this limitation ... attempting instead to eliminate (or at least truncate and minimize) it, rather than embracing it and consecrating it as an equally valid aspect of nature. That is true both in terms of waves and particles having equal and separate complementary types of existence — something which some Quantum theorists deny, as we discussed earlier in the chapter. It is also true in terms of the fundamental premises of 2FT — that the general categories of certainty and uncertainty have equal and separate validity as representing the dual nature of reality ... a complementary balance which is not in keeping with the spirit of the Uncertainty Principle. Another way to phrase this latter point is to say that in the established model, uncertainty is essentially treated as an exception to the general rule of certainty: the latter has a privileged status as providing the normal baseline, with the former only being invoked to explain deviating circumstances.

One way of describing the pattern we are talking about is to speak about process-like phenomena being distorted and frozen into a structure-like mold. Indeed, we suggest that such a tendency is indicative of a larger problem. We are talking about a broad, structurally-oriented mind-set which has long dominated much of physics' thinking and models; this is a matter we shall discuss at greater length in Chapter 9. Obviously, what gets lost with this distorted approach are not attributes of structure, but attributes of process; and nature comes out looking more discrete, structural, and particle-like than it actually is.

It seems clear that this was the spirit of what was happening with the Uncertainty Principle. For example, note that both Heisenberg and Born insisted "that the 'wave nature of matter' meant nothing more than the inherently random behavior of entities that are really particles." (Holton and Brush, p. 499) Jammer cites Landé as declaring that, "[I]t is the task of the quantum theory to reconcile the contradiction between the two classical concepts of particle and wave by proving the equivalence of the descriptions of physical phenomena in either terms." (Jammer, 1974, p. 453) But look at how Landé developed such a model: "In contrast to Schrödinger's *unitary wave interpretation* and Bohr's *dualistic wave-particle interpretation* he called his approach the *unitary particle interpretation ...*" (Jammer, 1974, p. 460). One of Landé's books was titled, *From Dualism to Unity in Quantum Physics* (Jammer, 1974, p. 463)!

We have so far been talking about the first overall answer to the question of how the Uncertainty Principle might be understood to describe the uncertain aspect of existence — as a method for isolating and thereby quantifying the basic uncertainty factor itself (such as the wave), and thence transforming it into what is effectively its opposite of certainty (such as the particle). But now we want to turn to the second interpretation we mentioned above, whereby the Uncertainty Principle would be understood as describing both partners in a more equitable sense — such as the interaction between them — and not primarily one in terms of the other.

Applied to the wave-particle duality, this second approach would mean that we are no longer pursuing the analysis based on associating the uncertainty factor with wave, and certainty with particle — because if the Uncertainty Principle describes the interaction or relationship between them, then it would be no more reflective of the one than the other. Rather than applying to either of the component partners (corresponding with certainty and uncertainty, respectively) separately, the Uncertainty Principle would represent some kind of combination between them — such as a conversion factor between them as to which partner will predominate in any given situation.

For example, how well can we measure both wave and particle phenomena, in some given situation — or how readily can we switch between them with precision? In the second interpretation terms that we are now discussing, the Uncertainty Principle would represent some kind of bridging factor which joins the two complementary portions ... such as describing the "ball" of energy which moves back and forth between them. We mentioned such a flipping or triggering factor in the Complementary Stages subsection. According to this approach, the Uncertainty Principle would provide a mathematical basis for quantifying such interaction relationships — for all conjugate variables.

In principle, it is hard to know where such a transition or interaction factor would end, in terms of the boundless scope of the present model of complementarity. Would there be a way to apply some form or equivalent of the Uncertainty Principle to all levels of complementary relationships in nature — such that it would act as a general conversion factor between the *X*-Force of Structure and the *P*-Force of Process, in whatever specific manner they manifest themselves for any given situation? This would presumably include complementary partners at every dimension of analysis, in the present model.

A related way of thinking about the issue is in terms of the basic relationship between the two forces themselves, along with all their subsidiary manifestations — that the two partners are never equal to each other, and that one is always dominant in relation to the other as recessive. Thus, 2FT is built upon the premise that all relationships in nature are of a figure-ground complementary character — whereby one entity is focus and sharp, relative to the other entity as background and blurred. Dominance and recessiveness may switch between them, as we keep repeating — but the dominance and recessiveness, the sharpness and blurness, will always be there ... in one form and direction, or the other. This would explain why "no physical situation can arise which exhibits simultaneously and rigorously (sharply) both complementary aspects of the phenomenon." (Jammer, 1989, p. 363)[48]

Now let us take a different tack, as we try to summarize these two interpretations of the Uncertainty Principle — by bringing them together with the aid of the other main theme of this work, that of hierarchy. In particular, we are interested in the relationship between a higher level whole and the lower level parts. At the same time, this approach also provides a useful framework for understanding the contrasting theoretical positions of Bohr and Heisenberg in their approach to the wave-particle pattern, which we described earlier in the chapter.

Again, Bohr's position was to emphasize the duality — the either/or aspect of the phenomenon, which was the basis for his Principle of Complementarity ... relative to which he treated the Uncertainty Principle as only secondary. By contrast, Heisenberg highlighted what we call the unitarity — the both/and aspect of the phenomenon, which was the basis for his Uncertainty Principle. As we have mentioned more than once, Heisenberg considered the Principle of Complementarity to be not just secondary — but an unfortunate distraction, if not an outright encumbrance: neither conjoint partner (e.g., wave or particle) had any real significance, and their very appearance was largely the byproduct of a flawed concept system.

Here is a complementary chart, which is intended to incorporate these two respective positions:

X-View	*P-View*
Principle of Complementarity	Uncertainty Principle
dichotomous either / or	continual both / and
wave vs. particle	wave-particle (e.g., "wavicle")
different lower level parts	same higher level whole

Once again, Bohr's position took account of Heisenberg's — whereas Heisenberg's position excluded Bohr's. Thus, "Bohr accepted the indeterminacy [or uncertainty] principle as a useful quantitative limitation on the extent to which complementary (and therefore partial) descriptions may overlap" (Holton and Brush, p. 499).

It is clear how we would approach this dispute, and the two respective Principles themselves. In terms of the above conceptual scheme, Heisenberg's overall position would be interpreted as embracing only the higher level unitary whole — which would also explain his dismissal of the lower level complementary parts. In particular, the Uncertainty Principle could be understood in terms of this higher level whole — which would be consistent with the latter interpretation above, that it represents a combination or translation device here. That is, the higher level differential would act as an

48 The issue can also be cast in terms of a completely different topic that we have not spent much time on in this chapter — namely, symmetry ... or more specifically what we call "false symmetry". In particular, the very expectation of sharp vs. sharp conjugate variables rests itself upon a presumption of perfect symmetry as applying in nature, whereby both conjoint partners should manifest themselves and their function in a way that is simultaneously and mutually precise.

interchange factor between the lower level complementary parts ... or even as a way of quantifying the limiting factor of such an interchange combination.

One can also argue that it is of no great consequence which of the two interpretations one adopts, as to how the Uncertainty Principle should be primarily understood — either as a one-directional factor for converting (e.g., wave-like) uncertainty into (e.g., particle-like) certainty, or as a unifying or conversion factor for joining them together. The end result is much the same, in ignoring the two-sidedness of nature. By virtue of such one-sidedness, the effect of the Uncertainty Principle would be to reduce the dual nature of reality into a new sort of "singularity" — here, based upon Planck's constant as describing the ultimate discrete (and particle-like) quantum, i.e., a minimal unit of action or energy in the universe. Needless to say, we reject any such conclusion: conjugate variables like particle and wave — or certainty and uncertainty — are each real and distinct aspects of nature, just as they also exist in relation to one another.

In closing this topic, here is how David Bohm depicts the two sides of nature that we are describing, in the way he contrasts "laws of cause" and "laws of chance":

> [T]he causal laws and the laws of chance *together* are what bring about the actual development of things, so that either of them alone is at best a partial and approximate representation of reality, which eventually has to be corrected with the aid of the other.... [T]he categories of necessary causal connection and chance contingencies are seen to represent two sides of all processes.... [O]pposing and contradictory motions are the rule throughout the universe, and this is an essential aspect of the very mode of things. (Bohm, 1957, pp. 28, 29, and 149)

> [W]e are led to regard these two kinds of laws as effectively furnishing different views of any given natural process.... [W]e do not assume, as is generally done in a mechanistic philosophy, that the whole of nature can eventually be treated completely perfectly and unconditionally in terms of just one of these sides, so that the other will be seen to be inessential, a mere shadow, that makes no fundamental contribution to our representation of nature as a whole. (Bohm, 1957, p. 143)

Correspondence Revisited

The final topic in our discussion of Quantum Theory is a return to the Correspondence Principle, which we introduced earlier in the chapter. There is not much more we have to say about the basic premise itself, so much as with the point we already mentioned — that the crucial Quantum Theory constructs of complementarity, uncertainty, and the observational whole are generally taken to describe the micro world alone, but not the macro world of our ordinary experience. Indeed, the idea would be that it is unthinkable how such patterns could possibly manifest themselves in our everyday existence, for that would seem to violate our usual sense of causality and the way we order our lives. So we fence off these newer ideas from our usual experience of macromatter — in spite of the fact that this violates the spirit of the Correspondence Principle, namely that the new laws and constructs (Quantum Theory) should incorporate the old ones (classical mechanics) as a special case.

Needless to say, from the perspective of the present model there would be no micro-macro split, because the overall patterns of nature are treated as being the same at all levels of the hierarchy. There are not separate micro and macro realms, with different constructs and sets of rules for each — any more than there are special levels of nature, such as are associated with physics' fundamental forces.

Indeed, the very idea of a micro-macro split would seem to result from a human-based perspective as to what is small and what is big. Given the hierarchical framework of the present model, one level of nature is inherently no more significant than any other level — and the very notion that the level we humans occupy somehow makes sacrosanct our vantage point (e.g., in differentiating small from large) itself represents a narrow and self-centered vision on our part.

We have already discussed the issues of observation and uncertainty, and it should be clear that we treat the underlying patterns as general characteristics of nature as a whole. Likewise with complementarity itself, which is the core precept in the present model. For his part, de Broglie "proposed that the wave-particle duality manifested by light might be a fundamental characteristic of *all* entities (electrons, protons, oranges...)" (Hecht, p. 267)

As to why complementarity is not treated as a pervasive phenomenon that applies just as much to what is called the macro realm as it does to what is called the micro realm, we discussed some of the main reasons in the last section of the opening chapter; and we need not repeat those points here. But there may also be particular reasons why some aspects of complementarity are more apparent in the micro than the macro world. For example, we have noted that small things are on a different time scale than large things — with the result that they may have a higher frequency and faster change rate between one complementary state and another ... thereby making such patterns more apparent.

Then, of course, microphenomena are much more readily manipulated than are macrophenomena, in terms of size and power — and so any macro intervention (e.g., an observation) which introduces a significant amount of energy into the situation can readily alter the outcome ... thereby making such patterns more easily demonstrable in the micro world than the macro world. As von Neumann said, for example, "[I]n the atom we are at the boundary of the physical world, where each measurement is an interference of the same order of magnitude as the object measured, and therefore affects it basically." (Von Neumann, quoted in Jammer, 1974, p. 266)

But as to our general premise that complementarity applies to all phenomena, micro or macro, we need only recall once again the basic framework we are working with — that there are two fundamental forces, which manifest themselves in terms of very broad characteristics that go well beyond particle and wave: Something will be dominant or recessive; it is either static or moving; a phenomenon is either matter (and particle-like) or it is medium (and wave-like); everything is part of either an inside or outside relationship; and a given entity is either stable or changing. We maintain that the essential characteristics for each of these dimensions are complementary in character, nesting in form — and universal in scope.

It was the genius of Niels Bohr to have discovered that complementarity was a crucial way of depicting the world; and he ultimately developed a very expansive view of this principle, and felt that it was the proper and pervasive description of nature in general. We are doing our best to implement that thesis in the present work.

Chapter 3

Atom & Co.

Although we have taken issue with various aspects of Quantum Theory's position about the role of observation in our examinations of nature, we heartily embrace the essential point that what we see depends on how we look. Put in terms of the present model, reality is a combination of being and seeing — and the two modalities cannot be divorced from one another.

Of course, the usual way of thinking about the observation or seeing function is in terms of the physical framework which is involved in any situation, such as would bring about a wave or particle outcome. But we contend that this category should also be understood on a more abstract level as encompassing the whole conceptual frame of reference which is involved, in any given instance. One thing this means in particular is that our theoretical preconceptions are an important part of what we bring to any situation, and so they play a significant role in the phenomena we find there. The result is that the things which one studies "are seen through the lens of theory." (Holton, p. 104) Indeed, "[I]t is possible to understand and use observations in physics today only if the scientist has, from the very beginning, a 'well-structured image of the actual connections between the events taking place.' " (Holton, p. 90)

The two sets of factors we are talking about — both the physical and the theoretical, as comprising the broader category of observation — become especially relevant when we are dealing with the world of the very small. We emphasized this point in relation to physical factors at the end of the previous chapter: micro states of being are especially sensitive to our macro modes of examination — simply in brute force physical terms, in that the latter can easily overwhelm the former. Now we want to highlight the point in regards to theoretical factors, as also being especially salient in our explorations of the micro world. Consequently, any conclusions we reach as to the nature of phenomena in the micro realm are particularly dependent on the conceptual framework we bring to bear. In the micro realm, we are dealing with very subtle states of being — in relation to our macro selves — which we maintain can just as well be interpreted in more than one theoretical way.

This is the case even when we are dealing with very well-established constructs. For example, Holton quotes Deutsch as saying that, "An electron [or any particle of modern physics] is clearly not an object with the general properties of a ball which we would see if we had a sufficiently good microscope, or feel impinging on our hand if our nerves were a little more sensitive. We are not *forced* by direct sense perception to use this image." (Deutsch, quoted in Holton, p. 90, with Holton's bracketed insert) Or as Morris puts it simply, "[N]o one has ever seen an electron." (Morris, p. 192)[1] Likewise, "the so-called internal structure of the hydrogen atom is *in principle* inaccessible to direct sensory perception." (Deutsch, quoted in Holton, p. 89) The important point is that, "[I]t is at least conceivable that physicists might have chosen a quite different way of looking at" such phenomena. (Holton, p. 92)

Toulmin addresses the issue in speaking broadly about...

> the fundamental concepts of science. We see the world through them to such an extent that we forget what it would be like without them: our very commitment to them tends to blind us to other possibilities. Yet a proper sense of the growth and development of our ideas will come only if we are prepared to unthink them. (Toulmin, 1961, p. 101)

Of course, for most people such quandaries do not arise — given that the main entities which are said to exist in the micro world have been well defined for a great many years now, and are no longer subject to varying interpretations as to their basic character or role in nature. Rather, the main task has been to further elaborate the scope and precision of the highly complex rules and specifications for describing and predicting the behavior ... of these established phenomena whose existence is taken for granted. As a result, the very idea of calling into question any of the basic components and their function in the conventional model of the micro world would seem to be a dubious proposition at best.

But that is part of our task now, in accord with our guiding precepts. For we maintain that the essential features of the established paradigm, here being in the micro realm, have been built up in a largely historic, incremental, and ad hoc fashion — rather than evolving in accord with some coherent theme. We contend that the phenomena in question could just as well have been construed in the latter terms, but that is not how modern physical theory developed. Yet seeking such a coherent outcome is precisely the mission of the present work.

Much of our discussion in this chapter will be centered on the construct of the *atom*, which has long been viewed as the basic building block of nature. We will present a few of the highlights of the orthodox model in the first part of the chapter, and then in the second part we will reexamine such issues from the perspective of the present model.

1 Similarly, "[A] 'pure' photon can never be distilled out of this complex network of transmutations. For the vast majority of particles..., the distinction between 'real' and 'ghost' becomes blurred." (Davies, 1983, p. 163) "A 'photon' seems to become isolated from the fundamental unbroken unity *because* we are studying it. Photons do not exist by themselves." (Zukav, p. 72) More generally, "[N]one of the subatomic particles, quarks or otherwise, are really particles in the common meaning of the word. Indeed, they may not even be 'things' at all." (Davies, 1983, p. 161)

The Conventional Approach

The atom existed as an idea long before physics came along to apply it to anything concrete and give it specific referents. In its early form, the atom was thought of as the smallest unit of nature, which could be broken down no further; and eventually that label was applied to a particular entity — that which we still call the "atom" today. In time, however, it was determined that the entity which physics had christened as the atom was not such an ultimate unit, but that it was itself composed of other things. What developed was the idea of a nucleus core in which the atom's mass was concentrated, surrounded by very light orbiting electrons. More specifically, the nucleus was said to be composed of individual, positively-charged *protons* — which were typically offset by an equivalent number of negatively-charged *electrons*.[2] That essential description continues to apply today, even though other aspects of the model have changed. Thus the atom's basic existence is explained in terms of the principle that opposite charges attract one another, thereby keeping the electrons bound to the protons.

But in many respects the primary purpose of a model of the atom is not only to explain the existence of individual entities, but also to provide a basis for explaining the interactions between them. There are two main facets of those interactions which are important: the joining together of individual atoms into collections called *molecules*; and the nature of the interactions themselves, which provide the basis for a whole set of phenomena like *electricity*, *magnetism*, and the combination between them called *electromagnetic radiation*. Taken all together, the product and character of the interactions between atoms is explained in terms of the *electromagnetic force* of nature.

As far as electrical and magnetic interactions, there is the field construct which describes both. Specifically, electric fields surround electric charges, and magnetic fields surround magnetic poles. In either case, we can think in terms of "lines of force" as characterizing the patterns of attraction and repulsion in these fields. The relationship between the strength of either force and the distance from the source thereof can be described in terms of the well-known inverse-square law. One apparent anomaly is that magnetic fields also surround electric charges which are in relative motion, vis-a-vis some observer — and the usual way of interpreting that outcome is to say that magnetism is simply a motion-based version of electricity, as opposed to being a separate phenomenon in its own right.[3] One significant difference between the two is that electric fields originate and terminate on specific electric charges, whereas magnetic fields do not.[4] This situation is apparent in the fact that there are no "magnetic monopoles" in nature.

2 "[A]ll atoms in their natural states contain equal numbers of protons and electrons" (Marion, p. 105; Hecht, p. 240).

3 "[M]agnetism arises out of motion.... [It] is actually a facet of electricity, a relativistic aspect born of the effects of motion." (Hecht, p. 444) Or, "[T]he magnetic force is really a manifestation of the electric force when the charges have relative motion." (Marion, p. 263)

4 "[M]agnetic field lines have no beginning and no end." (Marion, p. 267)

Electromagnetic radiation, such as visible light, is explained in terms of an interaction between the two field types. In particular, the electromagnetic wave is depicted as being a "transverse" wave (which we shall describe later) that consists of oscillating electric fields and magnetic fields, moving "in leapfrog manner" through empty space. In this regard, Rogers says that, "Electromagnetic waves seem to have potential energy in their electric fields and kinetic energy in their magnetic fields." (Rogers, p. 389) And of course, the wave-particle duality means that we are not only dealing with *electromagnetic waves* — but with the particle version thereof which are called *photons*.

Now let us go back to the electron. One way of thinking about this phenomenon is in terms of a discrete, particle-like entity; but the situation is considerably more complex than that. For one thing, there is the wave-particle factor we just mentioned, which also applies to the electron.[5] But the complexity also applies in the more specific terms of trying to identify individual electrons of whatever form, and their configurations in particular types of atoms. The details that are involved relate to a complicated set of quantum parameters, according to which each electron of an atom can be identified by a combination of several different quantum numbers. Most of these specifications, about the electrons which surround different types of nucleuses,[6] come from observation of "atomic spectra". These spectra indicate the vibration (or, emission and absorption) patterns of the atom, and are thereby said to reveal the precise structure of the overall electron arrangement.

Some of the most important characteristics which differentiate one type of stable atom from another are represented in the organizational scheme called the *periodic table of elements*. The table was originally devised by Mendeleyev and others in the late 1800's, as a way of finding some order amongst the various types of physical matter in nature. Each of the elements can be identified with an "atomic number", which refers to the element's number of protons; and as we alluded to earlier, for whole atoms that means the same number of electrons. There are some 90 naturally-occurring elements (along with additional synthetic ones) which are represented in the table, and they increase one after another in mass. But the electrons increment in a cyclic fashion, based upon what can be construed as a series of orbit-like shells or layers which surround the nucleus. Each layer is usually filled in stepwise, adding one electron after another — until that shell is full, and then the next electron goes to the following layer ... progressing from innermost to outermost layers thereof.

The periodic table reflects these two parameters, within vs. between orbital layers. As it is traditionally presented, the different rows (called "periods") represent the different layers (or "orbits") — whereas the different columns (called "groups") represent different numbers of electrons within each layer. A simplified outline of the periodic table would look something like the following:

5 "[P]articles can be understood in terms of standing waves" (Zukav, p. 107) That is true if they are effectively static — whereas the moving particle can be understood in terms of traveling waves.

6 Our use of the word "nucleuses", rather than "nuclei" may seem odd; but it is a valid term — and it is important for us to differentiate nucleuses from nucleons, and so we do not want to use the third term of nuclei which might tend to confuse the issue for some readers.

Electrons within Orbits

Group No.

I II III IV V VI VII VIII+

Fewest electrons ← ———————————————————— → Most electrons

1 Smaller

2 ↑

3 Period/row No.

4 *Different Electron Orbits*

5 ↓

6 Larger

7

Take the example of period 2 elements — meaning that there are two orbital layers for electrons to occupy. In that case, the first layer would be entirely filled, and the different group number columns in period 2 represent the successive filling of the second layer. Group I elements would have one electron in this orbital layer, group II elements two … progressing to group VIII elements, which would have eight electrons filling all possible positions in this orbit. (One of the ways that this chart is simplified is that it does not reflect the fact that higher number of orbital periods have more groups, or possible number of electrons, within them.)

In fact, it is only the outermost layer of electrons which determines an element's chemical properties, and that is why the elements are broken down into groups.[7] Regardless of how many inside layers or periods are involved, every element which has the same configuration of electrons in its outermost orbital layer belongs to the same group. That is why each group is also called a "family", because of the characteristics which they share in common. For example, all group I elements react to water in a similar way; certain groups form ores, while other groups form salts. Some groups have elements which are more combinable with other elements, whereas others like the "noble gases" are not combinable. It is the sharing or exchanging of the outermost layer of electrons, between one atom and another, which provides the basis for bonding into molecules — and thereby the existence of all normal matter in our experience. Several

7 "Only the outermost group of electrons are involved in chemical changes." (Rogers, p. 716)

different types of bondings are said to occur; but there are two principal types, which are called ionic bonding and covalent bonding.[8]

While a whole atom would have an equivalent number of positively-charged protons and negatively-charged electrons, some atoms are not complete in this sense. When the numbers of protons and electrons differ from one another, the atom in question is itself described as being *charged*[9] — in that case meaning either a deficit of electrons in relation to the number of protons, called "positively-charged"; or a surplus of electrons in relation to the number of protons, called "negatively-charged".[10] When a balance can be attained by a transfer or sharing of electrons between the two entities, then they are brought together into balance and are thus no longer charged.

As we noted above, physics' understanding of the makeup of the atom continued to evolve over time, beyond consisting merely of the proton and electron. Perhaps the biggest development came when it was realized that certain properties could not be reconciled with the notion of a single nuclear unit, and it became necessary to introduce an additional particle to go along with the proton in the nucleus. But since the charge factor was already incorporated in the existing model, by virtue of the relationship between the positively-charged proton and the negatively-charged electron, the new nuclear unit did not require any charge value. It was therefore christened a *neutron*, as being neutral electrically.

Note that the electrical charge factor is almost the only distinction between the proton and neutron, along with a very slight difference of mass. And even though the neutron concept developed as something of an afterthought, the two units were eventually accorded a kind of equal status — as *nucleons*. "It is useful to consider the neutron and the proton as simply two different states of a single particle, the nucleon." (Preston and Bhaduri, p. 6; also Davies, 1979, p. 126) Proton and neutron, then, are "merely the charged and uncharged versions of the same basic particle." (Hecht, p. 336)

But this new state of affairs also posed a problem, as far as what it was that held the nucleons together inside the nucleus. For one thing, since the neutrons carried no charge at all, then they would have no reason to be bound to anything. Yet another charge dilemma was not actually new, and this concerned the fact that there could be more than one proton inside the nucleus. And while the binding of the atom as a whole could be explained in terms of opposite charges attracting one another (i.e., protons and electrons), the other side of the coin is that same charges repel one another (i.e., protons and protons). So what keeps multiple protons bound tightly together inside the nucleus — which is the case for all varieties of atoms except the most basic form of hydrogen (composed of one proton and one electron) — when they should repel one another?

8 These also go under the aliases of "heteropolar bonding" and "homopolar bonding", respectively (Frisch, p. 120).

9 "The charge of a body or region arises as a result of an excess or deficit of electrons with respect to protons." (Pitt, pp. 64-65)

10 Such charged entities are called *ions* — which is a generic label that does not differentiate whether we are dealing with an atom, molecule, etc. The terms "negatively ionized" and "positively ionized" are, then, another way of referring to an entity having an excess or deficit of electrons, respectively.

It seemed clear that there was no way of resolving this dilemma within the theoretical confines of the previously established electromagnetic force, and so the answer was to introduce a completely new force which went beyond that which the conventional electrical charges could explain. This was the genesis for the *nuclear force* (or *strong force*[11]), a very powerful but short range type of interaction which is operative only within the realm of the nucleus, and whose primary purpose is to explain what holds together mutually bind-resistant protons, along with bind-indifferent neutrons.

Yet the problems still didn't go away. For one thing, what happens to the electrical charge factors, even if they are not sufficient in themselves to explain the composition of the nucleus? Don't we still have to take them into account, as resulting in different strengths for different compositions of nuclear binding? That is, if one were to have a neutron and a neutron,[12] or a neutron and a proton, there would be no electrical interaction, and thus only the effects of the strong nuclear force would apply. But in the case of a proton and a proton, there would presumably be electrical interaction, as well — the two same-charged entities would repel one another — and the net result is that the effects of the strong nuclear force should be mitigated to some degree by that of the electric, or electromagnetic, force. Two protons should be *less* strongly bound than would be two neutrons, or than a neutron and a proton.

However, this does not happen — the electrical properties do not intervene: the proton-proton case would be no different in strength of binding than the other two cases. What seems to happen is that the electric force disappears entirely, within the realm of the nucleus. Yet why should this be? The answer is unknown, but one resolution is simply to decree the "charge independence of nuclear forces" (Cohen, p. 60; Preston and Bhaduri, p. 52). This principle of charge-independence states that the strong nuclear force "appears to be unaffected by the charge on a proton and is of the same magnitude between any two nucleons (proton-proton, neutron-neutron, proton-neutron)." (Pitt, p. 261)[13]

Finally, we must note some other peculiarities in the rules by which the strong nuclear force appears to operate. For one thing, it is not just any combination of neutrons and protons which comprise the nucleus: neutrons are never alone within the nucleus, either singly or in multiples — they are always there with protons. And protons are never alone with other protons. Indeed, the pattern is that there are usually very similar number of protons and neutrons, for most stable nucleuses (Preston and Bhaduri, p. 5) — although as the size of the nucleus increases, so does the proportion of neutrons to protons. Correspondingly, larger (and heavier) nucleuses are less tightly bound than smaller (and lighter) ones.

11 The "strong nuclear force" would be differentiated from the "weak nuclear force" — or simply "strong force" vs. "weak force".

12 Such a simple neutron-neutron combination does not exist, and we only cite it for the purpose of discussing the issue at hand.

13 According to other interpretations, the neutrons function as a vehicle for implementing the strong force, thereby counteracting (and offsetting) the electromagnetic tendency of protons to repel one another.

There are other anomalies, as well. Even though the nuclear force keeps the nucleons bound together within the nucleus, the standard description is that this force must also keep the nucleons from getting squeezed too closely together. Lastly, although the neutron is described as being neutral, it nonetheless displays magnetic behavior: this should not happen to an electrically-neutral entity, which is how the neutron is typically thought of; yet it does (Cf. Rogers, p. 585; Hecht, p. 447; Davies, 1979, p. 135).

Let us summarize the complicated set of functions which the strong nuclear force must fulfill. For one thing, it must hold together multiple protons, which according to the electromagnetic force don't want to be anywhere near one another. For another, it must also act upon neutrons, which have no charge and are thus not subject to the electromagnetic force — and according to which, there is no reason for neutrons to be together with (or apart from) any other entity. Then there are some odd particulars: for one thing, the nuclear force won't bind together combinations of protons by themselves, nor combinations of neutrons by themselves. Rather, the nuclear force seems to prefer generally equal numbers of protons and neutrons, at least up to a certain size, after which it likes to have more neutrons than protons. At the same time, as just mentioned, this special force must push apart either type of nucleon from one another if they get too close together. Finally, this is a very, very powerful force, immensely stronger than either the force of electromagnetism or the force of gravity; but it operates in a much smaller realm than the other two — i.e., only within the dimensions of the nucleus itself ... in which place the electromagnetic force ceases to operate altogether, although gravity still does.

As far as the manner by which the strong nuclear force actually operates, consider the explanation which was put forth in the 1930's by the father of modern nuclear force theory, Yukawa. According to his model, we are dealing with an "exchange force" that consists of a meson particle which carries charge back and forth between the nucleons, and thereby binds the nucleus together. The overall composition of the nucleus is not altered in this interaction process, although individual nucleons would undergo transient oscillations of state — going from neutron to proton, and back again. This difference can also be understood in terms of spin (sometimes called "isospin"), with the two nucleons rotating in opposite directions ... which orientation would then be swapped between them.

More recent interpretations are that the nucleons are composed of quarks, which come in three different "colors" (and provide a different explanation for nucleon binding) ... and so on. Indeed, one might speculate that matter appears to be infinitely divisible into smaller and smaller units.

There is one more entity that we want to describe now. This is a phenomenon that is directly associated with the electron — indeed, it is considered to be the mirror opposite of that entity. The idea emerged in 1930, as part of Paul Dirac's abstract "theory of holes", according to which the universe is said to be filled with an infinity of electrons — but when one of these electrons is somehow removed, then the hole would be filled by the oppositely charged *positron* (Hecht, p. 345). The larger context was that Dirac was engaged in developing a relativistic quantum theory, and " ... among the solutions to his equation, Dirac was puzzled by the existence of those with *negative energy*, which inevitably arise because of the additional space-time symmetry which is inherent in special relativity." (Davies, 1979, p. 105)

So just as the neutron was discovered later as a general (but non-charge) counterpart to the proton, so likewise the positron construct developed much later than the electron.[14] And although the latter pair are clearly charge counterparts, such an association is considered extraneous to the basic composition of the atom ... both in the sense that the primary charge relationship there is deemed to be only between proton and electron, and that the positron has no regular equivalent role to play at all in the atomic model as does the electron.

Yet the positron manifests itself in various situations. One example is the phenomenon known as beta radiation, in which there is a change of nucleon — with either a neutron becoming a protron, or v.v. The former case is called *beta-* decay, and also entails the emission of an electron along with a tiny entity called the neutrino; the latter case is called *beta+* decay, and entails the emission of a positron, again together with a neutrino. Indeed, the medical procedure called positron emission tomography (PET) is an observation technique that is based on *beta+* radiation: a substance that has been specifically prepared to emit positrons is ingested by (or injected into) a patient, for the purpose of studying some aspect of the patient's internal tissue. A similar technique is also used for research and industrial purposes.

This brings us to the topic of antimatter, for Dirac's equations predicted that it was not just the electron which had a charge partner — but that there was no reason to stop there: every particle was assumed to have a corresponding mirror-opposite counterpart. This idea was originally postulated by Dirac as an inference from his mathematical explorations, as we just mentioned. That is because the symmetrical foundation of the Special Theory indicated that all matter in the universe had precisely opposite counterparts — meaning that there are two basic categories in nature. On the one hand, there is the usual type of *matter* that we are familiar with, which is composed of entities such as electrons, protons, and neutrons; and on the other hand, there is the esoteric type, called *antimatter*, which is composed of parallel entities like anti-electrons (another name for positrons), anti-protons, and anti-neutrons. These two sets of phenomena are said to have identical mass, but would have opposite or "conjugate" characteristics in other respects — typically represented by charge, but also entailing other so-called quantum properties. One conclusion is that all the universe is supposedly composed of equal portions of matter and antimatter — of anti-atoms, anti-molecules, anti-galaxies, and so on.

The idea was that around the time of the Big Bang, such oppositely charged matter and antimatter phenomena were formed in essentially equal measure through a process called *pair production*. But while they were both formed in the first moments of the universe, the two sets of phenomena necessarily became separated — because such matter and antimatter pairs will effectively neutralize one another by mutually annihilating, if they are brought into close proximity. Sure enough, such opposite charge counterparts have been frequently created in laboratory research, although most antimatter portions only survive for incredibly brief moments of time. Nonetheless, such findings have only bolstered physics' confidence about the universality of this pattern — with the implica-

14 The positron was identified experimentally by Carl Anderson in 1932.

tion being that somewhere in the universe, all the antimatter actually exists which goes along with all the matter that we encounter in our own world.[15]

This ends our very brief survey of some of the main features of the conventional model of the micro realm. Next we turn to an alternative perspective on these same issues.

A Complementary Approach

The steep task we now face is how to recast the phenomena we have been discussing into complementary terms. The main barrier to achieving that end is how completely alien any such broad reconceptualization is to the usual way of construing these matters, given the persistent non-complementary character of so much of the prevailing interpretation of phenomena in this realm — which of course has taken on a reality of its own, as we noted at the start of the chapter. That is not only because of tradition and force of habit, but also because the conventional model has been so successful in providing a viable framework for representing the details of what is going on. Still, any theory is distinct from the reality which that theory is meant to describe; and the same data can be construed in different ways, depending upon which theoretical map is brought to bear. We shall examine those issues extensively in Chapters 10 and 11, especially the point that what makes one theory better than another for depicting reality largely depends upon what function the theory is intended to serve — and there is no single answer to that question.

Our discussion below can be broadly divided into two parts. The first simply attempts to explain how various established constructs about the micro world can be understood in complementary terms, although they are not normally construed in this way. Thus, we contend that positive and negative charge are complementary to one another, and likewise with north and south poles. Also, charge patterns can be understood in terms of electric fields, just as polarity patterns can be understood in terms of magnetic fields — and electric and magnetic fields are also complementary to one another. We shall also look at how our theme can be applied to the periodic table of elements, the phenomenon of scattering, and various concepts about waves ... along with the general idea of things coming together and things going apart in the micro world.

While much of the first set of topics is largely a classification and illustration procedure, the second set is more substantive — and more controversial, because it challenges the conventional ways in which the phenomena are themselves understood, arguing instead that they would be better reconceptualized in complementary terms. This is the way we approach the whole topic of antimatter, which we will argue is a misbegotten thesis in its own right that should be eliminated. Next up is a different way of thinking about the nuclear and electromagnetic forces, followed by a discussion about

15 Another interpretation revolves around the precise ratio of matter and antimatter that were formed in the Big Bang start of the universe — revolving around the "essentially equal" description just mentioned. The idea is that there may have been a slight disparity of matter over antimatter — resulting in a residue of matter, after the main portions of matter and antimatter annihilated each other.

how the atom construct itself needs to be reassessed, for various reasons that we shall describe.

As we proceed, we must reiterate the immense challenge we are facing here: The constructs and interpretations we are dealing with are among the most well-established in modern science — yet the underlying phenomena themselves are among the most obscure, because of the indirectness with which they are observed. The result, once again, is that it is extremely difficult in this realm to distinguish the essential patterns themselves from the highly complex models in which they are generally construed — to differentiate observed fact from inferred conjecture. Yet that is our task — to comprehend the underlying phenomena in a manner which does not require our peering through the definitively-framed conceptual lens of the conventional model ... whose very limitations we are trying to transcend in this work.

Electricity and Magnetism. Let us start with the topics of electricity and magnetism — basic phenomena with which all of us are familiar, and perhaps fascinated by, in our daily lives. Electricity and magnetism seem to have extremely mysterious qualities, which set them apart from other phenomena of nature. But our position is that the pattern is not unique, and that there is no electric force or magnetic force, as such — or even the grandiose synthesis of an electromagnetic force. What sets these phenomena apart is the particular level of nature's hierarchy at which they exist, and the specific way the pattern manifests itself there. Yet in general terms we are dealing with the same pervasive pattern of interaction which occurs everywhere else in nature. In this subsection, we will consider different kinds of electric and magnetic field phenomena, and in the next subsection we will talk about the kinds of materials which are generally associated with each category.

One way of understanding the complementary relationship between electric and magnetic fields is that they form a figure-ground (or hand-glove) pattern with one another. Such a relationship also manifests itself in the fact that the two fields have an opposite spatial orientation: they are perpendicular to each other.

How are we to interpret this pattern in specific terms of the present model's dimensions of analysis? To start with, we treat both field types as predominantly medium domain phenomena — but beyond that we have chosen to associate the electric field primarily with the inside realm or closed field, in relation to the magnetic field as primarily outside realm or open field.[16] Like other complementarities, though, these inside-outside designations may be switchable between the two field types, as they interact with one another in different situations. But however they are named and categorized, the duality between electric and magnetic fields is essential: they must not be reduced merely to some kind of singular combination, as is sometimes suggested.[17]

16 In these terms, the conventional production of electricity can be understood in terms of the interaction between outside and inside realms — the manipulation of the former producing the motion of the latter, which is understood as flowing electricity.

17 It has been argued that Maxwell's electromagnetic theory, which provides a mathematical basis for understanding these patterns, can be stated in a manner that does not separately specify electricity and magnetism, making any distinction between the two forces artificial.

There are other ways that our complementary dimensions of analysis are useful for understanding the dualistic patterns here. In the case of electric fields, one can think of charge and current in terms of static and dynamic relationships, respectively: the dominant and recessive partners of the former (static charge) would be positive and negative charge, while the dominant and recessive partners of the latter (dynamic charge) would be incoming and outgoing current — of which there can be both stable and changing versions, as in direct current (DC) vs. alternating current (AC). The static relationship counterparts for the magnetic field would be north and south poles — and we would suggest that the dynamic relationship counterparts would be incoming and outgoing radiation, which could also be of a stable or a changing variety. And even though we are talking primarily in medium domain (or wave) terms for these various sets of phenomena, there would also be matter domain (or particle) counterparts. But to elaborate the former, there would be both static (standing wave) and dynamic (traveling wave) versions.[18]

Returning to the duality point about the two field types, we now take a position which specifically contravenes the conventional interpretation that magnetic fields only coexist with electric fields if the latter are in relative motion. By contrast, we maintain that electric fields and magnetic fields always exist in relation to one another — in the same way as happens with other complementary partners. One way of understanding the issue is in terms of a point we mentioned in the first chapter (and will develop further in the next) — that stasis and motion only exist in relation to one another. But if stasis and motion only exist in relation to one another, then any interpretation about motion which does not take account of its stasis counterpart is defective; and by inference therefrom, one could also say that electric and magnetic fields also go together. One may be predominant (or manifest) at any given time, and the other only latent, but the two are a pair.

In fact, we can think about these issues in two different ways, which is consistent with a core theme of the present model. On the one hand, we speak often about two complementary states of "being" — which here manifest themselves in electric field and magnetic field terms. On the other hand, we also talk about two different ways of "seeing" as also corresponding with these two states of being. In the case at hand, an *X*-view would reveal an electric field while a *P*-view would reveal a magnetic field. Such a pattern is in keeping with the fact that the electric field itself does not have to be in motion for the magnetic field to manifest itself — the electric field can just as well be static, with the observer doing the moving!

Let us try to summarize some of the complementary patterns we have been describing with the following set of nesting charts — bearing in mind that we are primarily dealing with medium domain phenomena throughout:

18 Physics sometimes describes electrons in terms of standing waves (e.g., Andrews, p. 81; Davies, 1979, p. 68; Zukav, pp. 99-101). These are associated with "vibration" or resonance patterns, which we shall mention below.

Inside Realm				Outside Realm			
electric field				magnetic field			
Static Relationship		Dynamic Relationship		Static Relationship		Dynamic Relationship	
charge		current		polarity		radiation	
Dominant	Recessive	Dominant	Recessive	Dominant	Recessive	Dominant	Recessive
negative	positive	incoming	outgoing	south	north	incoming	outgoing

Electric and Magnetic Materials. Now let us turn to a topic we did not discuss earlier, which concerns the different kinds of material that specifically relate to electricity and magnetism — and here we will emphasize nomenclature problems with the conventional constructs. The former category is simpler and more straightforward, and is based on whether the material is better or worse at carrying the flow of electrical current through it. Materials which readily carry this flow of energy are called "conductors", while the complementary type of materials which resist such flow are called "insulators" (or "non-conductors", or "dielectrics").

Unfortunately, the situation becomes more confused when we turn to the topic of magnetic materials. Thus, all matter in the universe can be classified into into the categories of diamagnetic, paramagnetic, and ferromagnetic[19]. "Diamagnetism" refers to the tendency of a material to *repel* — or more specifically, to repel a magnetic field; and all materials in nature are said to be diamagnetic to a greater or lesser degree. "Paramagnetism", on the other hand, refers to the tendency of materials to *attract* — or more specifically, to attract a magnetic field. There is also overlap amongst these categories: thus, some materials which are good conductors are classified as paramagnetic, while others which are good conductors are classified as diamagnetic (e.g., gases).

Of course, the most common form of magnetic material is that which is "ferromagnetic", such as the horseshoe magnet with which we are all familiar. In contrast to the other two categories, one might think of ferromagnetic materials as being equivalently able to *attract* and *repel*. We suggest that it is for this reason that such materials seem almost alive to us: being very well-balanced, they can easily "lean" internally in one polar direction or the other, if there is a strong enough external force to so orient them. In the present model's terms, one can describe this material by saying that it is *self-complementable* — it can complement itself. Likewise, in hierarchical terms, this material is *self-nesting*: Such a description would be consistent with the notion of "magnetic domains", whereby the ferromagnet material becomes polarized when its component domains cumulatively orient into progressively greater and greater alignment. This process strengthens the force in one direction, and weakens it in the other. Note the correspondence of this interpretation with the general pattern of hierarchy that we keep coming back to — synthesis and sameness in one direction, discrimination and difference in the other.[20]

19 This last incorporates the sub-categories of ferrimagnetism and anti-ferromagnetism, which we shall not discuss here.

One can differentiate two subsets of ferromagnetic material, in terms of how fixed is the polar orientation. A non-permanent ferromagnetic material can easily shift its alignment to conform to a stronger magnetic field. This is the case with soft iron, for example, whose component elements have various random alignments. By polarizing and freezing its orientation, this non-permanent ferromagnetic material can be transformed into the permanent form of hard iron, with component elements then all being aligned in the same direction. The hard iron object then acts as a whole unit, producing one coherent magnetic field. Such permanent ferromagnets can be used to orient smaller or weaker versions of themselves, as well as to polarize the non-permanent variety.

By now it should be clear that the concept of magnetism is much more confusing than the concept of electricity. For one thing, the former term is used to describe both tangible materials and intangible fields, whereas the latter term is not so used (leaving aside the relatively minor term of "dielectric" that we mentioned above). Also, we noted the completely different categories of "magnetic" materials — even though most people's thinking is almost exclusively dominated by a single type thereof, that of ferromagnets.

Not only is there terminological confusion between fields and materials, there is also confusion with the latter between adjectives and nouns! Thus, we would all agree that the fixed-polarity (or permanent) ferromagnet is "magnetic". But what about the non-fixed-polarity (or non-permanent) ferromagnet — is it "magnetic"? Well, we hem and haw: it's a magnet, and maybe it's "magnetic" when it's polarized or magnetized — but otherwise not. So, then, according to one way of looking at it, "magnetic" only refers to ferromagnetic material which has been magnetized ... either permanently and fixed or not.

Okay, now take two such permanent ferromagnets (e.g., horseshoe-shaped magnets), which have clearly been magnetized. Put them together, such that the opposite poles lock onto one another. Is it still magnetic? The answer depends on how you want to look at it: in relation to each other, we would say they are magnetic — after all, they're pulling each other together. But what about in relation to other permanent ferromagnets outside, or to other magnetic fields? The situation has changed now, because the "closed" magnet is not magnetic anymore, in the sense that it tends *not* to exert an influence upon other extrinsic magnetic fields, outside of its own unity.[21]

20 In this regard, it is worth noting an odd wrinkle in the attraction vs. repulsion patterns of magnetic field phenomena. That is, it appears that the strength of attraction between two unlike poles *is stronger than* the strength of repulsion between two like poles. Thus McCaig notes that, "The laws of magnetic repulsion are similar to those of attraction. Two magnets with like poles approaching repel each other with the same strength with which they would attract each other *if* they retained the same strength with unlike poles approaching.... [But a]s two permanent magnets approach each other they may induce changes in their respective magnetizations. If they are attracting both magnets become stronger, but if they are repelling both become weaker." (McCaig, p. 248) Also note Hecht, p. 443: "Two magnets placed with like poles near each other will slowly repel domains and so weaken each other." Describing these phenomena more in terms of the present model, we might think of attraction in general as stronger locally than is repulsion — whereas repulsion is stronger distally than is attraction! This may be a very specific version of a more general pattern that we mentioned in Chapter 1 — that "attraction" can also be understood in terms of "local" ... just as "repulsion" can also be understood in terms of "distal".

There are other problems with the concept, as well, especially the fact that the term *magnet* is all too close in morphological form and ideational association to the word *magic*. Surely the child we all grew from sees something very magical in his or her toy magnet. That is understandable, for it does seem rather mysterious the way two magnets pull or push one another, and the morphologic connection with the name of the material can only contribute to the mystery. It may be more than coincidental that "hypnosis" was once referred to as *magnetism*. Might the "magnetism" term have a kind of hypnotic effect on our thinking — quasi-magical associations, which are not easy to dispel?

It is important to emphasize that we are not playing word games, in discussing problems with terminology here. For we are dealing with a particular phenomenon that is difficult to understand and explain, and thus can easily take on an almost supernatural quality. Therefore, it is essential that we have reasonably clear and straightforward concepts, which are not unnecessarily contaminated by irrelevant associations. Such concepts not only form the basis for scientific communication, but far more fundamentally they play a very formative role in our thinking — thereby representing the tools of our minds when we cogitate about these matters. As we shall discuss further in Chapter 11, it will be far more difficult to discern general patterns and develop good models if we are laboring with messy and confusing conceptual units and clusters to begin with. And that is clearly the situation here, with archaic constructs like magnetism and its offshoots.

Perhaps we would be better off with new terminology for some of these phenomena, which might help reduce the confusion we have been discussing. This could serve both to eliminate irrelevant associations, and to separate the field vs. material function.[22]

Periodic Table. As far as the periodic table of elements goes, there are complementary patterns which can be identified amongst the various groups; and the different periods or levels would be understood in nesting hierarchy terms.

Regarding the latter, we must first reverse the vertical orientation of the table, so that lower number periods are at the bottom and higher number periods are at the top. The result is a table which which fits naturally and meaningfully into the kind of larger hierarchy that we are proposing in the present model. Each period of the table would represent one level of the hierarchy. Larger masses would be in the upward direction, and smaller masses in the downward direction — again, synthesis and sameness upwards vs. discrimination and differentiation downwards. We noted that heavier nuclei are less tightly bound than lighter ones, and one might interpret this in terms of our general principle that lower levels of the hierarchy are more concentrated and higher levels more dispersed. Fitting these levels into a broader hierarchy, one would go

21 Asimov, p. 156: "A bar of iron over both poles of a horseshoe magnet cuts down the magnetic force outside itself to such an extent that it almost acts as a magnetic insulator."

22 In the case of ferromagnetic material, for example, we might be better off talking about "bimatter" — with the meaning that it could readily alter its orientation.

further upward to find nesting combinations of nucleuses — molecules ... just as one could go further downward to find more elementary components of nucleuses — starting with the two basic nucleons themselves, and progressing lower to find constituents thereof (e.g., "quarks").

One way that complementarity would apply is in terms of contrasting processes of structural changes which occur at the level of the nucleus. *Fusion* is the mechanism whereby two lighter (complementary) nucleuses join together to form a single heavier nucleus — once again, synthesis and sameness upward in the hierarchy. By contrast, *fission* refers to the mechanism whereby a (non-complementary) nucleus separates into two, approximately equal, lighter nucleuses — discrimination and difference downward in the hierarchy. (In both of these cases, additional energy is also released in the process, aside from the structural changes in the nucleus.[23])

Another way of thinking about complementarity here would be in terms of different characteristics which describe same-level side-to-side groupings. One obvious possibility is a *metal* vs. *non-metal* continuum, with *metalloid* elements laying in between. Materials which are made from metals are shiny, malleable, and have good conducting properties (for heat and electricity); by contrast, non-metals are not lustrous and are good insulators.

On the next page, we sketch how these various characteristics might be represented:

23 It is the chain-reactions of these two processes in concentrated materials that lie behind "atomic" (fission) and "hydrogen" (primarily fusion, but with some fission) bombs. By contrast, so-called "neutron bombs" ("enhanced radiation devices") are basically hydrogen bombs, but with an even higher fusion:fission ratio — thereby releasing less heat and blast, and more high-energy radioactive neutrons. Note that nuclear power is generated by fission approaches today, although much research is underway toward future fusion reactors. Fusion is also the mechanism by which the energy release of the sun (and all other stars) is explained.

(molecules)

bigger (*P*-Pole/*P*-Force)

higher periods (more orbital layers)

↑

X-Force ← ——————————————————————— →*P*-Force

Group I | Group VIII+

metals | metalloid | non-metals

conductors | insulators

↓

lower periods (fewer orbital layers)

smaller (*X*-Pole/*X*-Force)

(nucleons)

By the way, ferromagnetic materials tend to be found midway left to right in the table. That may be precisely where such a material should be if it had the capacity to both attract *and* repel — which it does.

Scattering. Next in our quest to identify complementary patterns in the micro realm, we turn to another topic we did not discuss earlier. A simple way of describing the issue concerns what happens when one entity runs into another entity. In dichotomous terms, the first entity (or its energy equivalent) can either be absorbed or not absorbed by the second entity. This distinction can be described in terms of the two categories of "resonant scattering" (Pitt, p. 337), and "non-resonant scattering" (Hecht, p. 481). Resonance is related to "absorption", and non-resonance to "non-absorption" (e.g., Hecht). Sometimes, the two processes are talked about in terms of "absorption" vs. "scattering" (Neilburger, et al.).

One way of thinking about resonance is to say that every structural entity vibrates or oscillates at a unique rate or frequency. We can think about the pattern in terms of our breathing motif — that every entity breathes or resonates at a certain tempo. Speaking of the atom, for example, Murchie says that, "[E]very atom is somehow made of durable waves that vibrate continuously in hierarchies of energy.... [A]ll matter tends toward its natural rhythms" (Murchie, p. 411). Or in Hecht's terms, "[A]s with any mechanical system, atoms have natural frequencies at which they will oscillate.... [or] vibrate most effectively, taking up the most energy, when driven by an oscillation that matches its own natural (or resonant) frequency.... Every material has such natural frequencies at which energy can be absorbed." (Hecht, p. 482). Standing waves can also be understood in terms of resonance. Sometimes, waves interact in a manner that can be described in terms of "sympathetic resonance" or "sympathetic vibration", whereby similarly phased and sized waves will mutually reinforce one another's activity ... both proactively and retroactively (Murchie, p. 387).

The two basic categories of "resonant scattering" and "non-resonant scattering" can be understood as complementary to one another. In the former case, there is absorption of the incoming energy without re-emission — at least not in the form the energy was received. The incoming energy is compatible with the receiving entity and its natural frequency, and so the energy is absorbed. Again, this category is described in terms of "resonance" responses, because the incoming energy is of the appropriate parameters (such as frequency) to be absorbed by the receiving entity, which then "resonates". This category can also be described as "inelastic", in the sense that there is no "bouncing", but rather absorption.

On the other hand, if the incoming energy is not compatible with the receiving entity's natural frequency, then the energy is "rejected" and it bounces off. Refraction of light is a good example, with the result being that the direction and speed changes — as also happens with "reflection", which "is really a special case of refraction" (Murchie, p. 433). In contrast to the above, this category can also be described as "elastic", in the sense that there is a tendency for the entities to "bounce" off one another, as would two colliding particles.

To summarize the complementary pattern here:

Resonant Scattering	*Non-Resonant Scattering*
compatible frequency	incompatible frequency
absorption ("inelastic")	rejection ("elastic")
sympathetic vibration	bouncing-refraction-reflection

Wave Dualities. Now we would like to mention several types of complementary wave patterns or categories which manifest themselves in different situations. Of course, these issues are not strictly related to micromatter, but it is convenient to discuss them here.

On the most basic level, there is the alternating dominance of the oscillating wave, which waxes and wanes in a buildup and discharge manner. Using other terms, "Waves involve a complex pattern of kinetic energy and potential energy, the one continually changing into the other ..." (Rogers, p. 389) And of course, there is the incoming and outgoing aspects of waves in general.

Beyond that, one can distinguish between two basic types of wave propagation, transverse and longitudinal. *Transverse waves* are more familiar to us, in the sense that the medium visibly moves up and down or side to side — but perpendicular to the forward direction in which the overall wave is traveling. An example would be the wave that moves up and down as it travels along a vibrating string, or the wave that moves upon the surface of a body of water. By contrast, with *longitudinal waves* the medium does not move up and down or side to side, as the wave energy travels forward. Rather, this kind of wave consists of an alternating bunching up (a kind of buildup called "compressions") and spreading out (a kind of discharge called "rarefactions") of energy, in the same forward direction as is the wave itself. As Hecht says about longitudinal waves, "It's the disturbance that constitutes the wave, and it's the disturbance that does the traveling and carries the energy." (p. 468) Sound waves are probably the most obvious example of such longitudinal waves.

Transverse waves can also be classified in terms of a complementary sub-category. This relates to whether the transverse wave's outwardly-oscillating component is orientated along a consistent angle or not ... or whether it tends to rotate around (even randomly), from one angle to another. When the angle is consistent — when the transverse components (of the forward wave) are flip-flopping sideways at the same angle — then the transverse component is said to be *polarized* along such a plane, or simply *plane-polarized*. Mechanical filters can be used to produce this effect with light waves (e.g., calcite crystal, or polarized glass such as is commonly found in sunglasses), and certain common phenomena can produce it as well (e.g., glass or water — or the atmosphere). But more typically, light (and other transverse waves) are *unpolarized*: the secondary component is not orientated along a single plane, but shoots outward all around at various planes or angles.

Another complementary subpattern relates to the forward-advancing *wave front* itself, and there are two main possibilities: either the advancing wave front can stretch out along a line as it moves forward — such as the usual ocean wave coming into an open seashore: this is called a *plane wave* front. Alternatively, the advancing wave front can spread itself out in a circular manner as it advances — this being called a *spherical wave* front.[24] Needless to say, the nature of the initial wave disturbance is crucial in determining the form of the resulting wave front: a point source will produce a spherically-radiating wave front, whereas a linear source will produce a plane-radiating wave front. Furthermore, the shape of even a spherical wave front will change with distance: a wave front may be spherical very near to the point-source, but very far away the wave front becomes "stretched out" into a linear or plane wave form. Conversely, any wave front, when viewed close up enough, reveals itself as an endless source of spherical waves, though these "mini-wave fronts" are soon rapidly overwhelmed, swallowed up, or averaged out into the main advancing wave front itself.

One interesting category of waves consists of groups of waves which may be *phased* or "coherent" in relation to one another — as opposed to being *unphased* or not coherent to one another.[25] An illustration of such a phenomenon can be found with the laser beam (Trefil, p. 335). This phased effect is produced by stimulating the material from which the waves emanate, so as to force emission of the energy in synchrony. The result is a beam of light waves which act very much as a unity: a highly-focused "superwave", if you will, which can be directed very precisely and which carry a good deal of force — by virtue of the cumulative, phased, and focused energy involved. Because these wave constit-

24 Such a wave front is properly labeled a "circular wave" front when it exists in only two dimensions.

25 The broader phenomena are also called constructive interference or coherent superposition, on the one hand; and destructive interference or incoherent superposition, on the other.

uents are in synchrony with one another, they can also be said to be static in relation to one another — even though, of course, they are moving in relation to an external referent. In contrast to normal unphased (incoherent) waves, the individual waves of a laser are not moving within — not traveling, but standing — in relation to one another.[26]

Antimatter. Now let us return to the topic of antimatter, which we discussed in the opening section of the chapter.

To put the matter most simply at the outset, our position is that the whole construct of antimatter is a misconception. And the culprits are the same ones which we shall be discussing throughout this book, especially in Chapters 9-11: of the illusion of perfect symmetry reigning throughout nature, of mathematics dictating theory, of artificial laboratory outcomes taking precedence over natural contextual observations, of arbitrary theoretical assumptions squeezing out common sense — and of the overall lack of a substantial and coherent set of organizing conceptual principles with which to guide our explorations of nature.

In the case at hand, the entire explanatory framework for the antimatter thesis is completely unsatisfactory. Thus, the primary means for creating antimatter is through the process of pair production, whereby equal amounts of matter and antimatter are produced at the same place and time. As we noted in our earlier discussion, the assumption is that a massive pair production process took place for the universe as a whole, around the time of the Big Bang. Having been created in this way, however, the only way the resulting pair would survive is for each member to somehow split apart from the other, presumably very quickly. Again, that is because if matter and antimatter are brought into close proximity, then their mutually-offsetting character will effectively cancel out each of the two entities by leading to the annihilation of both (along with the emission of electromagnetic energy). Yet what brought about the necessary separation in the beginning, to avoid the immediate devastation ... presumably of the entire universe? There is no compelling mechanism to explain how these processes would have occurred — of what would have brought about the unmixing of matter and antimatter, which are said to have been originally created together.

Then there is the disconcerting issue of what has happened to the antimatter portion, which we don't encounter in our world. As Close puts it, "So where did all the antimatter go? Well, that's a very obvious question which has been asked for a long time; and theories of it have been created and died somewhat more rapidly than protons do." (Close) One contention is that the antimatter has traveled to another part of the universe. According to this thesis, most antimatter, from anti-sub-particles to anti-atoms and so on, are said to exist only in distant regions of the universe (e.g., Zukav, p. 211). Yet

26 One way to think about this phenomenon is in terms of a kind of simulated (or artificial) structure. The individual waves form an integral whole, which is static within. For example, a hologram is such a "fake" structure, being based on coherent light standing waves. Again, both the normal structure, and the simulated structure of a laser, can move — but in both cases the movement is something which is ascribed from an external vantage point, and not something intrinsic within.

there is no evidence whatsoever of the existence of such large quantities of antimatter remaining from the Big Bang period.

So what could be going on? In analyzing the issue, we start by asking why the general construct of antimatter should even be taken seriously to begin with, given that its phenomenological origin and existence are so hard to understand and impossible to verify. Most instances of antimatter play almost no role in the nearby world we experience directly, or in the distant world we know about indirectly. That is, most of what physics calls "antimatter" does not appear to exist in natural circumstances here on earth, or in this immediate region of the universe; nor has a large residue of antimatter from the Big Bang been detected in signals emanating from any other region of the universe either. We are mostly dealing with a hypothetical set of phenomena, which can be produced for incredibly brief periods, under extreme and artificial laboratory conditions — but which has been found nowhere else in nature. We maintain that the perfect equivalence of lab-produced symmetrical counterparts attests more to our technological prowess than to our ability to relate to and describe the normal patterns of the natural world.

We shall consider the issues here in terms of the two primary themes of this work — complementarity and hierarchy. Regarding the former, what we are encountering is another instance of the symmetry problem which we keep stumbling upon — which we described in the first chapter as a "symmetry conviction" (or "false symmetry"). In this regard, we already noted that the Special Theory's assumption of space-time symmetry was an important factor in the genesis of the antimatter concept — as it was in other unfortunate theoretical developments, which we shall discuss later in the book.

Physics' search for perfectly symmetrical counterparts is pure folly — the very concept is an illusion. And we are not alone in our position here. For example, Morrison notes that, "It is generally believed that the matter-anti-matter mirror is not exact." (Morrison, p. 64) Indeed, we mentioned a variation on the main theme in a footnote above — which is that there was a slight disparity in favor of matter over antimatter in the original Big Bang period, and that the bulk of matter and antimatter mutually annihilated one another, with the present universe being composed of the originally-disparate matter remnants. But that is a very ad hoc explanation, which is inconsistent with the perfect symmetry theme from which the antimatter thesis emerged to begin with.

And this is why the antimatter thesis is flawed — because it is based upon a concept of perfect symmetry ... rather than upon the more subtle notion of complementarity, which does not entail perfectly equal partners. Our contention is that complementarity provides a much more viable template for understanding the underlying patterns — not only in regards to micromatter, but pervasively at every level of nature's hierarchy.

In regards to that hierarchy, the main source of the problem we are encountering is that of the conventional fundamental force paradigm, and the undue importance which is attached to one particular level of interaction here — that which relates to electromagnetism, with charge being the outstanding factor. But a more satisfactory approach would recognize that there are corresponding sets of complementary properties at every level of nature, each of which has its own unique characteristics ... in the same way as do properties that are attributed to the electromagnetic force. And given that such general complementary patterns apply to every level, then there would be nothing special about the set of properties which exist at one particular level.

For example, our position is that the two kinds of nucleons within the nucleus are complementary partners to one another, proton and neutron. This means that it is not some synthetic "anti-proton" which is the proper complement of the proton — but rather, it is the neutron itself: the neutron is "antimatter" to the proton! These two counterparts are perfectly compatible with one another in their natural complementary relationship — and they do not need a sterile arranged marriage, with some idealized partner that physics creates.

Of course, the typical objections to this particular interpretation would be two-fold. First would be a restatement of the point that proton and neutron are not, in fact, identical to one another — leaving aside the charge factor — but they are only approximately the same, since they have a slightly different mass from one another. Yet that is precisely the point: it is complementarity which describes the relationship of natural counterparts, not symmetry. And complementary partners cannot be perfectly equal to one another, as the false symmetry myth presumes.

The second objection would be the assertion that these two nucleons are not charge partners to one another, since the proton carries positive charge and the neutron no charge. But one can argue that even in those terms, they are still charge partners — because that is the primary factor which differentiates them: recall the Yukawa thesis, of charge being passed back and forth (in the form of the meson) between them. In any event, we will return to the complementary charge relationship between proton and neutron near the end of this chapter.

We have already stated that a characteristic like charge is overrated as representing a general criterion for paired relationships in nature, and now we can elaborate on why. For, not only do complementary counterparts exist at every level of existence, but the same complementary theme applies to every dimension of analysis, in our model ... which is further testimony to the fact that the underlying pattern here is so much broader than anything relating to electromagnetic phenomena like charge. The most obvious example for the pattern we are talking about would be that of particle vs. wave — which we would describe as representing complementary partners in terms of our matter vs. medium domain. But they are not alone, either: there are complementary stable vs. changing partners, inside vs. outside partners, stasis vs. motion partners, and dominant vs. recessive partners. All of these could be described with the "antimatter" moniker — except why do so with any of them?

None of these paired sets of characteristics should take prominence over any other — and what is true for the dimensions of analysis is also true for existence, generally: all levels and dimensions are equivalent, in terms of overall patterns and principles. Complementary partners are complementary partners, regardless of which dimension of analysis or level of existence we are dealing with. That is why the antimatter thesis has no role to play in the present work — and why it belongs more in the realm of science fiction than it does as science fact.

Even the terminology reveals how defective the antimatter concept is — not in the sense of "false symmetry" as we have been discussing, but rather in the sense of "no symmetry". This relates to an issue that we will examine more fully in Chapter 9, concerning physics' overall structural (vs. process) orientation and bias in its approach to nature — with categories like "subatomic particles", but not also an equivalent category of "subatomic waves". In the case at hand, the very notion of matter vs. antimatter thereby represents that same one-sided approach to nature: Why are we not also talking about medium vs. antimedium — or even waves vs. antiwaves?!

Now let us turn to what might be seen as the main exception to our general argument about antimatter not existing in the real world, that of the electron and the positron. Consider the simplest form of pair production as it applies to these entities: a photon passes near an atomic nucleus and is transformed into both an electron and a positron. But the combination can come about in other ways — for example, a positron in free space is said to quickly interact with an electron. The brief, bound-state intermingling which results is even treated as a virtual hydrogen-like atom (called "positronium"), but with the positron instead of the proton serving as the positive counterpart to the negative electron. The pairing is said to only last seconds, however, until the two entities mutually annihilate and transform into photons instead. By the way, note the two (complementary) directions of the processes we are dealing with: on the one hand, there is the *creation* mechanism of pair production; and on the other hand, there is the counterpart *destruction* mechanism of pair annihilation.

The fact that paired phenomena like electrons and positrons do exist in the real world, whereas most other hypothetical antimatter partners do not, suggests that there is something different going on here. Yet what that is ... is hard to say — especially in the face of the standard model of the atom, which topic we will explore near the end of the chapter. In any event, we must be frank in saying that this electron-positron relationship represents the single most problematic instance in the book of where it is difficult to make our case as to why complementarity, rather than symmetry, should apply. The best explanation we can offer is that there must be some manner or grounds by which the proper conclusion is that one or the other partner is predominant in any particular situation ... in principle. Such a complementary dominant-recessive relationship should manifest itself in one form or another, either spatially (i.e., within time) or temporally (i.e., over time). We will return to the relationship between electron and positron shortly.

In summarizing our position about antimatter, we maintain that the construct should be eliminated altogether. For one thing, it is erroneous, by virtue of its perfect "false symmetry" assumption. And for another, to the very limited extent that it exists at all (e.g., electron-positron), we presume that it can be ultimately understood in complementary terms.

Interactions Between Nucleuses. In this and the following subsection, we revisit the dual topics of interactions between nucleuses and interactions between nucleons — which we maintain is a hierarchical way of talking about the electromagnetic force and the strong nuclear force, respectively. We begin with the former.

In speaking about the electromagnetic force, what is it that connects one nucleus with another? In some cases it is said to be the electron, while in other cases it is said to be the photon. For molecular bonding it is the former, while for electromagnetic radiation it is the latter. Note an anomaly, in terms of complementarity, right at the start — in that there is a lack of parallelism in the way these respective phenomena are labeled: there is only one name, electron, for representing both the particle and the wave versions of the former set of phenomena — whereas we have two separate names, photon and electromagnetic wave, for describing the latter set of phenomena.

Leaving that point aside, how shall we characterize the difference between these two sets in the present model? We maintain that the distinction can be understood in terms of the inside-outside dimension of analysis. If the two nucleuses comprise an inside or closed field relationship with one another, then the interaction is described in terms of an *electron*. On the other hand, if the two nucleuses comprise an outside or open field relationship with one another, then the interaction is described in terms of a *photon*.[27]

Here is how we would represent these complementarities:

Inside Realm		Outside Realm	
electron		photon	
Matter / Particle	*Medium / Wave*	*Matter / Particle*	*Medium / Wave*
electron	electron	photon	electromagnetic wave

Of course, in both cases we are talking about a dynamic relationship (labeled either in terms of electrons or photons) that exists in relation to a static relationship (composed of nucleuses). Concerning the issue of molecular bonding, our approach is rather simple (if not simple-minded): we merely take the general position that two nucleuses will bind by way of electrons if they are somehow complementary to one another — whereas they will not bind (or they will unbind) if they are uncomplementary. We would also argue that a similar description would apply to the case of whether electromagnetic waves connect two nucleuses or not ... a point that we will pursue further in Chapter 9.

Interactions Between Nucleons. When we recall the kind of definitions and restrictions we discussed earlier to account for this set of phenomena, it seems like we are dealing with a patchwork of arbitrary rules and definitions — designed to superimpose order upon a set of constructs which have no intrinsically harmonious relationship with one another. Thus there is a special force of nature, known as the strong (or nuclear) force, which somehow binds together multiple nucleons that would otherwise have no reason to be near one another, or else should actively repel one another — and so the interaction takes no account of any electromagnetic charge factors which would otherwise be involved. Yet the strong force acts upon the two types of nucleons only in combinations that have a certain proportionality to one another — rather than just being of one nucleon type only, or of any ad hoc combination of the two.

How, then, would the present model deal with the basic phenomena which the nuclear force was invented to explain? What is the mechanism which binds the nucleus together? Our answer is that complementarity is the key.

27 Note that in the first chapter we said that one way of talking about the inside realm was as 2-dimensional, in relation to the outside realm as 3-dimensional. Consistent with such a relative distinction, electrons are said to have two possible alignments of spin ... whereas photons have three. (Murchie, p. 323)

When we look within the nucleus from this perspective, there is nothing special to be explained — no electromagnetic laws which are being violated or which must be overcome, no unique phenomenon which requires a special force to be accounted for. As we said above with nucleuses, so with nucleons — and with every other level of nature: partners bind because they complement one another ... just as they do not bind (or unbind) if they are not complementary. It is obviously a very simple precept, and we propose that it can be adapted to the complexities at hand — thereby obviating the need for a special force of nature to explain it here (as opposed to describing how the persistent general pattern manifests itself at this level).

Of course, for the established paradigm there are no such general forces of binding and unbinding that can be applied to every level of nature's hierarchy. Indeed, the conventional model does not even ascribe independent existence to any other level of matter beyond that which can be described in terms of the 3-4 fundamental forces. And even the independent existence of those specific ordained levels is a relative attribute — as is demonstrated with the bizarre quandary about how it is that nucleons could not be infinitely compressible or "squeezed too close together". The arcane reasoning requires a special subsidiary property for the nuclear force, such that what was a strongly attractive tendency at the dimensions of the nucleus becomes strongly repulsive at the dimensions of the nucleon. This repulsive "twist" in the nuclear force theory arises because there is no acknowledgement of the existence of a nucleon structure, as such. The seemingly obvious nucleon structural interaction effect is sometimes treated as if it did not actually exist: "[F]or simplicity, it is often taken as a *hard core*" (Cohen, p. 56).

"For simplicity" — how about "for reality"? Once again, the reason the conventional model is forced to deal with this kind of peculiar issue is because it does not acknowledge the full-fledged existence of independent structural entities, at all levels of nature's hierarchy. In this case, there is no recognition of the nucleon structure as fully existing in its own right, and in that straightforward way as accounting for the resistance to further compression of nucleons within the nucleus. Instead, the theory is forced to put forward a roundabout explanation for the very obvious qualities which it treats as only illusory.

In the present model, when two complementary entities join together and form a (same) higher level unity, the lower level units do not thereby disappear as (different) entities in their own right — they continue to exist as separate lower-level phenomena. Therefore, we need not resort to such mental contortions here: the nucleon — and every other level of structure — *does* exist in its own right; and the fact that it has discrete boundaries and surfaces which resist arbitrary compression (to a greater or lesser degree) is simply testimony to the fact that it represents an independent level of nature's hierarchy.

How do we account for the fact that the force holding the nucleus together is massively stronger than other kinds of interactions? We suggest an explanation that is largely based on the hierarchy factor. As we noted in the first chapter, there are "power" differences between levels of the hierarchy — with higher levels being weaker (and less dense) in relation to lower levels being stronger (and more dense).[28] One might also speculate that there may be some way of differentiating what might be construed as a predominantly structural phenomenon like the short-range strong nuclear force, in relation to what might be construed as a predominantly process phenomenon like the potentially long-range electromagnetic force.[29]

It is intriguing to think of the nuclear force in Yukawa's terms, mentioned above, as an exchange phenomenon which consists of mesons (and charge) moving back and forth between the component nucleons, binding them together into the nucleus. One way of understanding this relationship between the two types of nucleons would be to treat them as dominant and recessive partners (or collective groups) of a static relationship (i.e., whole nucleus), with the meson being the dynamic relationship entity (and charge) which oscillates back and forth between them. This would be like a pendulum action, with a kind of SHM (simple harmonic motion) alternating to and fro between the two types of nucleons, in a process that we might refer to as "complementary nucleon transformation" — with the meson reciprocal transfer of energy continually changing the interacting nucleons from one state to the other. In that case, the overall composition of the nucleus would not be altered, in terms of the numbers of protons and neutrons.

The Crown Jewel. We end this chapter and our discussion of the micro world by looking at what might be considered as the crown jewel of physics, in the sense of a single level of matter upon which so many of its theories pivot — namely, the atom. Thus it is that while in some respects physics might claim not to be "atomistic" in its approach to nature, the reality is that it is difficult to interpret its theories in any other way, so centered are they on this construct and realm of nature (cf. Jammer, 1974, p. 204).[30]

Yet here is the main outlandish question which will guide us in the discussion here: does the atom exist? Or phrased more precisely, could there be a viable theory of nature in which the construct of the atom plays no actual role? In other words, are there other ways of interpreting the basic phenomena which are (or are not) normally associated with the atomic model? The position we shall argue here is that the atom is an outmoded construct — an anachronism, an artifact from the past — which should long ago have worn out its welcome in our models of nature.

28 Physics talks about the issue of explaining different levels of energy that are associated with different fundamental entities in terms of a "hierarchy problem". We touch upon that topic in the final chapter.

29 Although note that just as the nucleons-into-nucleus interaction might be seen more as a short-range structural force, the conventional nucleuses-into-molecule interaction also appears to apply only within the "short-range" molecular dimensions (Hecht, p. 337). Also note that another special force is called the "Van der Waals attraction"(Andrews, pp. 141), which may be invoked to account for why charge-neutral molecules or atoms attract one another electrically! (Hecht, p. 288) Plus, just as the separate nucleon structures will repel one another if pushed too close together within the nucleus, so likewise will the separate nucleus structures repel one another if pushed too close together (Andrews, p. 132) — requiring independent ad hoc explanations in both cases, rather than in terms of principles like that of multi-level existence, which is implicit in the present model.

30 Sachs' point long ago is still valid: "[T]he majority, to this day, has accepted the atomistic approach." (Sachs, 1974, p. 9)

The construct of the atom serves three basic functions, which we have already alluded to. One is the classical notion of the ultimate, indivisible unit of matter — a proposition which has long since been debunked. Not only is the atom composed of a nucleus, as distinct from the electrons which surround it; but the nucleus, in turn, is composed of two separate types of nucleons — and other tinier phenomena are also presumed to exist within nucleons. The upshot is that not only have smaller units of matter than the atom been identified, in practice; but there is no longer any basis for clinging to the proposition that there is an indivisible building block of nature, in principle — let alone that the atom is it.

The second function of the atom is to serve as the locus for one of physics' fundamental forces: the whole notion of an electromagnetic force is built upon the concept of the atom. That force explains both electromagnetic radiation (e.g., light) as well as electrical interactions — and the latter provides the basis for chemical bonding, whereby electrons from one atom combine with electrons from another atom. We already noted that such bonding explains the existence of the molecule[31], and thence all the rest of normal matter. Leaving gravity aside, it is electromagnetism which is said to provide the physical basis for all that we commonly experience in our world. An enormous mission, of course, and it is all said to pivot upon the atom — the result being the continued hegemony of the atom construct as representing the essential building block of nature ... no longer in terms of being indivisible or ultimate, but now in terms of being fundamental or basic.

What this means in our hierarchical terms is that the level associated with the electromagnetic force has been elevated to the point which makes it difficult for a coherent conceptual theme to emerge here — a theme that is based on the proposition that there are general patterns of nature which apply at all levels ... leaving aside the specific ways that these patterns manifest themselves at particular levels. Of course, that approach is very different from the conventional view, which is essentially that all levels of nature can be explained in terms of interactions which only occur at a handful of particular levels.

We discussed the third function for the atom in connection with the periodic table of elements: the diverse kinds of matter which exist in the world are ultimately said to be reducible to a variety of distinct atom types — and so in that sense the atom remains the smallest unit ... of an element that can exist in nature. We question this thesis, and we do so on fairly basic definitional grounds: these properties should be applied to the nucleus itself. To understand why, we will propose some fundamental reinterpretation of how the pieces of the atomic puzzle fit together — and in so doing, we will pursue the issue just mentioned regarding why an atom-centric model makes the generality for which we are striving so difficult (if not impossible) to attain, simply in terms of how a multi-level analysis could be constructed here.

We start by pursuing the history angle again. First came the smallest-unit-of-matter interpretation of the atom — which was followed at some point by the realization

31 "[M]olecular structure can be accurately accounted for by the electromagnetic force alone." (Cohen, p. 7)

that the atom was a compound, composed of a positively-charged center (protons) and a negatively-charged periphery (electrons). Eventually it became apparent that protons weren't enough to explain the nucleus — an additional type of nucleon was required. But because the basic charge framework was already in place, the new nucleon didn't need any of that; and so it was christened a neutron. Later it was recognized that the electron also had a counterpart, in the form of the positron ... with the two carrying negative and positive charge, respectively.

The important point to emphasize for understanding the conventional approach is that both these later constructs were essentially *afterthoughts* — added on to a largely completed atomic model, as a way of resolving obvious theoretical anomalies. Neutron and positron are stop-gap concepts, which have never gained an equivalent theoretical status as that of their respective partners. In both cases, the initial pair of constructs (proton and electron) had already been established when the need for new players arose — and there was never any attempt to reinterpret the basic model from the ground up, such that the end result was a more harmonious and meaningful (not to mention, complementary) whole. Moreover, the resulting situation has not turned out as well for the positron as it has for the neutron: at least the neutron has the role of filler to play in the basic atomic model, but the positron doesn't even get that minor function to perform — it is only an occasional guest, whereas the electron is ever-present.

But the picture which results is also very awkward. For example, one of the puzzling aspects of the atomic model is that the proton is far heavier than its charge counterpart of the electron: its mass is some 1800 times greater; and in addition, the electron is found very far outside the nucleus. Still, proton and electron are said to share an identical charge property, in equal (but opposite) measure. Yet it remains rather baffling why nature would provide these two *totally different* types of phenomena — in terms of mass, position, and other fundamental characteristics — but which just happen to have *precisely equal* (though opposite) charge.[32] And if that were not peculiar enough, we then find the further coincidence that there are normally just as many negatively-charged electrons in an atom as there are positively-charged protons! It is a very convenient and universally-embraced recipe ... but something may be amiss here.

Indeed, we must ask a very simple question: Could there be anything wrong with this picture? The atomic model seems cobbled together and based on ad hoc chronological developments, which do not meld together into a larger coherent vision. And remember, we are not talking about directly-observed entities whose character and function are beyond doubt, but rather with inherently obscure micro phenomena whose description and interpretation are intrinsically bound up with the physical and theoretical frameworks we bring to bear upon them. What we are left with, then, are abstract constructs that are associated with certain patterns of recorded behavior, which have been defined in their all-too-familiar ways. The question is whether there are other more conceptually efficient ways of representing the underlying phenomena.

32 People are perplexed as to why these two charges are of the same size, this fact being "one of the unanswered questions in physics." (Zukav, p. 206)

As we have said repeatedly, this entire work is predicated upon the assumption that a unifying conceptual theme will provide a sounder basis for a theory of nature than will a series of historical overlays that respond to this or that empirical problem which emerge along the way ... and that an all-embracing conceptual theme will do so the best. That is the basis for the present work, of course, in which the overriding conceptual theme is complementarity. So what might the picture look like when we approach the atom afresh from this perspective?

It is convenient for us to begin by examining the construct of the positron: how shall we understand the way this phenomenon is said to make its occasional entrance on the stage of nature? Our answer is to reject that very premise — and to pursue the notion instead that the positron exists in the real world in an equivalent and complementary manner as does the electron. Yet what could this possibly mean?

Consider one intriguing interpretation which was put forth many years ago. This 1949 analysis by Richard Feynman describes positrons as "electrons traveling backwards in time" (Feynman, 1949, p. 752; cf. Capra, 1975, p. 169). Of course, that could be equally phrased the other way — that the electron is nothing else but a positron traveling forwards in time. In any event, we would argue that such an interpretation accords full and equal status to the positron, putting it on a par with the electron — which is the goal that we are aiming for here.

As to what could possibly be meant by the notions of forward vs. backward time, this is not the place to get sidetracked by such deep ontological questions; we defer such a discussion until Chapter 9. But for now let us consider whether there is a way that such a general bi-directional pattern could be identified here.

We propose talking about the issue in terms of some kind of common process phenomenon, which itself would have complementary aspects — that might be called forward vs. backward, or inward vs. outward. In this respect we would suggest that a very productive way of thinking about the pattern is in terms of our breathing motif — that every structural entity in nature breathes ... in and out. One might even say that it is breathing — or as we have previously suggested, process (i.e., a dynamic relationship) in to and out from the nucleus structure (i.e., a static relationship) — which is the basis for what physics calls electromagnetic interactions in the universe. The electron could be specifically understood in terms of this in vs. out breathing, in one of two ways. *Either* the electron is a collective (e.g., forward time) way of representing the summation of both in and out facets, whereas the positron would simply be an alternative (e.g., backward time) way of doing so; *or* we could stay within the realm of forward time — with the electron representing precisely one of those two facets, and the positron representing the other. For example, the electron might be the inward segment (akin to inhalation), and the positron the outward segment (akin to exhalation).

One way or another, however, we presume that the positron must exist in the real world in an equivalent and complementary manner as does the electron. Then instead of contrasting the electron with the photon, as we did in a chart above, we would contrast both the electron and the positron with the photon. By the latter description we just proposed, the electron would be the inward portion and the positron the outward portion — of an overall inside realm phenomenon, that might jointly be named the *elsitron* ... as opposed to the *photon* representing the corresponding outside realm phenomenon, presumably with its own in and out portions, however they might be named. Thus:

Inside Realm	Outside Realm
electron and positron (e.g., elsitron)	photon

Of course, there is one little problem, leaving aside the fact that the whole atom model — which works so well and accounts for so much — is based upon the electron, not the positron. That is, how would we deal with the fact that including the positron, as having an equivalent presence to the electron, would throw off the whole charge analysis of the atom? Thus, if the positron were seen as regularly fulfilling the positive charge function which offsets the negative charge function of the electron, then we are left with nothing to offset the positive charge of the proton!

Or is that the case — what about the neutron? This brings us to the unthinkable question of whether the neutron is not, in fact, neutral in the way it is normally presumed ... but negative in relation to the proton. To pick up with the theme above, the main idea would be that protons are the positively-charged structural portion of the nucleus which are associated with drawing in the negatively-charged process phenomenon called the electron — in relation to neutrons as the negatively-charged counterparts which are associated with giving out the positively-charged entity called the positron.[33] We suggest that this proposal would be generally consistent with the Yukawa thesis about energy or charge being passed between the two nucleons, within the nucleus.[34]

Obviously, the main problem we must now confront, in this highly conjectural discussion, is how it would be possible for the new thesis to apply — that the neutron could be construed in non-neutral terms, even though its very name presumes the contrary. Consider several points.

For one thing, there is the argument we keep coming back to — that the theoretical assumptions and physical conditions we bring to bear are very important in determining the outcome of our explorations, in the sense of how experiments are structured and interpreted ... as opposed to simply and manifestly revealing an independently existing empirical reality. Thus again, how we see is just as important as what we look at, in our determinations about nature — and as we have said, this is especially true in the micro realm where so much inference is involved. One factor in this regard is context — in that we should be talking about neutrons in relation to protons, as they stably exist bound up inside the nucleus ... rather than as independent "free neutrons", which have a very short life span. Unfortunately, the latter is far more amenable to study than the former. Yet how do we know that the neutron, which is unstable outside the nucleus but stable within the nucleus, might not be similarly neutral outside the nucleus — but engaged in charge interactions within the nucleus?

In this respect, it is worth recalling once again that the neutron concept didn't even exist at the time of the original atomic model formulation; the electron concept did

33 Alternative names for the two nucleons would simply be *X*-nucleon and *P*-nucleon for describing the proton and neutron, respectively.

34 On the other hand, this approach obviously conflicts with the conventional model of the basic hydrogen atom, as consisting of only one proton in its nucleus — which raises the question of whether there are other (e.g., oscillating) ways of interpreting that nucleus structure.

exist — and it was thus the obvious candidate for providing the requisite negativity, to offset the proton's positive charge.[35] Only later did physics come up with the construct of the neutron — but by then the basic model was already established: all the leading roles of the play had been handed out, and only supporting parts remained. Actors such as the neutron were merely brought in to fill the gaps, but the basic play called "The Atom" had already been written.

We have already noted one important anomaly: if the neutron is chargeless, then only the strong nuclear force should impinge upon neutron-neutron interactions — with no electromagnetic force involvement — and thus neutrons should bind together by themselves. But such neutron-bonding does not appear to happen, and Hecht notes that evidence for a nuclear attraction between neutrons is "inferential" (Hecht, p. 336) In addition, there is the serious little puzzlement we keep coming back to, about how an electrically-neutral phenomenon like the neutron could respond to magnetic influences, as it does — a fact which is very difficult to explain if one assumes the neutron is completely neutral.

And that brings us to another significant point, which is that the very notion of a completely neutral neutron is actually misleading — for it has long been supposed that while the neutron is neutral in an overall sense, there are pockets of positive and negative charge within it. Thus in 1947, Enrico Fermi proposed what became a consensus — that neutrons have a positive charge at the center but a counteracting negative charge at the outer edge. However, in 2007 Gerald Miller used new data and theoretical tools to determine that the situation is more complicated, in that the neutron has "a negative charge both in its inner core and in its outer edge and with a positive charge sandwiched in between." (Choi) Furthermore, even this conclusion is tempered by the acknowledgement that further explorations may reveal other insights into the neutron's charge character, and that "the neutron could prove even more complex as more data come in ..." Clearly, then, absolute conclusions about the neutron's charge state may not be so absolute after all.

Now let us return to the basic conceptual issue, which concerns the question of whether there is a distinct combination of phenomenological units at this particular level of nature's hierarchy — conventionally represented by a nucleus core and electron periphery — which is worthy of a special name and status ... the "atom". We reject such a proposition and maintain that what we are dealing with here is merely a convention, which is itself a vestige of an outdated atom-centric mind-set.

The unfortunate result is to cloud our thinking, in the sense of walling off the level here by making it seem unique — thereby obscuring more general patterns which can congeal into a coherent model of nature. Thus, for example, we presume that every level of structure (e.g., nucleus) has a corresponding process aspect (e.g., electron) that is associated with it. Yet for purposes of a consistent hierarchical analysis, we should either name every level in combination ways (e.g., nucleus + electron = atom), or simply

35 In this regard, it should also be apparent from the discussion that the approach we are adopting has no relationship to the conventional thesis about unlike-charges-attracting as providing the explanation for what holds the negatively-charged electrons in orbit-like position around the positively-charged protons of the nucleus.

in terms of the structural construct itself (e.g., nucleus). The latter is the case when we talk about nucleons and nucleuses and molecules and multi-molecules, etc. Obviously, the atom construct does not fit this pattern, because it singles out one particular level of nature's hierarchy for special treatment, thereby muddling the broader multi-leveled picture here.

Nor need the problem be construed in hierarchy terms, for what we are dealing with in the atom construct is the fundamental starting point in the conventional model of micromatter — the prepossessing image of an independently complete, basic building block of nature which lies at the conceptual center of the physical world. This "isolationist" approach is based upon the notion of the electron as belonging to the individual nucleus, rather than the "interactional" approach of the electron as collective property belonging to the next higher level unit (i.e., molecule) of combined nucleuses.

Another way of thinking about the issue is in terms of a tale of epiphenomena — which can be told in two different ways, in this realm of nature. For physics, it is the atom which is real, and its electrons (and especially the outermost electrons) which are determining — whereas the molecule (and all other higher levels of matter) are essentially just agglomerations of atoms — mere *epiphenomena* which have no fundamental reality of their own. In the present model, by contrast, the nucleus is no more real, and the electrons no more determinative, than is the case for any other level of nature; indeed, the atom — which represents a model of nature centered upon the nucleus and its surrounding electrons — does not exist. Nor do electrons explain the binding of nucleuses within the molecule; rather, the electron patterns would be seen as *epiphenomenal* results — relative shadows — of the primary interaction patterns between nucleuses themselves.

All in all, we contend that the atom construct is a distraction which confuses our analysis — not merely because it is superfluous, but because it is actually a serious conceptual hindrance.

Obviously, the discussion here has been highly speculative, in terms of how this realm of nature might be understood very differently from the way it is conventionally construed. And again, none of what we are proposing is meant to question or contradict the basic observations which have actually been made about the "atom", or the established findings regarding the configuration of electrons in the microrealm. We are, for example, not challenging the spectral patterns which have been observed, but rather we are offering a different explanation as to how these observations are to be primarily construed — and named.[36]

In the face of all the sureness which underlies the orthodox approach, we are still dealing with phenomena that require physical intermediation to observe and theoretical interpretation to understand. The main question is whether the established ad hoc, historically-generated set of determinations is the optimal choice, or whether we would be better served by a framework that is based on the kind of consistent principles which

36 As another example, could it be that within the nucleus as a whole there are nesting layers (or "periods") of nucleons which represent additional levels of structure — they might be labeled as "nuclons" — that somehow correspond with the layers of orbital electrons?

underlie the present work. The particular ideas we have suggested here may or may not be a viable way of implementing our guiding theme of complementarity, but that theme remains what can be called our "central core of non-negotiable concepts" (Polkinghorne, p. 16). It is our bedrock — it is the wisdom we turn to when the situation we are confronting makes no sense or seems arbitrary. Somehow or other, therefore, we contend that complementarity can be applied just as much to the phenomena which are associated with the "atom" as to any other.

This completes our discussion of the micro world. We next turn our attention to the more familiar ground of stasis and motion, to see how complementarity would apply there.

Chapter 4

The Relationship Between Stasis and Motion

Our task in this chapter is to examine some of the main concepts in the area of stasis and motion. We start by briefly presenting a number of seminal ideas from classical and Relativity physics. First up are Newton's Laws and the special regions of nature where they apply, and then we move on to the important Principle of Relativity and the different ways that it can be stated and understood. After that, we talk about the relationship between Newton's mechanics and Maxwell's electromagnetism, and how Einstein joined them in his Special Theory.

We critique these ideas in the next part of the chapter — both in a general logical sense, as well as in terms of the precepts of the present model. This is to lay the groundwork for the last section, where we propose a complementary framework for dealing with these issues.

The Basic Components

The conventional model of stasis and motion is built upon a number of crucial laws and constructs. Some of these only apply to the classical theory of Newton and his contemporaries, whereas altered versions apply to Einstein's Special Relativity; yet certain ideas are common to both approaches. As we shall see, the classical rules are still today considered to represent valid ways of describing how our normal world works.

Newton's Laws of Mechanics. Let us first mention Newton's basic laws of motion. *Newton's first law* is perhaps the most important, as it describes the common property we call *inertia*. Put succinctly, "Every body continues in its state of rest, or of uniform motion in a straight line, except insofar as it is compelled to change that state by forces impressed upon it." (Hecht, p. 92)

Newton's second law associates the strength of this stasis-motion tendency with the property of mass. "The measure of an object's inertia is its *mass*." (Marion, p. 49)

Such "mass never changes", but rather "is an intrinsic property of matter." (Marion, pp. 49 and 53) The second law determines this quantity of mass indirectly, by means of measuring the resistance to change: Whatever force is required to change an object's inertia (i.e., uniform state of stasis or motion) is a measure of that object's *inertial mass*. Thus, by knowing how much an object's state of motion is changed and the size of the force which produces that change, one can determine how much of this stuff of inertia is involved. We call the departure from a constant state of stasis-motion *acceleration*. "Newton defined the inertial mass of a body as a constant state of proportionality between the acceleration of the body (the effect) and the applied force (the cause)." (Sachs, 1971, p. 25). The famous formula which results from this second law is $F=ma$.[1]

Acceleration refers not merely to change of speed, but to change of direction, as well — the proper term which encompasses both attributes being called "velocity". The result is that linear motion at a constant speed is considered by physics to be non-accelerating, whereas circular motion at a constant speed is treated as acceleration. Consequently, some additional mechanism is required in order to account for those situations where the constant speed motion is not linear, but circular. Thus, the planet which orbits the sun, or the stone being held by a string and spun in a circle, are each said to be responding to a *centripetal* (or center-seeking) *force*, and the constant speed motion which results is called *centripetal acceleration*.

Perhaps the most well-known principle of classical physics is enshrined in *Newton's third law*, which says that *for every action there is an equal and opposite reaction*. Thus, when one body exerts a force upon a second body, the second body in turn exerts a similar (but opposite) force back upon the first body.[2] It is a simple proposition which has received wide application in a variety of areas, inside and outside of physics. Newton's third law is probably so appealing to us because of the lovely sense of balance and order it expresses.

Regions of Applicability. Newton's laws of mechanics do not apply everywhere in nature, as he knew at the time he proposed them. If objects are to act in the orderly way which his laws prescribe — in the case of the first law, without intervening forces; and in the case of the second, with intervening forces — then we must be able to control and exclude any other secondary forces, to begin with. Therefore, we must start with a baseline situation where extraneous forces do not impinge on the region as a whole.

Newton's mechanics are thus built upon the notion of *force-free regions* of the universe. The idea is that if one is sufficiently far-removed from any external body, then whatever forces are produced by the existence or activity of such a body will be essentially eliminated. In this way, the "true" nature of physical existence, interactions, and motion can be ascertained. These force-free realms are given the formal name of *inertial*

1 Speaking precisely, Newton did not provide the mathematical formula here — but rather, the description of the relationship. The formula was specified later by someone else.

2 Newton's third law also "leads directly to the fundamental principle of *conservation of momentum* and may, in a restricted sense, be thought of as equivalent to it." (Hecht, p. 99)

frames of reference, and such regions are also sometimes referred to as "preferred" or "privileged". By contrast, those regions of the universe which are not free of external influences — and where Newton's laws of mechanics do not apply — are called *non-inertial frames of reference*. Largely paralleling this distinction, Einstein's Theory of Special Relativity is restricted to the realm of inertial frames, whereas his General Theory encompasses both.

The Principle of Relativity: Four Formulations

The plot thickens now, as we move from Newton's laws of mechanics to a crucial general precept, in our thinking about the nature of stasis and motion. The Principle of Relativity is one of the most important theoretical tenets for physics' overall model of macromatter. That includes both the classical world of Newton and the more modern world of Einstein. Different interpretations of the Relativity Principle underlie each approach, and it is crucial to understand how these are all related.

We shall first look at a statement of the classical Relativity Principle, which concerns the basic indistinguishability of stasis and uniform motion. Second is a statement concerning the nature of motion — i.e., that it is only relative, and not absolute. Third is a related formulation which denies the existence of an ether or medium or absolute space.[3] Finally, we consider a grand declaration concerning "laws of nature" — what their scope and form is, and where they do and do not apply.

Stasis Equivalent to Uniform Motion. Perhaps the most prominent and meaningful formulation of the Relativity Principle is the concept which is essentially contained in Newton's first law. This *primary form* of the Relativity Principle in fact originates with Galileo, and has long been an important impetus to thinking about the nature of the physical world and thus how theory should describe it. Indeed, the other three formulations which follow could be understood as axioms to this basic postulate, which by itself is known as the classical or Galilean Relativity Principle.

As we discussed above, Newton's first law stipulates that in the absence of extraneous forces, an object will continue in its state of rest or uniform linear motion indefinitely — meaning that the object qualifies as (part of) an inertial frame. Moreover, this primary formulation of the Relativity Principle can also be understood to mean that there is no functional difference between those two uniform states. The most common way of exploring this assumption is by examining the internal and external states of the same entity, or system. If that system is either static or engaged in uniform motion, as determined from without, then strictly internal factors within the system cannot differ-

3 We should note that most explanations of the Relativity Principle do not differentiate the second and third aspects, as we are doing here. Instead the third interpretation, that there is no absolute space, is treated merely as a corollary of the second interpretation, that there is no absolute motion. We find it useful to separate the two issues.

entiate between those two external states. In other words, there is no way to establish, by means of strictly internal cues, which of the two uniform states an inertial frame system is undergoing externally. This also means that there is a direct correspondence between a system's internal and external states: if the system is judged to be an inertial frame within, then it will also be categorized as an inertial frame without.

For example, playing table tennis inside a large shipping container that is sitting upon a static ship is no different from playing table tennis inside the same shipping container that is sitting upon a ship which is moving uniformly ahead in calm seas. Properly speaking, the container occupants in the latter case have no basis for concluding that their container is in a uniform state of motion: the game of table tennis goes on inside, just as in the former case.

Indeed, we can formalize these table tennis kind of internal inquiries, to the point where such a system becomes a test-conducting laboratory for investigating the underlying premise. In the present work, we often label these explorations as *inertialness tests*. Thus, physicists might conduct tests inside the ship container — standard physics experiments where balls are rolled and objects placed on balance scales — and if the outcome of these tests conforms with Newton's laws, then the physicists declare that the ship-laboratory container qualifies as an inertial frame of reference. But then the ship changes direction, and makes a sharp and sudden turn: physicists repeat the internal experiments in these circumstances, but now the system fails the test, and thus it is deemed to be a non-inertial frame of reference. Such real or hypothetical inertialness tests are commonly cited as strong evidence in support of the classical Relativity Principle, which can be understood as dealing with the correspondence between the internal and external states.[4]

The primary formulation of the Relativity Principle clearly describes our common experience, in the way it equates stasis and uniform motion. But the other interpretations which follow represent greater degrees of abstraction — even though, on the surface, they may appear as innocuous or obvious extensions of this basic theme. It should also be noted that Newton subscribed only to the first interpretation, which is why it is called the classical one; the ones which follow are associated with Einstein and his Special Theory.

4 Note that whereas the usual discussion of these scenarios talks about conducting such tests internally, within the system (as we have just described) one could just as well conduct such experiments upon the system externally (as we have also just described) — both cases determining a common outcome, as to whether a system is, or is not, an inertial frame of reference. The "internal" test examines the relationship between the contents of a system and the frame itself — e.g., between the balls that we are rolling around to test Newton's laws, and the apparatus or container walls (or floor) which frame this ball action. The "external" test examines the relationship between the system itself and whatever the next larger realm is — in which case the container then becomes contents, in relation to whatever the even larger frame is (such as the earth). In this respect, one could talk about both internal inertialness tests and external inertialness tests. We haven't taken that path in this work — but one could just as well do so, theoretically. Instead, we will continue to talk about inertialness tests which are conducted inside a system — realizing that the same pattern applies to the next larger system as well ... which is the next hierarchical level, in the present model.

No Absolute Motion. A second way of understanding the Relativity Principle is that the notion of absolute motion has no meaning. Is a person, who is sitting on the surface of the earth, moving in any clear-cut sense? Well, that all depends on how motion is defined: if one looks at the local region immediately surrounding that person, then the answer is no — the person seems to be quite static. On the other hand, if one takes a broader perspective and becomes aware of the motion of the person and the rest of his or her local region (indeed, that of the whole earth) in the solar system, then the answer is clearly yes — the person is moving.

The conclusion which this second formulation leads to, then, is that we are not justified in speaking about motion in an absolute manner, but only in a relative sense — as compared to particular other things which are not in motion. This formulation of the Relativity Principle "abolishes the concepts of absolute rest and absolute motion. All we can know empirically is that one object is in motion relative to another." (Hecht, p. 141) For example, say that the person is sitting on one of two adjacent trains, which are both static at a railway station — and then one of the trains begins to move, so very gradually that the person could not feel it. If the person only has his/her train and the other train to look at, then it will not be clear which of the two trains is moving and which is static.

In fact, one can go further and say that even this degree of relativism is unsatisfactory — because it does not sufficiently conform to a larger principle of symmetry. That is, even after the train moves out of the station and onto its journey, it is not enough to state that the train is moving relative to the tracks. For who is to say that the train "carriage" is moving relative to the static track "embankment", rather than the reverse scenario — that the train is static, and it is the track which is in motion relative to the train?! Here is how Einstein makes this point, starting with the presumption that, "[E]very motion must be considered only as a relative motion ..."

> ... we can express the fact of the motion here taking place in the following two forms, both of which are equally justifiable: (a) The carriage is in motion relative to the embankment. (b) The embankment is in motion relative to the carriage. In (a) the embankment, in (b) the carriage, serves as the body of reference in our statement of the motion taking place. If it is simply a question of detecting or of describing the motion involved, it is in principle immaterial to what reference-body we refer the motion. (Einstein, 1961, pp. 59-60)

Again, then, Einstein is raising the question of what basis we have for determining that it is the train which moves in relation to the embankment — any more than concluding that it is the embankment which moves in relation to the train. In this regard, Dingle refers to "a cartoon, in the days when relativity was a popular sensation, in which a passenger calls out, 'Hi, guard, does Manchester stop at this train?' " (Dingle, 1972, p. 140) In a fully symmetrical world, this altered interpretation of reality would be no cause for alarm.

Indeed, Dingle goes even further, and argues that the most thoroughgoing interpretation of the Relativity Principle here would deny that *either* the train or the embankment is moving, relative to the other's stasis — that such a description is merely "a concession to ordinary modes of thought; on the Relativity Principle, the only proper description in each case would be 'when the relative motion occurs'." (Dingle, 1972, p. 216) According to this full-fledged interpretation, then, motion cannot be parceled out to either one or the other party, but can only express a net state of affairs between them: there is, or is not, relative motion between two entities.

No Absolute Space. The issue becomes broader still, when we recall another meaning of the classical inertial frame construct. Inherent therein was the notion that there were not only particular force-free regions, where states of uniform stasis and uniform motion would continue indefinitely. More broadly, it was assumed that there was a universal framework for such phenomena — a universe-wide standard of absolute space, relative to which absolute motion would exist.

Therefore, even presuming that motion is relative as between one object and another, that does not necessarily preclude a more substantive referent — as expressed in terms of the notion that there is a basic underlying universal medium or substructure, relative to which a more fundamental kind of absolute motion might nonetheless occur. So could there be a difference in terms of motion between one object and another object, as against motion between one object and the larger universe? And if that were possible, then such a category of motion, which is not dependent on various other obvious objects, entails some kind of general substratum — a universal reference system, relative to which such absolute motion could then be said to occur (Dingle, 1960b, pp. 121-122). Historically, such a substratum has been called the "ether" or medium, at least in relation to electromagnetic phenomena. This third formulation of the Relativity Principle, then, concerns the non-existence of such an absolute space.

In fact, it is often taken to be one of the most significant conclusions of modern physics that such an ether (or absolute space) cannot be discerned. As Eddington puts it, "It is impossible by any experiment to detect uniform motion relative to the aether." (Eddington, 1920, quoted in Dingle, 1972, p. 204) Furthermore, it is supposedly one of the great accomplishments of the Theory of Special Relativity to have completely abandoned the concept of absolute space, along with the ether medium. That position is represented in this third interpretation.

Laws of Nature. The final formulation of the Relativity Principle is the most abstract of all. Here, Einstein introduces interpretations and conjectures which are either completely distinct from the original experiential formulation, or else are the furthest removed inferences therefrom. And yet this last approach is also the most widely cited — and is seen by many physicists as the most valuable aspect of the Relativity Principle generally. Simply stated, it says, "The laws of nature are the same to any observer, independent of his motion relative to any other observer of the same phenomenon" (Sachs, 1973, p. 38). Indeed, it is this very formulation which is described as "the basis of the Theory of Relativity." (Sachs, 1974, p. 96)

Yet what can it mean to say that a "law of nature" can be applied directly to any system which is at rest or undergoing uniform motion? And if it means what the most obvious interpretation suggests that it means, then why bother to say it? On the other hand, if it goes beyond the obvious, then are we unnecessarily narrowing the sensible meaning for "law of nature"?

Consider the possibilities. In keeping with the most straightforward meaning, one might argue that "laws of nature" refer to general laws of principle which can be applied to some set of phenomena, or to regular and predictable cause-and-effect connections — theoretical relationships which have been tested so thoroughly that they finally get consecrated as "laws of nature". Certainly that description applies to all three of Newton's laws of mechanics, which are taken as describing inertial frames. E.g., the famous third law of Newton that "for every action there is an equal and opposite reaction" would presumably apply to any given inertial frame. This statement was accepted as a "law of nature" long before the present issue under discussion arose — and, of course, the third law still applies today ... even though neither it nor the other Newtonian laws qualify as "laws of nature", under the more restricted interpretation which Einstein applied.

Indeed, we can survey all the general principles which apply in ordinary experience, and they all fail to meet Einstein's lofty criterion, as "laws of nature". Clearly, then, we must be dealing with a very select category, since nothing reasonably accessible to us in our normal experience qualifies. We can now see why this fourth Relativity Principle formulation is not merely redundant — i.e., it is not just a restatement of the obvious. A "law of nature" is what Einstein stipulates it is, irrespective of our common experience and normal use of language. As Bridgman says of "law of nature" and the Special Theory's approach to this issue:

> A natural interpretation of the meaning would be that we are here stating a new observation about the properties of Galilean [i.e., inertial] frames. The implication would seem to be that "laws of nature" are something that we know about from previous experience, and that the theory has given a new insight with regard to the relation of these laws to Galilean frames. This insight is that the laws of nature are the "same" in different Galilean frames, a statement that can be checked by formulating the laws for different frames of reference and observing whether they are the same. But that is not what the postulate is actually saying. What it is saying is something about a "law of nature"; it is saying that unless the law is the same in the two frames of reference it is not a "law." (Bridgman 1983, p. 118)

In fact, what Einstein meant was something very specific. One of the main goals of the Special Theory was to try and integrate two different sets of existing laws and formulas, only one of which we have discussed so far. On the one hand, there was the abovementioned Newtonian mechanics, which is where the classical Relativity Principle (i.e., the first of the four formulations) applied. On the other hand, there was Maxwell's electromagnetism, where that principle or interpretation did not apply ... at least, not apparently. We will examine the relationship between these two categories more closely in the next section, and especially the perceived conflict between them. But we can summarize Einstein's position in terms of a famous set of formulas which lay at the heart of his model, and which we will discuss at great length in later chapters: "The whole content of the special theory of relativity is included in the postulate: The laws of nature are invariant with respect to the Lorentz transformations." (Einstein, 1961, p. 148)

Yet even this description is not broad enough, because Einstein's theorizing did not stop with Special Relativity, but went on to General Relativity. Consequently, when we get into this fourth "law of nature" interpretation of the Relativity Principle, there are actually two sub-categories — resulting in two different forms of this fourth version of the Relativity Principle. The Special (or Restricted) Relativity Principle refers to laws of nature which are applicable to inertial frames — incorporating both mechanics and electromagnetism; whereas the General Relativity Principle refers to all kinds of frames, inertial and not ... and to all kinds of phenomena in the universe.[5] Having noted the distinction, however, most of our discussion of this fourth "law of nature" interpretation will deal with the Special version thereof.

Before looking at these issues further, we need to digress now and consider the historical background that is involved.

Newton vs. Maxwell

At the turn of the 19th-20th century, perhaps the most glaring contradiction in physics was the one which seemed to separate Newtonian mechanics (which dealt with the behavior of common objects) from Maxwellian electromagnetism (which dealt with the behavior of phenomena such as light). Like Newton's laws, the equations of Maxwell had also been well-tested and widely-applied — although Maxwell's laws were quite new, as compared to Newton's. The problem was that while the laws of mechanics appeared to conform to the classical Relativity Principle, the equations of electromagnetism did not. This contradiction represented a kind of stalemate for physics. In particular, two conspicuous differences between the two sets of laws concerned how motions were to be combined across separate frames of reference, along with how the notion of absolute space or ether was to be dealt with.

Let us recall the situation. As we have seen, the classical Relativity Principle set forth an explicit definition of those situations which qualified as inertial or "preferred" frames, i.e., the frames had to be either static or undergoing uniform linear motion. In addition, the classical Relativity Principle can also be said to provide an implicit method for quantitatively relating such preferred sets of phenomena with one another, by means of what is called the *Galilean transformation*. The basic assumption was that, for example, an object's motion which occurred inside one uniformly moving frame could simply be added together with the motion of the frame itself, in order to determine the net motion of the object ... as compared to a second inertial frame which was static relative to the first. So in this sense, the classical Relativity Principle both defined inertial phenomena by specifying which realms the "laws" (i.e., of mechanics) applied to, as well as indicating that groups of such preferred phenomena could simply be combined arithmetically in order to describe any net motion.

To take a simple example, if one rolls a ball inside a static laboratory, then Newton's laws apply to the ball's motion. If one then performs this same operation inside

5 Healey describes the step upward from the special to the general theory: "[I]t is normally taken to be Einstein's great achievement to have eliminated the need to assume the existence of any such privileged state of motion in formulating a fundamental physical theory.... [N]o relativistically invariant laws describing any physical process may single out a preferred inertial frame: This is the essential content of the special principle of relativity." (Healey, pp. 149 and 190)

such a laboratory that is part of a uniformly moving train, how are we to determine the ball's motion in relation to a static observer on the ground who is watching the train go past? Of course, the classical Relativity Principle says that Newton's laws will apply both to ball-in-train and to train-on-track. Furthermore, according to the Galilean transformation, one can simply add the ball's motion (relative to the train) together with the train's motion (relative to the track) to see how fast the ball is moving relative to the track (e.g., relative to an observer standing beside the track).

But there was a problem in that this classical Relativity Principle approach, which applied to Newton's mechanics, did not seem compatible with Maxwell's electromagnetism. Thus, suppose that the ground-based observer was not looking at a ball which was moving within the train, but at a beam of light. At first thought, there would seem to be no difficulty there: why not merely add the light's speed on the train, together with that of the train itself, in order to determine the net speed of the moving light in relation to the static track — just as we did with the ball? The problem was that whereas this simple additivity worked in the case of the moving ball (Newton's laws of motion), it did not seem to make sense in the case of moving light (Maxwell's equations of electromagnetism). But why? Why did the Galilean transformation appear to apply to mechanics, but not to electromagnetism? The dilemma becomes clear when we recall how the phenomenon of light was understood by Maxwell and his contemporaries.

Briefly, all of empty space was thought to be filled with an invisible substance that we have mentioned, called the ether, and light was understood to consist of waves which vibrated upon this ether medium. Given that there was only a single universal ether medium, which had an homogeneous character everywhere, then there would be no way that they could be added together, in the manner of common objects. Thus the ether medium inside the train was the same ether medium as existed outside the train — and so one could not add light's speed in relation to the inside medium, together with the motion of that medium within the train itself, to determine the net speed in relation to the outside (i.e., static observer or medium): there would be only one overall medium, not two separate ones.

The single medium concept was embodied in Maxwell's electromagnetic theory, which was essentially reducible to Maxwell's equations — in which the notion of a fixed speed of light, c, is the pivotal factor. Thus, light would travel at a constant speed in relation to the ubiquitous and universal ether medium — or equivalently, in relation to an object (such as the just-mentioned observer) which was static in relation to that medium. But if one attempted to alter these equations to take account of any motion of the medium, by applying the Galilean transformation, then the equations became much more complicated, if not completely unwieldy.[6]

6 Rogers says that, "Electromagnetism is summed up in Maxwell's equations, for a stationary observer. Ask what a *moving* observer should find ... with the Galilean Transformation: then Maxwell's equations take on a different, more complicated, form." Again, "Maxwell's laws ... change to a different, clumsier form for any observer except one 'at rest' ". Rogers mentions the "discomfort that physicists with a sense of mathematical form got from the uncouth effect of the Galilean Transformation on Maxwell's equations." (Rogers, pp. 475, 564, and 476)

Yet the problem was not only one of additivity in terms of Maxwell's equations — for the very idea of a universal ether medium also seemed to contradict the classical Relativity Principle, according to which there is no discernible difference between non-motion (or rest) and uniform linear motion. This is the basis for the fundamental contention that no experiment conducted within an inertial frame could detect whether the frame as a whole was undergoing uniform motion or was completely at rest. And for the field of physics, as it existed at the beginning of the 20th century, there was no contesting this premise — its validity was unchallenged in the mechanical domain to which it applied. But Maxwell's electromagnetic theory seemed to be saying that there was an absolute standard of rest — that of the universal ether, in relation to which there would be absolute motion.

As if all this were not enough, another bee in the bonnet of turn-of-the-century physics was a crucial experiment conducted by Michelson and Morley in 1887, which attempted (but failed) to detect the effects of motion through this universal ether medium. That experiment will be treated at length in Chapter 6, but what we need to note here is that the results did not seem to fit with the standard interpretation of the Maxwell model that we have been describing. In many respects, this experiment set the stage for a totally different way of approaching the issues we have been discussing, as we shall see later.

Revisiting the Principle of Relativity

We take exception to many of the interpretations and conclusions that we have been describing, and in this and the following main section we want to critically examine the crucial paradigms which underlie them. In so doing, we will also be laying the groundwork for our alternative complementary framework, partly based upon forks in the road with existing concepts and theory. In this way, we can understand where the conventional approach goes wrong, and how the present model tries to get it right. This new approach will become much clearer in the last main section of the chapter, where we discuss the overall 2FT model on stasis and motion more fully.

Our topic in this section is the Relativity Principle, and the four formulations we described above. On the face of it, these different interpretations may seem clear-cut and convincing — they obviously have been so for the established paradigm, lo these many years. But as we begin to examine the underlying conceptual framework, we shall see that serious problems emerge. For one thing, they are all based upon a very dubious proposition, as to the circumstances in which the principle even applies — that of the "inertial frame of reference". Equally problematic are the principle's main stasis-motion premises. Finally, we sharply question the singular "law of nature" thesis, which Einstein so vigorously championed. Taken all together, we shall argue that the Relativity Principle represents a highly unsatisfactory framework for understanding nature.

The Myth of Force-free Regions. A crucial assumption of the Relativity Principle is that it describes regions of the universe which are unaffected by other force influences.[7] It is supposedly only in these pristine force-free realms — a kind of "physics heaven" — where natural phenomena flourish and manifest themselves in their essential form, and where fundamental precepts about nature properly apply … laws like Newton's mechanics, Maxwell's electromagnetism, and Einstein's Special Theory of Relativity.

Consider the peculiar circumstances that are called for here. The conventional interpretation is that if an object got far enough away from external influences, then it would be in the neutral land of empty space. And since this space is empty, there would be nothing to influence any object that was contained therein — which object, therefore, would be able to manifest its "natural" or inertial state, of either simple stasis or uniform linear motion … forever, in principle, since there would be nothing in such a void to alter it.

So much for myth. The fact is that no matter how far removed one object gets from other objects, what basis have we for presuming that such influences would ever completely disappear, however minuscule the resulting impact? Even the orthodox paradigm is forced to acknowledge that, "There are no places in the universe where an object can actually move totally free of all external influences, and surely the idea of infinite straight-line motion is quite unrealistic." (Hecht, p. 73) Furthermore, even if one were to assume that such fabled force-free regions actually existed, there is also the question of how an object would ever reach that transcendent realm — and how it would manage to arrive there either static or at uniform speed, having traversed the real world of motion-changing forces en route.

But there is still another problem, as to the meaning of the stasis-motion state of the object itself, in the inertial frame. Take the motion: in this realm of pure *nothingness*, how are we to understand the motion which is said to exist there? For if motion is truly relative, as physics argues, then what does uniform linear motion — or any other kind of motion, for that matter — even mean, in a world where there is nothing for the object's motion to be relative to? Indeed, if there is not something to provide a referent for stasis, then the motion cannot be said to exist at all. Yet by definition, if something else is there, then it will have effects on the surrounding region, and so our object would not be in a force-free zone of empty space. Moreover, this limitation applies not only in terms of some other static entity, which would be necessary for the motion to exist in relation to. But it would also seem to preclude any kind of observational mechanism, which would somehow witness or record the alleged motion.[8] For again, if it is a force-free zone, then nothing at all can be there to disturb the object — and we surely know from Quantum Theory that to observe is to disturb. All in all, we are dealing with a complete idealization. As Rogers puts it,

> Perhaps the ideal experiment is *unthinkably* difficult, requiring a single moving body infinitely isolated from all others which might disturb it. Then how could we observe its steady motion? Where would we be and where

7 Again, we are not dealing with General Relativity in this chapter, which would incorporate such other force influences.

8 As an aside, one might even argue that this limitation would also seem to rule out the object itself, in such an empty and force-free region — for any such object must be assumed to exert forces upon the surrounding region through which it is moving.

> would our mileposts be? Since it would be impossible to observe such motion if it existed, are we wise to talk about it as part of scientific knowledge? (Rogers, p. 108)

Similarly, Zukav talks about the assumption...

> that somewhere in the universe there exists a frame of reference in which the laws of mechanics are completely valid ... This frame of reference is called an "inertial" frame of reference.... [However, n]o one yet has found a co-ordinate system in which the laws of mechanics are valid.... Since their very inception, the poor laws of classical mechanics have been left, so to speak, without a home. No one has discovered a co-ordinate system in which they manifest themselves perfectly.... On the one hand, we have the laws of mechanics, which are indispensable to physics, and, on the other hand, these same laws are predicated upon a co-ordinate system which may not even exist. (Zukav, pp. 123 and 125-126)

For his part, Bridgman says that, "[T]he part of our world associated with 'empty' space has as much 'reality' as the part associated with 'matter'.... [S]pace is never empty ..." (Bridgman, 1983, p. 153) In this regard, note that Quantum Theory would also question the category we are discussing here — according to which, "There is no such thing in the real world as 'empty space.' It is a mental construction, an idealization, which we have taken to be true." (Zukav, p. 240; cf. Morris, p. 158)

Einstein also doubted the objective existence or integrity of the inertial system concept, where objects move in a uniform, rather than an accelerating, manner. As he puts it:

> The weakness of the principle of inertia lies in this, that it involves an argument in a circle: a mass moves without acceleration if it is sufficiently far away from other bodies; we know that it is sufficiently far from other bodies only by the fact that it moves without acceleration. Are there at all any inertial systems for very extended portions of the space-time continuum, or, indeed, for the whole universe? (Einstein, 1956, p. 58)

Note that this reasoning is part of Einstein's development away from the inertial frame circumstances where the Special Theory applies, and towards the broader realm of the General Theory. Thus he writes that, "[W]hen the gravitational field is included in the framework of the consideration, the inertial system loses its objective significance." (Einstein, 1961, p. 151) And late in his life Einstein argued that if there were no gravitational field then there would be "absolutely *nothing*, and also no 'topological space'.... There is no such thing as an empty space, i.e. a space without field." (Einstein, 1961, p. 155) In summary, Einstein "considered the concept of the inertial reference system to be a 'logical weakness' of the special relativity theory." (A. Miller, p. xiii) We shall pursue Einstein's argument in Chapter 5, when we examine gravity and General Relativity.

For now, though, consider the contradictions we are faced with here. To start with, it is a very dubious strategy to build a model about the real world of nature that we all experience, based upon a supposed physical domain of the universe that it is logically impossible for anyone (or anything) to ever experience, either directly or not. How much will the resulting theory be based upon reality, as opposed to some idealization — and how can we ever determine that? Worse, still, while it is damning enough that no requisite force-free region exists, it is even more problematic when we realize that the laws which are based upon such a *force-free* region largely apply quite well in the *force-full* region of the universe which we all know! Not perfectly, of course — for example, there is friction, on and near the surface of the earth, which will affect motion. But nonetheless, mechanical laws like Newton's are not just idle philosophizing or science fiction about some unknown make-believe region of the universe: they are very well-tested laws of physics, which we experience to a large degree every day. In other words there is a serious gap between what the theory claims to be referring to (abstract situations which do not exist) and what it actually applies to (concrete situations which do exist).

One result is that there is no clear consensus in physics as to what does and does not constitute an inertial frame of reference. Take the earth, for example, and the fact that the reason why the earth remains in orbit around the sun is explained in terms of gravity. According to conventional theory, then, is the earth an inertial frame or not? One would assume not: the earth is not located in some force-free region of the universe, far from any sources of gravitational influence … as the concept requires. Zukav says, for example, that, "[T]he laws of mechanics are not valid for a co-ordinate system rigidly attached to the earth. The earth is *not* an inertial frame of reference." (Zukav, p. 126) But by contrast, Hecht says the laws of mechanics apply to "all nonaccelerating (inertial) environments — like your living room …" (Hecht, p. 140)

We would argue that inertial systems exist precisely where the laws apply — and that to a large degree, an inertial frame can be found right here on the surface of the earth. So we need not go to the far reaches of the universe to find such inertial domains: they are everywhere — not in *force-free* regions, but rather in *force-balanced* regions. That is the case regardless of whether we are talking in general terms about this model's two fundamental forces themselves, or whether we are talking about the particular way they are manifested in some given situation — either in terms of the composition and character of an object or system internally, or in terms of the larger context and influences to which an object or system is subjected externally. There are always two countervailing force tendencies, and that is why the notion of "force-free" makes no sense in this framework; rather, the issue is one of balance or not.

Furthermore, we have a very specific category for representing that "balance or not", in the form of one of our dimensions of analysis. The stable-changing dimension represents the difference between situations where forces manifest themselves in a balanced and *stable* manner, as opposed to those situations where forces manifest themselves in an unbalanced and *changing* manner. We shall see that in many respects, this is how the conventional model's inertial and non-inertial frames can be understood in the new framework that we are proposing.

The different concept of force with which we are working — always applying, but in a balanced and stable vs. unbalanced and changing way — is hardly just a trivial play with semantics: it represents an entirely different philosophy of nature … not to mention that it also moves the loci for such "privileged" inertial frame regions from Nowhereland to Somewhereland. The very notion of a force-free region of the universe makes no sense, and has no role to play in the present model … even abstractly.

Absolute Space and Motion. The simple contention that "all motion is relative" is probably the most powerful single theme underlying any relativity theory. The basic principle can be grasped in terms of common visual experience: in order to determine whether something is moving, we intuitively compare it with something else that is not moving. It is the difference between the motion and the stasis which we identify, the argument goes, and not any kind of absolute motion as such: "[T]he only motion that we can ever determine is motion, or lack of motion, relative to something else." (Zukav, p. 123) Why, then, retain the concept of absolute motion at all, if it has no independent meaning? Anything can be seen to be moving, if only we identify a non-moving referent for the motion to be relative to. Thus, the only conclusion which seems tenable: all motion is relative.

It is a seductive argument. On the one hand, it is a compelling insight — but on the other hand, it is only a half-truth way of describing the situation ... and even then, it is not describing a pattern which is unique to stasis and motion. That is simply because in our model this pair represents only one among several sets of crucial complementary characteristics, and the same corresponding pattern applies to them all. We shall return to the broader pattern later in the chapter.

But for now we can say that the same principle also explains our rather obvious "half-truth" objection regarding stasis and motion, for even the phrasing of the basic assertion is distorted. To argue merely that "all motion is relative" is truly a one-sided proposition. "Big deal", one might respond: "So is all stasis!" That is, while it is true that motion only exists in relation to stasis, the other side is equally true and must be equally stated — that stasis only exists relative to motion. Therefore, "all stasis is relative," would be the proper complementary contention. Putting the two together we find that both motion and stasis are relative. But relative to what? To each other, of course. Stasis and motion are complementary phenomena, which only exist in relation to one another. The two always interact with one other in the real world, and neither exists alone in an independent and universal sense.

So we contest the assumption which is implicit here — virtually to the effect that stasis is the normal state of affairs in nature, and not requiring of explanation ... whereas motion is exceptional and requires a special mechanism to be accounted for. That may, in fact, be a reasonable inference to draw from our usual, stasis-oriented earth-bound existence — but this is hardly the basis for a conclusion about the normal state for nature as a whole. Indeed, if you think about it, such a one-sided proposition runs counter to the very essence of the classical Relativity Principle or Newton's first law: the state of inertia can be equally represented by either (uniform) stasis or (uniform) motion. Therefore, a thoroughgoing conceptual analysis should give no more precedence to explaining the motion aspect than it does to explaining the stasis aspect. This is another instance of the problem we mentioned in the first chapter, concerning physics' one-sided approach to so many phenomena in nature — which we label in terms of "no symmetry".

Yet beyond such lack of conceptual balance, the no-absolute-motion argument raises serious ontological difficulties. The most important problem is that if all motion is only relative, and if all stasis referents are equal in relation to which such motion is gauged, then what basis do we have for presuming that any object is undergoing uniform motion — or any other type of motion ... or non-motion, for that matter? Indeed, how can we ever draw any general conclusion whatsoever about an object's stasis-motion state?

This is a problem that is specific to Einstein's Special Theory. By contrast, in classical physics the referent for stasis is taken to be, simply, "absolute space" — as in both definite and universal. Indeed, one would not even have to specify any particular stasis referent in that concept system, because the motion is deemed to be absolute and thus the stasis referent of absolute space is implicit. Classical physics might have said: "We don't need to specify the referent, relative to which the system is moving uniformly, because that referent is always understood to be absolute space ... which may or may not be identifiable (or even detectable) in any concrete way. It is enough for us to decree that a system which is stable internally — i.e., it passes the inertialness test — is therefore undergoing uniform motion (or stasis) externally, relative to this absolute space. The motion (or stasis) is always the same no matter how you look at it, because the referent for it is absolute, and never changes."

However, this same rationale cannot be applied to Special Relativity, which claims to have abolished both absolute motion and the absolute space referent relative to which the former can be said to exist — either in a universal or in any other definitive sense. Einstein's model eschews the absolute space concept altogether: the idea is completely anathema to the Special Theory, even to the point of contradicting it to the core.

One major casualty of such a purely relativistic approach is the classical Relativity Principle itself, which we just described: how can we any longer assert that a system's (uniform) stasis and motion are equivalent to one another — and indistinguishable to an internal occupant — if there are no absolute referents by which we can gauge those stasis-motion states? There would be a multitude of referents, for either stasis or motion, which are indiscriminately strewn about the stage of nature — and one can freely take one's pick, in order to attain some given outcome.

The result would be that the internal-external correspondence we discussed earlier, as part of the first formulation of the Relativity Principle, would be broken — and we could no longer depend upon an inertialness test conducted within a system as giving any indication at all, regarding the system's external state.[9] That is because the implicit but crucial assumption behind the inertialness test paradigm is that there is a single and absolute outcome as to the internal-external correspondence — and thus definitive designation of the system as an inertial frame or not. Yet this proposition makes no sense unless we have a constant and consistent external referent by which to gauge the system's stasis-motion state ... and such a constant and consistent referent must be tantamount to an absolute space — in some form or other.

But if it's up for grabs what the system's external stasis-motion state is — if all judgments about a system's external motion (or lack thereof) are equally relative and arbitrary, depending on the particular referent (among many) which is used to make that determination — then why should there be any correlation between the internal

9 General Relativity would make just this point, in the form of the Equivalence Principle that we shall discuss in the following chapter.

and external states? By denying the classical construct of an absolute reference system for describing motion generally, we are left instead with a fairly chaotic set of reference systems in which one thing can be static or moving in any manner that is desired ... simply by selecting out the requisite referent, relative to which such varying determinations can be made. And so we could prove whatever outcome we liked — correlation between internal and external, or not ... as well as uniform or not — by picking whatever referent corresponds with the conclusion we preferred. We could thus either confirm the classical Relativity Principle and Newton's first law, or not — effectively undermining them as general descriptions of nature.

So which way do we turn? How are we to resolve this fundamental dilemma? Is there to be some kind of absolute referent in every case — something akin to the classical interpretation of the Relativity Principle (with its single, universe-wide referent), and the absolute inertialness test? Or, are there to be an infinite number of relative referents — as the non-classical formulations of the Relativity Principle assume?

Consider the fact that most people talk about a moving object as having a given velocity — and physicists do so as well. For his part, Bridgman argues that any such descriptions are mistaken: "Velocity *is* a relative concept, and whenever the physicist allows himself to speak of velocity with an implication of absoluteness he is either forgetting something or is tacitly implying something he has not taken the trouble to make explicit." (Bridgman, 1983, p. 28)

Such relativistic arguments are all very well and good — but is there a sound basis for the more common approach, of assuming a clear answer to such questions as how fast an object is moving? Are some referents more crucial than others in determining the outcome, for any given situation? If so, the obvious candidate for that something, relative to which a given velocity is taken to be fixed and absolute, must have many qualities which contradict the third interpretation of the Relativity Principle.

As we shift gears now, in seeking to understand what is going on, we want to be clear at the outset that the issue is not whether there were problems with the classical theory — but rather whether the Relativity solution was actually an improvement, in terms of bringing us closer to a viable model of stasis and motion. We contend that it was not, and that instead we have been on a long and wild goose chase.

In fact, we maintain that there is an argument for having it both ways, in terms of the respective absolute vs. relative frameworks that are involved here. But such an interpretation requires a different conceptual orientation than that of either of the two conventional models. Therefore, our solution is to reject the overly-narrow classical approach, of a singular absolute space as representing all of nature — just as we reject the overly-broad Relativity approach, of an endless array of possible referents for motion.[10] Indeed, the Relativity approach amounts to a random representation of reality, whereby the validity of the underlying relative motion precept has been taken to a ridiculous extreme — which has left us with no coherent framework for understanding

10 There are obvious parallels between the two approaches described here, and that between the Copenhagen and Many Worlds interpretation of Quantum Theory which we discussed in Chapter 2.

the patterns that are involved. Yes, there is a relative component to stasis-motion determinations — but this should not lead us into a world of conceptual chaos, or the abandonment (as opposed to the modification) of those established constructs which make sense and are workable.

Let us consider the main conflict that we are faced with here. On the one hand, Newton treated absolute space in terms of a singular, universe-wide phenomenon — in relation to which there was only absolute and definite motion. By contrast, the Theory of Relativity rejects any notion of a universe-wide absolute space referent — and since there was no such absolute space, then there was likewise no absolute motion to go with it. All we are left with is relative motion, in relation to a miscellany of relative (and arbitrary) stasis referents.

Our basic approach to these two diametrically-opposed positions is in the form of a simple question: why should there be *one* universe-wide referent alone as representing the absolute space concept, which Newton's theory assumed? Why need there be one referent which provides a standard for absolute rest in relation to everything, everywhere in the universe — or else the whole conception of absolute space (and so, absolute motion) is fatally flawed and needs to be abandoned, as Einstein's theory concluded?

We maintain that the problem lies not in the viability of the classical concepts themselves, nor in the patterns which underlie them, but rather in their application: that they must be applied in a singular universal sense — *i.e., in relation to everything* — or else they are false ideas (if not idols) in the first place, which must be discarded. It was as though the disappointment was so great at the inability of the conventional concepts to apply everywhere, indiscriminately, that they were simply dismissed ... as illegitimate concepts in their own right.

Our position is that one must distinguish between "absolute" in the sense of having a definite universal or universe-wide referent, which is the classical Newtonian interpretation — as opposed to "absolute" in the sense of having some definite local referent, which is precisely the new twist that we are introducing. So while it is true that stasis and motion have no constant meaning when applied *universally* between one particular entity and every other entity in the universe, they still have constant meaning when applied *locally* between one particular entity and its orderly background. The end result is that stasis and motion are complementary concepts which still carry absolute meaning, relative to any particular context.

Conventional theory does not make such a differentiation, whereas we maintain that it is an inevitable outgrowth of our hierarchical model — and this is the key to resolving the problem here, which Newton's and Einstein's models do not take into account. According to the new approach, the notion of a universe-wide phenomenon is only one way of implementing the absolute space concept — whereas the same concept can be applied to each of the subsidiary levels which are nested within that most general framework. Every level of nature's hierarchy can be described as "absolute" in some definite, local, level-specific sense ... and not merely in a singular, universe-wide, level-irrelevant sense.

Put in the simplest of terms, one can say that every entity has an absolute relationship with the next adjacent level, such that a lower level entity can be seen as moving in relation to the stasis of a higher level background. The next higher level can, in turn, be seen as a moving lower level entity in relation to the stasis of the still higher level background, and so on — resulting in a hierarchically-tiered, interlocking, and alternating relationship between stasis and motion. For example, in the case of a ball rolling on the surface of the earth, the former represents the lower level of absolute motion in relation to the latter as representing the next higher level of absolute space; and the earth, in turn, would be seen as the lower level which is undergoing absolute motion in relation to the still next higher level of the solar system as absolute space. In this nesting scheme, the choice is relative as to which level is selected in any given case — but the result is then absolute.

In these terms, then, there are all kinds of referents for the absolute space concept — indeed a unique referent for each and every (e.g., local) level of nature's hierarchy, from the universe as a whole on downward. Yet, the same basic concept of absolute space would apply to each of these nested levels, from the largest to the smallest. The only difference is one of size or scope — but functionally, they are all identical as absolute space frameworks and referents: absolute stasis applies to each of these referents, in relation to which there could be absolute motion. Put all together, this means that we can continue to speak about absolute space and absolute motion — in the relative, multi-leveled, hierarchical terms that we have just described.

The upshot is that there is a natural and coherent stasis-motion relationship between phenomena of nature. This also means that wherever one goes in the universe, whatever level of nature's hierarchy one is focused upon, there is a clear order of things. And while it will not be the same hierarchical order of things in some other situation (e.g., elsewhere in the universe), it is nonetheless absolute ... in a relative sense.

One can also approach the matter we are discussing by saying that there is no single universe-wide standard of "inertia", which supplies the underpinnings everywhere, for what physics calls an inertial frame of reference. Rather, we assume that every nested level of the hierarchy is stable or "inertial" unto itself, based upon its own local absolute space center of the universe, and level of the hierarchy — each of which is equally valid, in its own realm[11] ... with Newton's first law (and the classical Relativity Principle) being applicable in every case. This is also why the inertialness test remains a consistent and reliable criterion: there is a correspondence between the internal and external stasis-motion states of a system — because there is a specific and absolute background representative of stasis, which thereby serves as an absolute referent for motion in every situation.

Law as Eunuch. We have been discussing various problems with the Relativity Principle, in terms of its different assumptions and forms. Now we come to the last, or fourth, formulation that we described earlier — Einstein's thesis that the only interpretation which is ultimately viable is one concerning "laws of nature". Of course, our focus is on the Special or Restricted version of this principle, as we noted earlier, and

11 Note the general similarity to Bridgman's 1936 point of view: "There is no escaping the fact that it is *I* who have the experiences that I am trying to coordinate into a physical theory, and that *I* must be the ultimate center of any account which I can give. I and my doings must be specially set apart and, in perhaps the only possible sense of the word, constitute an absolute." (Bridgman, quoted in Miller, 1983, p. xlii)

not the General one — meaning again that we are concerned with inertial frames and uniform motion, as opposed to non-inertial frames and non-uniform motion.

We have already talked about the climate in which Special Relativity arose, at the turn of the twentieth century. On the one hand there was Newton's classical mechanics, and on the other hand there was Maxwell's electromagnetism. The problem was that there seemed to be contradictions between these theoretical frameworks — and the Special Theory's main goal was to find a way of resolving this conflict, by merging these two distinct sets of laws, in the form of a new type of transformation ... thereby transcending and superseding the sort of mechanical vs. electromagnetic domain limitations which seemed inherent in Newton and Maxwell's theories. The new unity which Einstein proposed became, in turn, the sole component of a new "law of nature" category ... which Einstein devised.

Let us first address the main problematic issue which seemed to differentiate the Newtonian and Maxwellian realms, concerning whether there is a way of differentiating stasis from uniform motion. According to Newton's first law, these two states are indistinguishable: the game of table tennis goes on undisturbed if the container frame in which the game takes place is either static or moving uniformly — and the (windowless) occupants of the container cannot tell which state obtains. This lends support to the other interpretations we have been discussing — namely, that there is no such thing as absolute stasis or absolute motion.

As we have noted, Maxwell's model appeared to contradict these ideas: light waves were understood to exist as traveling disturbances upon a static and universal ether medium — which medium was tantamount to absolute space. A moving object would therefore be undergoing absolute motion in relation to this fixed medium system — and thus a methodology seemed to exist for differentiating absolute stasis from absolute motion ... even though no such distinction seemed to emerge from the previously mentioned Michelson-Morley experiment.

Our position is that there is no actual discrepancy between the two systems; rather, the problem lies with the inflated significance which is attached to Newton's first law that we just mentioned. While the point is indisputable that the (windowless) occupants of an inertial frame container system are unable to determine whether that system is static or moving uniformly in relation to the larger environment, this should not lead us to the conclusion that the two stasis-motion states cannot be distinguished ... if not by internal occupants, then by external ones.[12] As we indicated in the previous subsection, we still subscribe to the old absolutes of space and motion — but as they been updated in our relativistic (i.e., hierarchical) form, in a manner that would apply to both classical mechanics and electromagnetism ... Michelson-Morley notwithstanding, as we shall see later.

At the same time, we contend that the two realms of Newton and Maxwell should not be melded together. We take that position on consistent grounds which we have

12 The basic problem can also be understood in terms of a kind of existence vs. knowledge question, such as we discussed in relation to Quantum Theory in Chapter 2, and will pursue regarding the issue at hand in Chapters 6 and 7.

already encountered, and will do so again at various times in the book: complementary aspects of nature must retain their separate identities, even though there are characteristics which are common to them both. This tenet applies to two different sets of characteristics that we are primarily dealing with in this discussion — two separate dimensions of analysis in the current model, stasis-motion and matter-medium. In the first case, the fact that stasis and motion states cannot be differentiated from one another in certain circumstances does not eliminate the distinction between them. In the second case, the matter and medium domains are likewise not merely reducible into one common phenomenon, despite their similarities. Simply because there are shared aspects which the two partners of each pair have in common does not mean there are not also other aspects which make each partner different from one another.

Thus in speaking about Newton's mechanics, we are largely concerned with common objects (and the matter domain) — whereas in speaking about Maxwell's electromagnetism, we are largely concerned with waves (and the medium domain). For example, there are physical differences in the way that speed factors would be determined for each case: In the matter domain case, motion of the source must be factored in for determining the object's fundamental speed in relation to the local "absolute space" background — whereas in the medium domain case, motion of the source makes no difference in determining the wave's fundamental speed in relation to that background. We will explain the reason for the distinction in Chapter 6.

Indeed, we make the same point about the two types of transformations themselves: we approach them in complementary terms, and we argue that they do not serve the same function. The Galilean transformation is not merely an inferior approximation which just happens to be satisfactory at low speeds, which is what Einstein stipulated — whereas the new Lorentz transformation utilized by Special Relativity is the true "law of nature". On the contrary, we maintain that the later transformation does not negate the earlier one — and that the notion of simple additivity still applies. Therefore, both transformations are valid principles which properly describe different aspects of nature, and they should not be blended together — or one eliminated in favor of the other.

It is the same general problem that we keep encountering in this work — namely, an attempt to collapse a duality of nature into a unitarity, which was one of Einstein's special trademarks. In each case, we are dealing with two properly distinct categories of nature, whose complementary characters have been obscured by their joining — and the result distorts nature, rather than elucidates it. Indeed, every instance of this hyper-reductionist pattern also represents another example of the "no symmetry" problem which we keep coming back to.

Once again, of course, we have been focusing our attention on what is called the Special (or Restricted) Relativity Principle, and its world of inertial frames. But the same objections arise on a broader scale when we examine the General Relativity Principle, and its unrestricted world of both inertial frames and non-inertial frames. Just as the former theory "specially collapses" dual aspects of nature into a singular one, so the latter theory "generally collapses" in the same way. Like the notion of a "wavicle" from Quantum Theory as combining particle and wave, any such mergings will only lead us astray in our attempts to understand nature.

The Woes of the Inertial Frame

We continue our attempt to untangle the web of confusion here, by looking at some of the serious problems with the inertial frame concept itself. In this section, we will examine four aspects of the issue. First, we shall argue that the construct is unsatisfactory because it overemphasizes one side of a two-sided pattern in nature. Second, we shall explain why the concept is based on a confusing mixture of two different categories of phenomena. Third, we shall see that the overall frame construct is often misapplied, by a failure to differentiate figure from ground functions. Finally, we will consider the bizarre conceptual gyrations which physics is forced into, because of a lopsided vision of the construct of inertial motion.

One-Sided Frames. The first problem is that the construct of the inertial frame of reference is inherently distorted in its conception and formulation. As we shall see, the culprit is the usual one — the lack of a complementary mind-set or approach, in a world where that pattern is pivotal.

To begin with, we have noted that physics treats the inertial frame as a privileged or preferred state — which is distinct from those non-qualifying and less desirable states that are merely categorized as being a negation or deviation thereof ... i.e., a *non*-inertial frame of reference. In this respect, the second side is treated almost as a throwaway, in contrast to the primary side: the two are not accorded comparable status. The mere negation of inertialness hardly qualifies as a complementary equivalent in its own right — it is little more than a mock duality, still clothed in unitary garb. So the first step in our reanalysis here is to recognize the duality, and see what kind of equivalent constructs are entailed.

In formulating the answer, let us back up from the frame construct and consider the essence of the issue, starting with the notion of *inertia* itself. How do we describe the properties of an object that is either completely static or moving uniformly, unaffected by outside influences? It would seem that one of the main characteristics of the inertia concept is the notion of *in-oriented*: in abstract terms we might say that the object is following the beat of its own internal drummer — and thereby maintaining its own internal momentum, in the form of some consistent stasis-motion state. But, then, in the vision of complementarity that we are espousing, there would have to be a full-fledged counterpart for such a notion or property of inertia: we propose the novel — and very short-term! — concept of *outertia*, as representing that which is *out-oriented* and subject to external influences.

Bringing these ideas back to the frame of reference issue, one solution would be to adjust the pair of names so that they are commensurate with one other, in these terms: we could talk about *in*ertial frames of reference, as opposed to *out*ertial frames of reference.[13] In this way, neither type of frame or region would be considered as privileged or

13 Instead of "outertia" and "outerial frames", one could say "exertia" and "exertial frames"; but the point is moot, since we adopt neither set.

preferred, since neither is ideal and the other commonplace: instead, both are pervasive throughout nature, completely equivalent and complementary counterparts with one another.

But this very naming scheme brings us straight to the next problem, concerning a blurring of two different sets of qualities.

Mixed Dimensions. In the present model, we have chosen not to incorporate the classical "frame of reference" construct — even in the improved, complementary in-oriented vs. out-oriented form just described. For there is also the other factor that we touched upon earlier in our debunking of the force-free concept — that inertial frames correspond with no-forces, and non-inertial frames correspond with yes-forces. We argued that the correct analysis is not no-force vs. yes-force, but rather balanced force vs. unbalanced force.

So the equivalence issue that we discussed in the previous subsection is only one of the problems with the frame of reference pair: the other main problem is that the frame of reference concept itself blurs together two distinct sets of characteristics, which must be differentiated from one another. Because of this confusion — along with the layers of unwelcome associations which the conventional constructs carry with them — we conclude that it is better to simply abandon the inertial and non-inertial frame terminology ... and incorporate their valid properties in a more coherent approach.

In the present model, the two sets of characteristics we have been describing represent two different dimensions of analysis — that of the *inside-outside dimension* and that of the *stable-changing dimension*. We contend that the latter is the primary factor which differentiates the frame of reference pair — and so we say that the inertial frame can be primarily characterized with labels like stable, balanced, and uniform ... whereas the non-inertial frame would be primarily described with labels like unstable, unbalanced, and non-uniform.

To be sure, we still use the shortened "frame" label as a convenient plain language tag for describing the inside-outside dimension for matter domain phenomena. These patterns will become clearer in the pages ahead.

Object vs. Function. There is still another problem that we need to address, which will come into play later in the book when we discuss certain Relativity scenarios — and this does not concern the different types of frames of reference, but rather misusage of the underlying construct itself.

We treat the frame of reference construct as referring to the ground or background of any situation, relative to which there must exist a figure or focus (individual or collective). Yet any particular *object* that we care to identify, of whatever scope, can *function* in either figure or ground mode ... depending on the circumstances. For that reason, it is a mistake to refer to some object as a "frame of reference", when it may or may not function in that manner in the particular situation being examined.

Here is why. Let us say that we call something a frame of reference, in relation to its contents. But then what happens when we have more than one of these "frames", which are interacting with each another? A fundamental premise of the present model is that there can only be one background phenomenon or frame in any situation — which therefore means that what was a frame in relation to its contents will itself become contents in relation to a larger frame. That interpretation is inherent in the hierarchy premise:

Lower Level	*Higher Level*
parts	whole
contents	frame

The crucial difference between figure and ground is lost by a naming convention that obscures the distinction between object and function. So we should never identify an object with a label which freezes that entity into one function, when it can just as easily take on the other. This confusing usage is another reason why we are abandoning the conventional frame of reference concept.

Fictitious Forces. Ad hoc and one-sided definitions are unlikely to yield coherent patterns of nature; instead they inevitably lead to further complications, if not absurdities. We now encounter another example of this problem.

Newton said, in his first law, that straight-line motion was the natural state of affairs, in the realm of inertial frames. By definition, then, movement which departed from such straight-line motion required an external force to produce the "unnatural" non-linear motion. Of course, as we can see in the heavens all around us, such "unnatural" orbital motion is hardly atypical. So how are we to explain such widespread circular motion patterns, given that they diverge from the linear?

We described the conventional answer at the beginning of the chapter: the inertial (linear) motion assumption leads to another conceptual twist, whereby the circular category in question is labeled "centripetal acceleration" — which, in turn, requires a special centripetal or "center-seeking force" to complete the reasoning. As Hecht puts it, "We need a force to account for the deviation from straight-line motion, and there it is — even the logic is circular." (Hecht, p. 120) Or Rogers: "[H]ow can a thing accelerate towards the center of a circle and yet neither move faster nor get nearer the center? This is still puzzling, but it now seems to be more a matter of wording than physics." (Rogers, p. 301)

The upshot is that since circular motion is not the natural state of affairs, it requires a special "centripetal force" to explain it. One might intuitively think that there would also be a counterpart "centrifugal force" to go with it — and yet such a parallel construct is rejected as being both unnecessary and not strictly correct. That is, according to the conventional reasoning, there is clearly a centripetal force — this being what holds a circling body in its orbit. But the notion of a centrifugal force is unsatisfactory because if the inward or centripetal force were to suddenly disappear in the situation, the circling body would not then make an abrupt perpendicular shift to a strictly outward path — as the centrifugal force concept would supposedly entail — but would instead veer off in a straight line tangent from its previous circular path. Moreover, since such straight-line motion is natural, according to Newton's first law, it would not require a force to explain it ... because such linear motion is "just naturally the way things are"!

Of course, it is all very well to invoke Newton's first law — but what about the essence of Newton's third law, which we interpret in complementary terms? If the centripetal force were suddenly eliminated in the situation, then so would be the equal and opposite centrifugal force — and thus it is not surprising that the previously orbiting body would simply move with its momentum directly forward.

Be that as it may, physics' bold conclusion is that the centripetal force is *real* — whereas a parallel construct like the centrifugal force is *fictitious* (e.g., Trefil, p. 138). The result is a particularly bizarre anomaly of the conventional model: the idea that there is a centripetal force, which describes the tendency of an object to move in towards the center of an orbit, and that this is a real force — whereas the countervailing centrifugal force, which describes the tendency of an object to move outward away from the center, is only a fictitious force. So much for a sense of symmetry in our models of nature, which is the point that we keep making.

We maintain that an inward-seeking centripetal force of attraction is no more (or less) real than is an outward-seeking centrifugal force of repulsion, in a meaningful and complementary model of nature. And while such "centripetal" and "centrifugal" terms may or may not be useful overall constructs in the kind of coherent model that we are striving for, it is hard to see how the underlying pair of tendencies would not go together. Therefore, to the extent that either one is a useful concept alone, then both must be invoked on an equivalent basis together. It is not the centrifugal force which is fictitious here, but physics' one-sided approach to nature.

In the present model, all of the terms we have been discussing can be used interchangeably, for they can all be understood to represent slightly different conceptual approaches to the same two force principles — an in-oriented *X*-Force of Structure, together with an equally-valid out-oriented *P*-Force of Process. A complementary chart might look like this:

X-Force of Structure	P-Force of Process
in-oriented force	out-oriented force
attraction	repulsion
centripetal force	centrifugal force

The Dual Nature of Stasis and Motion

In this chapter, we have mostly talked about Newton's Laws and the Relativity Principle, and we have critiqued the conditions and conclusions which comprise the resulting model of stasis and motion, in both its classical and modern forms. Definitions may change, disappear, or be added — but many of the underlying concepts are still arbitrary in terms of their fundamental assumptions, and do not make sense in relation to one another. The theory we are left with is convoluted, incoherent — and certainly unsymmetrical. It does not provide a satisfactory framework for dealing with the issues at hand, let alone fit with other crucial constructs about nature. It is our task in the present section to lay out some of the main features of the 2FT approach to stasis and motion, which we will develop further in the rest of the book.

In particular, we want to consider how the conventional stasis-motion issues that we have been discussing can be understood in the complementary and hierarchical terms with which we are working. Right off the top, of course, we must again highlight stasis and motion themselves as representing complementarity incarnate — even though that obvious relationship is virtually never identified or made explicit in conventional treatments. Yet the inherent inverse relationship between stasis and motion is as compelling a statement of complementarity as any could be. Therefore, we celebrate the universal acceptance of the basic thesis, concerning the relativity of stasis and motion — only, we talk about it in terms of the complementarity of stasis and motion.

Of course, stasis-motion represents only one of the complementary and nesting dimensions of analysis in the present model, and our task here is to see how the underlying issues can be organized and construed in terms of three of the other dimensions (leaving aside the matter-medium dimension, inasmuch as we only discuss matter domain phenomena in the rest of this chapter). Our main focus is on the inside-outside dimension, where we shall see that its two different basic kinds of topologies also entail two correspondingly different states of stasis and motion. Aside from that, we shall examine stasis vs. motion in terms of static and dynamic relationships — i.e., the dominant-recessive components of each. Finally, we shall look briefly at the stable-changing dimension, to see how some of these stasis-motion issues would be differentiated there.

Inside vs. Outside Realms. We have proposed that there is a relative difference of topology or orientation as between the two realms — and that in one situation or another the inside realm might be variously characterized with terms like 2-dimensional, flat, and horizontal — relative to the outside realm as 3-dimensional, curved, and vertical. We now want to see how this general pattern can be applied to the main topic here of stasis and motion. We begin with a rather startling, albeit obvious, difference between the basic "natural" motion states for each of the two realms. At the end of the subsection, we shall consider their stasis counterparts.

The issue could hardly be more fundamental. Indeed, based on the topics we discussed earlier in the chapter, it should be apparent that the conclusions one draws as to the character of natural motion are far more than a mere reflection of one's aesthetics: in many respects, the answer can provide the conceptual underpinnings for an entire theoretical analysis of nature. That is the case with Newton's laws, and in terms of the whole of classical physics which results … even continuing to the present day.

Physics says that uniform motion is linear in character: this is inscribed in Newton's first law, and it is not a matter of dispute. Yet interestingly enough, it was Aristotle and Galileo who first identified the uniform motion aspect of nature, which eventually led to the law of inertia. Galileo described this motion as running along a "Great Circle track" (Toulmin, 1961, p. 54). Of course, the usual position is that such a conclusion was flawed. Thus, "Galileo approached the idea of *inertia* — the idea that an object will continue in motion unless a force acts on it to stop it. He still did not have the idea entirely clear, however, for he seems to have thought in terms of motion around the earth, and his concept was that in the absence of a force, a body will continue in motion around the earth at a constant height." (Clotfelter, p. 83-84) Again, referring to Galileo's inertia breakthrough, "Despite this brilliant insight, he got lost…. [and] in the end he wrongly concluded that this perpetual motion applied only to circular systems." (Hecht, p. 72)[14]

It was Descartes who came along and *linearized* this conception of uniform motion (Hecht, p. 72) — a position which, of course, was later adopted by Newton: "The essential feature of Newton's laws of motion was the realization that in the real world, unrestrained motion more closely approximated a straight line than a circle." (Trefil, p. 135)[15]

Yet recall the idealized scenario which is associated with these conditions, as to where "unrestrained motion" would apply. We start with empty space, where there are no other forces or influences, and we add in a solitary object that is moving — supposedly uniformly, and in a straight line ... as well as indefinitely, since there is nothing in empty space to change this pure state of inertial motion.

Of course, we have already seen that such a thesis is not realistic, when we consider what such uniform linear motion would actually mean. If the object is really moving in a linear path in open space, then its distance from different external influences (of differing strengths) *must* be changing: its distance from some external influences is increasing, while its distance from other external influences is decreasing. Regardless of how strong or weak are these other forces, there is no logical way to avoid them. We know this, precisely, from the inverse-square law: different distances result in different force strengths. Or as Davies puts it, "All the forces we know to exist share the property that they diminish with distance." (Davies, 1980a, p. 55) And just as changing distances result in changing forces, so likewise, changing forces must mean changing motion states — in terms of slowing down, speeding up, and changes of direction. Obviously, then, the concept of uniform linear motion in open space seems patently absurd, in the context of physics' own force laws.

Needless to say, we are not abandoning the notion of uniform motion — much less the idea that it could exist in open space. But what conditions would be required for this phenomenon? Clearly, the clue to stable motion is stable force — which, by inference from the inverse-square law, means stable distance. And the only kind of motion which fits that no-change-of-distance requirement in the open or outside realm is circular motion — *around* some object-source of external force, not *towards* or *away* from it. Therefore, the kind of motion which is stable in open space is precisely that of uniform circular motion.[16]

To reiterate the two positions: Physics claims that at least in principle there is nothing to change an entity's state of motion when the entity is in an idealized force-free region of the universe; we have explained that for 2FT there are no force-free regions of the universe, either in principle or in practice — there are only force-stable, as opposed to force-changing, regions. Physics says that if there are no external forces acting upon an entity, then it will continue in its constant state of motion; 2FT says that external forces are always present, but that if the external forces acting upon an entity are stable,

14 Sachs argues that there is controversy about whether, in fact, Galileo opted for the circular approach — although the consensus is that he did. See Sachs, 1988, p. 10. In any event, Aristotle supposedly did so, as is apparent in the following footnote.

15 Copernicus was also of the old school. "The natural motion of the heavenly bodies was for him, as for Aristotle, universal circular motion. By contrast, Newton employs throughout his work a quite different paradigm of natural motion: it is, to be sure, still motion-at-uniform-speed, but it takes place in straight lines rather than in circles." (Toulmin, 1961, p. 103)

16 The main purpose of this discussion is the contrast between linear and circular motion — leaving aside the elliptical and not strictly uniform character of orbital motion in general.

then it will continue in its constant state of motion. Physics sees uniform motion as only occurring in principle, never in practice — there being no zones of the universe which are totally free of such external influences; 2FT sees uniform motion as occurring in principle and in practice whenever the external influences impinging upon an entity are stable — stable influences produce stable motion. Physics says "no force: constant motion" (Rogers, p. 108); 2FT says "constant force: constant motion."

Clearly, we have arrived at one of the most simple and dramatic differences of interpretation between the present model and that of contemporary physics. And we argue that the source of the discrepancy, as usual, is another one-sided representation of reality. Here, the conceptual trap revolves around the issue of which is the natural state of motion — circular or linear. Galileo said *circular*; Newton said *linear*. We say *both*. Why should there be only one type of natural motion? Nature is a duality, and there are two kinds of natural motion: both need to be taken into account.

Yet the dissent asks how this could be! Our practical experience here on earth seems to argue for Newton's approach: an object which is propelled into motion seems to continue in a straight line, unless something intervenes to alter that course; and, moreover, if we could correct for the inevitable intrusion of factors like friction, the object would presumably continue in its straight-line motion at a uniform speed, ad infinitum. How could anyone argue with what seems so self-evident?

Consider the perspective which leads to such a "self-evident" conclusion. We and the moving object are both localized on a flat earth terrain, and the *uniform linear motion* conclusion is perfectly valid in that context. Yet what if we were to back off and adopt a very different perspective? What if we were to look at the moving object from the vantage point of outer space — then what would we see? We would also see an object moving upon the surface of the earth — but the terrain would no longer be flat: it would be curved, in the same way that the earth itself can be described. We would therefore be observing *uniform circular motion*. Consequently, the correct answer about the object's dynamic state in relation to the earth is that there are two kinds of "natural" or uniform motion states — precisely because there are two kinds of topologies, and not just one ... inside flat and outside curved, with a different trajectory for each.[17]

There is no contesting the fact that circular motion can have a uniform or constant speed: For example, "The earth is moving at a fantastic speed on its trip around the sun, but this speed is constant." (Zukav, p. 123) Or, again, an orbiting object's "speed in circular flight is constant." (Hecht, p. 126) Yet while technically the "speed" may be constant, the overall "velocity" is not. As we noted in the beginning of the chapter, this is because "speed" does not take account of direction, whereas "velocity" does. By orthodox definition, then, circular motion represents a change of direction — and therefore a change of velocity ... and so it could never meet the uniform velocity test. That is why practical and obvious examples like those we just cited would be rejected by physics as not being revealing of "natural" motion, in the Newtonian sense. Circular motion would never be

17 Jammer cites "the Aristotelian differentiation between terrestrial motions in straight lines and celestial motions in circles ..." (Jammer, 1957, p. 80).

seen as uniform motion, in the way an inertial frame is defined ... as corresponding with linear motion.

Indeed, the simple complementary interpretation that we are putting forward would represent utter heresy (not to mention, folly), and as being completely irrelevant to the conventional model. Yet consider some of the peculiar consequences which stem from such perverse reasoning, in terms of the kind of convoluted and illogical arguments and concepts which that unbalanced vision leads to. Thus, at the very end of the last section, we talked about how circular or orbital motion would normally be defined as accelerated motion — centripetal acceleration, to be precise — and how it would be explained in terms of some specific center-seeking force which would apply there. In the case of the particular center-seeking force that might be associated with gravity, we shall return to that issue in the next chapter — but for the moment we can say that there is never any way of avoiding gravitational influences: they are always present, to one degree or another, no matter which practical examples one cares to cite. Even Einstein acknowledged this point, as we noted earlier in the chapter. Therefore, such "center-seeking" influences are pervasive throughout the universe.

Furthermore, the underlying reasoning is itself circular: the notion of a centripetal force was only invoked in the first place in order to explain why motion which is "naturally" linear is, in fact, circular. But if we don't assume that the motion is naturally linear to begin with, then we have no need to posit a special mechanism to explain its circular counterpart. The different treatment of centripetal vs. centrifugal forces itself illustrates the sort of theoretical acrobatics which one must go through as a way of circumventing one-sided (i.e., no symmetry) constructs.

Here is another way to think about the problem. Is the earth flat or round? We maintain that there is no singular answer to this question, in spite of the great strides we have made from the flat-earth vision of our ancestors. Yes, the earth is a sphere, and so its surface is curved — and this represents a broad, external *P*-view. But the other perspective is equally valid: it says that the earth is flat, and so is its surface — and this represents the narrow, internal *X*-view. Not only would either perspective be just as viable as the other, but a comprehensive vision clearly entails the two of them; one will not suffice.

Consequently, we argue that the idea of uniform linear motion represents only a limited perspective — that of the narrow, local, *X*-view within ... the inside topology. And again, the notion of uniform linear motion is perfectly consistent with the view that the earth is flat. That is nothing to be intellectually ashamed of, so long as we recognize the complementary perspective which should go with it — that of the broad, distal, *P*-view without and uniform circular motion ... corresponding with the outside topology. Both views and both aspects must be taken into account: an object that is moving linearly on the flat earth surface is also moving circularly on the curved earth surface.

Newton's first law is based on a lopsided vision of natural motion. It only deals with the kind of natural motion which corresponds with the flat inside topology; it ignores the equally-valid complementary kind of natural motion which corresponds with the curved outside topology. So while the first law can be construed as describing a complementary relationship between uniform stasis and uniform motion, that law should be applied to two different, complementary topologies — one for the flat inner frame, and another for the curved outer frame.

Hecht notes that although the uniform linear motion thesis "cannot be directly proved true in this universe, we can 'understand' a tremendous range of observed phenomena [with it], and that is its ultimate justification." (Hecht, p. 73). We argue that far more phenomena can be encompassed by gathering up all those examples which are explained by the uniform linear motion thesis, and putting them together as representing one side of nature; and then by contrast, gathering up all those examples of the uniform circular motion thesis, and putting them together as representing the other side of nature. The resulting pair, which joins both perspectives, provides its own "ultimate justification" — both in principle and in practice.

Before finishing with this topic, let us mention a particular device which uniquely illustrates the two kinds of motion we are talking about — in contrasting relation to one another. We are referring to what is called the Foucault Pendulum. Described very simply, there are two components. One source of motion is that of a large pendulum — a long, heavy ball or bob that hangs from a rope or chain that is attached to the ceiling, and which has been set into what is essentially a state of back-and-forth uniform linear motion.[18] On the floor below is a thin layer of sand, which the bottom tip of the pendulum touches and through which it creates a linear track. The other source of motion is that of the earth itself, which, of course, rotates in the form of uniform circular motion. The result of these two separate stable paths is that, over time, the straight line which is drawn underneath the swinging pendulum will be continually changing its orientation — it "slowly changes its apparent direction of swing." (Rogers, p. 231) — in accord with the circular motion of the earth.

One conventional explanation for the effect is to say that the pendulum is swinging relative to all matter in the universe. By contrast, we would suggest that the pendulum is merely swinging in relation to itself, such motion being self-defining — just as the earth is rotating in relation to itself. That is, both these motion states can also be described in terms of simple harmonic motion (SHM), each of which delineates its own stable system.[19]

Most of this subsection has been concerned with the two kinds of uniform motion which characterize the inside and outside realms — and now let us mention their stasis counterparts. Given that there are two different kinds of motion which exist in the two different kinds of topology, so there must also be two correspondingly different kinds of stasis which go with them. Still, accepting such a premise does not come easily: just as our usual inclination would be to think of uniform motion only in terms of one (linear) of the two kinds of states, even more so would we normally think of stasis in one and only one way — the simple and obvious kind of stasis that we see all around us.

We suggest a broader perspective upon such stasis phenomena. To start with, we assume that the same conventional interpretation for understanding the relationship between stasis and motion would apply in the case of the stable outer frame as would

18 Actually such motion can be thought of as an arc or ellipse, but that detail is irrelevant here.

19 Rogers discusses and illustrates the relationship between what is essentially (uniform) linear and circular motion, in his treatment of simple harmonic motion. Regarding the latter, if one views any circular motion with a perspective that suppresses the depth component thereof, then the result is also a pendulum-like simple harmonic motion. (Rogers, pp. 174-175).

apply for the stable inner frame: one entity is static, in relation to the other entity as undergoing uniform motion. The main difference between the two realms is the character or topology of the respective phenomena. And in the same way as we have argued that the conventional approach to uniform motion is lopsided by virtue of only dealing with what we call the flat inside topology, the same goes for stasis. Just as there is a parallel type of natural motion in the curved outside topology, so there must be a parallel type of natural stasis there. We maintain that the most obvious candidate for meeting these requirements is what we commonly call "spin". Therefore, we are talking about spin playing the role of stasis in the outside realm.[20] (Note that when we speak of spin here, we are specifically talking about the rotating center, as opposed to the periphery — be that the surface of the same object, or other separate objects.)

Here is how the pattern might be represented, in complementary terms of the way some object is moving in relation to the earth:

Inside Realm 2-dimensional flat		*Outside Realm* 3-dimensional curved	
Stasis	*Motion*	*Stasis*	*Motion*
flat stasis	uniform linear motion	curved stasis (e.g., spin)	uniform circular motion (e.g., orbit)

By the way, as far as the spin-as-stasis role we are talking about, one can argue that the spinning entity is not going anywhere in terms of changing position — not by virtue of its rotation — and in that sense it is not moving ... even though it is turning.[21] In terms of heavenly phenomena, one might think of spin as functioning like a stake in the vacuum!

Static vs. Dynamic Relationships. Thus far, we have been discussing stasis and motion in terms of the different topologies which characterize the inside-outside dimension, particularly in terms of two complementary kinds of uniform stasis and motion. But now we want to switch gears, or dimensions of analysis, as we consider the issue of stasis and motion in terms of their components.

To be sure, the wording gets tricky here — given the nesting nature of this model, and the fact (discussed in Chapter 1) that the same terms can be used at more than one level and so in more than one way. For example, we use the static relationship label to describe the situation where there is relative stasis between two entities — whereas we use the dynamic relationship label to describe the situation where there is relative motion between two entities. And in either case, we use the generic terms of dominant and recessive to describe the two entities in each relationship.

20 The pattern is illustrated by the spinning gyroscope, which establishes its own spatial orientation. As Post says, "The gyroscope is one of the most sensitive devices that responds even to the slightest deviation of its frame from an inertial frame. This direction preserving property of the gyroscope is widely used in navigation." (Post, p. 476) Such an attribute is also called "gyroscopic inertia", whereby a freely rotating body will maintain a fixed direction in space.

21 It is also true that there are different ways of parceling out stasis and motion here — and spin itself can be construed as motion, in relation to a static background.

But now we want to shift downward one level in our analysis, where other meanings come into play. Thus, we can talk about the static entity itself as having nested components or aspects, which we also label as a static relationship; and likewise we can talk about the moving entity itself as having nested components or aspects, which we also label as a dynamic relationship. And again in either case, we label the nesting components or aspects in terms of dominant and recessive entities.

With the first set of labels, it is the (stasis vs. motion) character of the relationship between the two entities which determines whether they are called a static or dynamic relationship; whereas in the second set, it is the nesting components of the stasis entity itself which represent a static relationship, while it is the nesting components of the dynamic entity itself which represent a dynamic relationship. Obviously, the former usage is more conventional and the latter not. Yet it is in terms of the latter (or lower nested dimension of analysis) that we now want to set the conceptual stage for some very novel interpretations — and while these will be discussed more fully later in the book, we can take some big leaps here.

We would put forth the proposition that when two entities interact, they do so by means of either a static relationship or a dynamic relationship. The static relationship case represents a direct interaction between the two entities, whereas the dynamic relationship case represents an indirect interaction between the two entities in the form of some phenomenon which moves between them. This means that we are now associating the quality of *stasis with direct contact*, whereas we are now associating the quality of *motion with indirect contact* — and we will use the distinctive term "intraaction" to describe the former case, and the standard "interaction" term to describe the latter. One way that we sometimes talk about the dominant and recessive entities which comprise a static relationship would be in term of sink (e.g., end point) and source (e.g., start point), respectively; while one way that we sometimes talk about the dominant and recessive aspects of a dynamic relationship would be in term of something that is moving forward or backward between the static relationship pair.[22]

Here are some nesting complementary charts to represent the analysis:

Static Relationship	Dynamic Relationship
intraaction	interaction
direct & static form of contact	indirect & dynamic form of contact
absolute space	absolute motion

Dominant	Recessive	Dominant	Recessive
sink	source	forward-moving	backward-moving
puller	pusher	attraction	repulsion

22 Some phenomena display a kind of self-complementing pattern — such as that of the abovementioned pendulum, which oscillates back and forth between the two static outer points of its cycle. Such simple harmonic motion (SHM) is apparent when essentially equal-force source and sink entities alternate with one another (like the seesaw). But of course it is not only the static relationship partners which swap mirror roles back and forth — their dynamic relationship counterparts also alternate, e.g. between forward and backward motion.

Stable vs. Changing Domains. We have talked about two types of uniform motion — with the linear form applying to the inside realm, and the circular form applying to the outside realm. But either type of uniform motion would describe the stable region, whereas a non-uniform counterpart would describe the changing region. Our task now is to see how such motion patterns might be depicted in complementary and nesting terms which are consistent with the analysis above.

Let us start with the inside realm, where we can talk about forward and backward motion in the broader framework of leftward vs. rightward component tendencies. For the stable region, these leftward vs. rightward tendencies would be essentially balanced. The result would be uniform motion that does not veer off into either of those opposing leftward vs. rightward directions, but instead is balanced between them by going in a straight line — uniform linear motion — either forward or backward. On the other hand, for the changing region these two countervailing (leftward vs. rightward) tendencies would be unbalanced — with the result being motion that veers off from its straight-line path and heads into either a leftward or rightward direction. Rather than stable and uniform linear motion, there will be a change of direction — either leftward or rightward.

Now consider the corresponding patterns for the outside realm, which we can talk about in terms of inward (centripetal) vs. outward (centrifugal) component tendencies. For the stable region, these inward vs. outward tendencies would be essentially balanced. The result would be uniform motion that does not deviate into either of these opposing inward vs. outward directions, but instead is balanced between them by going around in a circle — uniform circular motion — either clockwise or counterclockwise. On the other hand, for the changing region these two countervailing (inward vs. outward) tendencies would be unbalanced — with the result being motion that diverts from its circular path and heads into either an inward (centripetal) or outward (centrifugal) direction. Rather than stable and uniform circular motion, there will be a change of direction — either inward or outward.

On the next page are a pair of nested complementary charts for summarizing these motion patterns.

Changing Region

Inside Realm		*Outside Realm*	
2-dimensional flat		3-dimensional curved	
leftward forces ≠ rightward forces		inward forces ≠ outward forces	
non-uniform motion		non-uniform motion	
Dominant	*Recessive*	*Dominant*	*Recessive*
leftward	rightward	inward	outward

Stable Region

Inside Realm		*Outside Realm*	
2-dimensional flat		3-dimensional curved	
leftward forces = rightward forces		inward forces = outward forces	
uniform linear motion		uniform circular motion	
Dominant	*Recessive*	*Dominant*	*Recessive*
forward	backward	clockwise	counterclockwise

Now for some exceptions and extensions to these patterns. For one thing, the above directional analysis is not the only way of differentiating stable and changing regions. Rather than stable vs. changing *direction*, we could also talk about stable vs. changing *speed*.

Beyond that, of course, we are only dealing with stable vs. changing motion patterns — the dynamic relationship which characterizes some entity that is moving ... between the two partners of a static relationship. But the static relationship itself may change — for example, in terms of the dominant and recessive components thereof (e.g., source and sink) moving closer together or further apart, and thereby exercising a correspondingly stronger or weaker force upon one another ... as well as influencing the dynamic relationship which moves between them. Such a source-sink shift closer together or further apart can (at least sometimes) be understood as a shrinking or expanding of the space or medium, whose boundaries are delineated by those static relationship partners. We will pick up on this issue in Chapter 8, where we discuss a variety of important constructs and phenomena that can be understood in terms of this changing region category.

So much for a discussion of general stasis and motion patterns, as they are represented in the template of our new complementary and nesting framework. We will develop these ideas further in the pages ahead — starting in the next chapter, as we turn to the formidable topic called gravity.

Chapter 5

The Dubious Concept of Gravity

In our discussion thus far about stasis and motion, we have been trying to develop a conceptual framework that is useful for understanding a wide array of natural phenomena, yet which does not require special constructs that are unique to those phenomena. Such ad hoc explanations and nomenclature are the trademark of established science, with the resulting theory suffering in terms of meaningfulness, coherence, and breadth. There is no better example of this problem than that of the conventional concept of gravity.

In fact, one of the purposes of this chapter is to question the usefulness of the gravity construct as it is conventionally understood, given what we shall find is a surfeit of unnecessary and confusing conceptual baggage, along with the usual lopsided picture of nature at its core. We maintain that the gravity construct is defective in various ways — and that the underlying phenomenon would be better understood in broader, complementary terms.

We shall begin by looking at the two established models of gravity, those of Newton and Einstein, and consider some of the features which differentiate them. After that, we turn to the famous Equivalence Principle of General Relativity, which attempts to join together different sets of constructs and phenomena related to gravity. With this presentation of the established concepts behind us, we then begin to systematically analyze the issues in terms of the present model's themes of complementarity and hierarchy, with most of our attention going to the former. All of our dimensions of analysis will come into play, as tools to better understand a variety of dualities which underlie the phenomena associated with gravity.

The Conventional Approach

In this section, we first examine the two main models of gravity themselves, and then look at a subsidiary precept of Einstein's called the Equivalence Principle.

Newton vs. Einstein. There are two main theories of gravity which are prominent in physics today. The classical one is Newton's very practical law of universal gravitation, and the modern one is based upon Einstein's very abstract Theory of General Relativity. Although both frameworks share certain assumptions, the two models are also quite distinct, in that they are each based on very different premises and visions of nature.

As we discussed in the previous chapter, Newton's approach is built upon the notion of absolute space. Regions of this universal space which are not subjected to outside influences are called "inertial frames of reference", and it is here where Newton's laws of mechanics apply. By contrast, those regions of nature which are subjected to outside influences are called "non-inertial frames of reference", and Newton's laws of mechanics do not apply there. In their place is another principle, called Newton's law of universal gravitation. This thesis was based upon the proposition that all matter in the universe directly attracts all other matter in the universe, by means of a pervasive force of gravity, with a strength that is solely dependent upon a)the quantity of mass of each of the two objects in question; and b)the distance which separates those two objects. The famous formula which encapsulates these factors and represents the force of gravity is $F_g=G \times mM/r^2$.

The picture of gravity underwent major changes when Einstein came along in the early part of the twentieth century and presented his Theory of General Relativity. Of course, this was the successor to his Special Theory, which latter (like Newton's mechanics) was only intended for inertial frame regions. Indeed, it was partly Einstein's recognition of the inertial frame limitations of Special Relativity which led him to General Relativity. But the new approach did not just incorporate non-inertial frames: it actually eliminated the conceptual distinction between the two frame-types altogether.

Einstein's model completely eschews notions like absolute space and direct interaction, and presumes instead that gravity is strictly a field-based phenomenon: gravitational influences in General Relativity are spread indirectly and over time, at the speed of light, conveyed through the mechanism of an intervening *gravitational field*, or *space-time continuum*. Therefore, it is not a direct force which one mass exerts upon another that is the basis for gravity, but rather it is the topology of the field which intervenes between such masses — by virtue of the way each mass modifies the character (or "curvature") of the surrounding region ... and only indirectly by means of that field to another body. The focus in General Relativity has shifted from the objects themselves which interact, to that of the intervening space-time continuum between them.

Here is one way that is frequently used for envisioning the Einstein model. Think of a heavy rubber sheet which has been mounted and stretched out upon a frame, such that it is tight and well-supported all the way around. Now imagine various size balls or rocks at different places upon the rubber sheet, which have the effect of distorting the smooth surface in their proximity. Needless to say, the sheet itself is supposed to represent space-time, and the objects upon it are meant to correspond with heavenly bodies. To see how gravitational influences would be conveyed in this system, one can simply roll a lighter ball upon the rubber sheet. The rolling ball would tend to fall in towards the heavy objects, which have created valleys upon the sheet, thereby being tantamount to attraction.

Take the classical image of Newton watching an apple fall from a tree to the earth — and then concluding that the cause was a direct force of gravity which operated between apple and earth. According to the Relativity model, the apple does not fall toward the earth because of some mysterious and direct "action-at-a-distance" force of attraction between apple and earth, but simply because space-time is sufficiently "curved" in the area surrounding the earth ... such that the apple merely falls or flows quite naturally upon the contours of this space-time region, toward the earth. The resulting natural motion of apple in the direction of earth would mimic an attractive function between the two objects, but there would be no need to invoke any specific force of gravity to explain this behavior. The apple's motion is an implicit part of the topology of the space-time tableau itself.

Indeed, the Relativity position is that the Newtonian approach to gravity borders on being magical: the mechanism for that theory's "action-at-a-distance" direct force of universal attraction has never been satisfactorily explained, including the fact that these influences are supposed to be conveyed instantaneously.[1]

The space-time continuum can be classified as either "flat" or "curved": flat space-time can be understood to mean that there is virtually no gravitational force involved; whereas curved space-time can be understood to mean that a force of gravity is present.[2] Of course, the degree of curvature would correspond with the size of the mass that produces (or is associated with) the curvature: the heavier the mass, the greater the distortion or curvature (or "warp", as it is sometimes called) of what would otherwise be a flat space-time.

Actually, the above imagery and descriptions — including the rubber sheet illustration — do not even conform to the Relativity model, in that they are themselves based upon outmoded conventional constructs. That is because discrete objects do not actually exist in Einstein's scheme: The object itself is merely a concentration, or "energy knot", of the space-time continuum — rather than being a distinct reality in its own right. Thus, it is not for matter "to react to the curvature of space-time", as Geroch puts it: rather, matter *is* the curvature of space-time. Einstein himself described matter as "pulselike concentrations of fields, which stick together stably, and would act almost like

1 Jammer disputes this orthodoxy, and instead argues that Newton was very careful, and quite agnostic, on the mechanism by which the universal force of gravitational attraction would be conveyed. Jammer quotes several passages from Newton to buttress this interpretation, including the following from 1713, after the basic law was proposed: "I here use the word *attraction* for any endeavor whatever, made by bodies to approach each other, whether that endeavor arise from the action of the bodies themselves, as tending to each other or agitating each other by spirits emitted; or whether it arises from the action of the ether or of the air, or of any medium whatever, whether corporeal or incorporeal, in any manner impelling bodies placed therein towards each other." (Newton, *Principia* (second edition), Scholium following Proposition LXIX, Book I; *Principles*, p. 192; quoted in Jammer, 1957, p. 137)

2 In broader terms, while there was an earlier assumption that the space-time continuum of the universe is generally curved, a more recent interpretation is that the universe as a whole is flat. In contrast to this, there would be smaller-scale localized curvatures in the topology, presumably corresponding with the existence of matter there.

small moving bodies." (Einstein, quoted in Murchie, p. 493) So the description of individual objects existing upon the space-time continuum, and thereby causing the curvature, must be considered only in a figurative sense — in deference to our conventional modes of thought.[3]

Another distinction is that whereas Newton's theory deals with interactions between masses, Einstein's approach meant that all phenomena would be impacted — including things without mass, like electromagnetic waves:[4] anything that exists in the affected field would be influenced. Thus, for example, light that is traveling in outer space can be "bent" sideways by some massive object,[5] in a phenomenon that is called "gravitational lensing".[6]

Much of the complexity of General Relativity is contained in its guiding formulas, called field equations. This set of mathematical relationships connects "the curvature of space to the distribution of matter in that space.... [and] can be used to determine the structure of the universe as a whole." (Capra, 1975, p. 163). Geroch describes it in terms of the curvature of space-time geometry equaling the mass density of matter in space-time (Geroch, p. 166), based upon a factor of G. The G here would be understood as a conversion ratio — "curvature of space-time" and "mass density of matter" — and it also corresponds with the gravitational constant in Newton's formula.[7]

What makes Einstein's model of gravity so appealing to many people, in relation to Newton's model, is not so much that it yields more exact descriptions or predictions about certain phenomena. Rather, the General Theory's overall value is said to derive from its ability to provide a more sophisticated, elegant, profound — and ultimately satisfactory — vision of nature.

One way that these qualities are said to manifest themselves is in what might be called the conceptual economy of the model, because of the way various constructs are incorporated or eliminated. To start with, of course, space and time are themselves merged, as providing the substratum of the field or continuum: Instead of a 3-dimensional space, along with a 1-dimensional time, we now have simply a combined 4-dimen-

3 There are variations of interpretation here: "Space-time is curved in the neighborhood of material masses, but it is not clear whether the presence of matter gives rise to the curvature of space-time or whether this curvature is responsible for the existence of matter." (Whitrow, in Capek, 1988, p. 97) It is a chicken-and-egg dispute.

4 What about the particle counterpart of electromagnetic waves, photons? The usual interpretation is that photons have no mass, but this has not yet been definitively determined, experimentally — plus, there are definitional issues regarding the concept of mass which would come into play, as well.

5 Note that as early as 1704, Newton had also put forth the notion that gravity could cause the bending of light.

6 Such bending of light is one of various effects in this category. Another is called "micro-lensing", which can happen under certain conditions when the gravitational object is situated directly between the source of the light and the observer — in which case the light waves can actually be enhanced or magnified by the intervening gravitational object. Then, the gravitational object acts like an amplifier, in a similar way as an in-line booster can augment the power of a signal traveling through a wire.

7 This function is analogous to the role of c, in the mass-energy relationship. We discuss that topic in Chapter 7.

sional space-time. Furthermore, gravity is eliminated as a distinct force, by virtue of being naturally integrated as an inherent characteristic of the continuum itself.[8] Plus, not only is...

> There is no such thing as "matter" — matter is a curvature of the space-time continuum. There is not even such a thing as "energy" — energy equals mass and mass is space-time curvature.... There is nothing but space-time and motion and they, in effect, are the same thing. Here is an exquisite presentation ... (Zukav, p. 179)

Nor, in fact, do we need motion, either:

> In classical mechanics, motion was viewed as a displacement of the unchanging particles (with a constant mass and volume) in static absolute space. In relativity it is viewed in a Cliffordian fashion as a disappearance of a local curvature in one region and its reappearance in a neighboring region; this creates for our senses an illusion of a displacement of a passive material in an equally inert space. (Capek, 1988, p. 95)

But it doesn't stop there. Thanks to an ancillary premise called the Equivalence Principle that we shall discuss next, not only is the distinction between inertial and non-inertial frames of reference eliminated, but so is the distinction between acceleration and gravity ... as well as that between inertial mass and gravitational mass. All of this amounted to a grand and daring set of theoretical leaps (if not coups), which catapulted the model forward, conceptually.

On the other hand, there was a real epistemological cost for this massive reductionism — in that the theory is very difficult to understand, and many of its details are not readily comprehended, even to most physicists. As they say to people trying to make sense of Einstein's model, "Don't worry too much about visualizing a four-dimensional world. Physicists can't do it either.... it is not even possible to picture it" (Zukav, pp. 150 and 170). Indeed, "We have no direct sensory experience of the four-dimensional space-time, nor of the other relativistic concepts." (Capra, 1975, p. 156) We can only observe the *shadow* or "projections of four-dimensional phenomena, just as shadows are projections of three-dimensional objects." (Capra, 1975, p. 158)

By contrast, the Newtonian model is exceedingly simple, which also explains why it continues to be so useful: It is far easier to work with and understand, and it is perfectly compatible with most aspects of our proximal world. The upshot is that much as contemporary physics touts the grandeur and beauty of Einstein's Theory of General Relativity, Newton's model continues to supply the nuts and bolts for most pragmatic analyses in this realm of nature. Furthermore, while Einstein's elimination of a specific force of gravity has been lauded by many physicists as a step towards a unified model of nature, neither Einstein nor anyone else has ever come close to attaining that goal ... a point we will return to in Chapter 12.

8 "[G]ravitation in general relativity has not the character of a force. It is a property of space-time.... a non-Euclidean space-time of variable curvature without gravitational forces." (Jammer, 1957, p. 260)

The Equivalence Principle. Let us now consider some of the alluring scenarios which Relativity used to demonstrate how several important classical distinctions in physics have no validity, under the umbrella of what is called the Equivalence Principle.

One of these thought experiments is intended to demonstrate that the sensation of resting on the floor of an elevator, which is accelerating upwards, would be experienced by an internally-restricted observer in a manner which is no different from the case of common gravity that we all experience on the earth's surface. In another rendition, an internally-restricted (and presumably deaf) observer of an upward-accelerating rocket ship would likewise be unable to differentiate that sensation from one of common gravity, which a person at rest on the earth's surface would feel.

A bizarre extension of the elevator example would place the elevator chamber deep in empty space, far away from any obvious sources of gravity, but being pulled along ("upward") at a steadily increasing rate by a giant cable coming from somewhere. Again, the reasoning would be that the occupants would not be able to differentiate their accelerating situation from the usual gravitational one, whereby they are at rest on the surface of the earth: in both cases, they would be pressed against the surface of either the elevator's floor or the earth's surface.

By virtue of such illustrations, Einstein declared that acceleration is essentially the same as gravity. The result is that, "There is absolutely no way of distinguishing between uniform accelerated motion and a constant gravitational field." (Zukav, p. 166) Or more simply, "[G]ravity and acceleration are equivalent" (Zukav, p. 168).

Next consider a very different conceptual tack which the Equivalence Principle takes for dealing with the issues at hand. In this case, we start with the elevator existing in an actual gravitational field, yet suspended in place by its cable — except that in this scenario, the elevator would be thousands (or millions?) of floors up from ground zero, in an incredibly tall building (e.g., Zukav, pp. 162-166). The salient question is, what happens to the various occupants or components of the elevator if the cable is cut?

Obviously, the elevator would fall in the gravitational field. Moreover, it has long been an accepted fact that the quantity of stuff, or mass, of which an object is comprised bears no relationship at all on the rate at which that object will fall in a gravitational field. Galileo is credited with having discovered this principle, which legend tells us was deduced in experiments that he conducted at the Tower of Pisa in Italy. What Galileo found was that all objects would fall at the same rate (i.e., correcting for atmospheric resistance) irrespective of how heavy or light they were.

Therefore, both the elevator and everything inside it will all fall at the same speed, regardless of their differing masses. The result which seems peculiar is that because they are all falling at the same rate, the various objects in the elevator would all maintain a constant spatial relationship with one another — i.e., all the component objects would remain mutually static: everything would appear just to be floating, even though they were actually falling. As a consequence, the occupants within the elevator, who have no window to look out of, would have no reason to think they were falling in a gravitational field.[9] Rather, they would conclude that they were in a force-free realm of the universe, where Newton's laws of motion would be perfectly valid — as would be confirmed if they conducted inertialness tests within the elevator. As a result, the occupants could rightly claim that their elevator represented a valid inertial frame of reference!

Of course, such a conclusion would be a mere illusion, according to the "rightful" perspective of an external observer who had a larger picture of what was really going on. This wise, broad-based external observer would conclude precisely the opposite as that of the more narrow internal observers, whom we just described: we are clearly dealing with an elevator that is falling in a gravitational field — hardly what one would normally think of as an inertial frame! The contradiction is clear: according to one (internal) perspective the frame is inertial, but according to the other (external) perspective the frame is not inertial.

Einstein's dramatic conclusion was that such incompatible scenarios make the underlying conceptual distinctions untenable. This is the premise of the Equivalence Principle: at least over limited distances, an inertial frame of reference is the same as a frame of reference which is falling in a gravitational field — i.e., a non-inertial frame of reference. Therefore, the only proper resolution of this paradox is to conclude that there is no substantive difference between inertial and non-inertial frames — and so the distinction between them should be eliminated altogether. "Anything like an 'inertial coordinate system' that can be 'wiped out' (Einstein's words) by the assumption of a gravitational field hardly deserves to be called absolute." (Zukav, p. 164). The upshot is that two fundamental distinctions about nature have been eliminated by the Equivalence Principle: that between acceleration and gravity, on the one hand; and that between inertial frames and non-inertial frames — or inertia and gravity, generally — on the other.

Now let us turn to one final equivalence which Einstein puts forth here, and this regards the related concepts of inertial and gravitational mass. We mentioned the former concept early in the last chapter, where we described it as the quantitative measure of an object's resistance to acceleration. Another way of measuring an object's inertial mass is to set it on a balance scale, as offset by another object of known mass on the other side. By contrast, gravitational mass — also called "weight" — describes the pull downwards on an object by some gravitational force (e.g., on a common spring-loaded scale). Gravitational mass therefore "measures the amount of stuff to be pulled by gravity, and also the amount of stuff for pulling gravitationally" (Rogers, p. 119).

The notion that there are two fundamental types or aspects of mass was a long-established principle of classical physics, going back several centuries (Zukav, p. 167). But the curious thing was that they were equal to one another — that an object's inertial mass was measured to be the same as its gravitational mass:

> Inertial mass and gravitational mass are equal. This explains why a feather and a cannonball fall with equal velocity in a vacuum. The cannonball has hundreds of times more gravitational mass than the feather (it weighs more) but it also has hundreds of times more resistance to motion than the

9 Presumably we are not to consider the sensory experience of falling which would be felt by the occupants.

> feather (its inertial mass). Its attraction to the earth is hundreds of times stronger than the feather, but then so is its inclination not to move. The result is that it accelerates downward at the same rate as the feather, although it seems that it should fall much faster. (Zukav, p. 167)[10]

However, it was a quandary as to why this relationship existed — until Einstein came along, that is, with his Equivalence Principle, which was said to finally offer an explanation for it ... or, at least, provide a basis for the non-difference. A presumption behind the Equivalence Principle was that if inertial frames and non-inertial frames were really different, as classical and Special Relativity physics maintained, then such a difference should manifest itself — for example, in terms of differing amounts of (inertial vs. gravitational) mass in each realm. This is not the case, of course, and so the Equivalence Principle offered its own solution to the longstanding puzzle: inertial mass and gravitational mass do not differ from one another because the two concepts represent the same phenomenon — a single, unitary sort of mass ... rather than the dual type.[11] Put another way, inertial mass is the same as gravitational mass — because (as we just discussed) the respective "acceleration" and "gravitational" forces, by which such quantities are determined, are themselves fundamentally indistinguishable from one another. Therefore, it was the same quantity of stuff which interacted with the external force, regardless of whether that external force was labeled "acceleration" or "gravitation".

This completes our presentation of the two main models of gravity. We will present a very different perspective on these issues in the remainder of the chapter.

10 Other descriptions explain the common rate of fall by saying that the strength of the gravitational force varies, depending upon the mass of the object: "[T]he force of gravity on a body depends on its mass" (Hecht, p. 123); "This is a remarkable property of gravitation, that Earth-pulls are proportional to the *inertial* masses of matter pulled.... [T]he Earth's attraction varies as the mass of the victim" (Rogers, pp. 276 and 318). Therefore, objects with heavier mass are pulled by the force of gravity proportionately more strongly than are lighter objects. In all cases the force of gravity varies accordingly, so that the end result just happens to come out right (Whew!) — such that all different mass objects end up falling at the same rate.

11 Note that Dingle offers an interesting historical perspective which contrasts with the conventional approach, regarding Einstein's Equivalence Principle discovery and the relationship between inertial mass and gravitational mass. Dingle notes that Newton identified the relationship between mass, distance, and a gravitational force. "[I]n the usual presentation of Newton's theory by those (including Einstein himself) who purport to show its inferiority to Einstein's, this discovery of Newton's is totally misrepresented as a defect. It is said that Newton's theory includes two kinds of mass — *inertial* and *gravitational* — which are mysteriously identical — a fact which the theory is at fault in leaving unexplained. But Newton's theory does *not* contain two kinds of mass: he defines only one, the so-called inertial mass — 'it is this quantity that I mean hereafter everywhere under the name of body or mass', he says, and his magnificent discovery that this quantity plays a major part in determining the actual accelerations of bodies observed in nature is totally misrepresented by the assertion that gravitation requires a second mass which happens, in a magical way, to be identical with the mass that measures inertia." (Dingle, 1972, pp. 196-197)

A Complementary Approach

Our position is that there are serious problems with both Newton's and Einstein's models, along with the pivotal construct of gravity which explicitly or implicitly underlies both. Neither approach is consistent with other forces or models about the world, and neither theory is based upon a coherent overall vision of nature.

The Equivalence Principle itself is a mixed bag. Its main downside is in the typical overly-reductionist way that Einstein attempted to sidestep the meaningful complementary patterns of nature, and reduce them to meaningless unitarities instead. On the other hand, its main benefit was in the realization that the basic patterns underlying gravitational phenomena also applied to non-gravitational phenomena, as well — a point that is fully consistent with the present model, as we shall see below.

We also maintain that there is also a certain amount of logical trickery here, in some of the Equivalence Principle scenarios — which is reminiscent of another such instance that we encountered in the previous chapter, where we talked about the case of the gradually starting train alongside a static train at the station: Without sufficient external cues, it is difficult for an internal observer to determine which train is starting to move and which remains static. There we were talking about inertial frames, whereas here we are talking about both inertial and non-inertial frames. But just as the anomalous aspects of the train scenario did not provide sufficient justification for obviating the overall difference between stasis and motion, so the confusion here should not obscure the fundamental complementary patterns which a coherent conceptual framework would provide.

Regarding the basic question of the modality by which gravity operates, we must ask how much of an advance it was to move from one locus of explanation (Newton's objects) to another (Einstein's terrain)? In Newton's model, the two objects interact directly — whereas in Einstein's model, the first object curves or warps the space-time which envelopes it (or is itself produced by a curvature or warp!), and thereby indirectly influences the second object. Yet why is an indirect mechanism inherently better or more straightforward than a direct mechanism? Why is the substitution of one transmission mechanism for the other necessarily an advance? Because the notion of a force has been subsumed by the notion of a field or continuum? Have we really resolved anything by speaking about the motion of one object as "naturally" flowing within or upon the contours of the terrain, which have been shaped by another object?[12] Why is that necessarily an improvement over simply saying that the second object's motion is specifically caused by the first object (or by an interaction between the two objects)? Because we are using seemingly-sophisticated terminology like that of a "flat" or "curved" "space-time continuum"? We will return to this issue later.

12 "If certain scientists substituted humps and hills for forces, then they have just substituted one superstition for another." (Stace, p. 172). Note the similarity between this General Relativity maneuver, and that of Special Relativity's substituting of an electromagnetic field for the medium, as we shall see in the next chapter.

Our discussion ahead will be mostly oriented upon a number of crucial dualities which should be applied to any interpretation of gravitational phenomena, and which dovetail perfectly with the present model's dimensions of analysis — all of which will come into play. That process begins with the first two topics, in which we set forth the framework for understanding gravity in terms of the inside-outside dimension and the stable-changing dimension. Then we challenge the basic assumption that gravity represents a force of attraction alone — and argue that it must be understood just as much in terms of repulsion, as well. Next we consider how a model of gravity should incorporate the paired phenomena of spin and orbit, whereas those phenomena play no particular role in existing theory. After that, we look at how gravity can be understood both in structural and process terms, with the former corresponding to surface contact relationships and the latter to dynamic ones — especially the conventional construct called "free fall".

Following this we turn to the Equivalence Principle's conclusion about the relationship between gravitation and acceleration. We will then discuss the other two related dualities which Einstein denied — that which differentiates the essence of the inertial and non-inertial frame constructs, along with that between inertial and gravitational masses. Next, although most of this chapter deals with matter domain rather than medium domain phenomena, we will briefly consider the role of gravitational waves, as they complement the more usual category of gravitational objects. Finally, we consider how Newton's concept of gravity can be understood as being equally viable in its essence as is that of Einstein. At the end of the chapter, we will also touch upon the nesting, or hierarchical, aspect of nature — which must be taken into account in any understanding of the gravity construct.

Inside vs. Outside Realms. We need to set the stage for our discussion at the outset, by delineating where the phenomena that are described by gravity do and do not apply. In this regard, let us reprise a nested complementary chart that we presented in the previous chapter (bearing in mind the definitional points about spin and orbit which we made there, and will reiterate below):

Inside Realm		Outside Realm	
2-dimensional flat		3-dimensional curved	
Stasis	Motion	Stasis	Motion
flat stasis	uniform linear motion	curved stasis (e.g., spin)	uniform circular motion (e.g., orbit)

It should be clear that in this model gravity is treated as an outside realm phenomenon, a point we will address further below.

Stable vs. Changing Gravity. The next complementary pattern which we want to identify is that gravity can be understood in either stable or changing terms. This contradicts the conventional assumption, which is that gravity is only a change phenomenon.

Not only is such a complementary interpretation inevitable in the present model, but it is readily apparent in the universe. Thus, any gravitational system where the components maintain an essentially constant relationship with one another is an example of stable gravity (as in the chart above) — in contrast to the changing gravity case where the interacting partners do not have such a relationship. Simply speaking, an object circling the earth represents stable gravity, whereas an object moving closer to or further from the earth represents changing gravity. It is the issue we spoke about in the previous chapter: constant distance means uniform (or stable) force, whereas changing distance means varying (or changing) force.

The spin-orbit pattern that we mentioned above is therefore a stable phenomenon, where the larger object is spinning relative to the smaller object as orbiting. By contrast, if the smaller object is moving closer to or further from the larger object, then we categorize the situation as a changing region phenomenon. Bearing in mind that we are dealing with outside realm issues in both cases, the stable-changing pattern might be charted in this way:

Stable Region		Changing Region	
uniform		non-uniform	
Stasis	*Motion*	*Stasis*	*Motion*
spin	orbit	spin	fall-rise

Gravitation as Attraction vs. Repulsion. Now we want to engage in a more extensive discussion, regarding another basic attribute: Gravity is almost always thought of as being *only* a force of attraction. Thus, the earth is said to exert an attractive pull on the apple — and the apple, in turn, is said to exert an equal-and-opposite attractive pull on the earth.

Once again, we reject the conventional interpretation. In this subsection, we will try to explain why the one-sided attraction-only thesis is unacceptable, on the grounds that there is nothing inherent in the phenomenon we call gravity that leads to such a conclusion: it is simply a convention. The pattern associated with gravity only appears as uni-directional because physics has chosen to define it that way, and we have all gotten trained and accustomed to this way of construing the phenomenon.

To start with, we must ask in principle why such an anomaly would exist: The other fundamental forces of physics have both attractive and repulsive aspects, so why should gravity be an exception? Not only does such a one-sided assumption run counter to the two-sided pattern of physics' other fundamental forces, but it is also completely antithetical to the whole two-sided basis of the present work.

We have said that the *X*-Force of Structure and the *P*-Force of Process always go together, in some form or other as they manifest themselves in any given situation. Attraction and repulsion epitomize each of the two forces, respectively — and so, likewise, they always go together. In principle, there will never be a situation where there is only attraction or only repulsion: there must always be a complementary relationship between these two basic aspects of nature.[13]

One way of understanding the point is in terms of simple motion tendencies — that functionally speaking, *attraction or movement towards* one place (e.g., the sink) effectively amounts to *repulsion or movement from* another place (e.g., the source) — and in these terms, one entity's attraction is another entity's repulsion. Another way that the dual pattern can be described is to say that in every case of gravitational interaction, the larger partner would be seen as attracting ... in relation to the smaller partner as repelling.

Thus, we maintain that there is no compelling reason to describe the earth-apple interaction in terms of mutual attraction — even though, once again, such a description of gravity is deeply ingrained in our thinking.[14] One could just as well say that while the earth is attracting the apple, the apple is resisting and repelling the earth's pull: we would then have a force of repulsion from the apple to offset and complement the force of attraction coming from the earth — a pattern, by the way, which is totally consistent with the essence of Newton's third law about equal and opposite actions and reactions.

Yet how do we answer the obvious retort: Earth and apple do stay together, and therefore can be properly described in terms of attraction — not repulsion. Wouldn't that indisputable reality justify the attraction-only conclusion, regardless of any linguistic quibbling about the component tendencies? So (the reasoning would continue) while it may be very entertaining to bend the issues here, in conformance with the "complementarity agenda" — according to which gravity is attraction for the one side (e.g., the stronger entity) doing the pulling, in relation to which the other side (e.g., the weaker entity) is pushing or repelling back — nonetheless the end result of the force of gravity is clear: attraction reigns, and this is what binds together earth and apple.

Consequently, it is argued that there is a good reason as to why gravity is thought of in terms of attraction and binding, and not also in terms of repulsion and unbinding. After all, that's how the gravity construct came about in the first place: to explain why things come together — particularly in relation to massive heavenly bodies, such as why the apple falls to the earth. Not only earth-apple, but earth-moon, and sun-earth — all such relationships are said to be bound together by the universal force of gravitational attraction.

The problem is that this counter-argument is a one-sided distortion in its own right. Thus, why should we specifically ascribe gravity to those overall situations where the force of attraction is predominant over the force of repulsion, and not the reverse? Why should our theoretical framework be built upon an explanatory principle of an overall binding force of attraction, as opposed to an explanatory principle of an overall unbinding force of repulsion? Things do not always come (or stay) together in the heavens; they also go (or stay) apart.[15]

13 Properly speaking, we should be talking about attraction vs. repulsion for inside realm phenomena, and contraction vs. expansion for outside realm phenomena — a distinction that we noted early in the first chapter. However, for ease of discussion, we often violate that distinction, and talk generically about attraction vs. repulsion for either the inside or outside realm. This simply makes it easier to dovetail with conventional usage.

14 Some physicists recognize the convoluted nature of the conventional approach: The earth is "pulling down on you; you, in turn, must be pulling up on the Earth, however peculiar that may seem." (Hecht, p. 99) Note that some observers do describe the pattern in more balanced terms. Thus, gravity is sometimes explained as, "The resultant effect of the gravitational attraction and centrifugal repulsion between any body of matter and a planetary body ..." (Pitt, p. 172) Murchie says that, "gravitation commonly amounts to a repulsion as well as an attraction." (Murchie, p. 575)

This is a hierarchical model, and the dual attraction-repulsion pattern manifests itself in different ways. For one thing, the overall tendency, which we cited in the first chapter, towards larger things in one (e.g., upward) direction of the hierarchy and smaller things in the other (e.g., downward) direction of the hierarchy might be understood in terms of an overall attraction and repulsion, respectively. Then there is the notion that two complementary partners at a lower level join together at the next higher level — as in thesis-antithesis → synthesis. Thus at the lower level, the larger object might be seen as attracting the smaller object as repelling; but they would join together at the next higher level in an overall attraction — which is just what gravity is supposedly describing.

But in this model, the pattern doesn't stop at any one level, such as the one we just referenced. For the combined entity that supposedly represents overall attraction at that higher level will itself interact with some other entity at that level — in which case the combined entity may take on a repulsion function in relation to an attraction function of a larger counterpart at that level.

This is not merely abstraction. Thus, we can say that the earth is attracting, relative to the moon as repelling — and of course, the combination can be seen in terms of an overall attraction. But that same earth-moon combination would then be seen as repelling, in relation to the larger sun which is attracting. And so on: the complementary attraction-repulsion pattern is persistent. Yet the conventional paradigm only recognizes the attraction aspect, while ignoring the repulsion aspect.

So why should the attraction outcome be treated as more significant than the repulsion outcome — exclusively so, such that it is consecrated as a fundamental force of nature? The implicit assumption seems to be that the baseline state of affairs is that things do not come or stay together, and therefore a special mechanism is required to explain the departure from that norm — to wit, gravity. In this way, unbinding would be the normal state of affairs, which does not require a force of nature to explain ... whereas binding would be the exception to the rule, and thus require a special force of nature to account for.

In some respects, it is the same unbalanced pattern that we saw in the previous chapter, where the centripetal force of attraction is taken as real, but a countervailing centrifugal force of repulsion is not. Likewise with gravity: it is deemed to be a force of attraction only, and not of repulsion. The logical imbalance is obvious, and obviously unsatisfactory ... and we have painted ourselves into a conceptual corner.

Let us now turn to the much larger overall issue — not concerning particular entities, but the universe as a whole. For if gravity did, indeed, exist by itself as a force of attraction alone, then why hasn't the universe all come together long ago into one large

15 Consider, for example, those situations where an attractive force of gravity is depicted as being the source for a momentum-building propelling effect — in the opposite direction. E.g., a man-made satellite is directed in the vicinity of a large planetary mass, but not close enough that the two objects bind together: the result is that the satellite loops around behind the large body, and then flies off in essentially the reverse direction, but with increased momentum — being functionally repelled away from the large mass! Even so, such a catapulting process is still usually characterized in terms of the standard attraction-only model of gravity.

mass? Evidently something is holding things back, some countering force which keeps everything from collapsing into the one-sided gravitational embrace![16]

Obviously, one viable candidate for countering the force of gravitational attraction is the expanding universe itself. In these terms, the force of gravity can be thought of as a universal force of contraction which draws the universe together — as opposed to the universal force of expansion which spreads the universe apart ... which latter can be understood as an extension of the Big Bang itself:

> The universe is thus the product of a competition between the explosive vigour of the big bang, and the force of gravity which tries to pull the pieces back together again. In recent years, astrophysicists have come to realize just how delicately this competition has been balanced.... [T]he observed structure of the universe seems to depend very sensitively on the precise matching of explosive vigour to gravitating power. (Davies, 1983, p. 179)

Of course, one might applaud this nod to complementarity — even though it is only implicit here, whereas any such explicit characterization would surely be resisted. And yet if the expanding universe represents the tendency that offsets the attractive force of gravity, then why is the former not accorded an equivalent role in physics' models of nature? Thus, why is the universal force of gravity treated as a fundamental force of nature, but the universal "force" of expansion is not? Needless to say, it would be a peculiar kind of complementarity when one partner would be treated as a fundamental force of nature which applied both universally and locally — whereas the other partner is neither fundamental, nor pervasive ... but is merely the remnant of a one-time prior event that applies uniquely to the whole universe alone.

Ad hoc, anyone? Something clearly underlies the process of expansion — which should represent a fundamental force, in physics terms, since it is an action that is caused by some phenomenon, and which cannot be explained in terms of the other fundamental forces ... irrespective of whether that expansion process is considered to be a one-time Big Bang event, or not. Moreover, adopting a general "fundamental force of expansion" explanation, as applying to all levels of nature, would represent a consistent complementary analysis — which could be applied both to the universe as a whole, as well as locally for particular interactions. Then we would no longer be dealing with an attraction-only phenomenon, but a complementary attraction-repulsion pair — however they are labeled: either a pair called gravitational attraction/contraction and gravitational repulsion/expansion; or of gravitational attraction/contraction and non-gravitational repulsion/expansion.

Two caveats should be noted, in regards to the expansion issue. One concerns the implicit assumption that we are only dealing with a one-time universe, and not a universe which could alternately expand and contract over time, in an oscillating manner: such a recurring universe would inevitably entail complementary forces or functions of attraction/contraction and repulsion/expansion that are precisely like those which we

16 What Rogers describes concerning repulsions between atoms should equally well be applied to large-scale matter: "The repulsions must be there or, with attractions alone, matter would not just hold together strongly: it would collapse inward." (Rogers, p. 145)

are proposing here. And if such complementary patterns would be understood to apply at the level of the universe as a whole in such ongoing terms, then it would seem hard to justify their not also applying at all other (lower) levels. We shall discuss the alternating universe interpretation in the final chapter.

The other caveat is physics' own vacillating love affair with a premise that Einstein put forward many years ago. One year after his General Relativity theory was presented, Einstein added a "cosmological constant" correction factor to the original field equations — whose "physical meaning is that of a force, completely unknown to science, which would counteract the effect of gravity between the galaxies — a kind of cosmic anti-gravity machine." (Calder, 1979, p. 148) As Davies describes it, "The new term concerned has the physical effect of a universal force of *repulsion* at all points in space. It was by balancing this cosmic repulsion with the gravitational attraction of matter that Einstein was able to construct a static model universe" (Davies, 1977, p. 169) — as a way to maintain the "quasi-static distribution of matter" (Calder, 1979, p. 148), "to eliminate universal expansion" (Kaufmann, p. 107).

Later, this static universe view was supplanted by newer evidence, put forward (especially by Hubble) some years after, that the universe was actually *expanding*. Einstein's "great antigravity force of repulsion" (Murchie, p. 215) was thus seen to be unnecessary, and so it was abandoned. The very notion of such a cosmological constant force of repulsion was later described by Einstein as "the biggest blunder of my life" (Einstein, quoted in Murchie, p. 215) which he "bitterly regretted" ever having introduced (Calder, 1979, p. 170). Not surprisingly, even today, "pure-minded relativists ... detest the Cosmological Constant, and the unknown force it represents." (Calder, 1979, p. 152)

Or do they? For, which was the big blunder — the original idea or its later abandonment? Modern speculations speak of a "supergravity," in which the "cosmological constant reappears.... corresponding with a novel force in the universe." (Calder, 1979, p. 170). Other recent theoretical approaches have also reintroduced the essence of such a cosmological constant, or some other universal force of repulsion — including the notion of a "negative gravity", which only comes into play over enormously large distances; so the idea has never gone away.[17] More recently, such a negative gravity phenomenon has been described in terms of the mysterious sounding, amorphously meaning, and typically misnamed construct called "dark energy". In any event, regardless of how it is labeled, we are dealing here with an example of how mainstream science sometimes lurches from one arbitrary theoretical vision of nature to another — because their models lack a coherent conceptual framework, a common and meaningful theme.

We need to make two final points before ending this subsection. First, it is not acceptable to maintain that the issue we have been discussing is fairly trivial, in the sense of representing another kind of semantic technicality, such as the "linguistic quibbling" we mentioned earlier in the subsection. According to such a more casual approach, one should be free to understand gravity either in a one-sided (i.e., attraction)

17 Note that a different notion of physics, also called "anti-gravity", would *not* perform this repulsion role — in that it relates strictly to the other-worldly realm of "antimatter", where it plays the same attraction-only role as it does in the real world which we all experience. We discussed the notion of antimatter in Chapter 3.

or two-sided (i.e., attraction-repulsion) sense — and that any argument, such as we have been engaged in, amounts to a trivial sideshow. We reject such a dismissive attitude, and maintain to the contrary that the issue is crucial.

For one thing, pivotal constructs like gravity form the tools of our minds: as we shall discuss in Chapter 11, such definitions largely determine how we think about phenomena — and so how they are defined makes a difference. Related thereto is the fact that we are searching for coherence and unity, and thus we must be concerned not merely with how we understand gravity, but with a larger model of nature. In this regard, there is clearly no conceptual advantage in singling out gravity, as representing a non-complementary phenomenon — when it can just as readily be understood as complementary. On the contrary, there is the significant disadvantage that it inhibits the development of a unified vision of nature, by treating this particular fundamental force as one-sided, while others are treated as two-sided. Defining gravity in a manner which makes it intrinsically different from other phenomena means that we must fence it off and treat it as something unique. That, of course, is the way gravity has been dealt with by physics, and one can contend that this is partly why the force of gravity does not fit with other forces in their theories. We must find a coherent underlying vision, if these apparently disparate phenomena are to coalesce into a unified framework.

The second main point is that it might be argued that Einstein's General Theory avoids the one-sided issues we have been discussing by eschewing a force of gravity, as such — be they attractive or repelling characteristics. General Relativity's space-time topology approach might be understood as incorporating all such directional factors, thereby eliminating concepts like attraction and repulsion altogether. Since such influences are conveyed indirectly through the intervening field, then only their net impact would matter, thus bypassing any description of entities as attracting or repelling one another. In response, we would point out that the rubber sheet model is intended to demonstrate how an object's warping of the surrounding region amounts to attraction, with no parallel property corresponding to repulsion. And in any event, we shall critique the notion in a later subsection that an indirect approach (a la Einstein) is necessarily superior to a direct approach (a la Newton) — and in the latter case, complementary tendencies like attraction and repulsion must come to the fore. As to the former framework, one might argue that such a "net sum" approach might be considered as another mechanism for collapsing an attraction-repulsion duality of nature into a net unitarity!

In closing this subsection, our conclusion is very simple: Whichever set of conceptual labels one chooses, there is a force of binding and a force of unbinding — and this dual pattern applies to every level of nature's hierarchy, from the top on downward. We shall return to the hierarchy pattern at the end of the chapter.

Gravitation as Spin vs. Orbit. We have said that attraction and repulsion are complementary and relative tendencies which are always present in some form or other — and we also noted earlier that spin and orbit can be seen as describing stable gravitational relationships. Now we would like to bring these together, in asserting that relatively speaking spin is associated with the larger body that is doing the attracting, in relation to orbit being associated with the smaller body that is doing the repelling.[18]

These two tendencies can also be described as centripetal vs. centrifugal forces, respectively.

X-Force of Structure	P-Force of Process
larger entity	smaller entity
spin	orbit
stasis	motion
attraction	repulsion
centripetal force	centrifugal force

In highlighting the spin-orbit pair, we are emphasizing a complementary function which is essentially ignored by the conventional paradigm, in either Newton's or Einstein's models — whereas we maintain that the pattern is significant. For example, the relationship between planets and sun might just as well be static, as far as the established vision goes. In terms of how gravity is thought to function, the planets might as well be standing still around the sun, rather than orbiting — and the sun itself might as well not be spinning!

Let us be clear about what we are proposing here — which is a correlation, rather than a causative link. We are not in a position to argue the latter, because there is no apparent evidence at present which supports such a link — and any such explanation is vehemently dismissed by the established paradigm. At any rate, it is worth reiterating that when we speak of spin here, we are talking about the rotating center in relation to the outlying region — regardless of whether the latter is the peripheral surface, or peripheral objects.

There are obvious patterns of nature which demonstrate a causative link between spin and attraction. One is apparent in the phenomenon of spinning *whirlpools* in water, such as with a bathtub drain, where there is a vortex force that pulls in nearby objects. Another manifestation of a vortex action was actually predicted by General Relativity, in terms of the phenomenon called *frame dragging* — whereby the rotation of a central object pulls along the surrounding "space-time" region; and experiments have confirmed such effects.

Structural vs. Process Gravity This subsection differentiates two qualitatively distinct aspects of the phenomenon we call gravity, and looks at one more closely ... while the other will be discussed further later in the chapter.

A conventional way of speaking about motion in a gravitational field is in terms of a construct called *free fall*. The idea is that if there are no constraining barriers, such as we all experience on the earth surface, then an object will simply fall freely in response to the earth's gravitational attraction. On the other hand, the term is also used to describe the experience which astronauts undergo when they are in high enough orbit around the earth such that they are unencumbered by surface interactions.

18 As we noted in the previous chapter, the fact that the same object can exhibit both behaviors is beside the point here — which is a relative one as between two objects.

We would argue that this is a most unfortunate term, which can only mislead us as to the nature of the underlying phenomenon. Neither the "free" nor the "fall" designations are particularly apt, and together they lead to quite irrelevant — but very compelling — imagery. *Free* may evoke something like the flawed force-free inertial frame scenario, in which an object moves in a region that has no external influences — whereas the activity in question here specifically applies in a gravitational field, where there are external influences ... so it is not free.

But at least it is falling, right? Not necessarily — why should that be the case? *Fall* suggests a specific orientation, as in "falling down" — which is not accurate. For one thing, the object could be just as much rising up as it is falling down, and for another it could be going sideways.

More generally, physics' tendency to treat free fall as some sort of idiosyncratic phenomenon is very misleading, the idea being that it is a unique activity which is different from more common sorts of motion. This is not the case, and so it would be better to dispense with such distracting terminology as "free fall". There is quite enough to decipher in terms of the essential issues without getting carried away by blissful (and illusive) images of floating on clouds in space, or whatever. Thus to say, as physics does, that free falling is "the normal state of affairs for an object in a gravitational field" is like saying that motion is the normal state of affairs in a gravitational field. Such an explanation is not helpful, for it is no more true in general than it is here: motion is no more normal than non-motion, or stasis — both are normal states of affairs.

It seems clear that what we are dealing with is another instance of an ad hoc construct being introduced to fulfill a necessary function, for which the larger model does not provide a coherent framework. Thus, there are gravitational situations which can be described in terms of surface interaction, and those which cannot. Yet the conventional framework is based upon confusing constructs which do not simply and clearly differentiate the two different factors that are involved here.

For example, the construct of gravity is sometimes applied only to the surface interaction — with the result being that gravity may seem to disappear without it: "[G]ravity is locally transformed away by the acceleration of free fall" (Davies, 1978, p. 96). At other times, the gravity label is still evoked, even without the surface interaction — as in this description of astronauts in orbit: "They were not gravityless but they were effectively weightless — freely falling." (Hecht, p. 109) Or, "Even in free fall, gravity is the only force still acting." (Hecht, p. 125)

We maintain that a complementary approach to these phenomena is far more satisfactory — both for understanding what is going on with the specific issue at hand, as well as for fulfilling the requirements of a larger coherent model. Specifically, we assume that there are two fundamental categories of gravity that are involved here, which can be essentially understood in terms of some version of stasis and motion — such as would correspond with the notions of *weight* and *free fall*, respectively.

In the present model, we can talk about one of the two categories in the form of a static relationship that we describe as *structural gravity*, while the other is in the form of a dynamic relationship that we describe as *process gravity*. Which reigns in any given instance will depend on the two interacting entities' relative mass, motion, and distance. For example, an object which is sitting on the surface of the earth (or flying in an airplane at a usual altitude) would obviously be described in terms of structural gravity.

But the spacecraft orbiting at a higher altitude would be described in terms of process gravity. Thus:

Structural Gravity	Process Gravity
surface interaction	no surface interaction
static relationship	dynamic relationship
gravitational mass (weight)	gravitational motion (free fall)

Then there is the stable-changing dimension, which is very important for this analysis. The stable case of structural gravity would be that which we have just described, of two objects pressed up against each other, whereas the changing case might be two objects crashing into each other. Correspondingly, the stable case of process gravity would be an object undergoing uniform motion in relation to another object, whereas the changing case would be an object undergoing non-uniform motion.

One of the main flaws of the free fall concept then becomes apparent, because while the term suggests a singular phenomenon, the reality is dual: there can be stable motion of one object around another object, or there can be changing motion of one object in relation to another object (either "falling downward" or "rising upward"). These two states could also be described in terms of uniform (circular) motion and non-uniform motion, respectively.[19]

So the free falling object could be moving in orbit around the earth at *stable speed, distance, and force*; or it could be moving with *changing speed, distance, and force* — either *towards* the earth (as in falling from the top of the Tower of Pisa, or in an elevator that is falling inside a fantastically tall Equivalence Principle building), or *from* the earth (as in moving upward and outward into space). Of course, the main stable-changing distinction we are talking about here is essentially that between what is conventionally called an inertial frame of reference and a non-inertial frame of reference — although obviously the former term would not normally be applied to gravitational situations, as we are doing here. We shall return to that issue in the second subsection below, but for now:

Stable Region	Changing Region
inertial frame	non-inertial frame
stable free fall (orbiting)	changing free fall (falling-rising)
uniform circular motion	non-uniform motion

Much of the discussion here has concerned the motion portion of the stasis-motion issue, in terms of free fall; later we will look more closely at how the stasis counterpart can be understood.

Acceleration vs. Gravitation. Our task in this subsection is to try and understand how the phenomena which underlie the constructs of acceleration and gravity would be related to one another. Yet pursuing such a strategy is not easy, for we

19 One could also talk about the two as horizontal free fall vs. vertical free fall, but we don't want to overuse those two orientation terms.

are dealing with two sprawling and disjointed constructs that do not readily fit with the kind of complementary and nesting framework that we are trying to develop.

To be clear at the start, however, we reject Einstein's conclusion regarding the equivalence of acceleration and gravitation: his position was that there was no compelling reason to differentiate the two constructs, because the essential phenomena with which they are each associated can be understood as representing the same thing. We do not accept this interpretation, for we maintain that there is a viable basis for distinguishing between two separate categories here, in spite of certain characteristics that they share in common.

The most common way that we would differentiate the two is in terms of our inside-outside dimension of analysis — with acceleration applying in 2-dimensional, flat, horizontal, sideways, and linear circumstances ... in relation to gravity applying in 3-dimensional, curved, vertical, up-and-down, and circular ways. Thus:

Inside Realm	Outside Realm
2-dimensional	3-dimensional
localized and narrow	globalized and broad
flat / horizontal / sideways / linear	curved / vertical / up-and-down / circular
attraction and repulsion	contraction and expansion
acceleration	gravitation

Applied to the most obvious example of our earth life abode, we can say that acceleration applies to what we call the inside realm and gravitation to the outside realm. Based upon such factors as topology, orientation, and scope, we tend to think about acceleration as a narrow and localized phenomenon, in relation to gravity as a broader and more globalized phenomenon — which is apparent from the fact that we construe gravity in terms of a field, but not so with acceleration.

Unfortunately, the correspondence is not so straightforward, because of other meanings that are involved. Take the simple example of the rocket ship which blasts off from the surface of the earth: the blasting off part is typically understood in terms of acceleration, while the earth surface part is understood in terms of gravitation. Another problem is that the acceleration construct is so broad as to include all motion that is not of the uniform linear sort — which is an analysis that we do not accept, as we discussed in the previous chapter and will return to shortly. Finally, the outside realm of the present model would encompass a wide array of phenomena, throughout the hierarchy of nature, which are not normally thought of in terms of gravitation — a point we will return to near the end of the chapter.

Now let us look at the non-parallel character of these two constructs, in terms of our stable-changing dimension of analysis. We have argued that there are both stable and changing versions of the phenomenon called gravity — as manifested in two obviously different types of free fall — and so the same pattern should apply to the phenomenon called acceleration. Yet, of course, such a proposition makes no sense on its face — the very idea that acceleration would not only be a changing phenomenon, which is how it is always understood, but also a stable one. Indeed, how would the stable version even be labeled? The best we can come up with is to invoke the term of "inertia" for describing the stable region counterpart of acceleration. Thus:

Inside Realm		Outside Realm	
Stable Region	*Changing Region*	*Stable Region*	*Changing Region*
inertia	acceleration	gravitation	gravitation

Next consider the uniform motion issue itself. On the one hand, we maintain that there is a stable inside realm version, called uniform linear motion — which of course is the only type deemed by the orthodox model to be "inertial", or what we call stable. On the other hand, we contend that there is a stable outside realm version called uniform circular motion — realizing that this category would be conventionally described in quite contrary terms of "centripetal acceleration". Here is our complementary analysis of what is going on, bearing in mind the point made above that outside realm phenomena range more broadly than that which we would normally describe in terms of gravity:

Inside Realm		Outside Realm	
Stable Region	*Changing Region*	*Stable Region*	*Changing Region*
uniform linear motion	non-uniform motion	uniform circular mot.	non-uniform motion
inertia	acceleration	gravitation	gravitation

Here is another point about the two confusing constructs we are discussing, and the random theoretical developments upon which they are based. The conventional model of physics obviously does not treat acceleration (or inertia, for that matter) as a fundamental force of nature — yet it does so with gravity. Why? The distinction makes no sense in the present framework. Indeed, given Einstein's Equivalence Principle treatment of the two categories as being essentially the same, then why is there still a fundamental force of gravity ... but not of acceleration — or why is it not a combined fundamental force of gravity-acceleration?[20]

What is the difference? Because Newton's laws of motion can be invoked to explain what we call an inside realm phenomenon, does this obviate the need for a fundamental force factor there? But Newton's law of gravitation can also be invoked to explain what we call an outside realm phenomenon — yet that latter phenomenon gets special treatment. Or is the explanation, as to why acceleration is not accorded "fundamental force" status, to be somehow found in the fact that it is obviously easier to observe or identify the specific "causal" agent which exerts the requisite change force there? Even so, why is that any different from gravity, where we can also identify the catalytic agent? Are the one set of patterns (gravity) somehow taken to be more mysterious than the other set (inertia or acceleration)? Or is the answer to be found in the fact that our causal explanations for the one (gravity) came later historically than did those of the other (acceleration)? In the end, how are all these determinations anything but arbitrary, and how can they ever possibly lead us to a coherent vision of nature?

20 Again, one can argue that there would be no fundamental force of gravity at all in Einstein's General Theory, such a property being incorporated into the topology of the space-time continuum. But leaving that model aside, physics still talks about fundamental forces of nature, one of which is gravity — and it is in this context that we raise the question.

Of course, the conventional interpretation is that no force explanation is required at all to explain one of the above categories, that of inertia, since it is presumed to represent the baseline category in which no forces are involved. But as we have discussed, such an abstract conception notwithstanding, it is generally acknowledged that in the real world this is a false assumption, and that there is no force-free realm of the universe. This is why we interpret the "inertia" category (along with its stable outside realm counterpart) as representing balanced forces, rather than no forces.

But let's go back to the Equivalence Principle itself and the way it treats these two constructs as being essentially the same, along with the two sets of underlying phenomena which they represent. We contend that the primary culprit here are the bizarre thought experiments themselves which are put forward to demonstrate the basic thesis. What essentially happens is that certain interactions are taken out of their normal context, where the difference between acceleration and gravitation would be obvious, leaving us with only artificial stubs of phenomena that appear to be the same. Expressed more specifically in terms of a distinction we made in the previous subsection, what is demonstrated in these thought experiments concerns the indistinguishability of surface interactions that may be related to either acceleration or gravitation (the latter of which we call "structural gravity") — while completely ignoring the precursor factors which clearly differentiate acceleration and gravitation (the latter of which we call "process gravity").

Not surprisingly, we contest the orthodox conclusion and claim that it amounts to another attempt to collapse a complementary aspect of nature into a unitary one. As to the general problem of confusing and misleading thought experiments, this is a subject we will address in the second half of Chapter 7.

Inertial vs. Non-Inertial Frames. Next we consider how the Equivalence Principle was used to eliminate the distinction between inertial and non-inertial frames of reference. Of course, the merging of these two realms is also the mission of the General Relativity version of Einstein's adaptation of the Relativity Principle, which we called the fourth interpretation in the prior chapter. According to the general form of that thesis, a true "law of nature" is one which describes all types of phenomena, regardless of whether they would be categorized as representing inertial or non-inertial frames.

To start with, we would not begin to pretend that the inertial and non-inertial frame constructs are problem-free: quite the contrary, as we discussed in the last chapter, especially under The Woes of the Inertial Frame section. But that is a very different matter from saying that the underlying construct is not viable in some revised form, and especially that the distinction between two such complementary categories is not essential. Indeed, the very distinction which Einstein embraces, in his flat vs. curved space-time continuums, can itself be understood as an illustration of the underlying difference between inertial and non-inertial frames — which Einstein rejects.

But what about the falling elevator scenario which seemed so persuasive, in our presentation of the Equivalence Principle above? How do we account for the fact that internal observers within the elevator container would describe their situation largely in terms of an inertial frame — whereas external observers would see this same elevator as falling in a gravitational field, thus representing a non-inertial frame?

Let us review that scenario. Recall that the inertial frame determination there is based on the fact that the elevator and its components will be "floating" in what is essentially a stable static relationship with one another — but the crucial detail which must be highlighted is that this determination will only apply "over limited distances". Of course, "limited" is a relative factor, depending on the overall force and distance involved, as between the elevator container and its components — and as between both, and that of the extrinsic force entity towards which the elevator is falling.

Indeed, this is precisely where the argument breaks down. For the inertial frame determination will only apply in the short term, even for the internal observers. That is because the elevator contents will change their relative positions as the elevator falls in the gravitational field. While such changes may seem trivial for the case of a small elevator container over a short distance, it would be more obvious if we were dealing with the spread-out components of an extremely large container, over a limited falling distance — or if we were dealing with a very large falling distance, for either a small or large container. In that longer-range case, there would be a more obvious tendency for the contents to converge (contract) substantially, as they approached the source of the "gravitational field" that they were falling towards (e.g., the center of the earth).

So we contest the conclusion as being an illusion! Rather than contradicting one another, the internal and external perspectives would reach the same conclusion as one another — if there were enough distance involved that the changing nature of the situation were allowed to become apparent ... that we were dealing with a non-inertial frame.

We reject Einstein's interpretation here altogether, as being based upon a faulty analysis. His essential argument, again, was that since an inertial frame system can appear to exist within the confines of a gravitational field, this then meant that the former construct was defective ... thus leading to his elimination of the distinction between the two frame-types. But in the previous chapter, we disputed the whole force-free underpinnings of the inertial frame construct, and argued that it was a false premise: yet this did not mean that the distinction between the two frame-types was flawed, so much as that it was misunderstood. Our position is that since external (i.e., gravitational) forces would exist everywhere in the universe, to a greater or lesser degree, then the only conclusion which makes sense is that what is commonly referred to as an inertial frame can exist in no other way than amongst gravitational forces. That is why the main difference between inertial and non-inertial frames should not be understood in terms of no-force vs. yes-force — as is indicated in Einstein's argument — but rather balanced-force vs. unbalanced-force ... or as we choose to call it in our dimensions of analysis scheme, stable vs. changing regions of nature.

Inertial vs. Gravitational Mass. Now we want to look at the relationship between inertial mass and gravitational mass, and how the Equivalence Principle attempted to merge these into one. In so doing, we will also be looking more closely at the stasis (or "structural") counterpart for the motion (or "process") aspect of gravity that we described in an earlier subsection. Of course, the basic reason why the two mass-types are said to be no different from one another is because the inertial and non-inertial realms of nature where they each apply are themselves said to be no different from one another, as we just discussed.

When we presented our analysis of the frame of reference construct in the previous chapter, we concluded that while the difference between the inertial and non-inertial frames could be understood primarily in terms of our stable-changing dimension, there was a secondary meaning in terms of our inside-outside dimension — and that latter distinction provides the main basis by which we approach the present issue.

Once again, we are talking about the difference between the inside realm and the outside realm, and it is in these terms that we can speak about a different category of mass in each realm — a kind of inside mass and outside mass, respectively. But, of course, there are more conventional ways of labeling these two complementary categories:

Inside Realm	Outside Realm
inertial mass	gravitational mass

Therefore, an entity's *inertial mass* refers to the inside realm "quantity of stuff" which the entity is composed of, while that entity's *gravitational mass* refers to the outside realm "quantity of stuff" of which the entity is composed. An inertial mass example could be something like an apple that is sitting stably in a local interaction with another apple — and maybe we'll get some good apple pie soon! — whereas a gravitational mass example could be something like an apple that is sitting stably on the surface in a more globalized interaction with the earth.

We have previously mentioned how these two types of relationships could be understood in terms of measuring scale devices. When we determine the inertial mass of an object, we may use a balance scale to compare the target object with another object in the same inner frame — and we are thus measuring what might be called the object's localized or "side-weight". Conversely, when we measure the gravitational mass of an object, we use a spring-loaded weight scale to compare the target object with another object in the outer frame (e.g., the earth) — and we are thus measuring what might be called the object's globalized or "down-weight". In both cases, we are dealing with a relationship that can be understood as being composed of dominant and recessive entities.

Are these two masses really different from one another, or are they actually the same? The answer must presumably be both — that on the one hand, it is the *same* "stuff" which participates in both types of interactions — but on the other hand, they are also *different* by virtue of what distinguishes the two realms or domains of which they are part. In this way, the same apple can have two different roles in these complementary plays. The apple which is sitting static on the earth surface can be part of an inner frame relationship (inertial mass) and an outer frame relationship (gravitational mass). Yet the two inside and outside plays are not being performed to the same audience, as it were. One is inside-based and the other outside-based.

Before closing this subsection, let us dovetail the "structural gravity" points we have made here with the "process gravity" points we made earlier, now labeled and charted more broadly to also include inertial and acceleration counterparts:[21]

Changing Region

Inner Frame: *Structural Interaction*	Inner Frame: *Process Interaction*	Outer Frame: *Structural Interaction*	Outer Frame: *Process Interaction*
accelerational mass (e.g., crashing)	non-uniform motion	gravitational mass (e.g., crashing)	non-uniform motion (e.g., changing free fall)

Stable Region

Inner Frame: *Structural Interaction*	Inner Frame: *Process Interaction*	Outer Frame: *Structural Interaction*	Outer Frame: *Process Interaction*
inertial mass (e.g., balance scale)	uniform linear motion	gravitational mass (e.g., weight scale)	uniform circular motion (e.g., stable free fall)

Gravitation as Matter vs. Medium. The most common way that we think about gravity is in relation to the massive bodies which populate the heavens, including our own earth; and the effect those bodies have upon common objects, such as those that we experience on and near the earth surface. But objects are only part of the makeup of nature ... there are non-objects, as well. This brings us to the dimension of analysis that we call matter vs. medium.

Physicists and astronomers often talk about a strange and obscure set of phenomena known as "gravity waves". These waves are said to carry a gravitational field disturbance from one region of the universe to another, by reverberating outward in a wave-like manner at the speed of light.[22] The conventional assumption is that the cause of such gravity waves is some kind of cataclysmic event which occurs elsewhere in the universe. Thus, Wald states that, "Astrophysical events which are likely to produce large amounts of gravitational radiation are (1)supernovas or other gravitational collapse phenomena..., (2)accretion of a star into a large black hole at the center of a galaxy or star cluster, (3)coalescence of neutron stars and/or black holes in a close binary orbit." (Wald, p. 119)

In terms of our dimensions of analysis, we would understand gravity waves as outside realm medium domain phenomena — but we want to add in two other dimensions, as well: the stable-changing dimension and the stasis-motion dimension. For one thing, then, gravity waves would not only be associated with the kind of dramatic activities just cited, which are changing phenomena — but they would also exist in an ongoing manner in association with stable phenomena. Beyond that, we would argue that just as gravity waves can be of the traveling sort (as are normally conceived), they should also be under-

21 One could also talk about a parallel pair called "structural acceleration" and "process acceleration": the latter would correspond with a change of velocity — which would lead to a surface interaction between one object and another that would correspond with the former.

22 Note that Poincaré proposed the idea, of gravitational waves propagating at the speed of light, prior to Einstein's General Theory.

stood in terms of the static kind — meaning that there would not only be traveling gravity waves, but also standing gravity waves. The latter would presumably represent the medium domain counterpart for the matter domain's absolute space; as such it would amount to something like an ether medium phenomenon.

As far as Newton and Einstein's models of gravity, the distinction can be understood here in matter and medium domain terms — with gravitational objects corresponding to the former and gravitational waves to the latter. But there is a fundamental way that the two models differ from one another, and this is the issue we will consider next.

Direct vs. Indirect Gravity. One of the differences we discussed in the previous chapter is that static relationships could be understood in terms of direct contact, and dynamic relationships in terms of indirect contact. Now, it is true that we tend to think of direct contact as being restricted to situations where the partners are obviously situated right next to each other. But we propose a very different interpretation, such that one meaning for direct contact would be much broader than anything we are used to.

Thus, for example, from our common human perspective we might describe one pair of entities as being immediately adjacent to each another, and another pair of entities not so — and yet we suggest that such a perception is too limiting. What we are proposing is that two entities may interact as part of a static relationship — and be engaged in direct contact — irrespective of whether they appear to us as being located next to one another or not.

We maintain that the essential distinction we are making here, between static and dynamic relationships, completely parallels the primary philosophical or metaphysical dispute which differentiates the Newtonian model of gravity from that of Einstein. In particular, as we discussed earlier in the chapter, Newton's model is built upon the premise of "action-at-a-distance" — or direct and instantaneous contact between one gravitational object and another. We noted that Einstein's approach treats such linkages as akin to magic, and his model replaces them with a mechanism of indirect contact which is transmitted over time (and, in particular, at the speed of light), by means of an intermediate gravitational field. Not surprisingly, we reject both of these options as being insufficient unto themselves, and instead adopt *both* — but not in any random or mélange sense. Rather, the two are treated as complementary and fundamental mechanisms of nature, both of which must always apply, in one combination or another. We will discuss the broad and even startling implications, of the dual interaction template that we are proposing, at some length in Chapter 9. For now, however, here is a complementary chart of these medium domain patterns:

<u>*Static Relationship*</u>	<u>*Dynamic Relationship*</u>
action-at-a-distance intraaction	traveling interaction
direct and instantaneous contact	indirect and over time contact
Newtonian gravity	Einsteinian gravity

The Hierarchical Nature of Gravity. Most of the issues that we have been discussing thus far in the chapter have concerned the complementary nature of gravitational phenomena. Now we want to bring the hierarchy factor to the fore, again focusing on matter domain phenomena.

According to the standard interpretation, gravity is a singular and universal force — which manifests itself either directly or indirectly between one entity and another, throughout nature. *All* phenomena in the universe are supposed to attract *all other phenomena* in the universe, indiscriminately — to one degree or another, depending upon mass and distance. At the same time, gravity is said to be really significant as a force only in relation to massive objects, which we have repeatedly noted — i.e., large heavenly bodies. The usual explanation for this is that the force of gravity is inherently so weak that it only manifests itself in a significant manner when it comes to very large bodies and their ability to accumulate into something powerful ... that which is otherwise an essentially weak force, as compared to the other fundamental forces of nature.

Nonetheless, that detail aside, the universality principle seems very odd. The other fundamental forces operate at specific levels: electricity and magnetism are said to apply at the level of the atom, while the nuclear force operates at the level of the nucleus. Yet gravity is said to operate between *everything*: nucleuses would attract flowers, molecules would attract books, elephants would attract stars, Mars would attract apples — and so on, ad absurdum.

Yet how do we know, and what does it mean to say, that everything interacts gravitationally with everything else? All that we can observe is that discriminate relationships can be described with this law, not that all relationships in the universe can be so represented. Granted, one can never observe everything — and thus one is always drawing generalizations in science, to some degree. But it is the underlying logic that we are contesting here, with the all-pervasive gravity assumption. We maintain that what is missing therein is the very notion of hierarchy — concomitant with the idea that such interactions occur in an orderly manner, through nesting levels of phenomena.

Of course, one might argue that the hierarchical factor we are propounding is already implicit in the existing paradigm, and so is taken for granted. This would be apparent in the fact that the strength of the phenomenon called gravity varies with distance, as exemplified by the notion of different gravitational potentials applying at different altitudes or distances or regions — and that these might be understood in terms of different hierarchical levels. But such an unspoken subtlety will not help us to discern the simple and pervasive patterns we seek.

So let us try to elucidate the hierarchy factor, as we propose that it would be applied to gravitational phenomena. We start by recalling our discussion in the first chapter, where we talked about the upward vs. downward mechanism of hierarchy — as well as the same vs. different characteristics that go with them: Two different lower level complementary parts merge together into the same higher level whole.

We have already indicated that the general concept of gravity can also be understood in these hierarchical terms. Thus, we have described gravitational mass in terms of a static relationship — and such gravitational mass can be understand in hierarchical terms. The lower level parts would be understood as two different (dominant and recessive) entities and their respective masses; the higher level whole would represent a synthesis or cumulative nesting of these parts. For example, the apple forms an outside realm relationship with the earth, and together they form a whole — which can be construed in terms of the earth-apple gravitational mass. Looking upward, we can say that together earth-apple forms an outside realm relationship with the sun, and together they form a higher whole. When we talk about this new unity of sun-earth-apple, we can say that they form an outside realm relationship with something higher, and thereby form another whole nesting level. And so on.

In many (if not most) cases, we do not have specific names for labeling the higher level whole, such as the combination of earth and apple — or more significantly, earth and moon. But in general terms, it makes sense to think about the combined gravitational mass as representing a third entity, which corresponds with the higher level whole — as distinct from that of the two component entities which comprise the lower level parts. Indeed, we can understand the basic gravity formula in this sense: Instead of saying $F_g = G \times mM / r^2$, we can think of $M_{III} = G \times mM / r^2$, with the two lower level parts corresponding with the right-side m and M. The force of gravity here is the attraction or binding strength of this new higher level combined (third) mass_{III} ... composed of repulsive (m) vs. attractive (M) partners. This hierarchical relationship could be understood in terms of the dominant-recessive dimension components (M and m) joining together into the next higher level (mass_{III}).

In fact, we should also be describing the corresponding dynamic relationship in terms of this combined whole — with the two lower level mass objects orbiting the higher level common center of mass between them (e.g., see Murchie, p. 57). Thus, for example, at this higher level, both the earth and the moon orbit their common center of mass — akin to a phantom entity, M_{III}, in its own right — and not merely the moon circling around the earth.[23] Given the huge disparities of mass in the typical gravitational relationships that we commonly experience, a slightly shifted center-of-mass distinction may not seem important — but it is significant in illustrating the nesting hierarchy pattern which is central to the present model.

We can stretch our thinking now, and see how this same principle might provide a basis for explaining phenomena like the Cavendish study. Originally performed in 1798, this experiment consisted of varying the proximity of two pairs of lead balls, in relation to one another. The dependent variable was a pair of small lead spheres (each weighing 0.8 kilograms), suspended by string from either end of a freely-rotating arm. The independent variable was a corresponding pair of much larger lead spheres (each weighing 167 kilograms), that were fixed upon a platform. As that large ball platform was moved closer to the small ball set, the latter pair tended to shift very slightly towards the former.[24]

Is it possible that the change of position by the small ball set could be understood simply as a "correction" phenomenon? That is, the small ball pair was undergoing uni-

23 "The moon does not make its orbit around the Earth's center; but Moon and Earth together swing around their common center of gravity ..." (Rogers, p. 328) As Rogers notes, this means that the earth-moon center of mass is 3000 miles from the center of the earth — still 1000 miles underneath the earth surface!

form circular motion (e.g., largely around the earth's center) until the large ball pair intervened and altered the center of mass around which the small ball pair was orbiting. The change which was thereby introduced by the repositioning of the large ball pair required, in turn, a correctional shift by the small ball pair so as to regain equilibrium — a new state of uniform circular motion around the newly-positioned center of mass ... rather than the small balls moving towards the big balls, per se.

One might apply the same analysis to other more natural shift phenomena. Two obvious examples are the earth's tides and the planets' orbital perturbations. In the former case, the tides of the oceans are said to be largely caused by the orbiting of the moon. But another perspective is to say that the fluid oceans orbit the common center of mass that is created by the combined earth and moon. The ocean's tides would be understood as correctional shifts, caused by the alteration of the earth-moon center of mass upon which they are oriented and about which they orbit.

A similar shift takes place in the orbits of the planets, when they come in close proximity to one another and their orbits are thereby "perturbed". Thus, Jupiter's (mostly) uniform circular motion around the sun is upset periodically by the proximity of Saturn; the altered center of mass (from what was, for example, sun-Jupiter, to sun-Jupiter-Saturn) necessitates a transitional correction in Jupiter's course, which consists of a shift back into equilibrium.

In all of these cases, we can say that when the target object is stable, then this is because it is undergoing uniform circular motion around a constant center of mass. When the target object changes its motion state, this occurs in order to reach a new position of stability in relation to the altered center of mass — thereby going from one equilibrium state of uniform circular motion to another.

Now let us return to the general issue with which we began this subsection. That is, we contend that neither the set of phenomena we think of in terms of gravity, nor any other phenomena, represent an amorphous type of relationship between *everything*. Rather, we presume that different entities in the universe influence one another in an orderly fashion ... which by our definition means that they are operating within a hierarchical framework. They are not just a jumble of interactions coming from everywhere, upon everything. And neither the Cavendish experiment, nor any other example that can be cited, fundamentally contradicts this premise — virtually by definition.

Nature is not just one big pot of soup. One thing in the universe may interact with all other things in the universe, but such interactions are hierarchically-organized — and in that way discriminate and orderly. In Newton's case, for example, his achievement was to demonstrate clearly that the nature of the force between the sun and the earth is no different from the nature of the force between the earth and the apple. All that Newton was then warranted in doing was to extend the inverse-square law from the particular case of the earth-sun relationship to any other hierarchically-related relationship. Instead, Newton's law of gravitation was understood as a universal principle

24 The results of this experiment were taken not only as support for Newton's universal law of gravity, but it was, in fact, the first time that the gravitational constant G had been established — by virtue of the fact that the two relevant masses were both known, and thus Newton's formula could be solved for the G factor.

which did not take account of different levels, and so was meant to apply in an indiscriminate manner between all bodies that possess inertial mass.

One of the implications of our hierarchy approach is that there would presumably be a different version of "gravity" for every level of interaction. Of course, if every level of interaction has its own version of gravity, then we are obviously dealing with a much more abstract concept than we are used to. For this not only means that every level from earth on *upwards* to the very top (whatever that is, assuming there is a top) has its own version of gravity; but also that every level from earth on *downwards* to the very bottom (whatever that is, assuming there is a bottom), also has its own version of gravity.

Such generic gravity is tantamount to no gravity, in some respects — but it is certainly not equivalent to the old-fashioned gravity we commonly think of, nor even the modernized Einstein version. We are simply talking about outside realm interactions, as they exist at one or another level of nature's hierarchy: we suggest that this is the broader category which gravity represents, irrespective of the particular horizons and perspectives upon which we earth-based humans are centered.

Let us describe some of these outside realm patterns, as they manifest themselves in nesting terms. We can begin with the level of the earth orbiting the sun. Nested within that framework are objects which are orbiting the earth — be it those engaged in what we have called process gravity interactions such as the moon, or free-floating apples, orbiting satellites, etc. ... or those engaged in what we have called structural gravity interactions such as apples, people, and other objects which are sitting on the earth's surface.

Here is one example of an outside realm relationship that is sitting on the earth's surface and is nested within the next higher level outside realm relationship of the earth orbiting the sun. It consists of a person who is doing something very interesting. This person is holding a string, the end of which is fastened to the open top of a can — and the person is actually spinning the can around in a circle. But there is more, for inside the can is a little bug — and by virtue of the centrifugal force that is created by the can's circular motion, the bug is being pressed up against the closed end inside the can. If the bug were exceptionally intelligent, it might label the phenomenon that it is experiencing — e.g., during a walk on the inside surface of the can's end — as "gravity" ... although, of course, we smart aleck humans know better than that.

Another nested level, outside realm example would be that of a very large centrifuge device which is also positioned somewhere on the surface of the earth. This device is rotating, the effect of which is to push any contents — like people and apples — in a centrifugal direction until they are pressed up against the outside wall of the device (just as was the bug upon the can's end). The people could even stand up and walk on that wall just as if it were a floor (as did the bug upon the can's end). They might even characterize the resulting phenomenon as seeming very much like normal gravity — and, indeed, this scenario can be described as representing "artificial gravity".

The general complementary patterns which apply to each nested level of outside realm relationships would be the same. Thus, there will always be two countervailing tendencies or forces being exerted — inward centripetal attraction vs. outward centrifugal repulsion. In general terms, either could be associated with the dominant entity or the recessive entity. In situations which are called gravitational, the dominant entity will always correspond with (i.e., be the source of) the inward centripetal force of attraction, in relation to the recessive entity with the outward centrifugal force of repulsion. In analogous non-gravitational situations such as we just described, the reverse will hold — with the dominant entity corresponding with (i.e., being the source of) the outward centrifugal force of repulsion, and the recessive entity with the inward centripetal force of attraction ... even though that last reference to "attraction" may seem peculiar in this context.

The earth is pulling inward, in relation to apples and people pushing outward. The can is pushing outwards,[25] and the bug is pulling inward. The large centrifuge is pushing apples and people outward, while apples and people are pulling inward. Also in keeping with the reversal point just mentioned is the different position of the structural interaction: in the case of what we normally call gravity, the earth surface barrier which the person is pressed against is located in the inward direction, toward the center of the orbit; but in the other two cases, the surface barrier (of the can's bottom in one case, and the centrifuge wall in the other) is in the outward direction, away from the center of the orbit.

As to what these different situations would be named, it should be obvious that the "gravity" label doesn't work very well as a general descriptor. Nor would we accept a label of "acceleration" (or even "centripetal acceleration"), partly because we characterize situations like the above as stable, rather than changing. That is why we would argue for simply talking in basic terms about (stable or changing) outside realm relationships, at some particular hierarchical level.

Before closing, let us mention one more way of approaching the hierarchical theme, which also illustrates how the underlying patterns of "gravity" are not unique. Let us consider the essential identity between an important pair of formulas, whose relationship has always remained a puzzlement to physics. On the one hand, we have the formula for describing the strength pattern of a gravitational field, as manifested in the Cavendish experiment we just described; and on the other hand, we have the formula which describes the strength pattern for an electrical field, as manifested in what is called the Coulomb experiment.

The basic scenarios underpinning the Cavendish and Coulomb experimental devices are essentially the same (e.g., Rogers, p. 541; Hecht, p. 388). And again, the resulting formula which describes electrical interactions, known as Coulomb's law, is identical in form to Newton's law which describes gravitational interactions (Hecht, p. 235). That is, Newton's law for gravity can be expressed as $F_g = G \times mM / r^2$, whereas Coulomb's law for electricity can be expressed as $F_e = K \times qQ / r^2$. Needless to say, the electrical formula's "similarity with the gravity formula is obvious and striking though still not explained theoretically; they have the same form, but nobody knows why" (Hecht, p. 389). Our position is that the underlying pattern is the same, at every level of nature's hierarchy — regardless of what particular name we use to describe a given level, and regardless of what fundamental force label we attach to the pattern. That is also why we

25 Note that the can is pushing outwards in response to the hand (which is causing the spinning) pulling inward.

argue that distinct "fundamental force" categories need to be abandoned, including that of gravity ... a point that we shall return to in the final chapter.

This concludes our discussion of the gravity construct. We now move on to a whole other realm, as we begin our interpretation of light and electromagnetic phenomena. As such, we are moving backward in time — so to speak! — from gravity and the later General Relativity, to light and the earlier Special Relativity.

Chapter 6

The Great Light Hoax

Most of our discussion in the last two chapters was concerned with issues of stasis and motion. In Chapter 4, we looked at that topic primarily in terms of inertial frames and Einstein's Special Theory, whereas in Chapter 5 our focus shifted to non-inertial frames and Einstein's General Theory. In this and the following chapter, we look more closely at the earlier theory and the radical new concepts which go with it.

There are, in fact, two basic postulates upon which the Special Theory is built. The first concerns the Relativity Principle, which we discussed earlier; the second is the premise that the speed of light is always the same. What we shall argue here is that neither postulate is acceptable, and that they have left us with a distorted representation of nature which is completely unnecessary for dealing with the issues. By contrast, our goal is to present a coherent framework for understanding these matters, in the usual complementary terms.

Underlying much of the discussion is our premise that light is not a unique phenomenon, and that it conforms to more general patterns found elsewhere in nature. In some respects, the established paradigm would concur, by virtue of the way it treats the properties of light as being not at all unusual. In terms of light's wave aspects, for example, "diffraction and interference effects are common features of *all* types of wave motion" (Marion, p. 350). And in general, "electromagnetic [or light] waves ... carry energy in much the way that water waves and sound waves do." (Rogers, p. 389) But what we shall find is that such gestures of conceptual consistency are the exception, rather than the rule, in the way that Special Relativity treats the phenomenon of light.

This is particularly true when the topic concerns the speed of light waves, and this is related to the question of whether such waves require a medium. As we shall see, the notion of a medium as being fundamental to all wave behavior contradicts an essential premise of Einstein's model — which is that there is no medium for light transmission. We argue that in many ways, it is the very antipathy that Special Relativity holds towards the concept of a medium which makes the theory a highly abstract model, one that bears little relationship to our experience with nature; and this same non-inclusion of a medium concept also explains why the theory is, at root, impossible to understand and make sense of. And perhaps the primary reason why it makes no sense is because it lacks a crucial conceptual pivot: it omits — indeed, eschews — a real-world function such as is represented by the notion of a medium. The result is a big gaping logical hole right at the heart of the model ... leaving behind a trail of conceptual absurdities in its wake.

We start our discussion by examining two different versions of the light postulate — and we shall see that while one makes perfect sense, the other seems very bizarre. We then mention a couple of general experiments which bear on the speed of light issue. After that, we consider some of the methodological, even metaphysical, issues and limitations which must be faced in actually measuring light's speed. In discussing these matters, we cannot avoid focusing next on some fundamental questions about time, and its relationship to any determination about the speed of light. Following that we will describe the main ways in which we contend that light's speed is not constant. Up next is a very important century-old experiment, one that was (and is) considered decisive for dealing with the speed of light issue. Then we examine the medium concept more closely, and why it is so important for the issues we are discussing. After that we talk about how light can be understood in terms of the complementary framework of the present model. We close the chapter by considering the kind of theory-building strategy which has brought us to where we are, along with some of the confusing factors that have contributed to the general misunderstanding of the issues here.

But wait! Prior to talking about these subtle and complex issues, we need to get something out into the open which is inevitably lurking in the background...

A Prelude

Before pressing ahead with our main discussion about Einstein's Theory of Special Relativity, we need to step back and take a deep breath ... so that we can first consider some of the problems which loom in even having this conversation. After all, we are talking about *Einstein's Theory of Special Relativity!* — and the emphasis is intentional, in order to highlight the psychological charge that we are dealing with here. Merely to mention the name, of either the person or the theory, is to invoke an almost trance-like awe and veneration for the institutions that are involved.

Rightly so, many people would think — for these are well-warranted measures of respect that have been fully earned and should not be trifled with. Indeed, we are talking about a primary pillar of our culture and our society — of the whole world's culture and society, so widely shared are the basic sentiments about what this person and his theory represent. Thus it is that most of us have grown up, having been properly reared and educated, to think about Einstein as the quintessential symbol of the enlightened mind — the interlocutor extraordinaire between mankind and nature, with his Special Theory representing a profound insight into the fundamental workings of the natural world. It is quite jarring — or at the very least, discordant — to even contemplate having these icons examined in any kind of critical way. We are touching upon the shared sense of basic values and beliefs which holds any society (or group) together — and it borders on sacrilege (if not insanity) for anyone to challenge these core precepts: such impudence is to be condemned, and any heretics should be banished outright ... or at the very least thoroughly ridiculed, until they renounce this blasphemy and rejoin the fold.

It is putting the matter mildly to say that this leaves us in an unenviable position — for the result is to instantly demonize the present work in the eyes of many people, and to completely undermine its credibility. That naturally raises the question of whether it is worth letting ourselves get sidetracked — even derailed — by an issue which is hardly the main purpose of the present work. After all, our mission in this book is to present a new model of nature that is based upon complementarity, and not to take on the whole Einstein/Relativity mystique. So why jeopardize the former for the sake of the latter?

The problem is that there is no way to skirt the issue. On the one hand is the larger than life figure and his theory, along with the associated strategy for theory-building. On the other hand are the fundamental precepts of the present model and the theory-building strategy which we advocate. The path of least resistance would be to either abandon this project completely, avoid the Special Theory altogether, or mend our ways and find some means for reconciling our ideas with the master's model. The first option seems foolish, the second impractical, and the third impossible — for the two conceptual frameworks don't mesh, in virtually any manner. So this is why we have chosen to go our separate way — and continue arguing for the general precepts which guide the present work, as being just as applicable to the issues that are involved here as anywhere else.

As far as the standard assumption that it is folly to question Einstein's thinking or theory, because they are so deeply-entrenched and widely-embraced in our world, the same challenge would have obviously applied to every previous period in mankind's intellectual development, when one or another earlier theoretical framework was predominant — and the only way that the same pattern would not apply here is if we have reached some kind of conceptual pinnacle which will never be superseded. The fact that such an outcome is unlikely should become apparent in the pages ahead.

Our position is that in historical and conceptual terms, the universal acclaim for Einstein's work is misplaced. Regarding the former, we shall see that virtually all of the essential concepts and mathematics which are utilized in Einstein's model had already been laboriously developed by others before him — even though he usually gets all the credit for them; and this is true regardless of whether one contends that he also developed some of these ideas independently or not. Regarding the latter, there are several points to be made.

For one thing, the preexisting set of concepts and mathematics we just mentioned emerged within a theoretical framework that was evolving in the period immediately prior (as well as subsequent) to Einstein's appearance on the scene — and this included radical reinterpretations about the nature of time, which is commonly considered to be the greatest contribution of the Special Theory. What Einstein did was to incorporate these new ideas into a very different (and slimmed down) theoretical framework, one which sharply departed from the conventional treatment of such issues.

It may be surprising for some readers to learn that there is no factual support for Einstein's theory, as opposed to the theoretical framework which already existed without him. As to why Special Relativity won out in so many people's minds, we contend that the explanation rests as much upon emotional and stylistic factors as it is does upon rational and logical ones. We will touch upon some of these points in the latter part of Chapter 7, but for the moment we can say that the very features which are usually cited as making Einstein's theory superior (including about time) are precisely those which we regard as the chief flaws in his model.

In the present chapter, we shall examine one specific topic which makes Einstein's model so unique — the arbitrary precept about a constant speed of light, along with the incomprehensible notion that there is no medium for light waves travel in. While it is clearly possible to build a theory that is based upon such novel conjectures — the ideas themselves can neither be proven nor disproven — we maintain that what results is a warped vision of nature. That outcome is evident in the fact that the Special Theory has never made any sense, in a meaningful manner which can be integrated with our normal lives — and the main reason is not because of high-speed motion, which is the usual rationale that is put forward in this regard. Rather, the primary explanation is to be found in the simple fact that treating light as an idiosyncratic phenomenon violates the patterns we find everywhere else in nature.

This kind of gross departure from mainstream reality is not necessary for dealing with the issues at hand — and that is why we do not adopt the usual deferential, if not reverential, stance towards Einstein and his Special Theory. Not only do we not see his approach as admirable, but we believe that it has actually taken us way off in the wrong direction ... as we shall now begin to explain.

Confusion Between Two Light Postulates

Now let us look at the theory itself, and Einstein's radical assertion about the speed of light. The second postulate states that light will always have the same speed (in vacuum) wherever one finds an inertial frame of reference. But established dogma notwithstanding, we shall see that there is actually considerable ambiguity about what such an assertion really means.

There are actually two very different ways of construing this principle — one of which is perfectly tenable, and the other of which is very hard to comprehend. The issue which separates them largely concerns whether the source and sink (or sender and receiver) of the traveling light waves are themselves static or moving. The first interpretation can be called the *weak light postulate*, and it states that the speed of light is a constant factor which is independent of any motion of *the source* of that light. The second interpretation can be called the *strong light postulate*, and it puts forward the broader proposition that the speed of light is always constant, regardless of whether we are dealing with any motion of *either the source or the sink* of that light.[1]

The weak interpretation is quite clear and rather easy to understand — even if it requires a little reflection by those unfamiliar with the pattern. The guiding thesis is that the velocity of a wave is a property of the medium in which the wave travels, and this is why its speed is independent of any motion of the source. As Hewitt puts it in general terms, "The speed of a wave is not affected by motion of the source, but depends on the medium through which the wave travels." (Hewitt, p. 293) There would be nothing

1 Schaffner also refers to weak and strong interpretations, which he specifically labels as the ILV ("independence of the velocity of light from its source") and CLV ("constancy of light velocity hypothesis"), respectively (Schaffner, pp. 53-54). Likewise, Evans talks about "the original principle" of the constancy of the velocity of light, as opposed to "the general principle" thereof. (Evans, p. 72)

unusual about light in this respect — since the same pattern applies to all wave phenomena in nature (e.g., water, sound, etc.). Indeed, one could argue that there is not even a need for the weak light postulate at all, because it is not saying anything new or unusual about light which makes it different from any other wave phenomenon.[2]

To illustrate what is going on, let us consider a simple example — that of a water wave. For example, imagine a body of still water, such as a large pond. Now envision an object moving upon that water, e.g., a boat. This boat is the source of the disturbance which produces the water waves — but it makes no difference how fast that boat is moving, in terms of the resulting wave speed, because that speed is a property of (and is determined in relation to) the medium itself. Thus if we imagine a second boat which is at rest somewhere elsewhere on the pond, the water waves will arrive at the same constant speed, irrespective of the speed of the source boat.

Next let us introduce a second motion factor into this system — that of the receiving boat, which is now moving.[3] We now have two speeds to consider — one of the wave which is traveling in relation to the pond, and the other of the second boat which is traveling in relation to the pond. Take the basic case of the second boat moving straight towards the waves which are approaching it. In order to determine the speed at which the waves are received by the second boat, we would merely add the two speeds together — the speed of the waves (in relation to the pond), and the speed of the boat (in relation to the pond). Simply speaking, in the case of the sink boat's (augmenting) movement toward the approaching waves, this compounding will be additive — but conversely, in the case of the sink boat's (receding) movement away from the approaching waves, this compounding will be subtractive.

Let us introduce some convenient labels for describing the two speeds which are involved here. On the one hand, we can talk about the wave's *fundamental speed*, as being something that exists in relation to (and is determined by the characteristics of) the medium in which the wave is traveling. On the other hand, we can talk about the wave's *effective speed* in relation to the receiving object (e.g., the sink) — as combining the fundamental speed of the wave in relation to the medium, together with any speed of the receiving object in relation to the medium.[4]

The distinctions we are putting forward — between source motion and sink motion, as well as between fundamental speed and effective speed — are a viable way of describing all wave phenomena in nature. Consistent with this general pattern, the weak light postulate affirms light's constant motion or fundamental wave speed c in relation to the medium, as being independent of any motion of the source object. Yet the weak light postulate would also allow for motion of the sink (or receiver) object to be taken into account: the constant, fundamental speed of light (c) in relation to the

2 The weak light postulate would differ from an older thesis, called the *ballistic theory* of light. That approach would treat traveling light not as a wave in a medium, but as a particle — and in that framework, source motion would make a difference in the result.

3 For simplicity, we assume that the wave speed is faster than the boat speed.

4 One might instead speak in terms of *absolute* vs. *relative* speeds of light, respectively, although we apply those labels in a different manner later in the chapter.

medium would be combined together with motion of the receiver (v) in relation to that medium. The resulting effective speed, therefore, would be compounded: $c \pm v$.

While the weak light postulate conforms to wave behavior found everywhere else in nature, the strong light postulate proposes a pattern of wave activity found nowhere else in nature: it is persistently counterintuitive, and contradicts our conventional understanding of motion and speed. Thus, the strong light postulate makes no distinction between fundamental and effective speeds of light, because the latter does not exist in that framework. In contradistinction to all other wave phenomena in nature, the strong light postulate says that it does not make any difference whether, how fast, or in what direction the sink is moving — for the same unvarying speed of light will supposedly still apply.[5]

Again, no other phenomenon in nature (wave or non-wave) displays such extraordinary properties, and the result is to make light a completely unique phenomenon altogether — and not only unique, but quite incomprehensible as well. It is in these terms that Dingle says, "The passage of light is unpicturable ..." (Dingle, 1960a, p. 22). And it is for this reason that writers who present the strong light postulate often advise readers to suspend their common sense skepticism about the peculiar scenarios which result, and simply to accept what the theory says on faith.

Given its manifest implausibility in terms of the framework we have been discussing, the question naturally arises as to why the strong light postulate is even entertained as a viable conception, to begin with ... let alone how it can be universally embraced as a brilliant insight into the patterns of nature. The main answer is that Einstein was looking for simple and general laws that could be equally well applied to both static and moving systems. That appeared to be the case with Newton's world of mechanics and objects, there being no definitive way of determining whether a system was static or (uniformly) moving — since either result can emerge if suitable external referents are chosen. Yet the same pattern did not seem to apply to Maxwell's world of light and electromagnetism — and unifying these worlds was one of Einstein's goals.

To clarify the root of the problem and what was done about it, let us start with the universally accepted premise behind the weak light postulate — that light's fundamental speed is the same regardless of whether the source is static or moving. But by contrast, what about motion of the sink — and the weak light assumption that the resulting effective speed could vary? If light's effective speed could vary in accord with motion of the sink, then this would provide a definitive criterion for determining whether the sink is static or moving — which essentially means that we are talking about absolute motion, in relation to some absolute referent for stasis. This was the objectionable feature of the Maxwell model of light transmission here: that there was some underlying reference system — a medium — in relation to which light waves would move ... and also in relation to which a sink or observer's motion could be definitively detected. Clearly, if

5 "A feature of the Relativity Theory which seems to have aroused special interest among philosophers is the absoluteness of the velocity of light. In general velocity is relative ... But it is a curious fact that if I speak of a velocity of 299,796 kms. per second it is unnecessary to add the explanatory phrase, Relative to what? Relative to any and every star or particle of matter in the universe." (Eddington, 1928, quoted in Builder, 1958b, p. 473).

the Relativity Principle was to be applied to electromagnetic phenomena in the same way that it seemed to apply to normal objects, then only relative motion could prevail — meaning that there could be no absolute referent, and so there could be no way of differentiating source motion from sink motion. The strong light postulate was born, as an amalgam of the weak light postulate and the Relativity Principle.

Unfortunately, some of the argumentation in favor of the Special Theory is based upon a jumbling together of the two different light postulates. For example, Einstein explicitly stated the weak form at the beginning of his famous 1905 paper — "that light is always propagated in empty space with a definite velocity *c* which is independent of the state of motion of the emitting body." (Einstein, 1905). But then he implicitly introduced the strong form later on — again, by virtue of the first postulate (the Relativity Principle), with the weak postulate thereby becoming transformed into the strong form, as we just discussed.

Einstein's ambiguous treatment of the issue set the pattern for most everyone else. The result is that these two altogether different conceptions about nature become confused, rather than being properly delineated and applied in a coherent fashion. Thus it is often the strong version which is described, along with an illustration of the weak version ... as we shall see in the next section. We therefore disagree with Bridgman's weak light conclusion that "different authors do indeed essentially agree. The second postulate is that the velocity of light is independent of the velocity of its source." (Bridgman, 1983, p. 113) Likewise, Pauli also states that, "The velocity of light is independent of the motion of the light source." (Pauli, 1921, quoted in Schaffner, p. 56).

By contrast, most discussions implicitly or explicitly adopt the strong version — and it is normally described in the particular terms whereby the receiving object (or sink) is labeled the *observer* ... an important factor in its own right that we will return to later in the chapter. This is how Marion states the second postulate: "The velocity of light (in vacuum) is the same for any observer in an inertial reference frame regardless of the relative motion between the light source and the observer." (Marion, p. 398) Clotfelter: "[L]ight ... always appears to all observers, regardless of their motion, to move at the same speed." (Clotfelter, p. 134) Hecht: "Einstein's second postulate is known as the principle of the constancy of the speed of light: Light propagates in free space with a speed ... that is independent of the motion of the source (and of the observer as well)." (Hecht, p. 143, italics omitted) Rogers: "[T]he Relativity hypothesis is this: The measured speed of light (electromagnetic waves) will be the same, whatever the motion of observer or source." (Rogers, p. 482, italics omitted) Lewis and Tolman: "The second fundamental generalization made by Einstein he calls the 'the law of the constancy of light velocity'. It states that the velocity of light in free space appears the same to all observers, regardless of the motion of the source of light or of the observer." (Lewis and Tolman, p. 511)

Now for a few words about the approach we will be taking in addressing these issues. To start with, of course, we reject the later formulations of the Relativity Principle. Thus, we have embraced both absolute motion and its counterpart, absolute space, as being inevitable components of an orderly scheme of nature — even though we see them as being only relatively absolute, in the sense that they are local and multiple (and nested hierarchically) rather than being universal and singular. Furthermore, given that we reject the notion that light should be treated as a special phenomenon which transcends the usual patterns of nature, we embrace the weak light postulate as the only sensible foundation for analyzing these issues — according to which source motion would not impact the speed of light, but sink motion would.

In our efforts to understand the issues in the pages ahead, we shall see that the strong light postulate can actually be understood as entailing two distinct aspects: one is strictly a definitional issue, in the theoretical terms of the Relativity Postulate which we have described above; and it has no independent validation whatsoever. The other is something far more subtle and confusing, and it entails a kind of blurring together of different properties and meanings — which nonetheless leads to a certain kind of validation, as we shall see later on. By the way, for ease of analysis, we shall differentiate the two main referents — and generally talk about the weak light postulate in relation to any motion of the source, and the strong light postulate in relation to any motion of the sink.

Two Relevant Experiments

Let us briefly consider two examples of the kind of experiments which are sometimes cited in support of the Special Theory's position about the speed of light. One is fairly old, the other somewhat more recent; but both yield a similar result, and have the same (ir)relevance to the issues at hand.

The first might be considered an observation, rather than a manipulated experiment. It was put forth by de Sitter in 1913, and was cited by Einstein as dramatic and compelling evidence for his position. It concerns certain categories of binary or double star systems in the universe, which can be visually observed from the earth (or as conducted by Brecher in 1977, double star systems observed by earth-orbiting satellites).

We are concerned with the case of the two stars in a double star system which orbit a common center of mass between them. The argument is that the light emitted by each of the stars will have a different speed, at different stages of the orbit — when the star is approaching in the direction of the earth, as opposed to when it is receding away from the earth: the presumption would be that the star's approaching speed should be added to the speed of the emitted light in the former case, and subtracted in the latter. Were that the case, the resulting visual image as seen from here on earth would supposedly be a bizarre doubling up and then disappearance of the waves as the star alternately approached and receded, "as if a very trendy celestial television producer was playing silly games with the continuity." ("New Limit on Constancy of Velocity of Light", p. 422) Suffice it to say that no such bizarre and spasmodic view is observed — i.e., there appears to be no change of speed in the manner described.

The second experiment was conducted in 1964 by Alvåger and co-workers, and repeated since then. In it, sub-atomic particles are accelerated in the laboratory to speeds which are said to approach that of light. At this point the particle decays into electromagnetic radiation, or light, and the speed of this emitted radiation is found to be the same speed as that emitted by a stable, non-moving entity.

What have we learned from these two well-known experiments? Not much, in fact, regarding Einstein's radical proposition. That is because only motion of the *source* is involved, not motion of the *sink*. We therefore learn nothing which is unique about light from experiments like this: they merely provide support for the bland weak light postulate — they say nothing whatever about the strong light postulate, and certainly do not lend support to it. As such they are little more than yawns which accord no particular credibility to the Special Theory of Relativity, in this regard.

It is important to be aware of the kind of subtle and misleading reasoning that is typically involved in interpreting experiments of this kind — amounting to a kind of sorcery — which fool us into thinking that more is going on here than actually is ... thereby luring us into embracing the Special Relativity mind-set. First, we are presented with the second postulate of Relativity, typically in the vague form that "the speed of light never changes". We are then told, of course, how deep and extraordinary an insight into nature this is, by contrasting the behavior of *light* with that of *mechanical* phenomena. Yet this feint merely distracts us from one distinction which can be made here — not that between light and mechanical phenomena, but rather that between any *wave* and *mechanical* phenomena ... or as we would categorize it, between medium and matter domains.

Next we are presented with experiments like Alvåger's, and told that it "is a direct and striking confirmation of Einstein's [second] postulate" (Marion, p. 399), or in connection with Brecher's version of the double star observation experiment, that "[I]t is pleasing to know just how right Einstein's second postulate in the special theory of relativity is." ("New Limit on Constancy of Velocity of Light", p. 422) Neglecting to mention that such evidence only supports the unsensational *weak* interpretation of the second postulate regarding motion of the source, we are then assured how fully Einstein's conjectures have been supported.

What happens in these cases is that somewhere along the way, in an almost subliminal manner, the other situation (motion of the sink) is often subtly introduced or merely implied: we are left with the impression that the speed of light is constant, even in the case of a moving receiver — we are led to believe that the strong interpretation has been somehow proven by experiments like these. But that is not the case.

Measuring Light's Speed

Let's get down to basics: why don't we simply test the constant speed of light proposition, by directly measuring light's speed in the different cases we are talking about? Great idea — but how shall we do it? The obvious answer is to determine the speed of light in the same way as one determines the speed of anything else: measure the distance between two points and then monitor the time it takes for light to travel between those two points — voilà, one has calculated the time it takes for light to travel a given distance ... the speed of light!

So much for easy and obvious answers. There's no problem with this procedure, in principle — it's just that the details of performing such a measurement for a phenomenon that moves as quickly as light are not easy. Here is the rub: light travels very fast, and this requires extremely accurate determinations of the spatial distance traversed, as well as the temporal duration which the journey requires. For example, how are we to measure the distance involved? Obviously the greater the distance, the better — which is true for any accurate measure of speed, but especially so for dealing with the extraordinary speed of light. By all accounts, light has been calculated to travel in the range of some 300,000 kilometers (or 186,000 miles) per second, in certain situations. Therefore, if we are to gain any kind of precise measure of its speed, in the usual terms, we need a very large distance. Of course, one can reduce the span required by incorporating intermediate reflecting devices, so as to effectively multiply the distances involved — but that introduces its own dubious set of assumptions, as we shall see shortly.

There is no escaping the fact that any substantial distance is itself difficult to measure with the kind of precision which is necessary. Indeed, this determination of spatial distance is considered a severe limiting factor: "The precision of many of the measurements [of light's speed] has reached a value comparable with that of measuring the path length." (Froome and Essen, p. 136). Oddly enough, it turns out that some methods of determining the distance are based upon the reverse procedure — of using an assumed speed of light to infer the precise distance involved;[6] this light-determined distance is then used as a critical factor in measuring the speed of light! Physicists even defend the extraordinary circularity of these techniques without blushing: "It is no criticism of them therefore to point out that as a means of measuring *c* they have this fundamental limitation." (Froome and Essen, p. 136)

So be it: determining the distance factor is a problem, but what about the time factor? How are we to gauge the amount of time it takes to travel between two points? There are similar problems of circular definition here, as well. Here is how Brillouin puts it, in discussing both spatial and temporal factors — as being defined in atomic measurement terms:

> Let us mention here a real difficulty resulting from internationally adopted definitions. The unit of length is based on the wavelength of a spectral line of krypton-86 under carefully specified conditions with accuracy 10^8 and the unit of time is based on the frequency of a spectral line of cesium with accuracy 10^{12}. Hence, *the same physical phenomenon, a spectral line, is used for two different definitions: length and time*, and the velocity *c* of light remains undefined and looks arbitrary. (Brillouin, p. 5)

MacRoberts also notes that, "In modern atomic transition chronometry if both time and space were defined by the same transition the velocity of light would be tautologically constant." (MacRoberts, p. 367)[7]

Yet as far as the time measurement goes, why get so complicated — can't we simply use something like a very precise stopwatch? Normally such a determination involves noting the starting time, when the travel begins here; and the ending time, when the travel ends there. But here is the problem in the case of light: how are we to know what the time is when the light reaches the destination at the other end of its travel, assuming we remain at the origin with our clock?

6 "By knowing *c* the distance is obtained." (Bergstrand, p. 758) Or, "The modern methods depending on a baseline measurement are of course designed to operate in the reverse sense giving the lengths in terms of the velocity of light." (Froome and Essen, p. 136)

7 Murchie makes the same point, that "to define both space and time in atomic vibrations is in effect like defining 186,282 miles as the distance light travels in one second, and one second as the time it takes light to go 186,282 miles. Under such a ruling, no matter how fast or slow light ever got to be, the distance it would go in a second would have to be 182,282 miles — no more, no less — because of the definition." (Murchie, p. 508)

In terms of most common phenomena, we might answer that we can see when the entity in question reaches its destination; and we can thus note the time on our clock when the entity arrives there. Unfortunately, that is not so easy to do when we are dealing with a substantial spatial distance between the starting and ending points — given that we must be able to see the ending point, from the starting point where we remain with our master clock. Furthermore, the very act of seeing the distant event can be considered as a process in its own right, which takes a certain amount of time to complete. Therefore, we would need to calculate the period of time which that process of seeing takes — i.e., in order to get the information back to us from the distant point — so that we can factor it out when we determine when the test entity did, in fact, arrive at the destination.

Of course, for most everyday events, that process of seeing is so fast that it has very little significance, in relation to the much longer time periods at issue. But in the case of light, this factor is not so easily disregarded, and the problem becomes vastly more complex.[8] That is the case because the phenomenon we are measuring *is* light and its speed, and therefore the process of seeing the moment that it arrives at its destination would take a comparable amount of time as the transmission of the target phenomenon itself!

Well, is there some other way around this problem, of communicating the information back to the starting point, as to when the light arrives at its target — aside from our seeing the event directly, which itself uses visible light as the mechanism of information transmission ... along with also being the subject of our inquiry? Of course, the answer to that particular question is, yes: there are any number of other modalities which could be used for conveying the information back to us at the starting point, aside from using the very phenomenon whose speed itself we are attempting to measure. Somebody at the other end could pick up the telephone, and inform us when the moment of arrival is reached — yet, of course, that alternative information transmission process also takes some period of time. Likewise with any other information conveyance that one can imagine — from carrier pigeon or tortoise, to voice transmission or a BB-gun fired back through the open air: any of these information vehicles could be used, assuming we are precise in quantifying the duration which is required in order to transmit the ending point arrival information back to us and our master clock at the starting point.

But there is still another problem, which none of these transmission vehicles address — namely, how are we to know that the clock at the receiving end is synchronized with the clock at the starting point, to begin with? For no matter how we convey the arrival time information back to the starting point, this has little value if the two clocks are not reading identical times. It turns out that this synchronization issue is the real nub of the problem here — both to determine the time required for the light to travel from here to there, but likewise to determine the time required for this arrival time information (there) to be communicated back to us (here) ... regardless of how this process is accomplished. And there are no easy answers, either.

8 The speed of light is reckoned to be some one hundred thousand times faster than typical speeds in everyday life.

Perhaps it's time to switch gears, and look for other ways of approaching the issue. That is, the methodology which we have been discussing thus far represents attempts to measure the speed with which a phenomenon travels from one point to another point — from a source to a sink — in one direction only. As we have seen, the problem with that methodology is in determining the precise point in time when the phenomenon arrives at its destination, and then for conveying that information back to us at the start ... inasmuch as we cannot be at both local source and distant sink at once, and thereby make both determinations directly.

However, there is another way to deal with the issue, and that consists of a round-trip measurement: instead of trying to ascertain when the phenomenon arrives at some distant destination, why not simply let it be the carrier of its own information — in the sense of having it bounce off that now-intermediate destination and then come straight back to us at the starting point? Then, there is no longer a problem of registering the precise arrival time and choosing one or another mechanism to convey that information back to us at the starting point. All such knotty problems are thereby eliminated, and we can directly note both the point on the clock when the phenomenon departs, as well as the point on the same clock when the reflected phenomenon returns.

But little of value in life is so easy or free; and not surprisingly, there is also a drawback to this two-way methodology. For the issue we then have to deal with is whether the speed with which the phenomenon travels in the outgoing direction is, in fact, the same as the speed with which the phenomenon travels back in the return direction — i.e., whether direction of transmission is itself a factor that needs to be taken into account. And as one might guess, there is no obvious answer to this question.

Physics has a pair of names which can be applied to the two outcomes, in this case for describing the relationship between the respective speeds of the outgoing and incoming light. When the go-and-return speeds are measured as being equal, then light is said to travel *isotropically* in the two directions — whereas when the go-and-return speeds are measured as not being equal, then light is said to travel *anisotropically* in the two directions. Nor is this distinction a trivial one, one that is merely put forward to confuse the situation. On the contrary, as we shall explain later, differentiating these two categories has enormous significance for understanding the issues at hand.

In the end, there is no obvious way out of this morass: On the one hand, we have seen that the problems involved in a one-way test of light's speed make that approach well-nigh impossible. Indeed, some theorists claim to have proven that the one-way measurement of light's speed cannot possibly be ascertained.[9] On the other hand, we can measure the speed of light based on a two-way test — but then we can only conjecture as to whether the to- and fro- paths take equal time, and thus whether light is actually isotropic, in this respect. The only way to answer that question would be to measure each of the two paths separately, and then compare them; but obviously, that takes us back to two separate one-way tests, and all the accompanying problems therewith. That is, if we

9 See, e.g., Builder, 1959, p. 93: "Measurement of the velocity of light over a unidirectional path from one point *A* to another point *B* is, in principle, impossible." Also see, e.g., Bridgman, 1983; Podlaha, p. 376.

wish to compare the speed of light between here and there, and then between there and here — to see if they really are isotropic — then how are we to resolve the inherent one-way problem of measuring and conveying information about time?

We keep returning to this same one-way dilemma: How can we, *here* ... determine the time, *there*? Yet we must do so if we are to calculate when light arrives *there*, in order for us to compare that time point with when light left *here*. This is the fundamental temporal problem — the paradox — of measuring light's speed in any direct, one-way sense: determining what the time is on the other end, so as to compare it with the time on this end.[10] We know the latter, but there is no way of directly knowing the former. It is an inevitable limitation of nature, and ultimately there is no recourse but to adopt whatever assumptions and definitions we decide are the best way of dealing with the issue. And it is in precisely this inherently unprovable realm where the conceptual battle must be waged — for it will not be fought in the laboratory, where the answers cannot be found.

The Subjugation of Time

It should be apparent that there are serious epistemological problems, both of methodology and of definition, which are inherent in any attempt to measure or explain the speed of light. It was upon this amorphous terrain that Albert Einstein was able to gain traction for a totally different way of approaching the issues — and in the process, of turning our view of the world upside down.

Recall our discussion in Chapter 4, regarding the "law of nature" interpretation which he imposed upon the Relativity Principle. This approach was based upon the Lorentz transformation, which entailed a consistent speed of light factor. Einstein interpreted this in terms of what we have described as the strong light postulate, in its fullest general form.

Yet how was Einstein able to exploit the ambiguities we have been talking about, regarding the difficulties of measuring the speed of light, in order to reach such a conclusion? On the face of it, there's no good reason why the speed of light should always be constant, regardless of the situation — and the idea seemed to make no sense in relation to a moving receiver, as we discussed earlier in the chapter. The only solution would be to change the rules, by establishing some novel definitions for dealing with these matters.

Before getting into these questions, it is important to emphasize at the outset that all the crucial conceptual components of Einstein's model, such as we are describing here, were not new and not his ... but were developed earlier by Lorentz and Poincaré, a point we shall examine further in the second part of the next chapter. However, it could be argued that one of the salient differences is that the latter pair of theorists were seeking practical explanations for conducting measurements and explaining experiments,

10 "Thus the clocks cannot be synchronized unless the unidirectional velocity of light is known, and the unidirectional velocity of light cannot be ascertained without synchronizing the clocks. Thus when we speak of 'the velocity of light' measured in an inertial coordinate system we cannot be speaking of a measure of its unidirectional velocity." (Builder, 1959, p. 93)

rather than necessarily claiming to be describing the essential nature of reality — whereas Einstein displayed no such humility or reservation. In any event, it is Special Relativity that we are examining — and which won the day — and so we treat the issues in Einstein's terms.

We already discussed one aspect of the new program in Chapter 4, regarding Einstein's dismissal of the notion of absolute space — which we treat as a matter domain construct; and likewise here, for its medium domain counterpart, classically called the "ether". But it is the time concept which bears the brunt of Einstein's assault on the classical world view, and which most commentators credit as representing the most significant contribution of the Special Theory. And it is the time factor that we identified just above as the chief problem in measuring the one-way speed of light.

For his part, Einstein would concur with the ambiguities we wrote about in the previous section — that there was no definitive way of determining the end-of-journey time at a distant location, in order to register how long it took for light to travel from here to there ... one consequence being that there was no way of measuring the speed of light in one direction. And for their part, both Lorentz and Poincaré recognized "the impossibility of measuring the unidirectional velocity of light over a one-way path in the system." (Builder, 1958b, p. 464)

Einstein had a solution, which in a way would kill two birds with one stone — in that it would deal both with the time-at-the-other-end issue and the speed of light issue in one fell conceptual swoop, by defining the two in terms of one another. The result was a "convention" — a consistent, albeit arbitrary, choice among the different possibilities for dealing with the issues here. Einstein essentially defined the moment of time at a distant point (e.g., when the light arrived) in terms of however long it took for light to travel there. But of course that's not the end of the story, because the end-of-travel-time information still needed to be communicated back from the receiving end, in order to do the calculation. What better way than to have light itself do us the favor of conveying this information, by means of a return trip (as we described above)?! However, that still left open the crucial question we have made reference to, as to whether there was a difference between the speed with which light was measured to be traveling in the outgoing direction as in the return direction — which was obviously relevant to any determination as to whether the speed of light was really constant or not. Einstein's solution for dealing with this issue was simple: merely stipulate that the two one-way (outgoing vs. return) speeds were to be treated as equal!

Look at what has happened, with this methodology. For one thing, it singles out light itself as the modulator, as providing the benchmark for the passage of time itself. For another, it is predicated upon the presumption that isotropy reigns — that each of the go and return legs that are involved represents precisely one-half of the total transit time. The resulting interpretation is enshrined in Einstein's *clock convention* (also called the "simultaneity convention"): it amounted to a definition of a time point, over a distance, that is based upon light and its isotropic travel.

What we are dealing with is an arbitrary rule that gives us no information whatsoever as to whether light should be the unique arbiter of time, and whether the two respective one-way speeds of light are in fact equal or not. As MacRoberts puts it, "The definition of 'time' in special relativity is based on light propagation and the 'constant velocity of light' is a tautological consequence of the definition." (MacRoberts, p. 365) Or as Builder puts it more benignly: "It is therefore clear that the concepts of 'time' and 'the velocity of light' are not independent within the context of the restricted [or special] theory itself, since each is definable only in terms of the other." (Builder, 1959, p. 93).

In this respect, Einstein's arbitrary choice of a method is quite dubious, in the sense that it hopelessly confounds that which we nominally seek to measure (the speed of light) with the means by which we ascertain it (the passing of time): the Special Theory determines the time on the receiving end by using light — and then proceeds to "measure" light's speed in terms of the time which that same light has delineated! We learn nothing at all about the supposed constancy of light's speed in this process. As Ives puts it, "The assignment of a definite value to an unknown velocity, by fiat, without recourse to measuring instruments, is not a true physical operation, it is more properly described as a ritual." (Ives, p. 156)[11]

One way to think about the issues is in terms of a conceptual vehicle which Einstein promoted, called the *light clock*. This simple abstract device might consist of a rigid rod of some length, with a clock at one end and a mirror at the other. A ray of light would be propagated at the clock end, whereupon it would travel to the mirror end ... and then return to the starting point. One could also add a second mirror at the starting end, in which case the light would then bounce between the two mirrors at each end. What results can be described as a clock, with the oscillating light representing its ticks. As Einstein put it, " ... a light signal, which is reflected back and forth between the ends of a rod, constitutes an ideal clock." (Einstein, 1949, quoted by MacRoberts).

MacRoberts muses about this device being a perfect clock: "The concept that this is 'ideal' or 'natural' is found in various authors. What is especially ideal or natural is not explained." (MacRoberts, p. 369) Moreover, he makes the point that, "Such clocks obviously cannot be used to measure the velocity of light; they *are* the velocity of light...." (MacRoberts, p. 369) That is because light will always appear to have the same speed when one uses that selfsame (and isotropically defined) light as a standard for determining the amount of time it takes for this light to travel between two points! MacRoberts again: "[T]he definition of time becomes the tautology: Light travels a round trip over a measured course in the time taken for light to travel the round trip over the measured course." (MacRoberts, p. 368)

Here is how Bridgman describes the essentially circular logic that is involved here, such as in the case where the light clock system is moving:

11 Here is Margenau's somewhat opaque description of the maneuver: "In the Special Theory, light signals perform a metric-defining role by providing a criterion of simultaneity of spatially separated events. To speak of proving experimentally that the velocity of light is rightly or wrongly conceived as a constant appropriate to this role is to presume that we are in possession of a more fundamental metric which may serve as a basis of the proof. If the Special Theory is correct, we have no claim to a more worthy kinematic starting point. A condition on measuring the velocity of light is therefore that the instruments of measure be calibrated with respect to the velocity of light. The circularity implicit in such an experiment therefore prevents us from calling the hypothesis of Einstein an inductive generalization of experimental facts." (Margenau, 1978, p. 193). More simply, as Calder puts it, "In Einstein's universe the speed of light is more fundamental than space or time. Space is what light moves in; time is how long it takes to move." (Calder, 1979, p. 60)

> The postulate that the velocity of light is independent of the motion of the system now becomes the most banal truism if we reflect that we might make the beam of light whose velocity we are measuring the beam which is reflected back and forth between the mirrors of the clock. In fact, the truistic element in this situation is so overwhelming that we wonder whether we have said anything with any physical content at all when we say that light velocity is the same in all frames of reference. (Bridgman, 1983, p. 136)

As far as the isotropy assumption goes, Einstein would not dispute the fact that this is merely a convention — although he also claims that it is "[i]n agreement with experience ..." (Einstein, 1905, quoted in Builder, 1958b, p. 467). Builder contends that the reference to "experience" partly relates to some crucial experimentation by Michelson that we shall discuss shortly. Brillouin is critical of such a conclusion, arguing that "The Lorentz transformation requires an averaging with light beams propagating in opposite directions.... [T]his *rule is arbitrary and even metaphysical*. It can neither be proved nor disproved by experiments, it *assumes* that signals propagating east to west or west to east have equal velocities, while Michelson's experiments only measure the average of these two velocities. The sudden and unverifiable assumption is obvious." (Brillouin, pp. 68-69; see also Ives, p. 188). Christensen puts it succinctly: "[I]t is judged (or decreed) that the signal-reflection occurs halfway — temporally speaking — between emission and reception." (Christensen, p. 51)

The reasoning behind the isotropy assumption can once again be traced back to the Relativity Principle: anisotropy would violate the underlying premise — because (as we shall see next) if light were measured to travel faster in one direction than the other, then this would indicate that there was the equivalent of an absolute space ... and therefore that the no-absolute-space interpretation of the Relativity Principle was false. Such an outcome would be most inconvenient for the theoretical scaffolding which Einstein was erecting — including the singular "law of nature" thesis which would effectively eliminate any distinction between stasis and (uniform) motion in relation to light phenomena — so he simply adopted an arbitrary definition which precluded the possibility of making such a distinction. That was not so hard to do, because such an approach could not be proven wrong, due to the difficulty in performing one-way measurements; but, of course, it could not be proven right, either. Yet since it is just a convention we are dealing with anyway, wouldn't that be the easiest solution?

The Non-Constant Speed of Light

We now want to look at two different reasons for why Einstein's strong light postulate approach is unsatisfactory. The first concerns the issue we have just been discussing, regarding isotropy. The second concerns the alterations which are associated with the Lorentz transformation, along with the speed of light which goes with it. As we shall see, these distinct problems can be understood in terms of our two main themes — that of complementarity and hierarchy, respectively.

Starting with the first point, there is no dispute about the isotropy issue when the system is static. To understand why, let us recall the simple situation that we described earlier in the chapter. If a boat is sitting still in a pond, then the speed of an approaching wave is no different in relation to the boat than it is in relation to the pond — since the boat and pond are mutually static — and we are thus dealing only with the wave's *fundamental speed* in relation to the (pond) medium. However, if the receiving boat is moving (i.e., towards or away from the approaching wave), then the boat's speed in relation to the pond must be combined together with the wave's speed in relation to the pond — resulting in a different *effective speed* of the wave in relation to the boat.

Of course, the scenario we have described is a one-way affair, in that the wave is generated by a disturbance (e.g., a source boat) elsewhere on the pond: the wave travels in one direction towards a sink boat. In order for the scenario to become a two-way affair, the one boat would have to act as both source and sink — with the wave traveling outward from the boat to some intermediate barrier, whereupon it would "bounce" and then come back to the same boat. For example, let us say that there is a single boat which is sitting still in the pond. Then a stone is dropped into the water at the forward end of the boat. This creates a wave that travels outward to a barrier beyond, and then "bounces" and comes back toward the boat. Leaving aside any non-relevant factors, we can again say that the traveling wave will have the same simple (fundamental) speed in relation to the static boat as it does to the pond. But the main point is that the traveling wave will have the same speed in both its outgoing and incoming segments: it will be isotropic — and in particular the two-way wave speeds will be isotropic in relation to the static boat (as well as to the pond, of course).

That detail is crucial, because this outcome will *not* obtain when the boat is in motion — for then we must factor in the boat's speed in relation to the pond, along with the wave's speed in relation to the pond, in order to determine the wave's effective speed in relation to the boat.[12] Thus, if the boat is moving in the same (forward) direction as the outgoing wave, then the effective speed of the wave (in relation to the boat) will be diminished ... whereas once the wave hits the barrier and is traveling back towards the boat, then the boat is heading towards the incoming (return) wave, and the effective speed will be augmented. The *effective* speeds (in relation to the moving boat) are therefore different for the outgoing and incoming waves — and so they are anisotropic.[13]

Now we can return to the main issue, for the problem should be apparent: this same sensible overall analysis as we have been describing for water waves on the pond cannot be applied to light waves and systems they interact with — not because the same paired pattern we just described does not exist there, but simply because Einstein has adopted an arbitrary definition which precludes it. And this is precisely why the clock convention assumption about isotropy is so problematic: it is a definition which masks or cancels out any disparity between two different one-way speeds of light, by merely looking at the average of both. Again, this would not be an issue for a static system, because then the two one-way speeds would be the same — isotropy. But that would not be the

12 As we have previously discussed, any motion of the boat as the source of the wave makes no difference to the fundamental speed of the wave, this property being a characteristic of the medium ... for all types of wave phenomena. Also note that we are not addressing any non-relevant factors here, in order to focus on the main conceptual issue that is involved. For example, we mentioned in a previous footnote that the wave is assumed to be moving faster than the boat.

13 The *fundamental* speed of the wave in relation to the (pond) medium is the same (isotropic) in both directions, but that is not the question here.

case for a moving system, as we have just explained, because the two one-way speeds would be different — anisotropy. Yet the latter is simply disallowed by the theory, and only the former is permitted.

As to how the Special Theory is able to get away with this maneuver, we already know the answer — because conducting one-way measurements of light's speed are problematic, if not impossible. And that is why we said earlier that the answers cannot be found in the laboratory, but must be fought in the conceptual arena — which is the reason we are even having this conversation, to begin with.

An obvious way of framing the matter is to bring us back to that perennial issue of stasis vs. motion, and how we are to make such a distinction for any given system — and what the resulting conceptual framework would look like. Einstein's dismissal of a consistent underlying reference for stasis, in relation to which the system (and the light) are traveling — be it called absolute space or ether medium — meant that there were no clear grounds for making such a stasis vs. motion determination: all motion is relative, and every system could be seen as static or moving merely by selecting an appropriate referent. As far as the speed of light goes, Einstein's solution was to essentially make each system its own referent — meaning that every system would be treated as static in relation to itself … even when the system is ostensibly moving, as we shall see shortly.

One way of thinking about such an internal reference for stasis is to say that every system carries its own local mini ether medium, in relation to which light always travels isotropically — although, of course, the ether and medium concepts are taboo in Special Relativity. For the purpose of determining the speed of light, the system is treated as *not moving* — even when it ostensibly *is moving* — and so not surprisingly, there is only a fundamental (and thus constant) speed of light … there is never an effective speed that is any different. To call this a one-sided representation of nature's patterns is putting it mildly, a point that we shall return to shortly.

Next let us turn to the second main problem with the strong light postulate, and this concerns a system which is clearly described as undergoing motion … and high-speed motion, at that. The essence of the issue is simple, and has been demonstrated experimentally: the moving system displays properties which are different from its static counterpart, and these can be described as alterations of spatial and temporal parameters. Needless to say, there is nothing especially intuitive about this pattern, because it does not appear to resemble anything in our usual experience. The standard explanation is that these alteration effects only become noticeable when an object is moving extremely fast, and this is not something we are normally in a position to observe.

But there is one particular aspect of these motion-related effects which bears on the issues we are discussing — namely, that the parameters of space and time would be modified in just such a manner that any measurement of light's speed in the "moving" system would come out to be the same as would be determined by a static system. That is an uncanny conclusion in its own right, of course, but we shall see that there is actually considerable confusion as to what it really means — and while in some ways it is an accurate depiction, in other ways it is not.

As usual, the original concept for this innovative alteration approach did not come from Einstein. It had been previously developed by others, in what became known as the

Lorentz-Fitzgerald (or Fitzgerald-Lorentz) contraction — which contraction would apply to the spatial parameters of a moving system.[14]

But the principle would also apply to temporal phenomena: Thus, if one thinks about the issue in terms of the light clock paradigm, the spatial dimensions of the rod would be shortened by the motion ... which would thereby alter the metric of time in a corresponding manner as that of space (e.g., Builder, 1958a, p. 294).

Yet there were several differences in the way these ideas were construed in Lorentz's model, as opposed to Einstein's. For one thing, there was an explicit physical explanation in Lorentz's theory which caused the alteration of spatial and temporal dimensions — the friction-like interaction between the moving object and the static ether medium through which it passed. But we have seen that in Einstein's model, there is no medium and thus there is no physical explanation beyond the notion of motion per se as somehow producing the effects.[15]

Not only was there no physical explanation for the alterations in Einstein's model, but there was also the disturbing little problem of why we should even conclude that the system is moving, to begin with. Given that there is no absolute space or ether medium in the Special Theory which would serve as a referent for stasis, then in relation to what does the motion exist ... along with the spatial and temporal alterations which go with it? Again, there was an ether medium in Lorentz's model, which not only provided a physical substratum that would explain the alterations, but which would also establish a clear referent of stasis in relation to which the motion would exist. By contrast, in Einstein's model all motion is relative — and therefore any system can be said to be arbitrarily moving or static, simply by selecting an appropriate referent. Are the spatial and temporal alterations also supposed to be arbitrarily relative? We will return to this issue in the next chapter, along with a related problem: that is, even if we were dealing with the same other object referent, there is still no basis for concluding which of the two objects is static and which is moving in Einstein's model ... and therefore the spatial and temporal alterations would apply to both. Needless to say, such a proposition leads to enormous contradictions.

But now let us leave aside the two radically different theoretical frameworks of Lorentz and Einstein for understanding what is going on here — one of which makes sense and the other of which does not — and consider the quantitative nuts and bolts of the issue, for both theories essentially agree about that.[16] The reason is because the resulting spatial and temporal alterations which we have made reference to would be described in terms of the Lorentz transformation. This set of formulas quantifies the changes which a moving system would undergo, based upon its speed ... however that is

14 The Irish physicist George F. Fitzgerald put forward the initial proposal in 1889. The Dutch physicist Hendrik Lorentz was unaware of this idea when in 1892 he proposed an almost identical mechanism.

15 "[A]ccording to Einstein, it is *motion itself* that causes contraction, and, in addition, time dilation." (Zukav, p. 138) Note that Einstein explained the length contraction in terms of time dilation.

16 As we shall discuss in Chapter 8, there is a distinction between how Lorentz construed the time alteration factor, at least initially, as opposed to Einstein's (and Poincaré's) approach.

determined. As we noted above, the space and time dimensions of the moving system would be altered in just such a manner that would still yield the same speed of light result as for a system that was static.

Here is how Einstein describes the outcome, in terms of the light clock paradigm: "We now assume that the clocks can be so adjusted that the velocity of propagation of a light ray in vacuum — measured with the aid of these clocks — will everywhere be equal to a universal constant *c*." (Einstein, 1907, quoted in Builder, 1958b, p. 469) In other words, the spatial and temporal metrics — and in particular here, the latter — would be altered as a result of the motion, with the rate of time "clicking" changing accordingly ... so as to yield the same constant *c*.

Note that there is no reason to assume that the underlying issue concerns some unique or magical property which has been discovered about light, per se — although the usual descriptions seem to indicate that, in keeping with Einstein's grand pronouncement on the subject. The reason is because the same pattern would apply in principle to all kinds of phenomena: any speed is simply a combination of spatial and temporal factors — and if those components change (in appropriate proportions) then the ratio between them will still come out to be the same, regardless of whether the topic is the speed of light or the speed of anything else. It is as though we are dealing with a shrunken Alice in Wonderland world in which the metrics have changed, but all in the same proportions.

Indeed, one can argue that the particular issue we are talking about here amounts to a virtual artifact, which can be thought of as akin to a fairly minor bookkeeping procedure. Such a position is consistent with the fact that the fine print of the conventional descriptions often refers to the apparent or measured speed of light always being constant.[17] For example, the numeric factor (as represented by *c*) would always be the same, in regards to the issue we are discussing — but that is only because the resulting speed ratio (of distance and time factors) is the same.[18]

Yet we contend this does not mean the two speeds are identical in relation to one another. That is because we are dealing with two different sets or levels of space and time parameters, with the moving system being contracted relative to the static system — and the result being that the two "same" numeric speeds are phenomenologically different.[19] One way of describing the two meanings we are talking about is in terms of a distinction between *measured* speed and *actual* speed. The former may be absolute, in the limited ways that we are discussing; but the latter is only relative.

17 For example, Zukav says that, "*[T]he measuring instruments* change from one frame of reference to the other in just such a way that the speed of light always appears to be the same. The speed of light appears constant because the rods and clocks used to measure it vary from one frame of reference to another depending upon their *motion*.... A "constant velocity of light" is exactly what would result if moving measuring rods became shorter and moving clocks ran more slowly because a moving observer would measure the speed of light with a shorter measuring rod (less distance for the light to travel) and a slower clock (more time to do it in) than an observer at rest." (Zukav, p. 138). Rogers puts it this way: "[O]ur measurements are affected by our motion, and only *the speed of light, c, is invariant*. A broader view treats *c* as merely a constant scale-factor for our choice of units in a compound space-&-time, which aspect observers slice differently." (Rogers, p. 485)

Now let us try to bring the two issues together that we have been examining — that of isotropy vs. anisotropy, on the one hand; and that of measured vs. actual light speed, on the other. To set the stage, let us go back to the earlier Lorentz theory — from which we maintain that two essentially different analyses emerge.

On the most basic level, we have seen that the Lorentz model was built upon Maxwell's concept of an ether medium, in relation to which light traveled at a fundamental (and isotropic) speed c; and any motion of a system would also exist in relation to this medium — the result of the two motions together yielding an effective (and anisotropic) speed $c \pm v$. As Builder puts it:

> [T]he Maxwell-Lorentz theory supposed that light is propagated in a universal stationary medium (ether) with a velocity c characteristic of the medium and independent of its source. On this theory it would seem incontrovertible that the propagation of the light, relative to any inertial system in motion relative to the ether, would be non-uniform and anisotropic, i.e. it would be propagated relative to the inertial system with different speeds in different directions. This inference fits in with our commonsense view of motion. (Builder, 1958b, p. 459)

At the same time, we mentioned in Chapter 4 the mathematically disruptive effect which such a motion-based factor would have upon the form of Maxwell's equations, and Lorentz eventually came up with modifications thereto that eliminated these complications. One consequence of this new approach was that light could be understood to travel at the abovementioned same measured speed in relation to a moving object — albeit within the "rest frame" of that object — by virtue of the altered dimensions of space and time (quantified in terms of the Lorentz transformation) which were related to the motion. Nonetheless, whether we are talking about a rest frame that is associated with the ether, or a rest frame that is associated with any other object — either of which would determine a constant (isotropic) speed of light, c, in relation to itself — any such rest frame would still conclude that a different (anisotropic) speed of light, $c \pm v$, would apply in relation to some other entity that was moving. Builder again: "[M]easurements made in each inertial reference system [i.e., rest frame] of the velocity of light are compatible with isotropic propagation of light relative to that system and anisotropic propagation of light relative to every other such system." (Builder, 1958b, p. 462, italics omitted)

18 Note that this same analysis would also apply to situations where we are only monitoring or gauging two respective speeds, as opposed to technically measuring them — such as in the groundbreaking experiment we discuss in the next section.

19 One way of understanding the point we are making here is in terms of a far more general pattern, which we have called multiple meaning. We talked about this issue in the opening chapter, in describing how the *same* construct can have *different* meanings at various hierarchical levels. This is precisely what we are dealing with here (although we are getting slightly ahead of ourselves in mentioning the hierarchy factor now) — not just in terms of the general construct called the speed of light, but also in terms of the specific theoretical or mathematical factor of c, along with the space and time dimensions upon which it is based. We argue that the way these phenomena vary at different hierarchical levels can be understood in terms of multiple meaning.

In this regard, we contend that in spite of the transformation he devised, Lorentz was still able to embrace the plain additive vision of light's speed in relation to a moving entity — whereas Einstein accepted only the unchanging speed of light proposition which lies at the heart of Special Relativity. For Einstein, a singular concept of light served as the essential independent mediator of existence and arbiter of reality — and its abstract speed represented the central rhythm which the rest of nature danced to, including our metrics of space and time. But we contend that the opposite (additive) position must also be incorporated, as part of a coherent model of nature.

We now want to elaborate on what we believe was the prophetic nature of Lorentz's thinking about these matters, along with that of Poincaré — for these two great physicists are commonly criticized for being of two minds on many of these issues, in contrast to Einstein's widely-touted single-minded convictions. This distinction is especially apparent in these theorists' dual vs. singular conceptions of time, as we shall discuss in Chapter 8.

Indeed, we contend that all the crucial issues which are involved here can best be understood as dualities of nature. One side could be called here the *X*-view perspective, and we will talk about it as being an internal stasis-based analysis — such as with a static system: there would be a fundamental speed of light, c, and isotropy would reign. The other side could be called here the *P*-view perspective, and we will talk about it as being an external motion-based analysis — such as with a moving system: there would be an effective speed of light, $c \pm v$, and anisotropy would reign. Needless to say, Einstein would only subscribe to the *X*-view framework here — even when the issue concerns what is supposedly a "moving" object — whereas we contend that the thinking of Lorentz and Poincaré was broad enough to encompass both.

X-view	*P-view*
stasis-based (static system)	motion-based (moving system)
fundamental c constant (isotropy)	effective $c \pm v$ (anisotropy)

Our explanation for the non-constant speed of light would join together this complementary analysis with that of hierarchy — neither of which theme is compatible with Special Relativity. The most which comes out of Einstein's model, regarding these issues, is the standard conclusion that the static entity has normal space and time metrics, while the moving entity has altered space and time metrics ... leaving aside various anomalies about how these are determined. The problem with that approach is two-fold — the first of which concerns a blurring together of complementary states of stasis and motion, and the second of which does not distinguish between different hierarchical levels of phenomena.

To elaborate, let us apply the concept of hierarchy here by talking about lower vs. higher levels — at each of which there can be a static entity and a moving entity. The crucial clue to our analysis is that what is a moving entity at a lower level can be taken as a static entity at the next higher level.[20] In this hierarchical framework, it is not the static vs. moving entities which would have normal vs. transformed (or what we can call *baseline* vs. *altered*) space and time metrics, but the lower vs. higher levels — each of which would have both stasis-based and motion-based outcomes. Here is how we might chart these complementary and hierarchical relationships (with alphanumeric labels designating each of the 2x2 categories, for ease of reference in the discussion below):

Altered space and time frame / medium (Lorentz-transformed)

X-view (IIa)	*P-view (IIb)*
stasis-based (static system)	motion-based (moving system)
fundamental c constant (isotropy)	effective $c \pm v$ (anisotropy)

Baseline space and time frame / medium (Normal)

X-view (Ia)	*P-view (Ib)*
stasis-based (static object)	motion-based (moving object)
fundamental c constant (isotropy)	effective $c \pm v$ (anisotropy)

We have seen how Einstein's approach was based on a general definition which simply precluded the anisotropy or motion-based outcomes (*Ib* and *IIb*), thereby leaving us with only two of the categories (*Ia* and *IIa*) above. And now we can see how a state of motion gets slipped in through the back door, as it were, by virtue of being converted into a state of stasis! That is possible because of the clue we just mentioned — that what is a moving entity at the lower level (*Ib*) can be taken as a static entity (a la "rest frame") at the next higher level (*IIa*).

In terms of this analysis, what Einstein does is to jump from a baseline (lower) level of stasis (*Ia*) to an altered (higher) level of stasis (*IIa*), thereby conveniently ignoring the motion-based counterparts (*Ib* and *IIb*) in each case. In terms of light's speed, Einstein simply shifts from one stasis-oriented determination of c (baseline *Ia*) to another stasis-oriented determination of c (altered *IIa*). It is this relationship between two levels which is represented by the Lorentz transformation — and that is why the conventional model talks about the same constant speed of light, c, applying both to the static system and to "the rest frame of the moving system" ... thereby transforming the motion factor of the latter (*Ib*) into a stasis factor (*IIa*). The unsurprising result is that no $c \pm v$ speed is to be found.[21]

It is also important to point out the problematic role which the concept of the *observer* plays in promoting the Special Relativity interpretation. Thus it is argued that the speed of light will always be the same as it is determined by an observer who makes measurements within the system. The subtle impression which is conveyed is that such an observer is somehow endowed with neutral and objective properties, the determinations of which should be taken as paramount. The problem, of course, is that the only category of observer which is taken into account is that which is consistent with what we have called an *X*-view — and this always yields a constant c outcome, as we have been describing for both baseline and altered systems; the corresponding *P*-view perspective (yielding $c \pm v$) is just as valid, but no such observers are recognized.[22] Thus, the *P*-view

20 Note that the interpretation we are using for discussing these issues is based upon an orientation which is the reverse of how we usually characterize nature's hierarchy, with larger entities being higher and smaller entities lower. We have adopted the contrary approach here, because it seems to make the patterns easier to visualize. Obviously, the most important point is the hierarchy factor itself, regardless of how it is applied to any given analysis.

observer would see a differential speed of light in relation to some entity that is moving in relation to the medium — regardless of whether such an entity is a simple object, or a larger system in which an *X*-view observer is present.

Before closing this section, here are a couple of remarks about the theorists whose positions we have been referring to. For one thing, we are certainly not suggesting that Lorentz and Poincaré construed any of the issues in either complementary or hierarchical terms, as we are here proposing. We are merely trying to adapt ideas which we maintain were consistent with their thinking to our own themes — whereas Einstein's do not fit.

Second, it hardly need be mentioned that ascribing a stasis-based framework for Einstein's interpretation certainly seems peculiar, especially given Special Relativity's distinct association with high-speed phenomena. Moreover, in later chapters we will talk about how Einstein's whole approach of treating light as a traveling phenomenon, which takes time to propagate from one point to another, itself represents a motion-based orientation that ignores its stasis counterpart. In both cases, what we are objecting to is the one-sided treatment, even though it is a different side in each instance.

The Michelson-Morley Experiment

We turn next to one of the most prominent experiments in the history of physics, which is often cited as providing proof of — if not the very raison d'être for — the Special Theory of Relativity in general, and the strong light postulate in particular. This paradigmatic research was first carried out by Michelson in 1881, and it was later repeated in 1887 with even greater precision by both Michelson and Morley, by which name the experiment is generally known. It has continued to be replicated many times since, with ever-increasing accuracy. But the basic result never changes.

The essence of the Michelson-Morley experiment is very simple. Take two beams of light and send them in perpendicular compass directions, along the surface of the earth. At the end of each path is a mirror which reflects the beam back to the starting point, in a classic two-way journey. The distance between starting point and mirror is measured to be the same in both directions.

21 We should mention the fourth category in the above charts (*IIb*), which we have not incorporated into our discussion. This would simply entail some moving entity that is a higher level parallel of the moving entity which is represented in the lower level analysis (*Ib*). Nor is there any reason to stop with two hierarchical levels: The same pattern would presumably continue, between this higher level and the still next higher one — with this level's altered parameters thereby becoming (lower) baseline ones in relation to the next higher level's altered ones ... and so on. By the way, it should be clear that the distinction between measured and actual light speeds would correspond to the distinction between altered vs. baseline, respectively.

22 One can argue that the situation is similar to the Quantum Theory notion that only the narrowest and most focused (*X*-view) idea of the discrete quantum is the correct depiction of nature, as opposed to the broader (*P*-view) counterpart of a continuum. We will pick up on that argument in Chapter 9, but suffice it to say here that a localized *X*-view perspective can have no exclusivity regarding the proper role of the observer ... and that the counterpart *P*-view perspective must play an equivalent role.

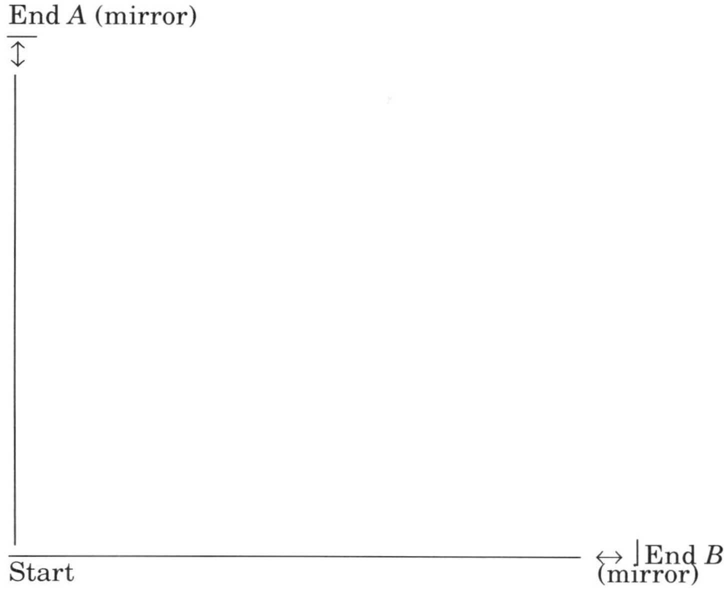

What do we learn from this famous experiment? What does it tell us about the constancy of the speed of light in general, or the weak vs. strong light postulates in particular? In some ways, the answer would be: not very much.

Oh, we forgot to mention the outcome — which is that there is no difference in the speed with which light goes and returns along the two round-trip paths. Yet, come to think of it, why should there be a difference? There is no apparent motion, and no other differential factors which would result in a speed disparity — so why should a difference even be expected … and what have we learned? There is nothing startling or revealing about this famous "null result", in the context in which we have just described it.

To be sure, we have intentionally presented the experiment in a very different manner from the way it is usually approached — although we have changed nothing about the essential content. Our purpose has been to show that in the most straightforward conceptual framework, there is nothing to be perplexed about, no paradoxical outcome — and certainly nothing decisive about the result, in terms of confronting fundamental theoretical anomalies. As we have presented it, there is no confusion, there are no problem issues … and the experiment would seem to have little purpose altogether.

Needless to say, this is not how the Michelson-Morley experiment is ever presented, for there would be little to talk about in that case. So having first framed this study in a simple and clear fashion, we now describe the experiment in terms of the conceptual overlay which normally goes with it, and from which its problematic aspects become apparent. That is, physics usually discusses the Michelson-Morley experiment in the historical setting of the late 1800's, where it was largely concerned with a particular issue — that of the role of the ether. So let us start anew, with a more typical presentation of the Michelson-Morley research. Curtain, please...

The Michelson-Morley experiment was predicated upon the classical assumption that light consists of waves which are propagated upon a fixed universal medium that fills empty space. This medium for light waves was given the unfortunate name of "ether" or "luminiferous ether" (sometimes with the classical spelling of "aether") — unfortunate, because of the very different and completely irrelevant associations which the term conjures up (e.g., as an anesthetic ether, in surgery). Light was understood to consist of vibrations in this fixed, yet gelatinous, ether substance. Maxwell fully subscribed to this ether model of light: the *c* which appeared in his equations was clearly understood to refer to the speed of light waves propagating upon the ether medium, in a matter-free region of the universe. His theory was based on the established principle that waves require a medium upon which to propagate: the medium is the "stuff" of which the wave pattern is built.

This ether medium was said to pervade all of empty space — including the space which surrounds the sun, and through which the earth moves. If light's effective speed can vary, as one would infer from classical physics theory, then light waves which move in a forward-and-back manner through the ether (in the same axis as the earth's own motion through the ether) would be judged to have a different speed than light waves which move in a sideways manner through the ether (in the opposite axis from that of the earth's motion through that ether).

We know that such a difference exists in our more familiar world. Thus, for example, a swimmer in a moving river takes longer to travel *up and down* stream than a swimmer who travels an identical distance sideways, back and forth *across* the river and its flow. The same pattern applies with an airplane which moves more slowly against, and then with, a wind ... than it does cutting sideways through the wind. In the Michelson-Morley experiment it was assumed that the earth is essentially moving through such a tangible (indeed, material) medium — and that this differential motion creates something like an "ether wind".

What the Michelson-Morley experiment did was take two beams of light and send one of them in a direction parallel to that of the earth's forward motion through the solar system, and the other in a sideways direction transverse to such motion. At the outer end of each of the two beams' paths was a mirror which reflected each beam back to the starting point. The distance between starting point and mirror was measured to be the same in both cases. Since the earth was moving through the ether, there should be a difference in the speed of the light beam running parallel to the earth's motion (and thus directly against, and alternately with, the ether medium), as compared to the speed of the light beam running transverse to the earth's motion (and thus crossing the ether sideways).

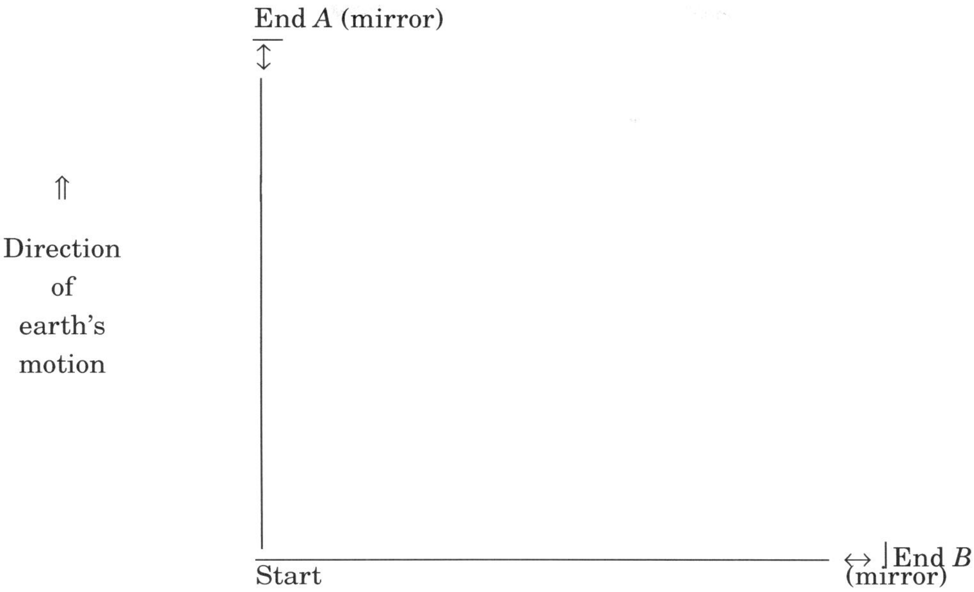

Unfortunately, for ether-based theorists of the day, the Michelson-Morley experiment did not result in any such differential speed pattern, or "phase shift":[23] there was no difference in how long the forward-traveling light waves took to complete their journey, as compared to how long the sideways-traveling light waves took to cover the same distance. Yet what could this outcome mean? How could one possibly understand this result? The theoretical dilemma posed by the "null result" of this experiment was both acute and deeply unsettling.[24]

We have already mentioned one theoretical way of explaining the outcome — in terms of the Lorentz-Fitzgerald contraction, whereby a moving object undergoes spatial and temporal alterations by virtue of its friction-like interaction with the ether.[25] The idea was that the earth's motion against the ether medium actually had concrete consequences: thus, the spatial dimensions of the apparatus would be distorted by its physical interaction with the material medium, such that the two different light paths would actually have different lengths. In fact, the ether-induced distortion would be of precisely such proportions as to perfectly offset the different (parallel vs. perpendicular) light speeds which were assumed — thereby making the two different light beams *seem*

23 We should note that any conclusion drawn from this and related experiments regarding the *speed* with which light travels is an inference, as speed itself is not really tested directly — so much as a differential in the phases of the respective waves which are said to travel in each of the two paths. But there seems no reason to challenge the basic premise of the research, in this regard.

24 According to Dingle's account, there were actually few physicists at the time who were concerned with this problem.

25 Later analysts since Lorentz have offered the same idea — that motion relative to the ether would produce physical effects. See the various papers of Ives and Builder; also discussions by Prokhovnik; Ruderfer; Grieder.

as though they were traveling at the same speed, even though they really were not. So the general idea was that differential motion in relation to the medium produced a contraction of dimensions and a slowing of clocks. As we have seen, the mathematical factor which describes the alterations is called the Lorentz transformation.

This contraction approach had the virtue of retaining the basic conceptual premise which lay behind the experiment — motion through an ether — and which formed the core of existing electromagnetic theory. On the other hand, this interpretation was not considered very satisfactory by physicists at the time, who found the reasoning convoluted, unconvincing, and arbitrary. As a result, the theory was still left in limbo, and there seemed to be a real crisis of understanding.

It was in 1905, during the period of this intellectual malaise, that Albert Einstein came onto the scene. In many ways, the conceptual void which was ostensibly left by the Michelson-Morley experiment provided the perfect entrée for his Theory of Special Relativity. In that model, Einstein offered an explanation which amounted to very radical surgery indeed. Although he may have been only peripherally aware of the Michelson-Morley experiment at the time that he wrote the groundbreaking paper which presented the framework for his new theory, his approach was very clearly relevant to the issues there, and offered a way out of those dilemmas.

Let us begin with Einstein's approach to the concept of an ether medium. There was one strong objection to such a medium model of light propagation — which Einstein seized upon: the ether appeared to possess no qualities by which it could be identified, in any concrete sense. It was tasteless, featureless, colorless, amorphous ... and generally thought not to be subject to direct experimental observation and manipulation. Was a hard-nosed scientist even justified in continuing to adhere to such a central concept as the ether, when its objective and demonstrable existence seemed beyond experimental verification? Michelson and Morley had conducted the most renowned experimental attempt to discern the ether — and yet they were stymied: no obvious or convincing signs of such a medium were found. Indeed, so devastating a blow were the null results of this experiment that it is sometimes credited — although not by its authors[26] — with dealing the death knell to the ether concept altogether (e.g., Zukav, p. 130).

Einstein's position was quite simple, and his theory put forward two revolutionary ideas about light. First of all, if there is no actual evidence for the existence of a universal ether — or any other kind of medium — then why not just abandon the concept altogether and build a model which does not require it? Better to base our theory on what can be proven, and not on intuitive assumptions which may be blocking the way to deeper insights. Second, why not simply presume that the speed of light is always the same, regardless of whether there is any motion of the observer or not? Enter the strong light postulate — not only as some abstract and restricted physical principle, but as we have seen, one having extraordinary implications for our notions about time and space, generally.

26 Michelson and Morley never embraced the no-ether conclusion that their experiment was said to prove, but rather the more careful and accurate conclusion about *non-detectability* of the ether. Builder's description is precise: "[I]t is impossible to detect uniform motion relative to the ether." (Builder, 1959, p. 88)

In fact, removing the ether was a logical prerequisite for establishing the strong light postulate itself, aside from any pragmatic arguments about detectability. That is, it seems impossible to imagine how the strong light postulate could exist together *with* a medium concept for light, in any full-fledged sense — just as it seems impossible to imagine what meaning the weak light postulate could have *without* a medium, as representing the substratum for the wave's propagation.

In any event, Einstein's approach certainly did away with the problems posed by the Michelson-Morley experiment! The very premises upon which that experiment were based suddenly evaporated, and there would be no reason to expect anything other than a null outcome there. It is not merely that light does not require an ether medium: there is also no reason to assume that the earth is interacting with any such medium. Since there is no ether medium for either light or the earth to travel in, then the parallel vs. perpendicular analogies have no validity; and since the speed of light is always determined to be the same, then any motion of the earth (and the apparatus) would be irrelevant to the outcome.

There was one little problem, though — for it was not just the ether premises of the Michelson-Morley experiment that were eliminated: the whole ether base, upon which all of electromagnetic theory was built, was simply swept off the table and dispensed with altogether. Maxwell's electromagnetic equations were saved, but their contextual underpinnings were gone — both in conceptual and physical terms. What remained was a phenomenon of nature, light, that seemed to have well-nigh supernatural properties: it was a wave, yet required no medium upon which to travel; and it had the same speed in relation to every object, irrespective of those objects' differential speeds relative to one another! No other phenomenon of nature displays such extraordinary properties. So what is going on?

The Necessary Function of the Medium

Many descriptions of Special Relativity trumpet the fact that, in Sachs' words, "It took the boldness of Albert Einstein in the first years of the 20th century, to recognize that the ether concept is indeed superfluous in the description of electromagnetic phenomena" (Sachs, 1973, p. 33). Instead, light was said to be an independent phenomenon, which required no substratum upon which to travel, but which rather existed in its own right: "[L]ight is always propagated in empty space with a definite velocity *c* ..." (Einstein, 1905, p. 38). Even today, the whole notion of a medium for electromagnetic waves remains anathema to established physics, and is officially renounced (e.g., Pitt, p. 125). Such an assumption was clearly rejected by Einstein's Special Theory — and most of modern science has not looked back ever since.[27]

Of course, one of the obvious problems with abandoning the ether medium and replacing it with nothingness concerns what it is that light is then supposed to travel at a constant speed in relation to. For, once the medium has been abandoned, what sense

27 On the other hand, Zukav writes that, "Quantum field theory resurrects a new kind of ether", in the form of "the featureless ground state of the field (the vacuum state)." (Zukav, p. 130)

does it make to talk about any speed — or even of motion or traveling, at all? This is true not only in the sense that waves require a medium, but more generally in that motion requires stasis: Every schoolboy relativist is sophisticated enough to know that the motion of one thing can only exist in relation to the non-motion of some other thing. But Einstein has eliminated the only obvious and consistent referent for non-motion here, relative to which the light wave motion can be said to exist. So what right have we to talk about light moving or traveling at all, at whatever speed, when the primary referent for that motion no longer exists? As Bridgman says about Maxwell's electromagnetic foundation, "The equations are meaningless, however, except in the context of a material frame of reference — identifiable points in space have to be assumed." (Bridgman, 1983, p. 150)

Here is Dingle's colorful description about the change from the meaningful model put forward by Lorentz to the meaningless model of Einstein:

> [T]here existed an ether with respect to which the velocity of a body had a definite, in principle measurable, value. Light consisted of vibrations in the ether.... What Einstein was proposing ... was to retain the finite velocity of light without the existence of any standard with respect to which that velocity had a meaning.... Light consisted of waves, with a definite length, frequency, and velocity, in nothing; it was like the grin without the Cheshire cat.... If his assumptions were granted he did save the equations, and when his theory ultimately made its general impact on the world, mathematics had so dominated physics that the non-existence of the Cheshire cat was regarded as a triviality; the grin remained, and all was well. (Dingle, 1972, pp. 155-156)

But the issue of an ether medium goes beyond Maxwell's model and the notion of wave speed in relation to nothing, for we are also talking about the meaning of the crucial Lorentz transformation. This set of equations are understood as quantifying the kind of spatial and temporal alterations which are said to occur in moving bodies. As we discussed above, both Einstein's and Lorentz's models make the same predictions about the nature of these alterations — each theory amounting to a particular way of deciphering the Lorentz transformation — but on very different grounds.[28] Thus, the Lorentz interpretation is based on the notion of a physical interaction between the moving object and the ether medium, whereas the Einstein interpretation is based on sheer motion alone, as between two objects.

The result is that, whereas Lorentz proffered the causally comprehensible idea that motion relative to a tangible medium results in physical modifications of space and time, Einstein proposed an "explanation" that was physically meaningless and causally incomprehensible. Dingle again:

28 We are not addressing one distinction between the two theories here, which concerns the "flippability" factor for the stasis-motion states of each entity ... an issue we examine elsewhere. Also, we restate an earlier footnote point regarding a distinction between how Lorentz initially treated the time alteration factor, as opposed to Einstein's (and Poincaré's) approach; we will return to this issue in Chapter 8.

> [Lorentz's] was a *physical* theory, not a *mathematical* one; that is to say, the proposal was that motion through the ether produced physical effects on bodies, and the mathematics expressed the physical results produced.... [I]t was Einstein, who had no qualms about abolishing the ether and still retaining light waves whose properties were expressed by formulae that were meaningless without it, who was the first to discard physics altogether and propose a wholly mathematical theory. (Dingle, 1972, pp. 165-166)[29]

Grieder makes the same point, about how Special Relativity "bars the way to a causal account of relativistic phenomena such as rod contraction and clock retardation." (Grieder, p. 41) Or as Zahar puts it, "[W]hereas Lorentz *explained* why a moving rod contracts, Einstein *bluntly asserted* that the speed of light is an invariant, an assumption from which Michelson's result trivially follows." (Zahar, 1976, p. 223) So while the Lorentz transformation made sense and could be logically understood in Lorentz's medium-based model, it became ad hoc and incoherent in the context of Einstein's no-medium theory.

Let us step back now and consider a different way of approaching the issue, regarding the abandonment of the ether medium construct. Specifically, the problem can be crystallized by asking what we are to do with classical constructs that make sense and serve us well in many ways, but which also seem to have their limitations — primarily in terms of needing more nuance in their meaning, which would also concern issues of detectability. Yet rather than wrestling with such subtleties in the former factor, Einstein took advantage of vulnerabilities in the latter. His solution was both bold epistemologically and bare conceptually: simply do away with these problematic constructs altogether, and replace them with a completely different (and slimmed down!) theoretical framework — yet one which certainly had its own limitations, both in terms of arbitrariness and in terms of coherence ... and the latter in both meaningfulness and fitting together respects.

We contend that although the classical constructs needed to be modified, Einstein's solution was far worse than the problem it tried to rectify. There is a better option — which may be less radical, yet which leads to more satisfactory results. That is to say, once we get past our pique at their imperfections, we should be able to see that the classical constructs really weren't so bad. More specifically, the solution is not to abandon such constructs, but rather to relativize them: they should be recast from an absolute to a relative framework — and then they will continue to serve us extremely well. Yet this more modest — and, we contend, more realistic — approach hardly fit with Einstein's model, either in terms of goals or content.

The issue we are facing here with the ether is similar to one we examined in Chapter 4. There we talked about problems with the absolute space concept, in relation to which absolute motion was said to exist. Both these constructs were to be dispensed with, according to the modern view, in favor of the notion that motion was relative to particular referents for stasis. The idea was that since there was no singular absolute

29 "Lorentz held that the universe obeys intelligible laws ... and Einstein held that the universe is governed by principles which can be given a mathematically coherent form (e.g., *all* laws are covariant)" (Zahar, 1976, p. 255).

space reference system which served as a criterion for absolute motion, then these constructs should be eliminated. At least this was the theory, but we argued that in practice the essence of absolute motion was retained. For one thing, the notion of a general state of inertial motion was not abandoned — even though the proposition lacked any conceptual base, inasmuch as Einstein had abolished the absolute space referent in relation to which such motion would exist. We also cited Bridgman (1983, p. 28), concerning his criticism of the commonplace manner in which physicists continued to cite an object's virtually absolute state of motion.

Now we have come full circle back to the same problem, but here concerned with light. In terms of the present model's dimensions of analysis, the discussion turns from the matter to the medium domain. Here we are facing a similar paradox, in that Einstein ascribes absolute motion (and so speed) to light — while denying the ether medium referent relative to which that absolute motion of light could have any coherent meaning. Thus, Newton's absolute space gave meaning to his absolute motion; similarly, Maxwell's equivalent ether medium gave meaning to his constant c. Yet in both cases, Einstein retains what is in effect uniform motion and fixed speed, while abandoning the primary referents relative to which such motion and speed have real meaning.

The difference is that in the realm of mechanics, Einstein at least *claimed* to be abandoning absolute motion, and replacing it with relative motion … which only existed in principle between one entity and another. The thrust of his position was that a motion state was specific to the particular entities involved, and had no general or universal meaning; thus it would be different between one referent and another. Yet in the case of light, this notion of relative motion got turned on its head — with the speed of light supposedly becoming absolute and referent-irrelevant.

We have argued that this is a false proposition — both in the sense of the unwarranted isotropy assumption, as well as in terms of what we have called a measured vs. actual speed of light distinction. The former is based upon a stasis-orientation, which defines away the corresponding motion-orientation that would contradict it; and the latter is based upon different sets of space and time metrics, which yield relatively different phenomenological outcomes.

Needless to say, the notion and function of a medium represented the complete antithesis of the strong light postulate: whereas Einstein was wedded to the latter, we are wedded to the former — because we believe that it is a crucial component of any coherent model of nature. Indeed, another obvious way of describing the relationship between wave and medium is in terms of the complementary figure-ground motif which we often invoke. Thus, the most straightforward way of visualizing a wave is as figure, in relation to the medium as ground. The classical approach implicitly recognized such a basic figure-ground pattern, in the sense that there had to be some sort of framework or substratum (i.e., medium) for light (waves) to exist in relation to.

In fact, even Einstein realized this — in practice, much as he would have rejected the idea in principle. He simply substituted a different name, for this inevitable functional component: *something* had to be posited upon which to "house" the electromagnetic waves. He called the new non-medium ersatz the "electromagnetic field" — as though such a substitution actually resolved anything fundamental. Even so, the new concept remained an abstraction in his model, one that he a)could claim was not really a medium, even though it acted like a medium; but b)was free to jump around with, conceptually, and assign at will — thereby implicitly attaching the inevitable phenomenological medium component to whatever system was most convenient for his analysis of the moment. In any event, the result is that physics contemptuously rejects the ether while retaining its essential properties, dressed up now in the newer and more sophisticated guise of the electromagnetic field!

> The invisible tentacles, the imperceptible effluvium, the mechanical ether are all now replaced by yet another construct, an unseen field spread out in the void of space.... Light was now to be seen as a self-sustaining electromagnetic wave propagating in the electromagnetic field rather than in the aether — a substitution of one construct for another. (Hecht, pp. 129 and 143)

Bridgman notes that, "The field gives a physical background, analogous to the old ether ..." (Bridgman, 1983, p. 150; see also Froome and Essen, p. 13). Parish similarly describes the new "field" as being indistinguishable from the old "ether":

> It is hard to see any difference. Both the ether and the field involve the surrounding space or the notion of an impalpable, invisible medium, itself inherently neutral as regards energy, mass or motion, none of which it possesses of its own account. Whether we call it an ether or a field, we shall always need something to link objects and the interplay of their forces together, for otherwise we should have to return to the idea of instantaneous action at a distance. (Parish, p. 116)

As Parish says, the concept of a medium supplies a crucial conceptual link, which we need if we are to bind together the various participants. And whatever name we give it, we need to call it something — this thing that acts like a medium — as a coherent stage which joins the different actors or entities and allows them to interact across a distance with one another. Plus, it is virtually impossible to conceive how a wave could exist without a medium. As Wolf puts it, "All waves are produced by vibrations within the medium holding the wave." (Wolf, p. 48) All waves, of course, except for electromagnetic ones!

Actually, we would suggest that there is a perverse way of understanding Einstein's approach here in figure-ground terms. Instead of the usual manner of thinking about light as a figure (wave) phenomenon, which exists in relation to some background (a la absolute space, or ether medium), Einstein essentially flipped nature's pattern — and eliminated the ether in the bargain: the new absolute and fixed base is light, which takes on a "phantom" background frame role. There is no need for the ether or any other medium as background — because light itself now (somehow) serves that background function.

In this way, the argument would go, eliminating the ether did not mean eliminating the only meaningful referent for light's motion to exist in relation to. On the contrary, it is light itself which now represents the new and fixed background centerpiece of nature, with everything else serving as varying figures in relation thereto. The result is that regardless of what real speed differences may exist — both between the two anisotropic one-way speeds of light for a moving object, and between the distinct relative speeds that would be associated with two separate objects' spatial-temporal worlds — Einstein's theory either defines away or distracts us from such outcomes, in the service of the constant speed of light myth. And after all, there is no way to prove that his contention is wrong ... any more than that it is right. The fact that it makes no sense is considered neither here nor there.

For his part, one can argue that Einstein himself later came around to the view that we needed a medium in order to understand what was going on — sort of! Thus, in contrast to the generally accepted view, Einstein's position seems to have evolved from one that was wholly negative to one that was either neutral or positive — moving from rejecting the ether altogether,[30] to treating it as either being superfluous or as actually being essential. Here is how Builder describes the change, speaking in terms of the Special (or Restricted) Theory:

> The belief that the ether hypothesis is incompatible with the restricted theory undoubtedly had its origins in the opinion to this effect so emphatically expressed by Einstein in his early papers (1905, 1907). Einstein, however, did not present any argument or evidence in favour of this opinion, and it is obvious that it had no logical relation to his formulation of the theory. In any case, he later (1920) explicitly admitted that, "More careful reflection teaches us — that the special theory of relativity does not compel us to deny the existence of the ether — on the other hand, there is weighty evidence in favour of the ether hypothesis. To deny the ether is ultimately to assume that empty space has no physical qualities whatsoever." (Builder, 1959, p. 90)

Ruderfer also notes that in this 1920 lecture at Leiden, Einstein

> stated that the existence of an ether is not denied by the special theory and is, in fact, demanded by the general theory of relativity. (Ruderfer, 1962, p. 325).

Hsu makes the same point about Einstein's position here, stating that, "He believed the existence of such an ether to be a conceptual necessity for understanding physical phenomena in general." (Hsu, p. 64)

It is interesting to note that in this same Leiden lecture of 1920, Einstein also seemed to acknowledge the utility of other non-observable phenomena which may not be readily demonstrated experimentally. In discussing classical ideas about rotation, Einstein says:

> In order to be able to look upon the rotation of the system, at least formally, as something real, Newton objectivises space. Since he classifies his absolute space together with real things, for him rotation relative to an absolute space is also something real. Newton might no less well have called his absolute space "Ether"; *what is essential is merely that besides observable objects, another thing, which is not perceptible, must also be looked upon as*

30 On the other hand, Zapffe quotes Einstein from 1909 as describing the first postulate of Special Relativity in ether terms : "The Principle of Relativity states that all natural laws belonging to a system of coordinates K', in motion relative to the ether, are identical with those belonging to a coordinate system K at rest relative to the ether ..." (Einstein, 81st Naturforscher-versammlung, Salzbethuslyrg, September 1909, quoted in Zapffe, 1977a, p. 21).

real, to enable acceleration or rotation to be looked upon as something real. (Einstein, 1920, pp. 16-17; italics added)

Unfortunately, even later still, Einstein (writing with Infeld in 1938) reverses himself once again and condemns the ether notion. In that work, we are advised to "forget the ether completely and to try never to mention its name." (Einstein and Infeld, p. 176). But in a postscript to their article arguing that Einstein did not reject the ether, Achuthan, et al. note that this later work was based on the Special Theory of Relativity, and in any case, "The discussion in this book [was] very perfunctory compared to this [above-mentioned 1920] Leiden lecture ..." (Achuthan, et al., p. 284).

A Complementary Model of Light

Obviously, we approach the phenomenon of light very differently in the present work from the way it is normally construed — in broad complementary terms that go well beyond the conventional wave-particle duality, which we discussed in earlier chapters.

To begin with, we can cite several general and abstract ways that the pattern of complementarity applies to light, as it does to all other phenomena — which we can express in terms of our dimensions of analysis. Thus, light will have a constant character for a stable region, but a varying one for a changing region. For the next dimension, there will be a relatively different type of electromagnetic phenomenon for the inside realm (e.g., electric field), as opposed to the outside realm (e.g., magnetic field). Nested next is the matter and medium dimension — the conceptual home of particle and wave, or absolute space and ether medium. The following dimension is that of static and dynamic relationships, which will play a prominent and innovative role in our explorations of light phenomena in Chapter 9. Finally, we assume there are two fundamental factors at the end of this conceptual chain, and one of the primary ways that we can talk about these dominant and recessive entities is in terms of sink and source, or inward vs. outward directionality.

But now let us get very specific, in speaking about different speed of light outcomes that might be applied to the Michelson-Morley experiment. We begin with Einstein's own description of how the results might be construed in relation to two separate systems. In the first case, the observer is said to be moving together with the experimental apparatus, along with the earth upon which both are grounded. In the second case, the observer would be viewing the whole show from a solar system vantage point. Here is the way he contrasts the results, in a later postmortem for understanding this experiment: "Thus for a co-ordinate system moving with the earth the mirror system of Michelson and Morley is not shortened, but it *is* shortened for a co-ordinate system which is at rest relatively to the sun." (Einstein, 1961, p. 54)

We concur with this analysis, except that we would reverse the stasis-motion labels which are applied — meaning that the first scenario would be better described in terms of stasis, given that no differential motion is involved between the Michelson-Morley apparatus and the earth system ... whereas the second scenario would be better described in terms of motion precisely because of the differential motion that is involved between the apparatus of Michelson-Morley and that of the solar system.

The reason why we began the Michelson-Morley section by introducing the experiment in such simple and straightforward terms as we did is precisely because it describes the first scenario: this is the most obvious approach — with the apparatus system and the earth-based observer who is examining it being static in relation to one another. Within the terms of this situation by itself, there is no Michelson-Morley anomaly to be resolved — no "shortening" to be explained, nothing which must be quantified in terms of the Lorentz transformation, and no reason at all to invoke a special speed of light postulate.

Yet the situation is very different when the larger picture is brought into play. One way of describing this is in terms of the classical assumption about a fixed and singular universe-wide ether medium background, which envelopes and surrounds all matter in the universe ... and through which the earth, as well as all light waves, are moving. According to that framework, there would be differential motion — not between the earth-based observer and the earth-based apparatus, but rather between the earth-based observer and the ether medium in which the traveling light waves are "housed". The problem, of course, is that we earthbound beings are not part of a solar system ether-based perspective which would reveal such differential motion; and the fact that such an outcome does not present itself becomes the basis for Einstein's approach — which is to simply reject the ether medium concept altogether, given that its existence cannot be demonstrated in a fundamental experiment like this.[31]

We contend this is a very shortsighted and quick fix solution, which does not help us to understand what is going on here. For one thing, we are unwilling to abandon a crucial premise — that waves only exist in relation to a medium — solely for the sake of expediency. On the other hand, while we still continue to embrace the notion of a medium for light waves — be it called "ether" or anything else — ours is a more nuanced approach that is in keeping with the notion of complementarity that we are espousing.

31 Here is another way of thinking about the earth-apparatus based system, in terms of whether it proves anything about an external medium beyond. We would suggest that the Michelson-Morley experiment could be understood as the equivalent of an "inertialness test" (which we described in Chapter 4), conducted within the earth-apparatus system container — and with light waves, rather than common Newtonian objects ... in which case, it might be better labeled as an "isotropy test". Thus, if the system passes the inertialness test, then it will also pass the isotropy test — which latter is the case for the first Michelson-Morley scenario we have been describing. Nonetheless, does this conclusion tell us anything at all about the existence of an extrinsic medium, through which the earth-apparatus system could be moving? To answer that question, let us return to an illustration we used in Chapter 4, where we talked about a ship that was either static or moving upon tranquil waters. Of course, classical mechanical experiments conducted inside the ship will give no indication as to the ship's stasis-motion state vis-a-vis the waters without. Next do this experiment inside the ship: set up two linear water canals, parallel and perpendicular to the ship's forward motion. Set water waves in motion within each canal "medium". Time the waves: one will find that the speed of the waves will be the same in both directions. Based on the Relativity reasoning, one would have to conclude from this null result that there is no basis for assuming the existence of a water medium outside in which the ship may be moving — or even that water waves need a medium. Are such conclusions warranted? We think not — for any intrinsic results are quite mute about the nature of any extrinsic stasis or motion, or existence of an external medium; and they certainly do not demonstrate that water waves need no medium!

Rather than a single manifestation of medium, as in the classical paradigm, we maintain instead that there are a whole series of nested mediums — one for each level of nature's hierarchy.[32] In the case at hand, there is a background medium which exists at the level of the earth — in which case there would be no relative motion between the apparatus and this medium.[33] Consequently, there is no reason to expect any particular effects which need to be resolved — in terms of contraction, Lorentz transformation equations, and light speed. But this is not the case when we incorporate the adjacent level of the hierarchy, and include a stance which is consistent with the second scenario described above. For then we would be dealing with relative motion — between the apparatus and the solar system.

Let us try to describe the whole picture in complementary and hierarchical terms — just as we proposed some pages back — with *X*- and *P*-views applying both to a (lower level) baseline orientation and to a (higher level) altered orientation.[34]

At the lower level we can talk about the solar system as having a baseline set of space and time metrics, and we can understand the situation in terms of what is essentially the classical ether medium concept — except that instead of talking about a single universal ether medium, in our nested level model we are talking here about a solar system ether medium. With this solar system vantage point, we can either talk in stasis-based *X*-view terms of a fundamental (and isotropic) speed of light, c, in relation to the fixed ether medium of the solar system; or we can talk in motion-based *P*-view terms of the Michelson-Morley apparatus which is moving at a speed, v, in relation to the solar system ether medium (as in our second and main presentation of the experiment); however, we describe this latter situation in terms of an effective (and anisotropic) speed of light, $c \pm v$.

In relation to this, what we can call the higher level earth system would be understood as having an altered (Lorentz-transformed) set of space and time metrics. Now we are dealing with an earth-based ether medium, which we can talk about in stasis-based *X*-view terms of a fundamental (and isotropic) speed of light, c, in relation to the fixed

32 It is noteworthy that Silberstein long ago ridiculed the ether model, at least partly because it would necessitate, in Builder's words, a separate "ether associated with every one of the infinity of possible inertial reference systems. This is clearly absurd and inconceivable, so that the ether concept breaks down and must be discarded completely." (Builder, 1958b, p. 459). Our position is very different: there must, indeed, be separate "ethers" for different systems; this is neither "absurd" nor "inconceivable" — it is inevitable, and it is an inherent part of a hierarchical model of nature. Note also that the hierarchical approach we are advocating is consistent with some modern interpretations of Lorentz's model, according to which there is not a single universal ether medium, but rather a series of local gravitational fields as providing this function.

33 The concept of nesting mediums, including one that exists at the level of the earth, bears some practical similarities to an interpretation of the Michelson-Morley experiment known as the "ether drag" hypothesis. The idea there was that the earth drags with it a layer of the universal ether, such that there is no differential motion between the earth and that medium. One can argue that the notion of a "layer" has some resemblance to our notion of a "level", but the two conceptual frameworks are really quite different.

34 We restate the point of a prior footnote, regarding a reversed order of things in this analysis (for presentational purposes) — with the larger entity representing the lower level baseline, and the smaller entity representing the higher level alteration.

ether medium of the earth system — in which case the Michelson-Morley apparatus would not be moving in relation to the earth (which is how we described the scenario in our opening presentation of the experiment). But some other object could be so moving, in which case we would talk in motion-based *P*-view terms of an object which is moving at a speed, *v*; and light would be moving in relation to that moving object at an effective (and anisotropic) speed of light, $c \pm v$.[35]

Of course, we have argued that the actual speeds of light, *c*, are relatively different, as between the two (lower and higher level) systems — even though they are measured to be the same. Furthermore, as mentioned earlier, one can extend this analysis to the still next higher level (as just mentioned, entailing another moving object), shifting the relative baseline vs. altered designations upward accordingly.

Here is how the two adjacent hierarchical levels that we have been discussing might be charted:

Altered space and time frame / medium of Earth (Lorentz-transformed)

X-view	*P-view*
stasis-based (static M-M apparatus)	motion-based (moving object)
fundamental *c* constant (isotropy)	effective $c \pm v$ (anisotropy)

Baseline space and time frame / medium of Solar System (Normal)

X-view	*P-view*
stasis-based (static object)	motion-based (moving M-M apparatus)
fundamental *c* constant (isotropy)	effective $c \pm v$ (anisotropy)

Once again, Einstein adopted only the two *X*-view perspectives for dealing with these issues — yielding normal vs. transformed space and time metrics — whereas we contend that both *X*- and *P*-view analyses are consistent with Lorentz's thinking. And while experiments neither confirm nor contradict either the one-sided or two-sided frame-

35 We will describe some examples of such moving objects in the next chapter. But for the moment, we can also mention ones which are related to a couple of other experiments in which obvious light speed differences are demonstrated, in a way that they are not in Michelson-Morley (for us on earth). Both of these use light beams which travel a one-way out-and-around path along a circuit — rather than being reflected back again in a two-way manner, as in Michelson-Morley. One experiment was conducted many years later by the same Michelson, together with Gale. In that 1924 experiment, light was sent in two opposite directions around a very large rectangular path. In contrast to the Michelson-Morley experiment, there *was* a phase shift in the waves — indicating a difference in light speed, or time necessary to complete the journey. Another important study, conducted by Sagnac in 1913, demonstrated the same thing. That research is conceptually very similar to Michelson-Gale, except that a)Sagnac was on a much smaller scale than Michelson-Gale, and b)the Sagnac apparatus itself underwent rotational motion. Ives notes that Michelson-Gale's light beams travel along a large one-way path "having segments of different latitude and velocity on the earth's surface." (Ives, p. 179) Ives also maintains that the rotation factor is not critical to the Sagnac outcome — and that both experiments (and their differential outcomes) can be accounted for in terms of the Lorentz-Fitzgerald contraction, just as Michelson-Morley can be.

work, we contend that our experience is far more consistent with Lorentz's approach than Einstein's — which is why the latter's model represents such a logical outrage, given that it completely dismisses the *P*-view perspective that is so familiar and meaningful to most of us.

Now for a few words about other variations of light speed which would be associated with a hierarchy theme — specifically in terms which reach outside the Special Theory's supposed purview, as entailing "inertial frame" situations that are ostensibly free of gravitational influences. Of course, such conditions have obviously not been met, as is clear from the fact that all demonstrations of the theory's principles have been conducted in circumstances where gravity is demonstrably present. But the main issue we want to look at now concerns *changes* of such gravitational influences, and how these would impact what is supposedly a *non-changing* speed of light.

On the one hand, Einstein himself made statements which are consistent with the former, rather than the latter. In this respect, the constant *c* thesis is abandoned by the General Theory of Relativity in the sense that it acknowledges *various* speeds of light in different regions of the universe: "[T]he law of the constancy of light *in vacuo* ... cannot claim any unlimited validity.[36] A curvature of rays of light can only take place when the velocity of propagation of light varies with position." (Einstein, 1961, p. 76). Sometimes this curvature is described in terms of gravity bending, or slowing down, light.

But on the other hand, this rather obvious acknowledgement is contradicted by the orthodox position, which remains wedded to the doctrine that the speed of light never changes! Thus, we are told that the constant speed of light is said to reign throughout the universe (e.g., Marion, p. 411). Or again, one of the curious aspects of so-called free fall in a gravitational field is that everything therein is said to fall together, all at the same ever-increasing rate, irrespective of how large or small its mass — everything, that is, except light! According to the standard approach, light will supposedly *not* fall at the same changing rate as everything else. Why? Because the Special Theory's constant speed of light postulate does not allow such variation of light speed.

The same interpretation is applied to other phenomena in the universe. For example, there have been instances of extragalactic expansions which appear to be much faster than the speed of light (e.g., Cohen et al.). Similarly at the edge of the expanding universe that we are able to see, there may be an "event horizon", beyond which the universe may be expanding faster than light: "The light from galaxies beyond that distance will never reach us because they move away from us faster than the speed of light." (Capra, 1975, p. 182) Yet after citing examples of astronomical bodies moving apart at 10

36 By the way, the very notion of a vacuum means "the complete absence of matter altogether" — which of course brings us back to the suspect notion of "empty space" that we considered in Chapter 4. In fact, there is no such thing as a pure vacuum in empty space (Davies, 1977, p. 1). Indeed, "[W]e can no longer think of a vacuum as 'empty'. Instead it is filled to capacity with thousands of different types of particles, forming, interacting and disappearing, in an incessant sea of activity. This is the quantum picture of space-time, one of violent fluctuations and interactions." (Davies, 1977, p. 136) Therefore, "[I]nstead of 'nothingness,' the vacuum is an infinite sea of particles of matter" (Sachs, 1973, p. 98) — so the vacuum obviously has physical properties. In any event, the main argument we are making here is that what is a vacuum in one space-time or gravitational region is a different vacuum in another.

times the speed of light, Clotfelter states that, "Physicists and astronomers are extremely reluctant to believe that anything is moving at a speed greater than *c*, for special relativity, a theory which has been verified in many ways countless times, assures us that such speeds cannot be attained." (Clotfelter, pp. 321-322)[37]

To understand the arbitrariness of how the issues are normally approached here, consider the very different way that physics explains what happens when light enters a transparent "material medium": in such cases, it is clearly acknowledged that the speed of light changes, this being described in terms of the category called *refraction*. Yet here is the fundamental question: why should the changing light speed which occurs in different "material mediums" be treated any differently, in principle, from the changing light speed that occurs in different "non-material mediums" (a point we will pick up on in Chapters 8 and 9)?[38] The point is that a wave is part of a medium — and it is the characteristics of the medium which determine the speed of a wave therein. Consequently, light's speed will change in different mediums — regardless of whether the overall phenomenon is labeled free fall, variations in space-time curvature, changing gravitational potential, refraction, or whatever.

One way of reconciling some of these various positions regarding the constancy of light's speed in changing conditions would be in terms of the difference between measured vs. actual speeds that we have discussed in this chapter. Thus, it may be arguable that at least in some situations space and time parameters have changed, as between one situation and another, in just such proportion that the measured speed of light remains the same in both situations — even though the different conditions which are being compared would result in different relative speeds between them.[39]

37 Another example of variations in light speeds can be found in a distinction which can be applied to all kinds of wave phenomena — in terms of a differentiation between an overall wave group velocity, as opposed to the constituent phase wave velocity. "[T]he synthesized pulse travels with a speed (the *group velocity*) that may be quite different from the characteristic speeds (the *phase velocities*) of the individual waves." (French and Taylor, p. 59) When we speak about measuring the speed of light, we are typically referring to the measurement of the group velocity of light, whereas the phase waves thereof could move even faster, "at velocities ranging from the speed of light upward to infinity" (Murchie, p. 405; also Rogers, p. 742). Not surprisingly, such variation in group vs. phase wave velocities is said not to occur when light travels through a vacuum, in which case there is supposedly no difference between the group speed and phase speeds. Once again, the primary rationale for such a conclusion is Special Relativity's thesis that the speed of light in a vacuum never changes.

38 In a similar way, there have been various research projects in recent years which have manipulated the speed of light, by various means — to slow it down, or speed it up, from what the speed would normally be in that situation. Such demonstrations provide further evidence of the malleability of light's fundamental speed, depending on the conditions in which it exists — i.e., the character of the medium.

39 On the face of it, this interpretation would not seem to apply to the refraction case — except that one might argue that if there were a way to conduct the measurements within the framework of the refracted situation, that there would be space and time alterations which would also yield the constant *c* outcome there.

Lifting the Veil of Confusion

In this final section of the chapter, we want to summarize the main issues we have been discussing and explain how Einstein's model has led us astray. Indeed, it should by now be clear that in discussing the speed of light in the context of Special Relativity, the issue is not so much a number as it is a logic — and perhaps not even a logic so much as a leap (if not a lark) about theory-building: Can one take a particular phenomenon of nature, as it manifests itself under certain conditions, and define a new set of general constructs and rules about the physical world — according to which this phenomenon will always be central and one of its properties always the same ... even as circumstances change? We contend that this is precisely what Einstein has done — and that he has succeeded spectacularly, in terms of the resulting theory's prominence in science ... not merely in its own right, but virtually as the epitome of what a good model of science (and scientist) should represent.

Of course, we take a contrary position — but in so doing, we may as well deal directly with one accusation that could easily be lodged against us: namely, that we are merely being stodgy absolutists who are stuck in old ways of thinking, unable to recognize the utter brilliance of Einstein's thinking and the profound conceptual breakthrough which his relativistic insight has brought us. For example, why should one assume that the old-fashioned world — of fixed space and time metrics, and different speeds of light in relation to moving objects — should continue to have validity ... in relation to the new-fashioned world — of variable space and time metrics, and a constant speed of light that does not change in relation to moving things?[40] Einstein rejected the former and embraced the latter — and look at the dramatic advances which his model has led to, in terms of our ability to identify the deeper patterns of nature!

Needless to say, we contest that last conclusion — particularly in terms of what we claim is the deepest pattern of nature, that of complementarity ... which, of course, Einstein wholeheartedly rejected. Indeed, we have argued that the fundamental issues we are discussing can best be understood in complementary terms, whereas they make no sense at all in the one-sided and hyper-reductionist framework of Einstein's model.

There are other salient theory-building issues which are also relevant to this discussion; and although we will deal with these matters more fully in Chapters 10 and 11, they are essential for a better understanding of the problems we have been encountering here, concerning the fundamental precepts of Special Relativity. In particular, it is worth noting that Einstein's whole approach to theory-building was not so much a matter of discovery as it was one of invention — and that the principles which resulted therefrom were to be considered as convenient abstractions, rather than generalizations drawn from the natural world. Those are our words, but we will quote him at length about these issues in the later chapters. Suffice it to say briefly here that Einstein spoke

40 One can apply these two quotes to the former and latter, respectively: "Space and time are primary concepts, and the velocity of an object or a light signal is a derived concept after distance and time are defined." (Hsu, p. 58) "If the speed of light is in fact fixed, the measures of space and time we normally think of as fixed are not. Space and time are relative." (Wolf, p. 181)

about concepts of physics being "free creations of the human mind" (Einstein and Infeld, p. 31); that he felt perfectly justified in making crucial and arbitrary assumptions about the nature of light transmission based upon "a stipulation which I can make of my own freewill" (Einstein, 1920, quoted in Zahar, 1977, p. 207); and of the "purely fictitious character of the fundamentals of scientific theory" (Einstein, 1933, quoted in Holton, p. 251). We will argue strenuously against this kind of theory-building strategy later on, but it helps to explain what we are up against here.

As far as the results go, we are not questioning the proposition that it is possible to formulate a theory in the way Einstein did: to argue otherwise would be flying in the face of reality! Nonetheless, we would contend that one of the main objections to the resulting model is precisely that it violates one of the primary tenets which a good theory of science should possess — namely, that it be based upon general patterns which apply to all phenomena in nature ... as represented, for example, in the generalization factor mentioned in the prior paragraph. The problem is that any such general patterns get lost when one singles out a specific instance thereof, and transforms it into an ultimate and unique principle based upon convenient abstractions. This is precisely what the Special Theory has done, by virtue of being built upon the proposition that the patterns which light exhibits are different from other wave phenomena. We argue completely to the contrary — that light is not special, and that it conforms to persistent patterns which are found everywhere else in nature.

Indeed, why are we dealing with speed of light, as providing the central metric for defining our reality? Even if we were dealing with the fastest speed we know of — which is itself a simplistic way of describing the issue, as should be evident from the last part of the previous section — that still does not justify any such choice. In principle, any particular phenomenon could serve as the conceptual centerpiece for the rest of nature — speed of sound, speed of water, speed of an ant ... or speed of *anything*.[41] Yet any such selection is inevitably an arbitrary choice, and this is a principal reason why Einstein's model is unacceptable.

There is no compelling reason to select light for this function — let alone claim some unique association of light with time. Yes, light *does* trace the flow, or ride the wave, of time — but then so does every other level of process in nature. Time is connected with the flow of process, but it does so differently at every level. The upshot is that no one level of process bears any more unique relationship to time than does any other level. Any frequency, any speed, can be understood in relation to time; and there is nothing special about light in this respect. Thus, there is no more compelling reason to link time together with light than there is for uniquely associating time with any other level of nature's hierarchy. It is merely an illusion to think that there is some idiosyncratic, not to mention causal, relationship between light and time. Plus, doing so undermines and confounds the crucial concept of time itself, such as to greatly impair its utility as a pivotal cog in our thinking.

41 "Sachs states that *c* "cannot be unique to electromagnetic phenomena. Rather it must be taken as the speed of propagation of interaction (of any sort) between matter and matter.... a universal speed.... a connective parameter between communicating observers" (Sachs, 1973, p. 45; Sachs, 1974, p. 178).

If one finds cogency in the arguments we have been making, then the question naturally arises as to how we have reached this stage in our theoretical development, and why are we still stuck here. In answering, we maintain that there are a whole confluence of factors which have laid the basis for Special Relativity's success, some of which are really quite amazing. We will discuss several of these points at the end of the next chapter, including historical ones. But for now we want to examine the issue in the conceptual terms of the present topic, that of the speed of light. Our intention is to review various abstract factors which have contributed to the veil of confusion that we are dealing with here — especially concerning the joining together of separate categories of phenomena that need to be differentiated, the mixing up between what is actually known and what is merely stipulated, the obscuring together of different meanings for the constructs we use, and the elimination of phenomenological components that are essential for conceptualizing the issues.

In these various ways, one can argue that Einstein is playing by different rules than the rest of us … yet in an almost insidious manner of which we are unaware. It's as though we need some sort of "truth in labeling" here — for example, when a commonplace term like *speed* (of light) is used in a very idiosyncratic way … as in isotropic and measured only, and not also in terms of their necessary counterparts. We should be constantly alerted to such misleading usage, so that we don't get seduced by the outcome. Yet we suggest that this is exactly what has happened here — with the result being that our basic concept system has been co-opted and our thinking thereby hobbled, making it very difficult to clearly analyze what is going on.

In fact, the most basic thing we can claim to know is that the speed of light is constant under a certain set of conditions — but then, that is largely true about everything in nature; so this is no revelation. The problem is that all the details regarding these conditions amount to fine print, which are obscured by the theory and which very few people take cognizance of — whereas it is the headline which dominates everybody's thinking, even though the thesis doesn't mean all that much in the circumscribed terms in which it is elaborated.

One example of the kind of loaded terminology we are dealing with concerns assertions about what we just mentioned as light's "measured" speed. Right off the bat, we are encountering a set of misleading connotations here: Thus, one implicit reaction might be to think, "Oh, it's measured — well, that must mean it's valid." But, of course, this is not what the word means in these situations: no measurements have been conducted which confirm the constant speed of light across these various conditions. Rather, we are dealing with a conjecture about what might happen if one attempted to make measurements that are based upon certain assumptions.

Then there is that which cannot be measured — the presumption that the speed of light is the same in one direction as in the opposite direction — meaning that the *one-way* and *two-way* speeds of light are identical, as they would be determined in any of these situations. Yet such an assumption has never been demonstrated, and it turns out to be grossly counterintuitive in the case where a system is moving in relation to the medium through which the light is traveling.

No problem, for Special Relativity: simply *do away with the medium* and have done with the issue! In that case, there can no longer be any motion between the system and the medium — because there is no medium relative to which such motion can exist! Neat trick — but what sense does it make? All traveling waves in nature are a disturbance of the medium through which they are moving. Who cares? Special Relativity is special, and it simply decrees that light waves do not travel in a medium.

One way that we have talked about the issue is in terms of a distinction between source motion and sink (or receiver) motion — with the former being describable in terms of the *weak light postulate*, whereas the latter is also incorporated into the *strong light postulate*: these two separate categories are typically blurred in the standard analyses — mixing together the former precept which says nothing unusual and is readily comprehended, with the latter precept that is far more suspect. In a related matter, it is helpful to distinguish between a wave's *fundamental speed* in relation to the medium through which it is traveling, and its *effective speed* as also taking into account any relative motion between an object and the medium. Needless to say, eliminating the medium makes moot this crucial distinction.

As if all this stasis-motion confusion vis-a-vis a medium were not taxing enough upon the inquiring mind trying to make sense out of the Special Theory maneuvers, the situation only gets more muddled by being mixed up with a second aspect of the strong light quagmire. This concerns alterations of space and time that are associated with moving phenomena — the result of which just happen to yield the same speed of light outcome as would apply to non-moving phenomena. This unusual concept is a difficult enough idea to grasp in its own right — although we contend that it constitutes the only valid factor in the whole Special Theory analysis here. Yet even then, the conceptual picture is murky — because of the mixing together of distinct meanings for speed, as we have already mentioned: In terms of the issue we are now discussing, there is a difference between what we call the measured and the actual speeds of light, as they would be gauged within a given situation or set of circumstances (e.g., hierarchical level) ... rather than between two situations or sets of circumstances (e.g., hierarchical levels). While light's (two-way) speed may be the same for the former case, it would be different for the latter.

Then there is the whole issue of the observer, and the distorted role which this concept plays in the determination of light's speed. By focusing our attention on the disconcerting idea that the (internal) observer of a static system would measure the same speed of light c as the (internal) observer of a moving system — by virtue of each system having different space and time metrics, and however such stasis-motion states are ascertained — we are distracted from the equally valid perspective of an (external) observer who is witnessing the altered speed of light $c \pm v$ in relation to the moving system. We maintain that both perspectives and outcomes must be taken into account, in order to reach a satisfactory and coherent model here. Then there would be no basis for the one-sided conclusion that the speed of light is the same in relation to a static system as it is to a moving system.

Finally, it is essential that we clearly distinguish between what we actually know, as opposed to what has merely been stipulated or conjectured. As far as the speed of light itself goes, nothing has emerged over the years which actually entails any obvious (high-speed) motion of an observer, to begin with — let alone one who is somehow physically-capable of measuring the speed of light under those circumstances. And regardless, how would any observer be able to transcend the kind of logistical problems we cited earlier in the chapter concerning the difficulty, if not impossibility, of measuring light's one-way speed? We are merely dealing with hypothetical projections which are based upon certain theoretical assumptions and imaginary applications.

What has in fact been demonstrated is that moving entities appear to have slower clocks, as compared to static entities — as witnessed by static observers, not moving ones — in a manner which is consistent with the Lorentz transformation. We will pick up on this issue in the next chapter, but the point here is that because there is a *c* factor which is included in those equations does not prove Einstein's thesis. It can be argued, for example, that not only Lorentz but also Poincaré "regarded the universality of light speed as only a rule, but not as a basic principle or as an intrinsic property of nature." (Hsu, p. 59)

Einstein, by contrast, had no such hesitation, and was quite prepared to jettison existing constructs and ongoing logic in the service of a set of ad hoc rules. In this way, the Special Theory of Relativity attained its goal of establishing a convention about a constant speed of light — but at the horrendous cost of sacrificing a meaningful conception of the world: waves without a medium, space and time only handmaidens to the new sovereign *c* — with light henceforth acting as something ghostly, supposedly having the same speed in relation to everything ... irrespective of those things' differential motion vis-a-vis one another, or the relative nature of such "same" speed conclusions. These kind of scenarios are impossible, by normal definition; Einstein made them inevitable, by special definition — by an arbitrary interpretation that is based upon a one-sided framework ... not the kind of coherent whole which should provide the basis for our models of nature.

We contend that there is not a shred of explicit evidence which buttresses Einstein's position on these matters — in spite of the blizzard of propaganda to the contrary. Thus it is that Holton refers to Special Relativity's "two postulates for which there was and can be no direct empirical confirmation" (Holton, p. 250). And so while it is clear heresy to say, we maintain that replacing the classical model with that of Einstein did not represent a step forward; and that Newton's absolute space — or Maxwell and Lorentz's ether — centerpiece provided a far more substantial framework to build upon than Einstein's absolute light centerpiece. While the classical model had its limitations, there are solutions to those problems which are not random, which make sense, and which fit with other broad themes. That is the position we are arguing for in this book.

Chapter 7

Special Nonsense[1]

We have seen that light occupies a very special place in Einstein's world of Special Relativity. In its virtual supremacy above all else in the universe, light becomes the central phenomenon on the stage of nature. Indeed, in many respects, light *is* that stage of nature, for we maintain that light plays a comparable role in Einstein's model as absolute space plays in Newton's — not merely in terms of theoretical preeminence, but far more importantly in terms of phenomenological function. What we are dealing with is a fixed and rigid quasi-structural phenomenon, around or upon which the rest of nature must pivot.

Within the Special framework the main theoretical construct is that of light's speed: "Relativity Theory is concerned almost exclusively with only one of the properties of light, its velocity ..." (Bridgman, 1983, p. 149). But as we have seen, it is not really light's speed or velocity, in conventional terms, that is involved — because it is not a variable which can change, in that model: the only question is how much other aspects in nature must be altered or reconceptualized, in order to always retain the predetermined, constant *c* outcome.

The end result is that Einstein's light is treated as a totally idiosyncratic phenomenon in nature. As we have said, this is true not merely in that light resembles nothing else in *our* experience — whereby "speed" means a quantity of motion that can vary with circumstances, understood both in fundamental and effective terms. It is also true in that light resembles nothing else in *nature's* experience — in that it is characterized as a wave without a medium, having the same speed in relation to everything ... unlike all other wave phenomena.[2]

In fact, as we have noted, light is treated as being so unusual that we are told to abandon any hope of actually understanding it. And unfortunately, such a conclusion is

1 We should note that this chapter has the same title as one in the book by Zukav. The difference is that he uses the term in an ironic sense, as actually referring to the Special Theory's brilliance — whereas we use it in a quite literal way, as representing something less than that, to put it mildly.

2 It is quite an understatement to say that Einstein felt that "light is somehow basically different from other forms of motion." (Bohm, 1980a, p. 122)

inevitable, and we may as well give up trying to grasp what is going on: Special Relativity's light *cannot* be understood, in any kind of usual meaningful terms. The logical pieces are not even complete, by which light is said to exist; so of course the phenomenon cannot be comprehended within such confines. And in many respects, physics doesn't even try! As to the main theorist himself, Bridgman says that, "The explanatory aspect is completely absent from Einstein's work." (Bridgman, 1927, p. 165)

The result is one of the supreme ironies of modern science: light is presented in a manner which makes it utterly inscrutable as a phenomenon, while this self-same light is offered as a centerpiece for our overall understanding of the physical world. It is no wonder that light has taken on quasi-mystical properties in the orthodox paradigm: light has great power, but it cannot be comprehended.

Regarding alterations of space and time, each of Einstein's two theories puts forth separate proposals as to how these dimensions will change, as well as how these changes are to be interpreted; and the effects and their causes are said to be completely different in the two cases. Thus, Special Relativity proposes certain types of space and time alterations which are brought about solely by motion, and especially by high-speed motion — including the notions that mass increases and time slows down, under such circumstances. In General Relativity, alterations of space and time are said to be caused by gravitational fields or by the "curvature of space-time".

The General Theory interpretations are far less controversial and difficult to understand than those of the Special Theory. We maintain that part of the reason is because General Relativity essentially abandons the speed of light centerpiece, and returns to a vision of nature that is based upon space and time — even though it is an unsatisfactory picture of a combined space-time, as we shall discuss in the next chapter. Nonetheless, the mechanism by which alterations occur are at least comprehensible and broadly coherent in the General Theory — whereas the Special Relativity framework is totally convoluted and misleading.

We shall look at some of the Special Relativity alteration issues in the first part of this chapter, and we describe a few of the scenarios and experiments which have been imagined or conducted in this regard. And then, given that we do not think that Special Relativity provides an acceptable framework for dealing with these issues and does not represent a coherent model of nature, we turn in the second part of the chapter to a discussion of how the theory has attained its lofty status, and why it is treated as gospel that is sacrosanct even today. We will consider the historical setting in which the theory arose and what the basis for its success was, as well as some of the crucial reasons for why the model is an unsatisfactory representation of nature.

Alterations of Space and Time

There are several reasons for why the space and time alteration effects proposed by the Special Theory are far more controversial than those posited by the General Theory, leaving aside the former's implausible logical footing upon the strong light postulate. For one thing, no real causal explanation is put forward as to why these Special effects occur, and it may be that none is even permitted within the theory proper — aside from the unsatisfactory argument that they are simply the result of motion per se, which is somehow associated with the speed of light. For another, there are various anomalies and contradictions between theory and practice, and there is no consensus about how these are to be reconciled.

Regarding the types of space and time changes which are said to occur, perhaps the most notorious is the notion that time slows down with motion. That is, a moving clock will tick more slowly than a static clock, and the same is said to apply to any other oscillating or rhythmic phenomenon. Time "stretches" or expands when the entity undergoes motion: the interval between beats is larger, and therefore time "passes" and things "age" more slowly. The formal name for this Special Relativity effect is "time dilation", or the even more pseudosophisticated "time dil*at*ation" — although it is important to differentiate this from the similar General Relativity slowing-of-time effect, since that is said to represent an altogether different phenomenon (see next chapter).

There are also two kinds of alteration effects that are proposed for the spatial dimension, as resulting from motion. One is in terms of size — the dimensions of a moving entity are said to change, in a very particular manner: there is contraction, but only in the direction of motion, and not along any other axis. The other spatial change is an increase of mass with motion.

Whether we are talking about spatial or temporal alterations, two more details can be added which relate to the Special Theory's motion predictions. One is the presumption that, as the speed of a moving entity approaches that of the speed of light, the alteration effects will continue to increase — but within the distinct limit that such change-inducing motion can never attain the speed of light. For, if an object's motion ever did reach the speed of light, the following results would occur: the flow of time would cease altogether, the forward-moving dimension would be contracted to zero, and the mass would increase to infinity. These predictions provide the basis for a very specific contention of the theory — that no material entity can ever travel as fast as light, let alone faster, as we noted in the previous chapter. The other point concerns the presumption of symmetry which is built into the Special Theory — leading to the anomalous situation whereby both side of a stasis-motion relationship would somehow be altered in the same way, which is a proposition that we shall see ultimately makes no sense.

The Standard Scenario. We begin by examining a common description which is sometimes put forward, in which Special Relativity's space and time changes are discussed. In one specific form or another, this scene is often invoked in the form of thought experiments which are proposed within the Special Relativity framework. To be sure, the essential theme predated Relativity, and was an outgrowth of the classical vision. In this regard, recall that according to the Newtonian conception, there was a pure, force-free realm somewhere in the universe: this region was described as an inertial frame of reference. Moreover, that pristine force-free status would be replicated in various locales, each of which thereby qualified as an equivalent inertial frame in its own right: "[T]here exists an infinity of inertial systems, any two inertial systems moving relative to each other with constant velocity in a straight line ..." (Bondi, pp. 26-27). This classical model, of an endless array of separately-moving inertial systems, was "carried over without any alteration into relativity." (Bondi, p. 27)

Because the essential situation here is so commonly invoked for illustrating the fundamental principles of Einstein's model, we refer to it as the "standard scenario".

Here is one popular rendition of the basic plot: Two spaceships are both said to be undergoing uniform linear motion, and each thereby represents an inertial frame of reference. In the course of their travels, these two spaceships interact, e.g., by means of certain measurements which are conducted between them. This scenario is deceptively simple, and very easy to visualize — at least in abstract terms.

Yet it is the peculiar outcome of the real or imagined experiments that are associated with this scenario which provides support for many astonishing, and often bizarre, theoretical conclusions regarding the nature of reality. Indeed, one might argue that it is partly on the basis of these uncanny outcomes that the whole Special Relativity approach is treated with such deference — which is a point we shall return to at the end of the chapter. Among the potent topics which are impacted are the speed of light, high-speed motion, the nature of space and time and alterations thereof, and so on.

The mathematical factor which is used to make comparisons about these different characteristics is the Lorentz transformation. The usual description is that this transformation allows us to make comparisons between the two different systems which are involved in the standard scenario. As Zukav puts it, the Lorentz equations "translate what an observer in one frame of reference sees into what an observer in another frame of reference sees" (Zukav, p. 146). For his part, Einstein describes the Lorentz transformation as indicating "the space-time magnitudes of an event when changing from one body of reference to another.... so that the law of transmission of light *in vacuo* is satisfied." (Einstein, 1961, pp. 31 and 32)

Now for another view. We take exception to the usual conceptual composition of the standard scenario, in spite of how reasonable such a portrayal may seem on the surface. For while it is superficially plausible, the typical scenario serves as an example of how the vagaries of an abstract thought experiment can all too readily lead us astray — which is another point we will return to later in the chapter. We maintain that there are logical holes in both the premises which underlie the standard plot, as well as in the conclusions that it leads to.

To start with, one of the main problems has to do with the frame of reference construct itself, upon which the classical model and Special Relativity are based. We discussed that construct's conceptual weaknesses in Chapter 4, and we recall one of these here, which concerns a confusion between figure and ground. In particular, the conventional approach confuses *object* with *function* in the way it names the participants in a scenario like this — as in each of the two spacecraft being labeled as "frames of reference" which interact with one another. As we explained in our earlier discussion, there can only be one background for any given situation, at one time. We have indicated that the frame of reference term is another way of talking about such a background — and consequently, there can only be one such background frame of reference in any given setting ... not two, as in the standard scenario. Therefore, it is a mistake to be characterizing the two spacecraft in those terms, as each representing a frame of reference.

There are two main reasons why we are so adamant about this point, which might otherwise be seen as nit-picking. For one thing, the essential issues are difficult enough in themselves, without adding on any additional layers of confusion which serve to obscure the underlying patterns. And the other thing concerns precisely those underlying patterns — for we contend that a coherent model of nature is based upon consistent themes and patterns, which perform a crucial role in helping us to understand the world around us. One of the enduring themes and patterns of the present 2FT model is the complementary notion of figure-ground, as providing an invaluable template for analyzing phenomena in nature. Another is the hierarchical notion of nested levels.

Those two features come together, in terms of the issues we are discussing here. Thus, what amounts to a figure at a higher level is itself a background at the next lower level.[3] But we cannot have two backgrounds at the same level — meaning therefore that we are either dealing with two spacecraft as figures in relation to some larger background, or one spacecraft as figure in relation to the other spacecraft as (e.g., part of the) background. And the fact that either spacecraft could potentially act as (or correspond with) a background frame of reference does not change the fact that only one of two differentially-moving spacecraft could serve (or be part of) that function at any given time, in relation to the other. Clarifying the crucial figure-ground distinction is precisely why making the secondary object-function distinction is so important.

These questions matter so much because of the way that figure-ground patterns apply to motion and stasis. Thus in the present model, stasis would be understood in terms of the background in any situation — relative to which motion could be understood in terms of the figure. Therefore, the fact that any entity can be seen as static or moving is another way of saying that any entity can be seen in terms of either background (frame of reference) or figure: it depends on which level of the hierarchy we are talking about, and thus what is being compared with what.[4]

Let us return to the terms of the standard scenario and ask where the stasis is in the analysis. Clearly, we are back to a problem which has come up before, and will do so again: the spacecraft motion is treated as being absolute in the standard scenario — whereas in Special Relativity, motion is only supposed to be relative. Yet if the only two components which matter here are the two spacecraft that are each supposedly undergoing uniform motion, then what is it that they are each undergoing uniform motion in relation to? While Einstein rhetorically rejects the notion of a universal frame of reference, or absolute space — what, then, are we left with (as a stasis referent) which will join the moving spacecraft participants together here?

3 We emphasize a point here that we also footnoted in the previous chapter — which is that we are not being consistent in the orientation which we associate with these higher vs. lower referents. Again, it is the hierarchy pattern itself which is most important, regardless of how it is oriented in each case.

4 We should mention a counterargument to the point we are making here — which would insist that the original hypothesis is a sound one: that the two spacecraft do, indeed, each represent independent inertial frames of reference — and that this determination can be proven. For example, what if investigators went inside each of the two interacting spacecraft and conducted "inertialness tests" which showed conclusively in each case that the internal lab was inertial, in the classical Newtonian sense? If everything appears as uniform, in this regard, then why should we resist the obvious conclusion that both of the two spacecraft should be certified as genuine inertial frames of reference — with the corresponding conclusion about what the spacecraft are doing externally … undergoing uniform linear motion? We contend that such an explanation is a distraction from the issue — which is a relative hierarchical matter, rather than being absolute. It also restates the object vs. function issue, and that either spacecraft could function as an inertial frame of reference, in which the internal "inertialness test" would be valid. But such a determination is a relative one, and depends on which level of the hierarchy we are dealing with.

So uniform linear motion in relation to what? Such a claim is a pure, irrelevant abstraction — unless there is some referent for making the determination concrete ... about uniform, about linear, and even about motion. Therefore, we can dance around all we like, conceptually, but we will inevitably end up back in the same old place: there must exist an underlying spatial referent for stasis, in relation to which either or both spacecraft's motion exists. Call it "space", or even "space-time" — yet there is no avoiding the necessity for some background frame of reference to exist, in relation to which either spacecraft is undergoing motion.

Of course, the standard line is that any entity can see itself as static — in relation to which any differential motion that is involved can be attributed to the other entity: we are free to pick and choose our referents for stasis and motion, in whatever manner is most convenient for the analysis of the moment. But, then, why are we even talking about the standard scenario as though it represented two spaceships that are each undergoing uniform linear motion — since one could just as well throw up the cards and let them fall down in some other random way, leading to another totally different description here than uniform, linear, and motion?

Obviously, we reject this kind of casual and sloppy analysis: while the basic notion of relative motion is almost frivolously true, it does not change our contention that there is an orderly relationship among such factors. Unfortunately, Einstein's theory provides no coherent conceptual framework for understanding these relationships — in its sweeping rejection of absolute space, on the one hand ... coupled with its implicit embrace of absolute motion, on the other. Actually, of course, the theory rejects both these classical constructs, in principle — but in practice cannot avoid their equivalents, in some form. By contrast, the present model assumes a whole host of such absolute space and absolute motion frameworks, which are hierarchically-arranged, in a nesting (and thus relative) manner that is specific to every particular situation. As a result, the stasis and motion referents will vary from case to case — but in an orderly and comprehensible manner.

In this discussion, we have not even touched upon another problem which is inherent in the standard scenario, and this concerns the trajectory of the uniform motion which would apply to open space. We described the issue at length in Chapter 4, where we argued that any object that is moving in a linear path — such as each of the two spacecraft are supposedly doing in outer space — would inevitably be moving closer to some external force influences, and further away from others; and this would inevitably impact the object's speed ... meaning that it would not be uniform. Indeed, there is no reason to assume that we are even dealing with inertial frames of reference at all in these various Special Relativity situations in which gravity or acceleration manifestly apply, given the force-free terms in which that construct is properly understood.

All in all, the standard scenario is not very satisfactory: to pick up on a point we made in Chapter 4 and will pursue further at the end of this chapter, the universe is not built upon arbitrary referents for motion. A motion state is not random, in terms of whatever ad hoc stasis referent happens to be convenient for the occasion. We do not have independent and equal spacecraft and trains and solar systems and embankments and earths, all undergoing mutually random and uniform motion. Such a haphazard image of separate, uniformly-moving, inertial frames of reference crisscrossing one another in their linear paths only leads to conceptual chaos.

The Perplexing Notion of High-Speed Motion. One of the puzzling aspects of the distortion phenomena we are discussing is the question of why we are not generally aware of it. If motion alters space and time, then why do we not experience this on a regular basis?

The usual answer is that for the effect to be noticed, things would have to be moving incredibly fast — and we do not encounter this in our everyday world. Consequently, the distortion effects only seem strange because our normal realm of experience is confined to phenomena having fairly low velocity. "We live out our lives in a limited situation of low velocities ..." (Zukav, p. 147) Special Relativity, on the other hand, is concerned with phenomena having very high speeds — i.e., in relation to that of light. Just imagine what it would be like to be "people living in a world in which ordinary objects travel near the speed of light" (Geroch, p. 148)! If only the rest of us were able to transcend our mundane existence and experience these very high velocities directly, then we, too, would grasp the more lofty reality which the Special Theory has revealed. We shall see an outstanding example of this high-speed world in a coming subsection, where we are led into Einstein's own musings as a young man about what it would be like to travel at the speed of light.

Leaving aside the way such an approach tends to throw non-specialists off balance mentally — by trivializing our range of experience and thereby undermining our powers of reasoning and our capacity to analyze these affairs, as being too narrowly-based and not sophisticated enough, etc. — the implication is that there is a world of low velocity and there is a world of high velocity; and it is in the latter world where Special Relativity's space and time distortion effects are to be found.

But what does this mean? For one thing, it suggests that things are moving at some definite speed — a point that we have touched upon previously as being out of keeping with the Special Theory. Thus, are all of us who live on the surface of the earth part of a world at rest, with ordinary objects traveling at slow speeds? Yet we could also be said to be moving very fast — around the earth, around the sun, around the center of the galaxy, and so on. What, then, is the criterion for determining how fast an object is moving — what is the proper gauge of an object's speed? That is, relative to what does the moving entity have the high speeds which are required in order for the spatial and temporal distortions to become significant? There are two basic ways of answering this question — which concern the referent for stasis that is being used, in relation to which the (e.g., high-speed) motion exists: one answer is absolute and the other is relative.

To understand the choice, let us go back to the standard scenario. If there is a larger background frame of reference relative to which the two spacecraft exist, then we are dealing with an absolute space, stasis-based referent for absolute motion — such as high-speed motion. On the other hand, Relativity would maintain that either of the two spacecraft can function as the quasi-static referent for the other spacecraft's motion; then we are dealing with relative motion — and it is only the differential motion between the two spacecraft which determines the outcome. (By the way, while it is clear how the former type of motion could be construed as being uniform, that is not the case with the latter type of motion.)

This brings us back to the issue we discussed in the previous chapter, regarding the different ways of interpreting what is going on. For Lorentz and others, there was a real physical ether medium — which for our purposes is tantamount to an absolute space referent for stasis — relative to which (e.g., high-speed) motion would exist; therefore, the friction-like interaction between object and medium would produce concrete distortion or alteration results. For Einstein and his supporters, there was no concrete physical medium of any kind — no absolute space referent for stasis — relative to which the (e.g., high-speed) motion would exist; therefore, the transformation is strictly the result of motion per se, as it would exist in a relative sense between any two differentially-moving entities.

This difference of interpretation has dramatic implications. Thus, in Lorentz's theory there can be only one measure of speed, and thus a single description of space and time changes. But in Einstein's theory an entity can be seen as moving at any speed we choose — and thus there can be a whole panoply of different outcomes as to the space and time transformations which result — simply on the basis of whatever stasis referent we happen to select, relative to which the motion occurs.[5]

Yet at some point, do we have the right to ask what such relative motion actually means, and to what extent we are merely speaking about *perceived* (if not *conjectured*) motion and associated results? Another way of talking about the issue is whether we are dealing with physically *real* or merely *apparent* effects, regarding the space and time changes which are said to occur in these relative terms.

There is another point, which relates to the almost surreal status which light takes on in Einstein's model — and here we are specifically talking about the speed of a moving object in relation to that of light, as though there were some significant connection between the two which is a factor in the space and time alterations that are said to result. However, there is no actual basis for concluding that such space and time alterations have any actual (let alone causal) relationship with the speed of light, per se — just that an object is traveling very fast, in relation to some referent of stasis.

Finally, there is another very significant issue, which we made reference to in the previous chapter. That is, given the symmetrical basis of the Lorentz transformation, what grounds do we have for concluding that one entity is moving in relation to some referent of stasis — as opposed to the reverse? If it could go either way, then this would of course mean that the transformation effects are reciprocal — and that each entity could see itself as static (with normal space and time metrics), in contrast to the other entity as moving (with altered space and time metrics). The choice, between a one-way (asymmetrical) and a two-way (symmetrical) interpretation, is reflected in the two different theories: for Lorentz, there was an absolute referent for stasis, relative to which absolute motion (along with space and time changes) would occur; for Einstein, there were only relative referents for stasis, and so either of two entities could be seen as static and

5 "The relativistic equations of [Einstein's] electrodynamics are, in form, identical to the Maxwell-Lorentz equations. They differ in that the velocities occurring in them are defined as the velocities measured in the particular reference system being used, whereas the velocities in the Maxwell-Lorentz equations are defined as being relative to the ether, i.e. as absolute velocities." (Builder, 1958a, p. 283)

the other as moving (along with space and time changes). While the Lorentz transformation itself had a symmetrical formulation, Lorentz did not conceive the world as stasis-motion "flippable" ... whereas Einstein did.

This is the issue we turn to next.

The Twin Paradox. We now come to a problem which has long been a thorn in the side of Special Relativity — but one of its own making. We are talking about an anomaly that is built into the very heart of the model, by virtue of its symmetrical foundation. The issue is the source of endless debate — mostly based on various defenses of the theory which supporters put forward, and numerous explanations which have been proposed as to why there is no problem ... or if one exists, how it can be resolved. But we maintain that none of the contorted posturing and intellectual scrambling really matters, for there is no way of escaping the underlying theoretical dilemma.

The twin (or clock) paradox was first proposed by Langevin in 1911, based upon comments which Einstein made in his original 1905 paper on Special Relativity (MacRoberts, p. 372). Here is one imaginary scenario for illustrating the problem. There are two brothers — twins, to be precise. At some point early in their lives, one of the brothers goes off in a spaceship, and travels for some years at speeds which are appreciable relative to the speed of light. When the traveling twin eventually returns home, he is found to be much younger than his stay-at-home brother. In particular, the traveler is still a very young man — returning home to meet his brother's grandchildren! It's that simple: the twin who stays at home ages much faster than the twin who travels at high speed.

This thought experiment sounds clear enough, leaving aside the salient question of whether such a scenario could ever actually be implemented. Thus, it is hard to imagine what kind of vehicle would be able to travel that fast, not to mention whether a human being could survive such biologically-challenging circumstances. But let us leave such practical matters aside, in order to deal with the crucial theoretical issue. The question is, which twin is actually moving in relation to the other as static — and what basis do we have for making that determination? The answer may seem obvious to the casual observer, based on the description we have given; but this is not the way Special Relativity sees things.

The problem centers around Einstein's interpretation of the Relativity Principle and of the Lorentz transformation. In brief, if one object is undergoing motion, in relation to another object (or set thereof) which is the static referent for such motion, the symmetrical formulation of the theory means that these roles could just as well be reversed — with the "moving" object then being designated as static, and the "static" object as moving.

There is no clear way of parceling out these two phenomenological states within the confines of the theory, as between the two entities. The result is that either entity could be seen as moving, relative to the other entity's stasis — and there is no way of differentiating which one is which. Recall, once again, Einstein's conclusion about the train and platform: either the train can be said to be moving, in relation to the static platform — or the platform can just as well be seen as moving, relative to the static train.

This is the way Rogers describes how the Lorentz transformation would be applied in terms of a static vs. a moving experimenter:

> [T]he symmetrical form shows that absolute motion is never revealed by experiment. We can measure relative motion of one experimenter past another, but we can never say which is really moving.... [E]ach sees all the other's apparatus shrunk ... Each sees all the other's clocks running slowly ... In this symmetrical 'relativity' we see the same thing in the other fellow's laboratory, *whether he is moving or we are*. Only the *relative* motion between us and apparatus matters — we are left without any hint of being able to distinguish *absolute* motion through space. (Rogers, pp. 483-484)

Likewise, Davies says that, "[E]ach twin would see the other's clock running slow relative to his own" (Davies, 1980b, p. 465). Therefore, while we may think that the traveling twin is moving relative to the stay-at-home twin, Relativists can just as well argue the reverse — that the "traveling" twin can be understood to be static, relative to the "stay-at-home" twin as moving. The disturbing result is that, "The effect is reciprocal and contradictory: both clocks cannot be running more slowly than the other." (Calder, 1979, p. 106)

Now it is all very well to talk about these kind of scenarios in terms of abstract concepts of stasis and motion which one can flip at will — but here we are supposedly talking about changes or transformations of space and time parameters ... which apply to the moving entity, but not to the static entity. Yet if either twin can take on either role, in relation to the other, then whose space and time parameters are actually changing? In the case of the twins, which one is younger than the other? Neither? Either? Both?

There is no coherent answer to these questions. On the one hand, proponents will put forward various explanations as to why there really would be asymmetrical aging — with the stay-at-home twin aging more rapidly than the traveling twin. But then there is the other hand, according to which the overall symmetry conviction must not be abandoned: this means that the first hand argument is only one side of the coin — with the equally asymmetrical counterpart being that the "traveling" twin would see itself as static, in relation to which the "stay-at-home" twin is the one moving and aging less rapidly.

Special Relativity's reaction to this conceptual challenge could variously be described as comical and/or sad — perhaps depending upon one's theoretical predilections. On the one hand, there is mild bemusement, such that "the clock paradox has been relegated almost to the mathematical puzzle department of popular science magazines" (MacRoberts, p. 373) ... to something approaching indignant anger — as in "[N]ot 'paradox,' please!" (Terrell, p. 9) Bondi echoes this indignation: "[I]t was always ridiculous to call this a paradox; no paradox of any form is involved." (Bondi, p. 151). Then there is the remarkable pleading, "Why Pick on Einstein?" — this being the title of an article by Davies largely concerned with the subject. (Davies, 1980b, p. 463)

More seriously, the problem with Special Relativity's response is that while most theorists believe there would be asymmetrical effects here — that the stay-at-home twin would age much more rapidly than the traveling counterpart — there is no consensus as to why such differential aging would be one-sided. Nor is there any broad agreement as to why the model should be symmetrical in abstract theory, yet asymmetrical in real world practice. These contradictions are generally bypassed with some form of conceptual subterfuge, as we shall see shortly.

One brave and candid scientist who began to wrestle with these questions was named Herbert Dingle. We repeatedly cite Dingle in this book; but here, with this topic, we find the genesis of much of his critical thinking about these matters. By the way, it is worthwhile noting that at one time Dingle had been a strong proponent of Special Relativity — until he began wondering about the twins problem. In fact, Dingle was himself a physicist, and he even authored an important 1940 text on Special Relativity, which was respected enough to be cited in the very limited bibliography of the authorized English-language translation of Einstein's popular *Relativity* (Einstein, 1961, p. 158).

In the 1950's, Dingle began to ask why it was that the traveling twin would age more slowly than the stay-at-home twin. First tentatively, then more persistently, Dingle wrestled with that single "Why?" question. In the end, he was not able to find a good answer himself; and no one else was able to provide one either, that he found satisfactory. The essence of Dingle's concern was that the paradox could not be resolved within the strict symmetrical confines of the Special Theory.

As we shall discuss at greater length in the second part of the chapter, Dingle argues that there is a confusion between two different theories of relativity, that of Lorentz and that of Einstein — again, with the former rejecting a symmetrical interpretation of the Relativity Principle, and the latter embracing it. Dingle maintains that this provides the breeding ground for the dilemma here. He explains the development of the problem this way:

> Although the "paradox" is obvious to anyone who reads Einstein's paper, it did not at first attract much attention, for the reason I have already given, namely, that Einstein's theory was regarded as merely a recondite form of Lorentz's, and on Lorentz's theory it does not arise.... [O]n Lorentz's theory, "asymmetrical aging" ... would actually occur.... [T]here was no difficulty of disposing of [the paradox] because it [the paradox] followed from the postulate of relativity which that theory rejected. (Dingle, 1972, p. 187)

The general reaction to Dingle's argument was very negative — virtually to the point of treating the author as a traitor or a heretic. As MacRoberts puts it in describing the whole affair, "All in all, it has not been a dispute in which the scientific establishment should take much pride." (MacRoberts, p. 373)[6] Aside from the way Dingle was treated,

6 That is an understatement, for the reactions to Dingle's doubts and his probing questions do not reflect well on the scientific community. Some of this history is recounted in Dingle's eloquent book, *Science at the Crossroads*. The work makes fascinating reading, both in terms of the scientific issue at hand, as well as in terms of the attitude of the scientific community towards troublemakers — from within, no less! — like Dingle. He eventually rejected Einstein's model altogether, and the treatment which Dingle later received from other scientists was rather shocking; and again, it is quite illuminating in its own right, regarding how the scientific community treats radical dissenters. (see Dingle, 1972) By the way, Relativity theorist John Taylor. (1975) cites Dingle's book as an "interesting study in the psychology of scientists": one wonders who Taylor thinks is the interesting subject of this case study! We return to the issue of scientific conformity in Chapter 10.

some commentators even seemed to argue that Special Relativity is itself beyond criticism. Thus Davies writes,

> If a thought experiment appears to give a result in contradiction to relativity, a logical or mathematical error must have been committed.... It is not necessary to measure the time dilation effect directly to verify that the theory's principles are correct ... If relativity were wrong, our detailed understanding of much of subatomic physics would collapse. (Davies, 1980b, p. 465)

In his note on the issue, Terrell protests that, "It seems strange that Special Relativity should still be viewed as a controversial theory after countless experimental verifications ..." (Terrell, p. 9)

By contrast, we maintain that such attempts to wiggle out of the real dilemma here by claiming the general value of the theory are out of order, and concern a different question that we will address later in the chapter. In any case, those points have nothing to do with the issue at hand.

As far as the abstract theoretical point which we made reference to above — that the "traveling twin's" perspective on this whole state of affairs would be the stasis-motion opposite from its normal portrayal — there is really not much to say ... except to contend that it is merely an illusion. We shall refer to it later in the chapter as entailing an "upside down" way of constructing reality. Of course, factually speaking there is no way of proving what that other entity's perspective is, one way or the other. Doing so would presumably require us to develop an entire phenomenological framework that is built upon the orientation (if not the consciousness) of the traveling twin and its supposed perspective of itself as being static — why not moving? — in the circumstances we are examining.

In any event, our position is that the fully symmetrical model is implausible and cannot lead us to a coherent vision of nature — even if it is the most convenient way of interpreting the Lorentz transformation, thereby providing the simplest "law of nature" formulation which Einstein sought. As we shall discuss in Chapter 10 regarding the role of mathematics in our theories, we should not be misled into thinking that everything which is represented in mathematical statements actually exists in the real world — and we contend that the symmetrical form of the Lorentz transformation provides a perfect vehicle for such illusion.

We want to use the rest of this subsection for describing some of the theoretical arguments which have been put forward for explaining why asymmetrical aging would definitely occur in the abstract twins scenario. Regarding the practical question of what actually happens in analogous physical situations, we shall describe some of the experimental evidence in the subsection which follows.

As far as elucidating the asymmetrical aging which it is generally believed would occur in this scenario, Dingle talks about the lack of consensus in this regard and how it would be explained "on grounds that are almost unbelievably diverse. They involve Doppler effects, observations by external observers in countless varieties of circumstances, the influence of the rest of the universe, electromagnetic considerations, and more ingenious situations than one would have thought possible." (Dingle, 1972, p. 186) Indeed, a whole book was written many years ago, called *Time and the Space Traveller*

(Marder), which tried to compile in one place all the different explanations for why the asymmetrical aging effect would occur.

Actually, it is hard to avoid the conclusion that these are not always explanations so much as desperate rationalizations — and often very convoluted arguments — to try and salvage the conceptual integrity of a flawed theory ... as someone flailing in water, who will clutch at anything to keep from drowning. As Anderson says: "Perhaps no other problem in special relativity has had so many papers written about it, nor called forth more nonsense" (J.L. Anderson, quoted in Ruderfer, 1979b, p. 329). We maintain that these proposals amount to excuses or guesses for trying to circumvent the obvious conclusion, which is that the theory is built upon a fundamental contradiction.

Here is one interpretation that was put forward by Keating, among others — although he is also one of those who deny that there is any paradox involved. Keating maintains that we must distinguish between symmetrical time dilation on the one hand, and the asymmetrical twin (clock) effect on the other. The former

> tells nothing about the physical time differences that might exist between spatially separated clocks.... [T]he statement that "moving clocks run slow" is an artifact completely without physical significance.... It is purely a symmetric, definitional artifact of measurement created by the adoption of the Einstein simultaneity convention, a "bookkeeping procedure" ... [7] [On the other hand,] the twin clock effect is the natural phenomenon that a clock having a past history of acceleration indicates a total elapsed time less than that indicated by an identical clock having no history of acceleration. (Keating, p. 219A)[8]

Acceleration? Did somebody say, "acceleration"? Here, in the context of Special Relativity? But acceleration means non-uniform motion, and every schoolgirl relativist knows that Special Relativity does not deal with non-uniform motion — this being outside of its purview, and taking us over to the General Theory. So how can the established paradigm use acceleration to explain an effect which emerges from Special Relativity?

It is not only Keating who introduces acceleration as a symmetry-breaking explanation for the twin paradox; this represents by far the majority (and perhaps orthodox) position on the subject. Rosser: "Asymmetry comes in when a clock is accelerating relative to an inertial frame of reference." (Rosser, p. 352). Korenman: "[H]ow many times must it be said that the felt acceleration of the traveling twin is sufficient asymmetry to remove the paradox?" (Korenman; see also Davies, 1980b, p. 463). All these interpretations explain the asymmetrical aging of the twins in terms of the fact that the traveling twin was accelerated while the stay-at-home was not. Thus, the two twins are not equivalent, and the symmetrical Relativity Principle does not apply. Asymmetrical aging, therefore, has been shifted from Special Relativity to General Relativity for its explanatory framework!

7 MacRoberts makes the similar statement that time dilation "is not some mysterious effect of motion on 'time' but a normal consequence of the definition of 'time'." (MacRoberts, p. 370)

8 We have not quoted the last sentence of Keating's closing summary paragraph of this article, which might be construed as resurrecting the myth by referencing "velocity", rather than "acceleration", as explaining the asymmetrical effect. (Keating, p. 219D)

But not so fast: The Theory of Relativity's conceptual circumlocution does not stop there. For the equivocators retort: "Because inertial referent frames are not involved throughout, the analysis must be carried out quite carefully. A proper calculation ... can be made within the context of Special Relativity if appropriate care is exercised ..." (Marion, p. 403) Or simply: "Special Relativity ... leads unambiguously to the time discrepancy in the twin paradox" (Noonan, p. 15). Podlaha: "[T]he general theory of relativity is not necessary for the solution of the clock problem.... [T]he wrong opinion arises that the cause of the clock retardation effects is the acceleration." (Podlaha, 1979, p. 379) "[T]he accelerations are not the real *physical* cause of the relativistic effects." (Podlaha, 1978, p. 182). Or note Ljung's comment that, "General Relativity is used only in determining that an asymmetry does exist and hence in deciding which twin changes frames of reference" (Ljung, p. 13), after which Special Relativity and non-accelerated explanations would suffice.[9]

So the extreme position is that the whole analysis can be conducted within Special Relativity, although some theorists acknowledge that General Relativity must be invoked — but only as a momentary "tie-breaking" interlude for deciding which twin accelerates, thereby managing the paradox — after which one can immediately return to the haven of Special Relativity to complete the analysis. The upshot is that most Relativists continue to want it both ways — that the Special Theory remains undiminished in its ability to deal with situations like the twin paradox, even though many concede that one must reach outside the confines of Special Relativity (to the General Theory) in order to resolve the anomaly.

Then again, we encounter the astonishing claims by some analysts who dismiss the notion that General Relativity is even relevant to situations where gravity is not involved: "The unequal aging of accelerated and non-accelerated 'twins' has been observed" in various laboratory experiments. In these, "the transport systems have huge magnets that bend the beams with great accelerations, yet Einstein's factor [i.e., the Lorentz transformation] always works. Since the gravitational constant G never enters, there is no need to invoke General Relativity." (Williams, p. 11) Others make the more general claim that acceleration will not affect clocks: "The data indicate that the behaviour of a clock is independent of both longitudinal and transverse accelerations. Arguments that accelerations change the clock are not correct." (Roos et al., p. 245) Bailey et al. also report an experiment which found no effects of acceleration. Likewise, Ruderfer also claims to offer proofs which "merely affirm the prevalent view that acceleration per se does not directly affect the behavior of ideal clocks." (Ruderfer, 1979c, p. 333) Builder likewise cites "the generally accepted view, which is basic in the general theory of relativity, that the rate of a clock is not a function of its acceleration." (Builder, 1958a, p. 280)

Needless to say, these are very peculiar arguments — given that according to General Relativity, gravitational influences (or curvatures of space-time) would alter clock

9 Such miraculous mental contortions are accomplished by converting the accelerated system's parameters into those of an infinity of successive instantaneous inertial systems. See e.g., Builder, 1957.

rates, as we shall discuss further in the next chapter. And any attempt to completely dissociate gravity from acceleration makes very little sense, certainly in its utter disregard of the Equivalence Principle. Indeed, even Marion would concur on this particular point, based upon his discussion of General Relativity: "[C]locks run slowly in a gravitational field.... Because the effects of gravity and acceleration are indistinguishable, we must conclude that an accelerating clock runs more slowly than a clock in an inertial frame of reference." (Marion, p. 42)

This ends our discussion of the abstract twin paradox. We turn next to some of the concrete results which have been observed in the real world, to see what further light can be shed on these issues.

Experimental Results. We will very briefly mention several lines of actual experiments and observations which relate to the time dilation issue we have been discussing.

One noteworthy experiment was conducted by Hafele and the abovementioned Keating in 1972. The study began with three atomic clocks which were synchronized together at the outset, and were then separated: one remained at rest on the ground, while the other two were carried aboard passenger jet aircraft traveling in opposite directions around the earth. The assumption was that there would be two types of effects — those associated with gravity and General Relativity, but especially those associated with motion and Special Relativity. The overall conclusion was that both theories' predictions were said to be confirmed — and the results have been widely cited, as they would appear to closely parallel the twins scenario. Thus, the experiment would seem to resolve the dispute and offer "proof acceptable to everyone." (Terrell, p. 9)[10]

Unfortunately, there are at least two problems with such a conclusion — one of which is general and we shall address just below, and the other of which is particular here: the two oppositely traveling clocks did not show the same readings, as would be expected from the fact that they both underwent identical amounts of motion in relation to one another; rather, the earth's rotation also had to be taken into account, with one traveling clock moving along with that rotation and the other moving against it.

Another line of research involves cosmic rays that are entering the earth's atmosphere at exceedingly high speeds. Collisions with air molecules result, one byproduct of which are called muons — and the inference is that the resulting high-energy muons then also travel very fast, as they descend towards the earth's surface. The conclusion is that the high rate of speed produces time dilation, and thereby extends the life of the

10 Schlegel dissents from the conclusion that Special Relativity is even involved here, arguing that the motion of the clocks themselves "is in fact circular and hence accelerated, so we properly interpret the motional time acceleration as a general-relativistic effect." (Schlegel, 1973a, p. 293) He also maintains that "the relativistic time loss prescribed is *independent* of the speed at which the clock is moved" (Schlegel, 1974, p. 186). His conclusion is "that we can expect atomic clocks carried at any speed around the world will give the same effect." (Schlegel, 1974, p. 186) Thus, "the observed effects may be readily derived as consequences of the (general relativity) time rate of change with difference of the gravitational potential" (Schlegel, 1973a, p. 293), and so "the experiment was in all respects a test of general relativity." (Schlegel, 1980, p. 350)

muons, since they travel a far greater distance (downward) than would otherwise be expected.

Finally, we can mention observations which are associated with a technology that developed in the 1990's and which is becoming more and more prevalent in everyday life, called the Global Positioning System, or GPS. This makes use of a set of satellites which orbit our planet and are primarily intended to help identify precise locations on the earth's surface. GPS is often cited as offering unique verification of the premises of both Special and General Relativity.

Now for the disclaimers — and our dissent — about the conclusions which can be drawn from these or any other lines of experimentation and observation that might be cited. For one thing, there is the little detail, which we have mentioned before, that Special Relativity is supposedly only dealing with inertial frames — which properly speaking should be free of gravitational influences and entail only uniform linear motion … not non-uniform linear motion, and not circular motion; proper inertial conditions do not therefore apply.

While this point may be arguable, something else is not — and it concerns a point we have made before and must emphasize here: *what has been endlessly validated are the specifications of the Lorentz transformation, and not Einstein's particular interpretation thereof.* In this regard, none of these experiments could possibly resolve the twin or clock paradox itself, simply because the primary issue is not a *factual* one but a *conceptual* one. The problem exists precisely because of the contradiction between the symmetrical predictions of Special Relativity and the asymmetrical observations of the reality we occupy — and it is the latter which have been abundantly demonstrated, not the former.

In terms of the standard scenario, all the experiments and observations show one entity that is moving in relation to a larger background frame of reference that is static … and the latter may include a second entity which is being specifically compared with the first entity: it is the moving entity which would undergo space and time transformations, not the static entity … and they do not flip-flop back and forth in this regard — there has never been a physical demonstration that the two roles could be reversed, at the same time and in the same setting. As far as the phantom reality which the Special Theory muses about regarding the moving entity, there is not the slightest physical evidence in support of such a symmetrical mirror image stasis-motion representation and space-time transformation of reality: it is pure fantasy.

Albert-at-16. And now for some more fantasy! Let us look next at a particular instance of something traveling at the speed of light. We do this in the form of one particular mind game, which may have been Einstein's first thought experiment. He is said to have contemplated this scenario as a young man, and it supposedly laid the groundwork for his later theorizing. It is surely one of his most provocative images, and was an early sign of the "budding genius" to come. Here is how Sachs sets the scene:

> The initial impetus for the development of the theory of relativity in 20th century physics seems to have come when the curiosity of a 16-year old boy led him to ask a question about the nature of light…. Young Albert Einstein then wondered if it might be possible for him to see a light beam standing

> still, by traveling parallel to it.... In contrast with his expectations — which were based on "common sense" — Einstein discovered that, according to the form of Maxwell's equations (the most fundamental description of light that he knew of), if one should insist on the objectivity of these equations, then there are no frames in which light would be propagating at any speed other than $c=3x10^{10}$ cm./second! That is to say, even if one should be traveling at the speed c, parallel to a propagating light beam, he should still determine from his measurements that this light is moving away from him at the speed equal to c. (Sachs, 1974, p. 1)

It is difficult to know what to make of this imaginary scene. Thus, what does it mean to say that Albert is traveling at the same speed as light? Does he do that, but then light hopscotches ahead and travels outward again? How can he both travel along with light, and yet that very light is also moving away from him?

Let us say that several of us are sitting around on the ground in a huddle, young Albert amongst us. Then he suddenly declares that he will do something very daring — and thereupon he gets up and somehow manages to propel himself up to the speed of 300,000 km./sec. ... which also happens to be the same speed as we calculate for light itself. And yet that proposition conflicts with the way Maxwell's equations are treated by the strong light postulate — in the sense of light always moving at the same 300,000 km./sec. speed ... which is supposedly what Albert would continue to calculate, as mentioned above.[11]

We suggest that two different perspectives are involved here — a ground-based perspective and Albert's perspective. According to the ground-based huddlers' perspective, light is moving at a fundamental speed of c — yet Albert is also moving at that same speed: the net result would be an effective speed of zero, meaning that Albert and the light would be static in relation to one another. On the other hand, Albert's perspective would be based upon the strong light postulate, which a)does not allow for such a fundamental vs. effective speed differentiation, and b)uses altered space and time metrics, according to which light would be seen as always moving away at 300,000 km./sec.

According to this latter approach, no matter how fast an object is moving, light is supposed to always travel at the same speed in relation to it. In those terms, an object can never travel at *any* portion of the speed of light — since the speed of light will always be the same, relative to that thing: One is always traveling at *zero* portion of the speed of light, because that speed never changes. Even if something were to travel at an exceedingly high speed (however that is determined), the speed of light would supposedly be the same as it was before — and just as much out of reach as ever. Like the carrot which leads the donkey, the relative speed of light would not only never be attainable, it would not even be approachable!

Some descriptions of this experiment seem to entail the Albert-based perspective with light continuing to travel away at high speed, such as the Sachs quotation we

11 Note that in this regard, Holton cites Hoffman as indicating that, "[I]t is hard to see why the situation described in the thought experiment should be considered surprising or impossible *if* one applies a Galilean instead of a Lorentz transformation to Maxwell's equations. Somehow, Einstein appears already to have tacitly been thinking that Maxwell's equations must remain unchanged in form for the observer moving along with the beam ..." (Holton, pp. 376-377)

started with. Other descriptions seem to entail the ground-based perspective, according to which there would be no differential motion between Albert and the light — such as with the following two descriptions: " ... Einstein tried to imagine what it would be like to travel as fast as a light wave. He saw, for example, that the hands on a clock would appear to stand still, since no other light waves from the clock would be able to catch up with him until he slowed down." (Zukav, p. 122) Likewise, Bohm cites Einstein as posing the question, " 'What would happen if one were to move at the speed of light and look in a mirror?' Evidently, one would see nothing because the light from one's face would never reach the mirror." (Bohm, 1980a, p. 122)

Let us contemplate further the light-as-static approach which is depicted in the ground-based perspective. We contend that this is a legitimate scenario to ponder — and so we might wonder what it would be like to essentially translate our worldly experience into light's worldly experience. This could be understood in the present model simply in terms of shifting our orientation from one to another level of nature's hierarchy — to the level at which light is static.

Since exploring a world of fantasy about such matters seems to be *de rigueur*, anyway, let us suppose that there are incredibly tiny things which move very fast in relation to that light as static — maybe even things which move at the equivalent of some 300,000 km./sec. in relation thereto! Let us further suppose that photons can also perceive and conceptualize — in which case they might accord some special name to the respective entity which is moving very fast in relation to their own stasis. Of course, it would not be named "light" or "photons", but the phenomenon could well serve some analogous function in the photons' world as photons do in ours. The point is that while we may not know what those other fast-moving things are, we have no basis for assuming they do not exist. Thus, no matter what level of the hierarchy we are dealing with, it seems plausible to presume that there will be things which move very fast ... relative to that level as static. As we have said before, there is no reason to think that light is unique in the hierarchy of nature.

But let's get back to the young man's own fantasy. Of course, there is the obvious practical reason why the fast-moving scenario of Einstein makes no sense, and that is biological: he could not survive those conditions. So one answer to Albert-at-16's wonderment is that it won't work — and we might tell him: "If you were traveling at that speed, then you'd no longer be the curious and upstanding genius you were (or became) when you wrote the Special Theory." Things which travel so fast are not going to be Albert Einsteins: they'll be an entirely different phenomenon.

Now let's try another tack, as long as we're in the realm of far-fetched scenarios which make no sense and have no relationship to reality. That is, if we really were to take the Special Theory's first postulate seriously, about the relativity of stasis and motion, then why not consider reversing the roles here? For according to the symmetrical interpretation of stasis and motion, if there is differential motion between two entities then we have no more basis for concluding that one entity is moving in relation to the other entity's stasis ... than the reverse scenario.

So why need we assume that light moves at 300,000 km./sec., in relation to Albert? Why not assume the reverse — that light is static, and he is moving at 300,000 km./sec.? By this account, perhaps young Albert could have saved himself some energy in his imagination: maybe he was, already and always, moving as fast as he thought light to be

— except that such motion is in relation to light as static … according to such a phenomenal, figure-ground stasis-motion reversal! Perhaps in this light-static world, light envies us and the kind of experience we have — and wishes it could experience life in our sublime and enlightened high-speed state … rather than in its own dull world! Do they in that world have some genius who can help them realize such fantasies?

A silly proposition, you say — to swap moving light, with something which appears to be static? Perhaps, but no more so than many conjectures which emerge from Special Relativity. You say that such an interpretation violates the second postulate, which states that light will always move at 300,000 km./sec. — so therefore we have no right to reverse perspectives and imagine light as static in this way. Yet isn't playing around with stasis-motion roles part of what the Special Theory is doing all the time — and part of the scene which young Albert was himself exploring here? And if all such speculation is out of order, then why are we entertaining this whole ridiculous scenario to begin with?

Trains and More. Next we turn to some other thought experiments, starting with a famous one that Einstein first proposed (Einstein, 1961, p. 26). It is a variation of the train vs. embankment scenario that we have previously described.

Imagine a train which is moving along a track (or embankment, or ground) at velocity v. Two bolts of lightning strike at each end of the train. The ostensible question here revolves around whether the two lightning flashes occur simultaneously or not. Simply put, Einstein argues that the judgment of simultaneity will be different for a mid-point observer who is static on the ground than for a mid-point observer who is moving on the train.

How can this be? Einstein reasons that if the two beams of light (emanating from the twin lightning flashes) reach the mid-point observer on the ground simultaneously, then these same two beams of light will reach the mid-point observer on the train in succession. That is, the static ground-based observer sees two beams of light coming towards him from two opposite directions (i.e., from where the front and the rear of the train are located) at the same time. By contrast, the moving train-based observer will see the beam of light coming from the front (toward which the observer is heading) before seeing the beam of light coming from the rear (from which the observer is receding). (Einstein, 1961, p. 26).[12]

Consider a second train example. We mentioned this in Chapter 4, where we talked about the difference between a ball which is traveling (e.g., down the aisle) of a uniformly moving train and a beam of light which is doing likewise. The issue concerns how fast each entity is moving, for both train-based and ground-based observers.

So how are these scenarios to be understood in the present 2FT framework? Our approach is to frame these situations in terms of the general framework of complementary and hierarchy that we discussed in the prior chapter … in which case different observers would have different options.

12 It can be argued that Einstein undermines his own argument about simultaneity with this interpretation. See, e.g., Evans.

We will analyze the situations here in terms of two adjacent levels of the hierarchy, each associated with a different level of (ether) medium — described here in terms of a lower baseline level that corresponds with the earth or the ground/embankment, and a higher altered level that corresponds here with the train. Thus:[13]

Altered space and time of Train system (Lorentz transformed):

X-view	*P-view*
stasis-based (train observer)	motion-based (moving ball)
fundamental c constant (isotropy)	effective $c \pm v$ (anisotropy)

Baseline space and time of Earth / Ground / Embankment (Normal):

X-view	*P-view*
stasis-based (ground observer)	motion-based (moving train)
fundamental c constant (isotropy)	effective $c \pm v$ (anisotropy)

In the case of the lightning flashes, the ground-based observer would have two options. One is a stasis-based *X*-view of light that is traveling at a fundamental speed of c in relation to the observer and the ground/embankment (the two being equivalent) — in which case, both fore and aft beams would arrive simultaneously, since any motion of the train has no bearing on the outcome and can easily be ignored or eliminated. On the other hand, this same observer can have a motion-based *P*-view of light that is traveling at an effective $c \pm v$ speed in relation to the moving train — faster for the forward beam $c+v$ that the train is approaching, and slower for the rearward beam $c-v$ from which the train is receding — thus the two beams would be seen as interacting with the train sequentially.

Next let us switch to the other level and the train-based observer — even though such a perspective would not necessarily apply there, as we will indicate shortly. But to start with, we are dealing with a level of medium that exists inside the train, in relation to which the observer is static. Yet there are still two options. On the one hand is the stasis-based *X*-view, which would determine that light is traveling at a fundamental speed of c; and to the extent that the externally-triggered bolts of lightning would manifest themselves inside and at each end of the train, they would be deemed to occur simultaneously. On the other hand, a motion-based *P*-view would gauge light's speed in relation to some moving object within the train — such as a ball rolling down the aisle — in which case the observer would determine that light travels at an effective speed of $c \pm v$ in relation to the moving ball.

As to the perspective point just mentioned, one could argue that an obvious possibility is for the train-based observer to see the lightning-caused events by looking outside the train at the show — e.g., by leaning out from an opening between train cars or maybe from the top of an observation car ... or simply by ingeniously imagining what a

13 This framework can also be applied to the previous subsections' topics, e.g. by substituting the traveling twin — or satellite, or Albert-at-16 — for the train in the above charts. (The phenomenologically static counterparts for each of these entities would go in the lower left category.)

"proper" overall perspective would look like. That would presumably mean that this train-based observer would actually be construing the situation in terms of the ground-based observer's *P*-view, in which case the two lightning flashes would interact with the train in effective $c \pm v$ terms mentioned above (along with the conclusion about non-simultaneity). Of course, one might go so far as to argue very generally that any observer is able to imagine the situation in terms of another observer's perspective, and not just the single one that we are identifying here. (See the footnote on page 329 for further remarks about this factor.)

Now let's get back to the second train example that we set forth above near the opening of this section, concerning the speed of the ball which is rolling down the aisle of the train, as opposed to the beam of light which is doing the same. Of course, in relation to the observer inside the train, the ball is moving at a speed of v and the light at a speed of c. But what about in relation to the ground-based observer? One way of answering the question is in terms of a basic difference between matter and medium domain phenomena that we first noted in Chapter 4: motion of the source must be factored in for matter domain phenomena, whereas it is not relevant for medium domain phenomena. In those terms, if we treat the train as the source here (boats were the big hit for such issues in Chapter 6!) then its motion must be included for the rolling ball, in determining how fast the ball is moving in relation to the ground-based observer; but that is not the case for the light wave, in which case motion of the train source does not impact the resulting wave speed.

Before ending this section, let us consider another line of even more abstract thought experiments, which supposedly demonstrate how the constant speed of light manifests itself in seemingly inexplicable and incompatible ways. In particular, what is supposedly being illustrated here is the difference between material waves and mediums like water, as opposed to the more intangible waves and mediums of light.

This is how Marion presents the basic plot (Marion, pp. 399-400). Suppose there are two independent systems which are undergoing relative motion between them, and suppose further that one of these systems is static in relation to a body of water.[14] At the instant when the other (i.e., moving) system is situated directly overlapping that of the first (i.e., static) system — with both thereby overlooking the body of water — a stone is dropped into the body of water, and waves travel outward from the point of origin at a given velocity. Observers in both the static and moving systems will treat these water waves as "belonging" to the system which is static in relation to the body of water. The implication is that the static system observer will see the water waves moving at a fundamental speed, whereas the moving system's motion must be factored in ... and this will result in a different effective speed for the water waves in relation to the moving system.

Now for the wrinkle: switch the scenario from water to light, and this will supposedly make all the difference in the world. For, now a totally different outcome is said to

14 Marion does not describe the two systems here in terms of static vs. moving, as we are doing for simplicity (not to mention, reality) in what follows. Rather, he treats the motion as being merely relative between them.

occur when the "moving" system is situated in an overlapping position with the "static" system and a light wave is emitted. In this case, "each observer sees exactly the same thing — a spherical light wave spreading out uniformly in all directions from his origin with the velocity *c*." (Marion, p. 400) That is because "there is no material medium involved — the light wave propagates through empty space. Because there is no matter to be identified as 'belonging' to one system or the other, each observer views the light wave relative to his own coordinate system ..." (Marion, p. 400) Since there is no underlying medium in relation to which motion can exist, each system can claim to be effectively static in relation to the light — even though there is relative motion between the two systems — and therefore no need for there to be an effective speed of light $c \pm v$ that is any different from the fundamental speed of light *c*.

Needless to say, we reject this conclusion, for it contravenes the whole argument we have been making that light waves require a medium just as much as do any other kind of waves in nature, and it is mere sleight of hand to presume otherwise. Furthermore there is a hierarchy of nested mediums, as we have been describing. In terms of the orientation we have been presenting, the lower level would correspond here with the static system ... in relation to the higher level as corresponding with the moving system. In these terms, the static system would *X*-view light traveling at a fundamental speed of *c* in relation to itself, whereas the static system would *P*-view light traveling at an effective speed of $c \pm v$ in relation to the moving system. This is the most relevant conclusion which applies to the scenario we are discussing — in spite of the fact that at another level the moving system can *X*-view itself as static with light traveling at a (different actual) fundamental speed of *c*, and a corresponding *P*-view of light traveling at a (different actual) effective speed of $c \pm v$ in relation to some object which is moving at that level.

So we contend that there is no difference between the water wave situation and the light wave situation. The problem is the symmetrical framework of Special Relativity, and the flip-flopping of stasis-motion factors which is built in. We will return to this issue at the end of the chapter.

Real vs. Apparent. One of the most significant controversies which is related to any theory about high-speed motion that is based upon the Lorentz transformation is the question of whether the proposed alterations of spatial and temporal metrics are real or only apparent, which is a question we alluded to earlier. As we shall see, the issue is not an easy one to clarify, partly because it can be interpreted in various ways.

To start with, different answers may emerge even within the established paradigm. On one end of the spectrum would be the orthodox position, which is that anything which is postulated and not disproven in regards to these matters is the correct answer ... meaning that the space and time alterations can only be treated as real.[15] At the opposite end of the spectrum are those theorists who argue that we are only dealing with apparent or pseudo-effects.[16] There is also a third approach which might be seen as a compromise — whereby one dimension's alterations are treated as being real while the other dimension's alterations are only illusory or apparent.[17]

It might be argued that the modern consensus is in favor of treating the temporal alteration as real, but not the spatial one. And, indeed, this is bolstered by the evidence

— where there is ample support for the slowing of clocks in moving systems, but none which demonstrates spatial contraction. Of course, the former is easier to show than the latter: temporal dilation may be deduced from outside the moving system in question, in terms of altered life expectancies or clock tickings — but there is no way to get inside the fast-moving system in order to conduct spatial measurements therein. And even if one could, the presumption would be that any measuring sticks would be equivalently distorted as that which is being measured — thus yielding no discernible change in spatial parameters.[18]

There is a further point about the spatial alteration issue — which is that it overlaps with an additional effect that can be described in terms of aberration. In both cases, we are dealing with a phenomenon that is said to apply to fast-moving phenomena, such that they are compressed or contracted as a result. At least part of this additional effect would be attributed to a local distortion that is caused by motion between the object and the observer, and is explained in terms of a visual rotation. Thus Calder says that, "Many accounts of relativity say, quite incorrectly, that a passing spaceship appears unnaturally squashed or contracted along its length. It *does* appear foreshortened but only in accordance with the entirely natural perspective of an object seen from an angle" (Calder, 1979, p. 113). Zukav makes the same point: "According to the special theory of relativity, a moving object appears to contract in the direction of motion as its velocity increases. Terrell has presented mathematical arguments which demonstrate that this phenomenon is something like a visual illusion ..." (Zukav, p. 142)

Another way of thinking about the real vs. apparent issue is in terms of the symmetry factor which is inherent in the structure of the Lorentz transformation. This, of course, is the basis for the twins paradox — and the argument here is that the asymmetrical outcome would be explained in terms of the fact that only one of the twins experiences real transformations, whereas any reverse conclusion is only apparent and is thus an illusion. Such an interpretation would be related to a more general differentiation that could be made between absolute and relative motions: absolute motion in relation

15 "Relativistic effects are not optical illusions nor are they scientific hocus pocus. Relativistic effects are *real*." (Marion, p. 403)

16 Thus, Whitrow: "[I]nstead of assuming that there are real structural changes in the constitution of matter due to motion, Einstein's theory only involved apparent changes, that is changes relative to the observer" (Whitrow, 1972, p. 111). Likewise, Lewis and Tolman: "The distortion of a moving body is not a physical change in the body itself, but is a scientific fiction." (Lewis and Tolman, p. 517) Sachs represents another theorist in the "apparent" camp. (e.g., Sachs, 1974)

17 E.g., Palacios claims that the length contraction is real, whereas the time dilation interpretation is added simply to achieve an invariant speed of light. Builder also argues that only the assumption of length contraction is required, while "clock-rate reduction ... is not an independent hypothesis." (Builder, 1958a, p. 294) See also Prokhovnik (Prokhovnik, 1963, p. 195), as well as Lorentz himself. Ives also argued that only length alteration need be assumed, whereas time alteration could be understood essentially as a byproduct thereof.

18 In fact, what is often put forward in the spatial alteration treatments is the bizarre proposition that an observer inside (e.g., what we see as) a fast-moving system would conclude (based on its own premise of self-stasis) that it is the other system (e.g., our own) whose spatial parameters have been altered by motion — and expanded, no less, rather than being contracted (e.g., Marion, p. 402). Such is the contorted reasoning that is involved in these discussions.

to an underlying substratum would be understood as producing real physical effects, whereas merely relative motion between one entity and another would be construed as producing only apparent changes. Of course, such an interpretation would not be consistent with the purely symmetrical and relative nature of motion that is depicted in Einstein's theory, as we have seen; nor, conversely, is it consistent with the non-symmetrical absolute motion basis of Lorentz's theory or the present model.

Finally, here is one more way of approaching the issue — which is consistent with Einstein's model — that is actually based on the proposition that we are simply dealing with two different perspectives, if not reciprocal conceptions of existence. According to this idea, each of the two interacting entities would see its own judgments about itself and the other entity as being real — in relation to which the contradictory determinations that are based upon the other entity's perspective (about both entities' metrics) would be seen as only apparent. As we have seen, this is a common way in which the high-speed motion scenarios are described. Note that in this purely relative way of construing motion, each entity's perspective is based upon construing itself as static in relation to the other entity as moving ... yielding "proper" vs. "relative" length and time determinations, respectively (labels which we shall return to in the next chapter).

Summing Up. What we have found, with the Special distortion issues we have been examining, is a combination of paradox and anomaly. We start with a series of sensational claims about the effects of motion in distorting space and time. Yet when we look more closely, various problems arise. To start with, there is the point that all motion is supposed to be merely relative and thus there are an unlimited set of essentially random stasis referents that can be selected as benchmarks — leading to a hodgepodge of conclusions about the motion-based effects, which do not lead us to an orderly conception of nature. Related to this is the question of whether the alterations are real physical changes, or merely appearances — in addition to whether the effects are temporary or permanent ... i.e., whether they are reversible or not (which we have not discussed). Finally, there is the symmetry conviction, and its assumption that if one entity is static and the other moving, then either can be construed to occupy either role — or both can, which is impossible and of course has never been demonstrated; and therein lies the paradox that we find in scenarios like the twins.

Of course, for the once-equivalent two partners to experience differential outcomes requires that they experience differential conditions — meaning at the very least non-inertial (or accelerated) motion by one of them. That is precisely the conclusion we mentioned above regarding the twin paradox. Yet how are such factors to be accounted for in a theory which is supposedly built upon inertial frames and uniform linear motion — conditions which do not properly obtain in the circumstances that are cited in any of the actual scenarios we mentioned above as supposedly validating the Special Theory?

What is clear is that there is no denying the viability of the Lorentz transformation as describing the various outcomes. But to repeat again, validating the Lorentz transformation and validating the Special Theory are two separate issues, especially given that the latter was a later historical overlay upon the former. And in any event, regardless of their chronological relationship, it would be difficult to deny that Einstein's model represents a particular conceptual framework for interpreting those equations — and that alternative approaches are available, as we shall discuss further in the next section.

We maintain that the Special Theory is unacceptable, both because of the various anomalies we have cited and because it offers no explanatory vehicle or mechanism for understanding how its conjectures would be implemented — and it does not lead us to a coherent model of nature; on the contrary, the Special Theory is built upon two postulates which can never make any sense. By contrast, we shall see in the next chapter that the General Theory retains its credibility in dealing with these sorts of issues, by providing a consistent and meaningful framework for understanding such space and time alteration effects.

But for now, let us look further at the historical background, and various factors which account for the high esteem in which the Special Theory is held — and why that faith is unfounded.

The Triumph of Special Relativity

We have been very critical of the epistemological framework of the Special Theory. On the most crucial of issues, the theory is contradictory, convoluted, ambiguous, arbitrary, or absurd.[19] It is impossible to fully integrate into our minds in any coherent sense, in spite of all our efforts to edify ourselves and let our thinking evolve over a whole century now. The theory's special combination of magic, power, and dogma leave us fairly helpless to comprehend nature; and the propaganda is so intense that we are blinded, as in a tremendous sandstorm. All in all, we contend that Einstein built his model upon conceptual leaps and end runs around nature — and we are still paying the price for that shortsighted strategy today.

But, hold on: this is a very sweeping indictment, which flies in the face of accepted wisdom! For one thing, physicists have always known that Special Relativity strains credibility as a conceptual vehicle for understanding nature, and yet they have embraced it nonetheless. Therefore, if the theory was so manifestly untenable as we claim, then there must have been compelling reasons for its adoption anyway. There must be a good explanation for why Einstein's theory eventually took hold some years after its inception, and why it is still embraced almost universally today.

In this section, we will discuss the basis for Special Relativity's success, by describing various factors on both sides of the case. We start by considering some of the arguments in favor of Einstein's approach. Then we examine the issue from a historical perspective, followed by a look at certain psychosocial and cognitive factors which have played a part, including a discussion of some of the dubious thought experiments which have been put forward to advance the theory. Next we consider the important mass-energy equivalence, as representing a powerful component of the model's appeal. Then we look at some general arguments against Einstein's model, as well as discuss the the-

19 Bohm's critique of Relativity is more measured: "[T]he theory remains generally unsatisfactory, not only because it contains what at least appear to be some serious contradictions, but also because it certainly has a number of arbitrary features which are capable of indefinite adaptation to the facts ..." (Bohm, 1980a, p. 134)

ory's relationship to the Lorentz transformation. Finally, we attempt to summarize the problems we are facing, by framing the issues in terms of the two categories of symmetry that we often cite — false symmetry and no symmetry.

Among the just-mentioned social aspects is the fact that the general public is usually exposed only to Einstein's lofty reputation and theoretical successes, and not to his theoretical failures. That lack of balance is part of what has made him a larger than life figure, a point we will return to below. But the end result is that Einstein's ideas are typically presented in a manner which puts them beyond question — and tends to make people dismiss contrary positions at the outset, as we mentioned at the beginning of the previous chapter.

In this regard, and before proceeding with the critique below, it is worth mentioning several areas where it is widely acknowledged, or at least strongly arguable, that his ideas were simply wrong. We discussed one in Chapter 5, that of the cosmological constant, which Einstein himself regretted having ever put forward ... leaving aside whether that factor is making some kind of a comeback in recent times. A second is that Einstein was very much a determinist, and he resisted the notion of chance as having an important role to play in nature (mentioned in Chapter 2). A third is in his intractable opposition to duality (as we will discuss further in Chapter 9). A fourth is in his failure to embrace the reality of a phenomenon that he sought to discount, in the form of the Einstein-Podolsky-Rosen experiment (again see Chapter 9). A fifth is in his treatment of the dimension of time as having only a static, and not a dynamic, essence — along with his contention that time only existed as part of a space-time combination (both of these points are addressed in Chapter 8). Finally, of course, Einstein never succeeded in his quest to develop a unified model of nature (see Chapter 12). Our purpose in mentioning these points here is to try and establish a little more of an even playing field: Einstein was a human being, many of whose ideas had merit ... while others did not. All should be subject to discussion.

The Great Breakthrough. Let us begin by cutting to the quick and listing some of the characteristics which Einstein's supporters might cite as to why the Special Theory has been so successful: a minimum of basic postulates, and an economical use of abstract constructs; a maximum of reductionism, and an elimination of extraneous ideas and complicating factors; mathematical utility; elegance of argumentation; conformity with principles of symmetry; compatibility with observations ... what more could anyone want?!

In these terms, it seems impossible to question the utter brilliance of Einstein's work. In one fell swoop, various tasks were accomplished: The troublesome construct of the ether was eliminated, the null results of experiments like Michelson-Morley were accounted for[20] — and the Lorentz transformation was derived in a seamless logical manner, whereas it was seen as an ad hoc postulate in other theories like Lorentz's own.

20 "If the Michelson-Morley experiment had not placed us in greatest embarrassment, nobody would have perceived the relativity theory as (half) salvation." (Einstein, 1908, quoted in Holton, p. 479)

Electromagnetic and mechanical phenomena were merged, in a way which preserved the form of Maxwell's equations and their constant c essence, while also combining them with Newton's laws of mechanics. The Relativity Principle was now seen to apply to both, in the form of the Lorentz transformation which now superseded the Galilean transformation. This also provided the basis for Einstein's "single law of nature" approach, which we discussed in Chapter 4.

Einstein's new model was also a dramatic break from convention, and in this respect Einstein was a revolutionary — not just in terms of a particular scientific theory, but in terms of our deepest conceptions about nature and reality. In fact, it might be said that the main contribution which Einstein made was precisely his ultimate "revolutionary and bold" dismantlement of the absolute space and time concepts, and especially the latter.[21] Max Born, for example, remarks that,

> Although I was quite familiar with the relativistic idea and the Lorentz transformations, Einstein's reasoning was a revelation to me.... For me — and many others — the exciting feature of this [1905] paper was not so much its simplicity and completeness, but the audacity of challenging Isaac Newton's established philosophy, the traditional concepts of space and time. (Born, 1969, quoted in Schaffner, p. 58)

Max Planck et al. mention the

> devastating results of [Einstein's] new interpretation of the time concept, which have repercussions on the whole of physics, and above all on mechanics and thence deep into epistemology. (Planck et al., quoted in Schaffner, p. 58)

Tucked into the Special package was the use of seemingly sophisticated new concepts to replace ordinary old ones. Einstein ridiculed or dismissed those classical concepts — even though there was not much functional difference between many of the old and new sets, aside from the latter being more elastic and esoteric — such as the old absolute space or ether medium being replaced by Einstein's new space-time continuum or electromagnetic field. Finally, the unorthodox theory-building argument which he put forward, about inventing arbitrary solutions to intractable problems — which we alluded to in the previous chapter and will address at length in Chapter 11 — was seen as being validated by his model. As far as ordinary logic and common sense go, their losses were seen as minor casualties (if not outright victories) in the larger war ... of creating and establishing Einstein's brave new world of Relativity.

Having set forth some of the main arguments in favor of Einstein's model, let us now put the issue into historical perspective, beginning with the crucial Lorentz transformation itself.

21 Holton talks about the issue of whether Einstein's work was a culmination of previous contributions, or represented a dramatic break in its own right (Holton, p. 191) — along with the question of whether Einstein should be labeled a revolutionary or not. He says that, Einstein "dismissed all talk of revolution when applied to his own work or to that of other modern scientists." (Holton, pp. 197-198).

Historical Background. The celebrated set of equations which can be described collectively as the Lorentz transformation is one of the most well-documented relationships in physical science. It was largely developed by Lorentz, and to a lesser extent Poincaré, primarily as a means of reconciling Newtonian mechanics with Maxwell's electromagnetism.[22]

We have seen that Maxwell's equations are based upon a constant *c*, as describing the speed of light in relation to the ether medium. If one subjected these equations to the Galilean transformation, by varying the resulting *c* in relation to a moving object — thereby transforming it into an effective $c \pm v$ — the equations become extremely complex and change their form (as described in a footnote on page 124). Aside from that distasteful element, the outcome seemed to conflict with the Relativity Principle, by establishing the ether as a reference for absolute rest or stasis ... a premise which seemed contradicted by the Michelson-Morley experiment. The new approach amended Maxwell's equations slightly, by providing "portability" or "convertibility" of Maxwell's formulas between different systems in relative motion — thereby preserving both the symmetrical form of those equations and the basic concept of a constant *c* upon which they were based. (Cassirer, p. 372). The result was that "Lorentz generalized Maxwell's theory to make it apply to moving as well as static systems." (Dingle, 1972, p. 132)

It was not just the Lorentz transformation which was presented prior to Einstein's entrance on the scene — but virtually all the essential concepts of relativity were laboriously developed and presented by Lorentz and Poincaré in the period before 1905. As Lorentz put it in 1909, "[T]he chief difference [is] that Einstein simply postulates what we have deduced with some difficulty and not altogether satisfactorily, from the fundamental equations of the electromagnetic field." (Lorentz, quoted in Holton, p. 322)[23] Builder is far from alone in asserting that, "Poincaré and Lorentz, by work extending over the period 1892-1904, succeeded in establishing the restricted [i.e., special] theory of relativity and thus reconciling" the Relativity Principle and the principle

22 Properly speaking, it was not even Lorentz who first conceived these equations, as he himself later acknowledged (O'Rahilly, p. 325). That was done in 1887 by Voigt, who came up with a transformation of coordinates that he applied to sound wave phenomena. Lorentz noticed in 1895 that what is called the first-order of Voigt's transformation preserved the first-order of Maxwell's equations. Larmor developed the second order equations in 1897. In his 1899 presentation of the transformation, Lorentz discussed the variation of mass with velocity arising from it, and he showed (as had Larmour) that the Fitzgerald-Lorentz contraction was a result thereof. In 1900, Larmour argued that the time dilation effect was a consequence of Maxwell's electromagnetic theory. In 1904, Lorentz presented the transformation in a form which preserved Maxwell's equations in vacuum. Poincaré is responsible for refining the formulas to account for the group property and invariance, correcting for the current density. For his part, Zapffe goes back much further, in arguing that the Lorentz Transformation is simply modelled after a very general physical relationship — "nothing but a cosine function — the relationship of a hypotenuse to the leg of a right triangle, such as delighted ... Pythagoras.... [T]his factor is one of the most common to be found in all of physics ... [for] Nature abounds in cosine functions." (Zapffe, 1977b, p. 777 and Zapffe, 1980, p. 485). Bridgman makes the point that the Lorentz "equations are not immediately concerned with light as such, but are concerned with the unique mathematical properties of a particular velocity *c*." (Bridgman, 1983, p. 149)

of the constancy of the velocity of light (Builder, 1958b, p. 463). Some observers like Whittaker (1953, cited in Builder, 1959, p. 91) credit the theory of relativity to those predecessors, and not to Einstein.

Notwithstanding the commonality of shared concepts between them, it is probably better to think about two different theories here, if not three. But leaving Poincaré aside for the moment, Dingle contends that one of the main reasons for Special Relativity's success is due to its nominal similarity to the relativity theory of Lorentz — even though Dingle himself considers Lorentz's ether-based theory to be non-relativistic. Thus,

> [T]he name "relativity theory" should be applied only to a theory that regards motion as a purely relative phenomenon.... Lorentz's theory was not, strictly speaking, a relativity theory at all; that is to say, it did not regard the relative motion of two bodies as, with equal validity, divisible between them in any of the various conceivable ways; each body had its own absolute motion — i.e. its motion with respect to the ether — and, although we had not discovered how to find what that was, it was nevertheless real. (Dingle, 1972, pp. 166 and 187)[24]

Dingle cites Einstein's ability to present his theory in a sufficiently vague way, such that it was subject to varying interpretations in different situations, which thereby contributed to the confusion between the two theories.

> The general confusion ... has proved the chief means of preserving Einstein's theory, in spite of its obvious untenability, because of the freedom which it allows of switching from Einstein to Lorentz and back as occasion makes convenient. (Dingle, 1972, p. 179)

In this regard, Dingle came to particularly deplore the proclivity of physicists to blend these two very different theories, the one ether-based and the other nothing-based:

> [P]erhaps the most serious aspect of the comparison of Lorentz's theory with Einstein's was the fact that both were called "relativity" theories, for this ... has led to a confusion that I think has been the most effective agent in allowing Einstein's theory to persist so long in spite of its manifest impossibility.... [T]he ideas associated with the name [of relativity] in the minds of physicists became an incompatible mixture of Lorentz's and Einstein's — a fact that preserved the theory from disproof, since any attack on the relativity aspect could be met by an appeal to Lorentz's non-relativistic ideas, and criticisms of those could be disposed of by a reversion to relativity. Thus, for example, the "FitzGerald contraction" was variously regarded as an actual physical effect and as a mere appearance, according to the needs of the occasion. (Dingle, 1972, pp. 166 and 233)

23 Lorentz goes on to acknowledge Einstein's more parsimonious approach: "By doing so, he may certainly take credit for making us see in the negative results of experiments ... not a fortuitous compensation of opposing effects, but the manifestation of a general and fundamental principle." (Lorentz, quoted in Holton, p. 322)

24 Holton reaches the same conclusion, with a different explanation, in arguing that, "Lorentz's fundamental assumptions are not relativistic" (Holton, p. 200). That is because Lorentz maintained a concept of "true time" and real simultaneity, as we shall discuss in the next chapter.

> The physical differences between the theories are profound: here are a few. Lorentz ascribes the contraction of rods and slowing down of clocks to an *ad hoc* physical effect of the ether on moving bodies; Einstein ascribes them to an *ad hoc* modification of kinematics at high velocities. Lorentz's theory is impossible without an ether; Einstein's (because of its relativity postulate) is impossible with one. Einstein's theory makes a velocity greater than *c* logically impossible; Lorentz specifically restricted his theory to a "system moving with any velocity less than that of light", and, from the nature of its effects, it must break down well short of that velocity, just as Boyle's law breaks down well before the volume of a gas shrinks to nothing; it makes the "light barrier" no more necessarily impassable than the "sound barrier". Einstein's theory merges space and time into an unimaginable "space-time"; Lorentz's leaves them independent, as in ordinary understanding. (Dingle, 1972, p. 232)

Dingle's point on the critical question of the ether medium is important — that the Lorentz theory retained it, while Einstein's theory abandoned it. The result was that whereas Lorentz's approach remained a physically meaningful theory, Einstein's was essentially a purely mathematical model ... effectively renouncing a comprehensible link with physical reality. Whereas Lorentz attempted to explain Maxwell's ether-based system of equations, "Einstein was satisfied to 'explain' it in terms of things that we understood nothing of — in other words, not to explain it at all." (Dingle, 1972, p. 155) Thus, for example, Dingle talks about the fact that:

> ... the velocity of light in empty space is *c*. Velocity with respect to what? The development of the theory shows that it means "velocity with respect to anything at all" ... [Y]ou can choose any standard of rest you like, and the velocity of light with respect to it will always be *c*. It is impossible to picture this, for if we have two bodies in relative motion, we cannot picture a third body having a finite velocity which is the same with respect to both. We must content ourselves with stating the result of an experimental measure of the velocity of any beam of light with respect to any body at all. The theory therefore demands that we give up the attempts to picture, not merely what it is that moves, but the process of motion itself, the passage from point to point as time goes on.... The theory forbids us to form a picture of *any motion at all*. (Dingle, 1960a, pp. 20-21).

Of course, Dingle is speaking in logical terms — of light velocity as being potentially variable ... which is not the case in Einstein's system, as we discussed in the previous chapter. Still, even this background is not sufficient to explain what brought about the eventual embrace of Einstein's model. In this regard, there are some who argue that it was the gradual acceptance of Special Relativity by prominent physicists such as we mentioned above (like Max Planck and Max Born) which started the tide turning in favor of Einstein's model, as opposed to Lorentz's. Others maintain that it was the validation of predictions which emerged from General Relativity which established Einstein's status, and thereby the acceptance of his other (Special) theory.[25] A similar argument would continue to apply during the many years since, with the two theories being seen as mutually enhancing one another in a single irresistible package, both brought to us by a highly esteemed individual, as we shall discuss shortly.

But now let us return to Poincaré and his thinking on these issues, for his theory was not the same as that of Lorentz. In fact, it was so close to Einstein's that it is reasonable to ask why Einstein should get (virtually) all the credit, rather than Poincaré who came first. Thus, one can make a good argument that there was not much daylight between them on most of the central precepts of relativity. They were both committed to the Lorentz transformation and a symmetrical interpretation thereof, including identical transformations of space and time that were based upon relative motion; likewise, both theorists argued against absolute space and absolute time; and both of their models utilized Lorentz's local time concept — which dovetailed with both theorists' arguments that distant simultaneity was not self-evident and needed to be defined ... and was defined in both models by a procedure which Poincaré devised.[26]

So how did we end up with the situation where turn of the 19th-20th century physics is usually depicted as being stuck in an intractable theoretical quagmire, until Einstein galloped in on his white horse to save the day?

Some of the answers are substantive and some stylistic. Regarding the latter, a common criticism of Poincaré is that his theory was developed in a pragmatic fashion, as partly responding to the exigencies of experiments;[27] and that it was clumsy, by virtue of including other fundamental precepts besides that of relativity itself (e.g., concerning conservation and thermodynamics). More generally, Poincaré is accused of being the grand conservator who was more of a gradualist and a reformer, and who refused to make a clear break by completely abandoning the foundation of classical physics and fully embracing the new ideas — and that instead he had one foot planted in the old line camp and one in the new (e.g., Holton). The argument is made that Poincaré was almost there, but couldn't quite make the conceptual leap which was necessary for the radical change of thinking that went along with the new paradigm (e.g., Galison).

25 Dingle claims that, "[F]rom 1904 until the eclipse observations of 1919 that brought Einstein's *general* relativity theory to everyone's notice, 'the theory of relativity' meant, to almost all concerned, Lorentz's theory: Einstein's, if it was known at all, was regarded merely as a more obscure form of a theory that belonged to Lorentz." (Dingle, 1972, p. 167) Zahar makes a similar argument, although he identifies a different catalyst: "[B]efore the advent of General Relativity the scientific community ... regarded the theories of Lorentz and Einstein as observationally equivalent.... In my view Einstein's relativity programme superseded Lorentz's in the empirical sense in 1915 with its explanation of the precession of Mercury's perihelion. This explanation requires the *general* theory." (Zahar, 1976, pp. 271 and 262; see also Holton, p. 350.) Regarding this issue, note that the underlying equation which properly describes Mercury's orbit was first presented and published by Paul Gerber, a number of years before Einstein's work.

26 As early as 1898, Poincaré had explicitly questioned the operational meaning of the simultaneity concept, and he proposed a standard technique by which one could synchronize distant clocks, using light signals as a means of communication between the two points. The initial idea seems to have been based upon telegraphic signals, but the wireless proposal soon followed. In 1900, he applied this procedure to moving clocks (Galison, p. 207). It is also worth citing two other credits to which Poincaré is due: as we mentioned in a Chapter 5 footnote, he predicted the existence of gravitational waves propagating with the velocity of light; and as we shall mention below, he also discussed the mass-energy relationship prior to Einstein.

27 Lorentz's approach was even more criticized, as being ad hoc and entailing a kind of trial and error.

In substantive terms, the essential difference between them is apparent in their respective treatments of non-detectable phenomena: Einstein simply dispensed with the associated constructs, whereas Poincaré was less willing to completely abandon longstanding constructs which served vital functions — even if they could not be readily demonstrated experimentally — presumably on the grounds that nature may be more complex than is readily observed ... and warrants a model which is coherent, and is not merely the slimmest way of reconciling the data with reality.[28]

For example, it is said that Poincaré's position was more equivocal than Einstein's regarding the concept of the ether: the latter renounced the idea altogether — whereas Poincaré might be said to have just put it to the side, by virtue of being undetectable; yet it remained lurking in his thinking and argumentation. (In fact, there are anomalies in both theorists' statements over time, regarding the ether concept, as we have discussed.) Likewise, both models entailed a postulate of relativity — but it can be argued that Einstein's was more a matter of principle, whereas (again) Poincaré's was more based on practicality ... regarding the undetectability of the ether and thus absolute motion in relation thereto. Also associated with the undetectable ether would be the concept of "true time" which we will discuss in the next chapter.

The universal assessment is that Einstein's bare-bones ontological approach was the correct one, along with his clever alternative way of logically deducing the same practical outcome as Poincaré — as though the issue was a derivation puzzle contest, rather than a fundamental description of our common reality. The fact that Poincaré utilized his skills as one of the world's greatest mathematicians to assiduously develop all the fundamental ideas over a number of years (even if in piecemeal fashion) is not considered germane. Einstein was spared of all this work, since all the conceptual pieces had already been devised by the time he arrived — leaving him free to consolidate and repackage these ideas in a new and streamlined framework ... one which we contend was based on pure abstraction and conceptual shortcuts.

Leaving theory aside, other factors also came into play — like personality, age, and health. For example, Einstein might be described at this time as young, brash, and rebellious (e.g., Galison, p. 46) — in contrast to Poincaré as (some 25 years) older, more reserved, and little inclined to self-promotion ... and whose declining physical health later in these years may also have been a factor. Also, Poincaré was inextricably linked to the established order — including the fact that his cousin was the conservative president of the French Republic — whereas Einstein was the radical upstart, ready to uproot the old model and way of thinking, and put a new one in its place.[29]

Given the points we have just made, it is somewhat paradoxical that it was Poincaré who was ready to alter his thinking in accord with new information, whereas

28 Einstein once wrote that, "The logically simple does not have to be physically true; but the physically true is logically simple, that is, it has unity at the foundation." (Einstein, 1938, quoted in Holton, p. 259) Unfortunately, we contend that the Special Theory is based upon principles which are consistent with the former assertion and not the latter. One can also argue that the "logically simple" can be understood as referring to mathematical statements, a topic we will discuss in Chapter 10. Finally, we interpret the notion of "unity at the foundation" as referring to coherence, which is an attribute we find totally lacking in the Special Theory.

Einstein was firmly committed to ideas which he saw as transcendent. An example which is sometimes cited is a 1906 experiment by Kaufmann, which appeared to contradict relativity. Poincaré's reaction was that the theory might need to be altered, whereas Einstein remained resolute — and he was later vindicated when flaws were identified in the experiment (Holton, p. 206).

Yet what is not always mentioned in this regard is the fact that Einstein's adamant commitment to theoretical precepts — or conceptual intransigence, to put the matter less benignly — can be just as much of a hindrance as it is a benefit. Being fervently committed to basic principles is all very inspiring — but it is not necessarily the best approach, particularly when those principles are arbitrary and incoherent. From the present model's vantage point, that is always the case when it comes to Einstein's broad and persistent predilection for reducing (balanced) complementary patterns to (distorted) singular forms. We will discuss the most important instance of this problem in the next chapter, concerning Einstein's treatment of space and time generally — and especially in terms of the local time construct which lay at the heart of Einstein's model ... as it also did for Poincaré.

But the difference is that Poincaré's thinking about these matters was more flexible and able to evolve — whereas Einstein's was stuck in an ideological rut. Indeed, we contend that what the established paradigm treats as grave deficiencies in the earlier models — which were supposedly rectified by Einstein's more sophisticated approach — were actually right on the mark or open to moving in the proper direction. By contrast, Einstein's rigid one-sided framework has led us on a very long and unfortunate conceptual detour that has left any sense of reality in shambles. Our audacious conclusion is that we would have been far better off if Einstein's program had not overpowered and superseded that of Lorentz and Poincaré.

What seems clear is that if Albert Einstein had never come onto the scene and made his mark upon these matters, we would still have the Lorentz transformations and all the phenomena they describe. Only, then, we could have been dealing with these issues in terms of reason and meaning — rather than squandering all those years, hopelessly bogged down with dogma and idolatry. Our understanding of the world would have been better for it.

29 It should come as no surprise that to the end of his life, Poincaré "only once deigned to mention Einstein's name in print, and then only to register an objection" (Holton, p. 249); Einstein returned the favor, in only once barely mentioning Poincaré's contribution over the years. In any case, Poincaré was not alone among the great physics theorists of the time who never embraced Einstein's work. For his part, Lorentz "gave Einstein personally every possible encouragement — short of fully accepting the theory of relativity for himself" (Holton, p. 249). Mach described relativity theory generally as becoming "more and more dogmatical" (Mach, 1913, quoted in Holton, p. 248) In fact, Einstein acknowledged that, "Mach rejected the special relativity theory passionately.... The theory was, for him, inadmissibly speculative." (Einstein, 1948, quoted in Holton, p. 250) Finally, Michelson "to the end of his days did not believe in relativity theory, and once said to Einstein that he was sorry that his own work may have helped to start this 'monster.' " (Holton, p. 249)

Social and Psychological Aspects. Next we consider some of the insidious social and psychological factors that have contributed to Special Relativity's success, issues which have no particular relationship to the content of the theory.

Some of the influences we are talking about apply to all theories, as we will discuss when we address theory-building issues more generally in Chapters 10 and 11. These include such factors as pressure, conformity, rewards and sanctions, as well as the use of arcane terminology. But we are treating Special Relativity as special, and so we need to consider what is unique here.

For one thing, Einstein put forward a model of nature which was incomprehensible, in a virtually-explicit sense: the conceptual pieces did not correspond with anything remotely corresponding to our ordinary experience, nor did the model's "logic" fit with our usual reasoning. Indeed, common sense was dismissed altogether, as being a complete distraction from the issues and phenomena which the Special Theory was describing — because such common sense was deemed to be outmoded and based upon relatively primitive ways of viewing the world. Some of the explanation for all of this supposed inadequacy, of course, had to do with speed — that Einstein's model came into play with high-speed phenomena, as opposed to the low-speed world in which we all live. Leaving aside the absolutist implications of such labels, the result was that these phenomena were touted as being virtually in the realm of the gods, well beyond the ability of mere mortals to comprehend.

Speaking of which, it should also be clear that we are talking about a very unusual individual, who has taken on god-like attributes: Einstein was an extraordinary sociocultural phenomenon and legend in his own right, which has obviously continued to this day. Of course, the fact that the man was a terrific salesman, whose timing was fortuitous — and whose colorful character made for fantastic "marketing" — also did not hurt the cause: Einstein had a great name, a great face, and a great persona. None of these seemingly trivial factors should be discounted in helping to explain the endless social acclaim for the exalted master and his uncanny insights into nature — which are not only beyond the ability of most people to understand, but which would be absurd (not to mention highly intimidating) to challenge. Such hero-worship phenomena are another interesting psychosocial aspect of the issues here — typified by the point we made at the start of this main section, concerning the mostly one-sided trumpeting of Einstein's successful theoretical positions, but not his failed or dubious ones.

Note, by the way, the self-serving aspect of this "iconization", for the discipline of physics in particular and for science in general — and it is no wonder that Einstein has become their patron saint. Promoting Einstein's status in the culture also becomes a vehicle for enhancing physics' and science's own status in the culture, thereby heightening the esteem in which these institutions are held and the support that they are given by society. By the way, we are not suggesting any kind of conspiracy or conscious campaign of manipulation in this regard, but something much more submerged and subtle than that.[30]

One consequence of the elevated treatment of Special Relativity is to essentially disarm any non-specialist observers or critics who examine these issues, by effectively discounting any powers of reasoning and analysis which they — we — bring to the table:

our concept system is broken, our logic flawed, our common sense irrelevant, and our experience inadequate. One immediate casualty is any sense of self-assurance people have in their own cognitive processes, because the conceptual and logical cogs in their thinking have been jammed. Add in what might be called the contamination of meaning, whereby the theory uses words in confusing or misleading ways — such as the supposedly constant speed of light potentially representing only the average of two different speeds, or in blurring the difference between what we have called measured vs. actual speed. The result is that our conceptual tools of analysis have become corrupted or crippled (if not hijacked), thereby further limiting our ability to comprehend the issues. Then there are the charming thought experiments, which we will focus on in the following subsection, that only pile on to our sense of bafflement and bewilderment ... and general incompetence for dealing with the issues.

Not only is there no way to understand the new definition of reality that was put forward by Special Relativity in terms of our usual concept system, but there is nothing to replace it with, either: we are provided with no alternative theoretical foundation which restores our conceptual bearings, in terms of a larger framework of meaningfulness that is rational and coherent, and in which the given phenomena and interpretation can be comprehended. And of course, we are dealing with enormously important matters which concern the very essence of our common reality, and not merely some obscure phenomena within the discipline of physics.

In the end, the only choice left for most people is to either submit and conform, or be viewed as obstinate and antediluvian. Either way, the sociological and psychological forces at work here have had an enormous impact. We will examine the problem further in the next two subsections, where we will consider the issue in abstract and concrete terms, respectively.

Conceptual Illusions. Having been effectively immobilized as legitimate participants in the analysis process, and as further evidence of our ineptitude in these lofty realms, we are also tantalized and misled by what we would call magic tricks — which step only has the effect of further eroding any remaining confidence we may have in our ability to grasp these matters. Another way of describing the issues we are concerned with is to speak about "abstraction run amuck" — for that explains a good deal of the problem with Special Relativity, and why it has been so difficult to contend with.

We are here referring to a series of imaginary scenarios or mind twisters — bizarre fantasies or perspectives that are dangled before us, in ways that are very difficult to understand, yet which nonetheless seem very compelling, especially in light of all the other confusing factors we have been discussing. Although they are not an unusual practice in the development of science, such abstract thought experiments (or *Gedankenexperimente*) have played a particularly important — and, we would say, conceptually subversive — role in the advancement of Einstein's model. The resulting

30 Speaking of conspiracy and campaign, note that we are not even broaching the topic of anti-semitism in this work, and the influence which such vile sentiments may have (had) on any (positive or negative) assessments that are involved.

manipulation of imagery is very subtle, in that it presents common phenomena in a manner that seems very compelling, because it is "right before our eyes" ... even though it makes no sense. It is difficult to realize that the conclusions which are reached amount to half-truths; for by now we are so unbalanced, overwhelmed, and confused that it is virtually impossible to see through the haze — and we have lost both our conceptual orientation and any faith that is left in our capacity to meaningfully analyze such phenomena and issues.

What typically happens in these situations is that we are first presented with a plausible scenario, which is so simple and obvious that we can all relate to it; but then a new and peculiar way of thinking about the underlying phenomena is introduced — which is impossible to comprehend. On the one hand, the new interpretation is very alluring and difficult to contest, within the narrow confines of the scenario and theory itself; yet on the other hand, it makes no sense.

In many cases, the peculiar extensions are so far removed from their normal context that there is little hope of properly evaluating them. The fantasies are often so abstract that they are difficult to pin down, in terms of their real-life applicability: they are based on (or include) phenomena or states that we can never know directly ... and which may well not exist. Thus, it is unclear how satisfactory these conjectures would be if there were actually some way to implement them. In other cases, the situations may be more familiar, but they are analyzed or manipulated in ways which make them seem very strange or incomprehensible. Such conceptual maneuvering may be consistent with the highly abstract framework which Einstein was proposing — but the results are not meaningful in more usual terms, and may actually be absurd. Yet because the phenomenological and conceptual holes are difficult (if not impossible) to identify, we are seduced into accepting the conclusions which go with them.

We discussed some of the issues in Chapter 5, where we considered several scenarios that were put forward as part of General Relativity's demonstrations of the Equivalence Principle — like elevators getting pulled through the universe, or falling from astronomically-tall buildings. We contend that Einstein's treatment dazzles us, rather than educates us, by failing to provide a coherent framework with which we can understand how these phenomena are related.

But our focus here is the Special Theory, where the problem is apparent in the confusion between stasis and motion. Take the classic example regarding which of two gradually-starting trains is in motion, and which is static — or in the full-fledged version, whether the train is moving relative to the static embankment, or vice versa. Of course, the captivating mental imagery of this train enigma is also reflected in scenarios we discussed earlier in the chapter, like that of the twins.

We are told that the lesson from these scenarios is that motion is relative in two very different ways: not only does motion only exist in relation to stasis, in general terms — but we have absolutely no basis for apportioning out the stasis and motion, in any given situation, except in an arbitrary manner! As a result, there is no basis for concluding that the train is moving relative to the static embankment, any more than that the embankment is moving relative to the static train — and any decision we make in this regard is strictly random. Therefore, motion is really relative, and we are living in a helter-skelter universe!

So just how are we to make order out of these scenarios, and regain our cognitive equilibrium? The main answer is one we emphasize repeatedly in this book, and discuss at length in Chapter 11 — that eventually we must somehow end up with a meaningful and coherent conceptual framework, within which the phenomena in question can be understood. In our case, as we keep saying, such "conceptual salvation" will be found in the twin themes of complementarity and hierarchy.

Mass-Energy Equivalence. It seems clear that the most dramatic concrete support for the Special Theory eventually devolved from the sheer physical power that is commonly associated with the famous equation, $E=mc^2$, which is usually treated as being a unique outgrowth of the model. That equation is certainly the theory's most famous trademark, for which Einstein is generally credited with having derived. Indeed, "[M]ost physicists, including Einstein himself, came to regard.... the idea that mass and energy are equivalent ... as the most important single result of relativity theory." (Calder, 1979, p. 12) Without this mass-energy formula, the theory would probably be much less compelling and be considered much less significant.

Indeed, it would not be an understatement to say that we have become mesmerized by the $E=mc^2$ formula, and the emotional impact of the awesome phenomena that it is associated with should not be underrated in explaining the Special Theory's influence. In actuality, the linkage between the formula and the bomb is not that direct — but given that we are talking about the most potent mechanism for controlling nature that mankind has ever devised, any association is important. As a result, the formidable power we are talking about not only impacts the world around us, but also the world inside us ... in terms of the theory's power to manipulate (if not intimidate) our minds.

But once again, although it is a mere footnote in history, Einstein was not the first to present such an idea — for the general notion of mass-energy equivalence can be derived from non-Relativity ideas (e.g., Lewis, 1908), and in that sense did not require his theory. Furthermore, Lorentz had in 1899 already discussed the variation of mass with velocity.

Ives analyzed the mass-energy relationship question, both in historical and logical terms; he faults Einstein's treatment on both counts. In regards to the former historical aspect, Ives points out how once again the groundwork was laid by Einstein's predecessors: the basic idea was implicit in work by Poincaré published in 1900, while a rather more explicit statement was presented by Hasenöhrl in 1904. Therefore, the claim that in 1905 Einstein was the first to derive "the principle of the inertia of energy as a universal law ... is incorrect; Einstein, in the work referred to (1905), did not give this derivation ... and his derivation was actually incompetent to give the result he announced." (Ives, p. 184) Ives credited Planck with providing the first correct derivation of the famous equation, in its full meaning. More informally, Ives wrote in a letter regarding Einstein's "discovery": "It is very difficult to deal with a man who 'announces' a guess or hunch, but does not admit it is a guess, and gives a phony line of argument, wrapped up in such a way that nobody sees through it. It adds insult to injury when the guess happens to be correct." (Ives, 1952; see Ives, p. 186) We will return to Ives' negative views on Special Relativity below.

A typical physics description of the phenomenon here would talk about the concept of *mass* (or "rest mass") in terms of "congealed energy", and the concept of *energy* in terms of "uncongealed energy". "What Einstein concluded was that *mass and energy are equivalent.... Mass* is equal to and convertible into and from *energy*." (Hecht, p. 333) As far as the third factor in the standard equation, "The c^2 comes in only because of the conventional ways in which physicists reckon energy and mass. One could just as well, and more simply, write $E=m$ and adjust your units of measurement to suit." (Calder, 1979, p. 14) March makes essentially the same point in stating that, "the velocity of light in the formula $E=mc^2$ ... is no more than a conversion factor." (March, p. 216)

As an aside, the present model would approach the mass-energy relationship in a very different way. That is, nature is an endless catalogue of such equivalences, and there is nothing unique about the connection between mass and energy in this respect. We would treat this as another instance of the relationship between the matter and medium domains, one of our dimensions of analysis.

Indeed, we would cite this relationship as another instance of the very pattern which Einstein not only never embraced, but heartily rejected: complementarity. In its place we find the usual way of treating such complementary relationships — and that is to embrace the singular whole, and deny the underlying duality. In the case of mass and energy, therefore, this would mean that, "The two seemingly distinct conceptions are actually manifestations of one single entity: *mass-energy*." (Hecht, p. 333) That conclusion is unwarranted: as we keep saying, no aspect of nature's duality can be merely collapsed into a unitarity — for such actions will merely distort and disguise the essential complementary pattern.

Contrary Points. We now turn to some specific arguments against Einstein's approach, starting with the theory-building criteria we mentioned earlier.

To start with, we maintain that the simplest theory is not necessarily the most reductionist theory — which is perhaps the main way that Einstein's model qualifies; nor should mathematical utility be determinative. Moreover, as we noted in the previous chapter, generality does not merely refer to the wide range of situations that a theory can be applied to: it should also refer to the wide range of phenomena that it describes — one consequence of which means that it should not single out a specific phenomenon as being unique. Lastly, coherence should not only refer to whether a theory is consistent within itself internally, as well as with this or that subset of external "facts" about nature — but whether the theory is compatible with the broadest array of external referents, as they are construed as part of a comprehensible and comprehensive concept system for representing the natural world. We contend that Special Relativity fails on all these counts.

To pick up on the last point first, proponents of the Special Theory would presumably counter that we need to be patient, and that the larger goal of coherence will eventually be reached, once we get our old-fashioned thought processes up to date and in order. Again, the argument is that Einstein provided extraordinary insights into the deeper nature of reality, and that a more satisfactory overall concept system will eventually emerge to reflect this. But, continuing the reasoning, it is the advocates of the old

Newtonian model who must get with the new program and concept system … rather than the new program conforming to the old.

This is a false choice, for none of what we are saying should be taken to mean that we are arguing for the intransigence or rigidity of whatever conventional concepts happen to be in vogue at any given epoch: our fundamental ideas about nature must always be subject to revision, as Poincaré demonstrated. These include our core concepts of space and time, which had clear limitations in the way they were classically understood — and they needed to be made less absolute and more relative. Thus there are variations in the topology of nature, in the General Relativity sense; and the specific parameters of space and time should be situation-specific, in that regard. But we maintain that the same pattern also applies to light and all other phenomena — and as just indicated, our generality criterion would preclude the possibility that the particular phenomenon of light should become the primary characteristic, around which everything else should vary. Space and time are, indeed, relative — but not to light as a transcendent and absolute centerpiece of nature.

Regarding a point about theory-building style that we mentioned in the previous chapter, we contend that ad hoc[31], deductivist edicts such as Einstein put forward do not provide a sound foundation upon which to construct a coherent model — and that invention cannot trump discovery in our attempts to understand nature. As we shall discuss in Chapter 11, such a mode of theory-building may fulfill the relatively narrow requirements that apply in some limited realms, but the results are unlikely to mesh with the broader patterns found elsewhere in nature. Again, the best evidence for the resulting failure can be seen in the fact that neither Einstein nor anyone else has succeeded in bringing together the various theoretical subsets which have long reigned into a unified overall model of the physical world.

As far as reductionism goes, we are certainly not suggesting that more, rather than less, fundamental precepts are desirable: economy of thought is a good thing, and our models should reflect this. But such reduction should not be done in a manner which either distorts or obscures the more general patterns found elsewhere in nature. It is for this reason that we repeatedly criticize Einstein's excessive reductionism in this book — in terms of eliminating crucial complementarities of nature by reducing them to unitarities. Thus,

> Einstein disliked the dualism of particles and fields. The fact that the laws of mechanics (which govern the motion of particles) obey the Principle of Relativity, while Maxwell's equations (which govern the behavior of the field) do not, makes the dualism much worse … [There were] two courses of action open to a unificationist like Einstein: he could maintain either that

31 There are different ways of using the "ad hoc" label (see, e.g., Holton, Zahar), and we would not pretend that our usage in this work is particularly sophisticated. We are using the term here to contrast Einstein's peculiar usages with what would be entailed in a more meaningful approach — which we contend are to be preferred. We discuss the issue further in Chapters 10 and 11. In any event, we would contend that — like beauty — ad hocness is in the eye of the beholder.

> the Relativity Principle applies *neither* to mechanics *nor* to electrodynamics or else that it applies to *both* at the same time. (Zahar,1976, pp. 243-244)

Of course, Einstein took the latter approach — whereas we reject the unificationist mission, and have suggested that the mechanical and electromagnetic realms should not be collapsed into some combined unity. Put once again in terms of the present model, there is a difference between matter and medium domains.

And returning to space and time, there is also a difference between them — and this is surely one of the most crucial distinctions between the pro and con evaluations of Relativity that we have been engaged in. In particular, Einstein's reinterpretation of time and simultaneity represent the conceptual core of the Special Theory, and what made it such a unique and stunning break from its predecessors ... and so alluring for some, as expressed in the Born and Planck quotes earlier. Yet we strongly contest the notion that Einstein's treatment of time was a step forward in our understanding of the world; on the contrary, we treat it as a gross, one-sided disfigurement of nature.

We shall discuss these issues in the chapter which follows, but we can take the present opportunity to highlight an extreme form of the hyper-reductionism which is so objectionable about Einstein's work — in terms of the way he collapsed together space and time. First, the dynamic essence of time should not be sacrificed to the static essence of space, with the neutered parameter of time being made a mere event-marker appendage of space; and second, it is also important to differentiate time from space, in order that (temporal) simultaneity can continue to have a meaning which spans different (spatial) positions, as in ordinary usage. We shall return to this point just below.

In closing this topic, consider the argument about the Lorentz transformation being derived more elegantly from Einstein's model than from his predecessors, entailing fewer explicit postulates and premises. That may be so, but such a conclusion does not take account of the various implicit presumptions that are inherent in Special Relativity, about light and about time — unfounded assumptions which are themselves arbitrary and incomprehensible, and which contravene other viable constructs about nature. So we reject the idea that somehow the Lorentz transformation got thrown in for the bargain, with Einstein's approach: the "freebie" wasn't really free: it cost way too much in terms of reality, to do it this way conceptually.[32]

The Place of the Lorentz Transformation. Now we return to a rather basic point — which is that in spite of all the allure which Einstein's model holds for many of its proponents, the fact remains that there is no idiosyncratic correspondence between the crucial Lorentz transformation and Special Relativity. That is, while these equations serve as the cornerstone of Einstein's theory, we have seen that they were not developed by — nor are they unique to — his model.[33]

Thus, amidst all the hype and hoopla about how well-grounded Einstein's theory is experimentally — and verified daily by thousands of laboratories across the world — it turns out that there is not a whit of research support for Einstein's scheme as opposed to other observationally equivalent interpretative frameworks: the main thing that has been substantiated experimentally are the quantitative parameters of the Lorentz transformation. In this respect, the Special Theory does not have massive experimental support: the Lorentz transformation does, upon which Einstein's theory is based ... even though his model now gets all the credit for the accuracy of the resulting predictions!

So, whole theories can be constructed upon the Lorentz transformation, and we have seen that they have been. Any such theory which is based upon the Lorentz transformation amounts simply to a particular interpretative or conceptual framework, in which those Lorentz transformation relationships can be couched.[34] That may be a hard pill to swallow for a science which claims to reject vague concept systems and just wants the "facts". But as we have seen previously in this work, and will discuss more extensively in Chapter 10, facts don't mean anything without an interpretative framework in which to embed those facts.

Indeed, one can even argue that the question of proper historical attribution should not be the main point of our discussion — for the issue is not so much the past, as it is the future. We argue that it is an open question as to how the Lorentz transformation is to be best construed — just as we maintain that Einstein's theory does not provide a viable conceptual framework for fulfilling this function.

Although we have focused our discussions on the interpretations of Lorentz, Poincaré, and Einstein's, other later conceptual schemes include Builder's and Ives'. In many respects, these later theories can be considered as outgrowths of Lorentz's ether theory. But none of these various theories should be taken as having been proven or disproven, including Einstein's. As Dingle states, regarding the Lorentz alternative:

> *[A]ll* the supposed experimental verifications of Special Relativity can with exactly the same justification be advanced as verifications of Lorentz's earlier and quite different theory ... This is so because both theories have the same mathematical structure and give indistinguishable physical interpretations to the symbols involved so far as the experiments so far performed are concerned.... If Einstein's paper ... had never been written, all the experiments now held to prove Einstein's theory would still have been per-

32 As Builder puts it, "[N]o one would be inclined to dispute the elegance and effectiveness of the restricted theory formulation. We are, however, concerned here with assessing the fundamental character of the theory and not with assessing its elegance and its practical advantage." (Builder, 1958b, p. 479) Ives says that, "The Lorentz transformations were obtained by Lorentz as a succession of *ad hoc* inventions, to reconcile Maxwell's theory with the results of experiments on moving bodies. By Einstein they were derived after a discussion of the nature of simultaneity, and the adoption of a *definition* of simultaneity which violates the intuitive and common-sense meaning of that term." (Ives, p. 112) Elsewhere, Ives says that, "Lorentz arrived at his formulae by a process of invention and accretion; Poincaré arrived at his by giving Lorentz's equations a mathematical going-over to make them fit his principle of relativity; Einstein derived the Poincaré equations by the use of a paradoxical and untenable 'principle'." (Ives, p. 160)

33 We cannot deny our good fortune in the fact that Einstein was not the first to develop or refine this crucial set of equations — nor any of the other basic concepts that are involved here — for otherwise the row we have to hoe would be far more difficult in arguing against his model.

34 d'Espagnat talks about "Einsteinian relativity, which from a certain point of view can also be considered as an 'interpretation' modelled on mathematical formalisms (such as Maxwell's equations, Lorentz's formulae and the work of Poincaré) in large part prior to it. The strangeness of Einstein's theory and its shock to common sense did not prevent its wholesale acceptance by physicists." (d'Espagnat, p. 108)

> formed and held with the same conviction to prove Lorentz's. (Dingle, 1972, pp. 166, 141, and 233)

This means, for example, that all the experiments, which supposedly provide evidence in support of Special Relativity's thesis, about high-speed motion producing this or that effect, could just as easily be understood as representing motion-related differential space-time effects that are caused by physical interactions with the medium — and not by high-speed motion in nothingness.

And when the inevitable response keeps coming back that this great theory "works" and has been endlessly validated experimentally, we must keep retorting, in turn, that this is not the case: only the Lorentz transformation "works" and has been endlessly validated experimentally — and once again, we maintain that Einstein's model represents a wholly unsatisfactory conceptual scheme for interpreting those equations.

In this regard, we can mention one further point about Herbert Ives, whom we just mentioned. It was Ives, along with Stilwell, who in 1938 conducted an important experiment that is typically treated as providing dramatic confirmation for Einstein's model.[35] Indeed, Einstein himself cited Ives' research as the first compelling evidence in support of Special Relativity, in the form of direct confirmation of time dilation. But as we have been discussing, the Lorentz-ether theory and the Einstein-etherless theory are virtually indistinguishable experimentally — and Ives strongly believed in the former. By contrast, Ives deplored Einstein's theory, and his professional life was partly devoted to a series of arguments and research efforts directed against Special Relativity.[36]

No Symmetry. Now let us look at the two main ways in which Special Relativity makes no sense — in terms of the symmetry concept, and specifically in terms of our two general categories of critique in this regard: too much symmetry and not enough symmetry. On the one hand is the perfect symmetry illusion, which we label as "false symmetry"; on the other hand is the other extreme which we call "no symmetry". We fault the Special Theory on both these counts — and maintain that herein lies the dual core of the problem with Einstein's model. We address the latter issue in this subsection, and then turn to the former in the next.

We begin by recalling that the single most important contribution which Special Relativity is said to have made revolved around the concept of time, and the idea that this dimension of nature needed to be radically reinterpreted. The idea of a universal or absolute time was to be done away with, along with the notion of simultaneity at a distance. Time was now seen to be only relative, with the delay of traveling light signals

35 See Ives, pp. 29-40. This experiment consisted of a rather complicated apparatus whose main feature was that it supplied "a very rapidly moving source of light" (Gill, p. 37), whose direction of motion could be established with precision. Parceling out any changing distance factors, the result was that the frequency of the light waves was seen to differ according to speed — the observer seeing a frequency which was smaller than the actual frequency emitted by the source.

36 Respected observers like Arthur Miller, writing in his introduction to the revised edition of a work by Bridgman, note that "Although Ives was a life-long critic of relativity, his criticisms were constructive and were offered at the highest levels of scholarship." (A. Miller, p. xxiii)

being a necessary limiting factor in any attempt to identify a common point in time between two events that are separated in space. Needless to say, this interpretation is prominently praised by most observers — whereas we treat it as one of the principal flaws of the model. And once again, the source of the problem is Einstein's unwillingness or inability to recognize the dual nature of reality — instead opting for a one-sided or unitarity approach to time which is unacceptable. We shall discuss the complementary aspect of the time issue in the next chapter, so we will not dwell on it any further here.

Instead let us look at the broader pattern of complementarity that we have been discussing — in terms of higher vs. lower levels of the hierarchy, each having both stasis-based and motion-based aspects that we have described in terms of *X*- and *P*-views. Inherent in each of these perspectives is the strong and weak light postulates, respectively.

Now let us focus on one particular issue, which picks up on the topic of the previous subsection — except that the argument we are now making is that it is a mistake to treat the Lorentz transformation as a transcendent description of nature. That is to say, we dispute the common orthodoxy about this matter — namely, that the Galilean transformation is primitive and archaic ... and only worth considering as a low-speed approximation to the real and sophisticated description represented by the Lorentz transformation. We argue to the contrary, that the Galilean transformation has just as pivotal a role to play as does the Lorentz transformation — for precisely the reason that they both describe two different types of relationships. Consequently, neither supersedes the other — but both have a vital role to play together in a complementary framework for understanding nature.

We contend that the Lorentz transformation is a viable description of the space and time alterations which exist between adjacent hierarchical levels, as we have been describing in this and the prior chapter. On the other hand, the Galilean transformation would correspond with patterns that exist within the space and time parameters which exist at one and the same hierarchical level — between what we have called the *X*- and *P*-views thereof.[37]

The complementary interpretation that we are putting forward here is completely consistent with the position which Builder takes about these matters. He differentiates Galilean "law-like physical statements", which apply within the same physical system; and Lorentz "transformation formulas", which apply between different physical systems (Builder, 1960). Regarding the former, he says that, "[I]n every statement of a physical law, in every physical equation, and in every description of natural phenomena, all the quantities referred to must be stated in terms of the measures of one and the same reference system." (Builder, 1958b, p. 472) The latter transformation formulae, on the other hand, tells us how data in one system "can be translated into equivalent data expressed in terms of the measures of another system." (Builder, 1960, p. 4.2) Builder goes on to argue that, "Transformation formulae are to be sharply distinguished from physical

37 Another way of construing the pattern is in terms of the Lorentz transformation as describing the higher level whole, which joins together two lower level parts ... each of which can be separately described in terms of Galilean transformations.

statements" (Builder, 1960, p. 4.2), the former being an almost abstract technique for translating one system's measuring units into those of another system's measuring units, while the latter are more concrete statements about natural relationships applying within the same system and its constant set of measuring units (see Builder, 1958b and Builder, 1960).

False Symmetry. We now turn to the opposite problem with Einstein's model. And the main way that the false symmetry issue comes into play is in terms of the relationship between stasis and motion, and the notion that these can be arbitrarily flipped between two interacting entities. We contend that such depictions do not faithfully represent the real world.

Yet why do the scenarios which Einstein puts forward seem so tantalizing, if in fact they are not sound? The main reason has to do with the abstractness of these situations — such as we discussed above with Conceptual Illusions — in which the component entities are presented in isolation. As a result, the role of context is effectively ignored as a significant ordering factor in nature. For our part, we maintain quite the contrary: context counts, and meaningful figure-ground relationships will apply.

Clearly, we are advocating for a form of absolute motion in relation to absolute space, or its ether medium equivalent. That said, we are not talking about the "flat" classical version of these concepts, but rather the "nested" hierarchical version that is inherent in the present model. Yet for these patterns to become apparent, it is essential that any analysis of the issues take place in a logical, orderly, and realistic framework ... with the resulting states and entities fitting naturally together into a cohesive whole. Instead, Einstein uses the pure symmetry assumption as a conceptual pivot and justification for randomly jumping back and forth between incongruous aspects, which represent different states and relationships for each of two interacting systems. The result is not a consistent analysis from either system's point of view — but rather a melange of analyses taken from both.

There is no need for such bizarre anomalies — for the larger nesting reality demands a break in the symmetry ... and in one direction only, practically-speaking. Einstein was therefore wrong in treating two systems that are in relative motion as being completely equivalent in their respective roles. Once that perfect symmetry is purged from the theory — as Relativity's own resolution of scenarios like that of the twins inevitably entails — then there is no paradox. And of course, actual experiments discern a reality which is orderly and not chaotic ... and not strictly symmetrical.

In figure-ground terms, then, the train and embankment are not symmetric — and so they are not randomly static or moving, vis-a-vis one another. If there is relative motion between them, then one of them is moving and the other is static — and there is a basis for determining which is which. That is because nature does not consist of a jumbled array of individual stasis and motion referents. In the larger hierarchical context, we are dealing here with one small train that is moving in relation to the much larger static background ... only a small part of which is the embankment itself. What we do not have, in such broader and natural context, are things that are static and things that are moving, in isolated and willy-nilly relation to one another — but rather, an orderly universe, in which small things generally move in relation to large things ... to put the matter most simply.[38]

One therefore cannot talk about whether the train or embankment are moving without considering the rest of the extended, nesting web of hierarchical interconnections, with which those particular elements interlock. So while it is true — and the half-truth is — that the individual roles could be theoretically swapped, there is a much broader nesting array (a set of hierarchical figure-ground relationships) which does not so readily reverse. For we are dealing with a larger collective background of stasis in this non-symmetrical relationship: the embankment is part of the earth, which is part of the solar system, etc. ... relative to which the little train is moving.

In terms of the stasis-motion dimension, one direction of the hierarchy (e.g., upwards) is towards the larger, static background — while the other direction (e.g., downwards) is towards the smaller, moving figure.[39] Yes, one could, in principle, reverse this orientation — and construct what might be called an "upside down" universe, whereby moving embankments are seen in relation to static trains. Furthermore, in principle, this upside down analysis is just as valid as the more usual "right side up" version, whereby moving trains exist in relation to static embankments. Either of these constructions is arguable, and in this way one could build up a consistent model of reality based upon either perspective — in either a right side up or an upside down universe. But while we acknowledge that it is possible to construct an argument in favor of the upside down perspective, we maintain that it is tantamount to claiming that the sun orbits the earth, rather than that the earth orbits the sun — with a continuing reversal extending up and down the rest of the hierarchy of nature. The orderly and nested hierarchy still applies and constrains such easy reversals — even if the polar orientations of stasis vs. motion could be theoretically reversed.

That nesting hierarchy is always there, whether we are explicitly talking about stasis vs. motion or not. The apple inside a room, for example, is not just floating amorphously somewhere in the universe: e.g., the apple exists in relation to the table, and the apple and table are sitting upon the floor; the floor is part of the room, and the room is part of the building, and the apple on the table on the floor in the room of the building are sitting upon the surface of the earth; the earth is part of the solar system — and thus, the apple on the table on the floor in the room of the building is sitting upon the surface of the earth which is part of the solar system ... and so on. No element in this web is randomly placed there: it must interlock with the nesting framework or contextual

38 Perceptually, we tend to see the smaller thing moving in relation to the larger thing or background (Murchie, p. 531). By the way, in the case of the trains at the station, internal observers may not initially know which of the two trains is moving and which is static — but that is the kind of existence vs. knowledge issue which has come up periodically in this book. There is a difference between them — with the former matter of existence being the real issue here, as between the two trains ... and the latter matter of knowledge being just an excuse for denying reality.

39 This is the conventional hierarchical orientation that we use in this book — even though we adopted a reverse orientation for much of the description in this and the previous chapter for ease of visualization, as we explained earlier. Once again, the main issue is the hierarchical pattern, regardless of how it is oriented for discussing any given topic.

background which essentially determines its function and thus its character. While in principle one could extract some piece of this whole, and examine its attributes out of their natural hierarchical context, that proves very little — except that nature is malleable, and we humans are sometimes too well-adept at focusing in on parts and ignoring wholes.

Stasis and motion do not generally flip-flop with each other in a random manner; nor is our choice arbitrary as to which state applies to which of two interacting entities. A given entity *does* move (or else it *is* static) in relation to some other specific entity and larger context — even though this same entity will take on the opposite state in relation to an adjacent level of the hierarchy: In particular, an entity which is moving in relation to the next higher level is itself static in relation to the next lower level. Thus, the passenger seated on the train sees itself as moving, in relation to the embankment which is static — even though that passenger sees itself as static, in relation to an object that is moving within the train. In this way, there is a stasis-motion order in nature.

But what the Special Theory effectively does is to turn this stasis-motion theme into a game, by distorting the underlying pattern. The fact that a given entity's stasis-motion function can change in relation to different counterparts, as just stated — coupled with the fact that in principle one could construct an upside down stasis-motion reversal in relation to the same counterpart — is then exploited by the theory to create a chaotic vision of nature, by suggesting that the resulting hodgepodge accurately portrays reality ... from which only the theory can rescue us.

We can talk about the illicit Relativity tendency here in terms of flip-flopping reality, or *frame hopping*. The theory jumps from one perspective and relationship to another, thereby pretending that the incompatible orientations can be mixed and matched into a valid description of reality. The resulting juxtaposition is, of course, impossible to comprehend — for good reason, because the states are physically incompatible, and they cannot coexist in the combination that we are lured into assuming.

As an aside, we also reject the interpretation that is sometimes put forward, and we have made reference to, that the differential motion which exists between two entities should be construed as a phenomenon unto itself, which cannot be properly allocated in an orderly manner to either one or the other entity, but which somehow represents a state of its own. Stasis and motion are a complementarity, and neither exists without the other.

A brief note about *perceptual illusions* is in order here. One can design or arrange — or encounter — certain anomalous physical (typically visual) situations whereby more than one reference setting seems tenable, as providing the background for some target figure phenomenon ... to the point of even seeming selectable. Indeed, one can argue that an abstraction like the Rubens faces-vase image represents precisely such an example. But we maintain that in the real world there is a clear figure-ground order of things. We made this point earlier in the chapter, where light (as opposed to water) was described as being arbitrarily assignable to either of two differentially-moving systems. We dismissed such a random approach in our previous discussion: Intriguing as such anomalous descriptions may be, they can easily be taken as almost surreal demonstrations of the relativistic argument — which we are here rejecting.

Not surprisingly, then, we take strong exception to the implications of Bridgman's point when he says that, "It makes sense to talk about the 'same' event in two different frames of reference. The event is prior to the frame of reference and is independently describable and knowable, regardless of the particular frame which happens to have been chosen." (Bridgman, 1983, p. 116) We can translate his terminology of "event" as figure, and "frame of reference" as background: the idea that one can freely switch the figure from one background to another — or even that figure exists independently from background — is heretical, in the present work. Figure and ground exist in relation to, and form an integral whole with, each other; in the real world one cannot arbitrarily select out figure and ground in any manner that one wishes. Reality is not a cut-and-paste game, as in the frame hopping pattern we described above — or even as the all-motion-is-relative mantra implies.

Incidentally, one might suggest that some of the problem we are dealing with here can be traced back to Einstein's elevation of the role of light in nature. We have noted that instead of the usual metrics of space and time as providing a background framework for understanding the motion (and speed) of light, that latter phenomenon itself takes on the role of background ... with space and time essentially becoming figures in relation thereto. And while such an interpretation may serve certain functions as part of a complementary analysis, it becomes a one-sided distortion when presented in its own right. The result is that since there is no broader context besides light itself which needs to be taken into account, then that provides a great deal more freedom to manipulate things conceptually, in the way one "stages" the situation — thereby permitting such strange figure-ground reversals as Einstein engages in.

Consider what the extraordinarily gifted artist M. C. Escher was able to do so skillfully in many of his graphics: he managed to mix together incompatible perspectives into a harmonious — and paradoxical — whole. Nonetheless, stimulating and brilliant as his work is, this blending together of elements from different perspectives into a quasi-coherent whole represents no particular perspective or reality context faithfully, but rather combines together contradictory perspectives and different realities. Artistically it is masterful; scientifically, as it were, it is illicit. And while it may seem strange to compare Einstein with Escher, one can say that some of the Special Theory's interpretation represents just such an illicit *scientific* mixing together of incompatible perspectives.

We need to be concerned about how hazardous this kind of abstract juggling can be to our conceptual health. By creating colorful examples and incongruous imagery, our normal expectations are violated and the definitional borders we are accustomed to become blurred: we get confused, if not bewitched, by the conceptual wizardry. After that, science fantasy is not far behind or clearly differentiable ... and we have no way to understand the world in which we live.

Chapter 8

The Relationship Between Space and Time

Our main task in this chapter is to look generally at how the parameters or dimensions of space and time are conceived differently by the Theory of Relativity, as opposed to the present two-force model. In the first section, we will briefly describe the Relativity concept of a unified space-time, and the standard arguments in favor of this interpretation; and then in the following section we shall put forward the 2FT alternative, on grounds that should be quite familiar by now. Our primary mission is to reestablish the separateness of these two fundamental characteristics, and rescue them from the jaws of Relativity, as it were — where they have been deprived of their distinct roles in nature.

We switch gears in the third section, when we try to bring together a variety of disparate phenomena dealing with spatial and temporal alterations — including transformations, Doppler effects, gravitational shifts, and refraction; and we see how these might be approached in terms of a broad organizing framework. In the last section, we summarize some of the issues we have been discussing, which surround our twin themes of complementarity and hierarchy.

The Relativity World of Space-Time

In this section, we will look at several aspects of Relativity's treatment of space and time — including the core concept of a unified space-time (including the "space-time continuum" and the "space-time interval"), as well as how the notion of time has been transformed in Einstein's model.

The Space-Time Continuum. We have made repeated references to Einstein's negative view of duality, as representing an unwarranted and unfortunate description of nature. We have talked about how, in either Special or General Relativity, he effectively reduced a variety of complementary phenomena and patterns into one — including stasis and motion, background frame and foreground figure, inertial and non-inertial frames of reference, inertial mass and gravitational mass, acceleration and gravity, Newton's mechanics and Maxwell's electromagnetism.

But perhaps Einstein's greatest collapsing act is to be found in the joining together of the two dimensions of space and time, by which we commonly order our natural world, into what is called the space-time continuum. We first discussed this combined concept in Chapter 5, where we talked about it in terms of General Relativity's theory of gravity. But now we want to consider it in a more fundamental sense, in terms of the essential logic and world view which comes from the merging together of these two central characteristics of reality.[1]

Einstein's position was that we are mistaken in thinking that there are separate parameters for space and time, and that the two only exist as a single space-time unity. This conclusion rests largely upon contributions made to the theory by Minkowski, whose famous edict is endlessly quoted: "Henceforth space by itself, and time by itself, are doomed to fade away into mere shadows, and only a kind of union of the two will preserve an independent reality." For Einstein, the result was "that the world in which we live is a four-dimensional space-time continuum." (Einstein, 1961, p. 55)

This four-dimensional world sounds very exotic and sophisticated, but what it actually consists of is rather simple. We can all comprehend the idea of a three-dimensional space, comprising the qualities of length, width, and depth. Well, to go from there to a four-dimensional space-time, simply tack on the parameter of time as an extra bonus, and there we are! True, the result does not mean very much conceptually, in the sense that it is impossible to imagine just what such a four-dimensional space-time would actually mean, as we noted in Chapter 5. Yet it yields some very intriguing mathematical results — and curious concepts to boot, as we shall see later in the section.

So we are now told that we should no longer think about time as distinct from space in nature: those independent concepts have to be eliminated. "Space and time lose their separateness, becoming a unified concept." (March, p. 146) One can regard "the distinction between the two as purely man-made, a peculiarity having something to do with the nature of life and man's consciousness." (March, p. 146) Invoking the Plato's cave analogy, Galison says that, "In speaking of space and time separately, physicists were ... contemplating nothing more than shadows projected onto three dimensions. The full and higher reality of a four-dimensional 'absolute world' would reveal itself only through the liberation of thought, and specifically through the insights mathematicians could provide." (Galison, p. 265)

Sachs is among those who treat the very ideas of space and time as being essentially just simplistic and misleading linguistic units...

1 One might argue that in some respects Einstein's theories can be separated in overall regards to the issues here — with the Special Theory primarily focused on time, and the General Theory primarily focused on space. Of course, this is a very unorthodox, if not outlandish, interpretation of a space and time construct pair that has long been treated as one. But we suggest that General Relativity deals primarily with the topology of space — transformed into "space-time" — and related issues of mass: we examined that topic in Chapter 5, where we talked about the concept of gravity. By contrast, Special Relativity deals primarily with issues of time, as we discussed in the two previous chapters.

> ... not more than subjective language elements to express the laws of nature for physical processes, rather than being physical processes themselves.... "[S]pace" is not more than the name for a set of ordered sequences of *words*, which one uses, in turn, to facilitate an expression of natural phenomena.... [T]o consider space and time as separate entities is strictly illusory.... [They] are purely language constructs, invented to facilitate an expression of purely physical effects — probably for want of a better language! (Sachs, 1974, p. 72, 30, 44, 64; see also Capra, 1983, p. 89)

Note the correspondence here with how the wave-particle issue was construed by Heisenberg and others in the Quantum Theory realm: in that case, the goal was to discount the wave-particle distinction, in particular ... and complementarity, in general. The argument there was that we were not dealing with two distinct categories of phenomena, but rather with one overall phenomenon *which could be construed* in terms of classical constructs like particle and wave ... even though that is not what physical reality supposedly consisted of. The explanation was that we were dealing with flawed language units, just as we are encountering here: for Quantum Theory, it was particle and wave; for Relativity Theory, space and time. In both cases, the same — we would say, defective — analysis is used: the evident complementarity is dismissed as a limitation of flawed classical concepts ... whereas the underlying reality is supposedly singular and not dual in nature.

The parallel goes further, for it is not just a merging of equals that is envisioned within each theory — but a one-sided outcome that results: in both cases, the structural side of nature wins out over the process side. In the case of Quantum Theory, that meant particle-like characteristics being preponderant over wave-like characteristics; and in the case of Relativity Theory, it is space-like characteristics that rule over time-like characteristics. It is the same one-sided pattern that we are encountering repeatedly, such as with motion being a variant of (or transformed into) stasis, and not the other way around. We shall discuss the general problem more fully in the next chapter, concerning science's tendency to over-emphasize structural or mechanistic explanations and orientations in our models of nature ... and to discount, or underweight, their dynamic or process counterparts.

In the case at hand, it is clearly time that suffers far more than space in the merging between them. While the concept of space is altered somewhat, time is devastated — both in terms of no longer representing a distinct quality in its own right, on a par with that of space, as well as in terms of losing its dynamic essence: time is relegated to being a mere appendage of an essentially static space.

What we end up with in the space-time unity is basically a spatial universe, with time thrown in almost as an afterthought. For in the Theory of Relativity, time takes on the role of strictly an event marker, to be superimposed upon spatial parameters. Einstein:

> [A]ccording to this theory time is robbed of its independence.... [for] the time co-ordinate plays exactly the same role as the three space co-ordinates.... It appears therefore more natural to think of physical reality as a four-dimensional existence, instead of, as hitherto, the *evolution* of a three-dimensional existence.... It is neither the point in space, nor the instant in time, at which something happens that has physical reality, but

> only the event itself. (Einstein, 1961, pp. 56-57; Einstein, 1954, p. 371; Einstein, 1956, p. 30)

As others put it, "[T]he mathematician's description of time turns out to be very similar to that of space.... Clearly time may be assumed to share many of the properties of space.... [T]ime can be regarded as a one-dimensional mathematical *space*." (Davies, 1977, pp. 3 and 11)

Of course, one can argue that in the context of the Special Theory, both space and time have been displaced and shifted to the sidelines, as being only relative ... in deference to the speed of light, which amounts to the sole absolute centerpiece of reality.[2] Time, in particular, has been made secondary to light, in that the latter is also the exclusive bearer of information about the former — to the point where light acts as a sort of master clock or arbiter of time in the universe.[3] Any notion of an independent concept of time has been eliminated, in Einstein's system.

There are two main areas where the nature of time has been drastically altered by Relativity. The first is in the abovementioned role of time being treated as strictly an event marker, and in that way as amounting to a mere overlay upon space. One consequence is that we have lost the notion of a dynamic, flowing time — what might be called the "river of time", as referring to a developing or moving phenomenon that we all experience in our daily lives — leaving us instead with only a static dating function, or locational ordering of events. Indeed, this is one of the great tragedies which we have been bequeathed by Relativity: the death of dynamic time. Einstein abolished this crucial process aspect of nature, and all that remained was some kind of lifeless space-time tableau (as illustrated by the rubber sheet model we described in Chapter 5).

The other big time change concerns Einstein's re-definition of simultaneity, in conjunction with the definition which was incorporated into Special Relativity about light and its speed being a limiting factor. One result is that the classical Newtonian concept of absolute universal time, formerly treated as sacrosanct, was now seen as defective and was discarded ... as had been its counterpart of absolute space. In this way, it was concluded that there is no absolute simultaneity (at a distance), but only a relative simultaneity as mediated by light signals.

Let us now look at these two issues more closely, starting with the loss of time as process, and then the eclipsed and narrowed view of simultaneity. After that we will talk more about the combined space-time concept (and the notion of a space-time interval), when we also summarize the impact which the ideas of Minkowski have had upon our view of nature.

The Death of Dynamic Time. Einstein wrote, in a letter to a friend: "You have to accept the idea that subjective time with its emphasis on the now has no objective meaning." (Einstein, quoted in Prigogine, 1980, p. 203) Geroch: "There is no

2 "[A]ll physical processes are governed by the speed of light." (Calder, 1979, p. 62)

3 " ... light, the motion of which is closely coupled with the fundamental workings of time in the universe" (Calder, 1979, p. 129) Or that c is "a natural seam of time," and that "c's constancy truly must be of the essence of time." (Murchie, p. 558)

dynamics within space-time itself: nothing ever moves therein; nothing happens; nothing changes." (Geroch, p.20) Stapp notes that "Einstein's static realm of readings ... is explicitly constructed to have no trace of the idea of process." (Stapp, 1982, p. 383). Henceforth one must reconcile oneself to "the physicist's static picture of time." (Davies, 1980a, p. 190) Whereas "The Newtonian view of space and time is a *dynamic* picture", in which space should be considered as a three-dimensional affair that "moves forward in a one-dimensional time", by contrast the Einsteinian approach is a "*static*, non-moving picture of space and time.... [R]eality is four-dimensional.... In this static picture, the space-time continuum, events do not develop, they just are." (Zukav, p. 150) As to the source of the confusion, it is our consciousness that is the culprit: "The illusion of events 'developing' in time is due to our particular type of awareness ..." Zukav, p. 220) Again, "the *passage* of time is merely to be regarded as a feature of consciousness that has no objective counterpart" (Whitrow, p. 125).[4]

Of course, such conclusions are themselves based on a peculiar set of abstractions which the theory has generated. The fact that we all experience an essential dynamic flow in nature is chalked up to some "peculiarity having something to do with the nature of life and man's consciousness", to repeat from the quote above (March, p. 144). Yet where would such consciousness or awareness have come from, if not from nature itself? Dynamic process is just as much a part of nature as is static structure, and merely because we are able to convert the former into the terms of the latter does not obviate the necessity for both. This is a persistent problem with the established paradigm, as we shall see in the next chapter.

There are those in science who realize that something is amiss here. Consider a variety of musings by Paul Davies, who is generally a passionate advocate and popularizer of the Theory of Relativity:

> It is one of the most perplexing puzzles in physics that the elementary conscious experience of time — the flow or motion of the present moment — is absent from the physicist's description of the objective world.... [T]he time which enters into the equations of theoretical physics seems to lack that essential quality of human psychological time.... [W]hatever we may experience mentally, time does not pass, nor does there exist a past, present and future.... The four-dimensional space-time of physics makes no provision whatever.... for a flowing time or for a moving present moment.... In scientists' theoretical analyses of nature, tense plays no part: there are only dates.... These statements are so stunning that most scientists lead a sort of dual life, accepting them in the laboratory, but rejecting them without thought in daily life.... (Davies, 1977, p. 3; Davies, 1974, pp. 21 and 3; Davies, 1977, p. 57; Davies, 1980a, pp. 14 and 187)

> [P]hysical science can make very little progress in the discussion of the now.... Unless this moving-now phenomenon is to be discarded entirely, it has to be admitted that we do not understand something about time, or about the mind, or both.... [I]t could be that for once our minds are more reliable than our laboratory instruments, and time really does have the richer structure that we perceive. In which case the nature of reality, of

4 There are those who dispute this interpretation. See, e.g., Capek, 1975.

> time, space, mind and matter, will suffer a revolution of unprecedented profundity. (Davies, 1977, pp. 221-222; Davies, 1980a, p. 199)

In some respects Einstein was merely carrying on a tradition, in terms of how he treated the phenomenon of time; and in this regard, the problem goes beyond Relativity. Thus, Prigogine speaks of, "Time, the Forgotten Dimension" (Prigogine, 1980, p. xi). He disparages the conventional models of physics as being essentially static: "[B]e it classical or quantum, time enters only in a quite restricted way" (Prigogine, 1980, p. xi) He goes further and says that, "This static view of the world is rooted in the origin of Western Science." (Prigogine, 1980, p. xi), noting that this "has been the prototype of the scientific approach." (Prigogine, 1980, p. 20) Yet the view of nature which this static vision reveals is unacceptable: "[O]f the changes that occur in nature, classical physics retained only motion. Consequently, as Bergson and others emphasized, everything is given in classical physics: change is nothing but a denial of becoming and time is only a parameter, unaffected by the transformation it describes. The image of a stable world, a world that escapes the process of becoming, has remained until now the very ideal of theoretical physics." (Prigogine, 1980, p. 3) We will return to this issue in the latter part of the following chapter.

Lost Simultaneity. A second way that our world of space and time has been disrupted by Einstein's model is in terms of the meaning of simultaneity. We have already discussed some of the underlying issues, in terms of the clock convention methodology which led to this new conception, and in terms of the consequences for the notion of speed of light. Here our focus is specifically on the concept of simultaneity itself.

Recall that one of the main issues we discussed earlier was how to determine the precise moment at which some distant event occurred — such as the arrival time of a pulse of light whose speed is being measured. Now we ask the question more generally: how does a given moment of time at some distant spatial position relate to some moment of time at a local point? Indeed, can there be a moment of time that is shared between the two locations? In particular, can we speak about an event happening here, at the same time as an event happening there? In other words, is there such a thing as simultaneity between two events which appear to be separated spatially?

Most people would probably answer affirmatively, in an instinctive sense: obviously there can be a moment of time which is common to both locations — there is no reason to think that time is merely local, in this respect. Yet such an intuitive assessment would be dismissed as outmoded by contemporary physics, because Relativity has supposedly shown us that there is no such thing as simultaneity over a distance.

Einstein approached the issue in definitional terms, essentially based upon Poincaré's arguments, which we described in Chapter 7. The reasoning was that there was no confusion regarding what simultaneity meant locally, in one particular region — as between whether an event occurred at the same time as a second event, or not — because we can *see* it. But he questioned whether there was the same degree of reality to the notion of simultaneity over a distance: he wondered how we were to understand what simultaneity meant when the two events whose correspondence was at issue were separated sufficiently that we could not directly and immediately observe what was happening to both. His position was that in such a case, the answer was not obvious — since we cannot know in any certain sense when something happens at some local point here, as well as at some distant point there, inasmuch as we cannot be in both places at the same time to make the determination: therefore, whereas local simultaneity is self-evident and indisputable, distal simultaneity can only be defined.

And not only *defined* — but by the procedure Einstein adopted, it was to be *denied* altogether! His argument was that because it takes a certain amount of time for information to be conveyed between two distant points, we therefore have no basis for concluding that two such separated events actually occur at the same moment — and since we cannot confirm it, we can just as well dismiss the concept altogether. Thus, Einstein's position was that spatially-separated events meant temporally-separated events.

Note that Einstein's approach here is similar to other existence vs. knowledge issues we have discussed — that things only exist which we can measure. We will discuss the general topic further in Chapter 10, in terms of the logical positivist doctrine. In any event, physics has generally embraced Einstein's position — that spatially-separated events cannot occur at the same moment; and this is true no matter whether we are talking about events which occur in the past, present, or future. In the case of the present, for example, the conclusion is that, "There is no universal now, but only a personal one — a 'here and now' " (Davies, 1977, p. 221). As Einstein put it, "Every reference-body (coordinate system) has its own particular time" (Einstein, 1961, p. 26).

The result is sometimes described by saying that, "There is no such thing as simultaneity of distant events." (Einstein, quoted in Capek, 1975, p. 612)[5] At other times, physics talks in terms of simultaneity being only relative, as between distant points in space — and it is in these terms that the phenomenon of light comes into play.

The reasoning is that we must have some procedure for correlating when an event occurs here with when an event occurs there, and that we need some mechanism for accomplishing that task. Yet what modality shall we use for this purpose? Of course, there is no single answer — and in this respect whatever choice we make is arbitrary. But given such flexibility, Einstein suggested that the phenomenon of light might make a good candidate for fulfilling this function. The argument was that we happen to know a great deal about light signals, anyway, and this information may prove very helpful.[6] Aside from that, of course, light also happens to be the fastest communication vehicle we know of — and in that respect, also, the best transmission candidate for conveying information about time, and hence about simultaneity, as regards distant events. And thus it

5 "The very essence of special relativity is an elimination of *absolute simultaneity*. There is no universal cosmic present, no *world-wide instant*, to use Eddington's term.... [T]he relativity of simultaneity is not merely relativized but simply *denied*" (Capek, 1988, pp. 99-100).

6 Here is Einstein's own description: "In order to give physical significance to the concept of time, processes of some kind are required which enable relations to be established between different places. It is immaterial what kind of processes one chooses for such a definition of time. It is advantageous, however, for the theory, to choose only those processes concerning which we know something certain. This holds for the propagation of light *in vacuo* in a higher degree than for any other process which could be considered, thanks to the investigation of Maxwell and H.A. Lorentz." (Einstein, 1956, pp. 28-29)

was decreed that light signals would henceforth be used to set distant clocks, in order to coordinate them with local ones.

It should be clear that what the theory has done is to introduce its own peculiar brand of simultaneity. Bridgman makes the same point, noting that, "Einstein adopted one particular method of setting distant clocks, which, because it was his choice, amounted to a particular 'definition' of the simultaneity of distant events." (Bridgman, 1983, p. 146)[7] The effective result was that such a definition of distant simultaneity would have to be constrained by the speed of light's transmission, since it doesn't happen instantaneously. Therefore, what was previously assumed to be an absolute character for (distant) simultaneity would have to be mitigated by the limiting — and newly absolute — factor of light's speed; and so it is proper to consider such distant simultaneity as existing only in a relative sense, as determined and delimited by the speed of light.

To recap: simultaneity can only be said to exist to the degree that we are able to measure it. Light would be an ideal agent in this operation, both because it has been studied in depth and because it is the fastest transmission vehicle we know of — and by virtue of this unrivaled speed, it is by means of light signals that we can get closest to (although not attain) absolute simultaneity. Thus, simultaneity is only relative ... to the speed of light.

Although all of this argumentation sounds rather innocent and neutral as regards to the underlying issue, there is actually a rather subtle manipulation that is going on here. That is because Einstein's interest in this matter was not primarily that of some detached and disinterested academic observer, whose goal was to provide a speculative and provocative counterpoint ... regarding the epistemological basis of time. For the fact is that Einstein's questioning, and subsequent redefinition, of the time concept was hardly serendipitous: he had a particular theoretical motive. For he knew that the time concept *had* to change, as a necessary step toward realizing his objective elsewhere — i.e., that of finding a unitary law of nature, which would allow him to successfully merge Newtonian mechanics and Maxwellian electromagnetism.

In order to attain that end, he needed to find a means of securing a constant speed of light for all systems, regardless of such systems' differential states of (uniform) motion. In this way, he would adapt Maxwell's theory to the Relativity Principle, by making the equations "frame-invariant", based on a constant c in all systems — and the concept of time (or here, simultaneity) would have to be changed in order to accomplish this. Along the way, such a strategy would incidentally account for apparent anomalies like the Michelson-Morley experiment, which we mentioned in Chapter 6 was bedeviling concerned physicists at the time.

It was a brilliant stroke, of course — but we argue that Einstein's approach did not involve the profound insight into the nature of time for which he is commonly credited. Rather, what we are dealing with is a strictly utilitarian maneuver that entailed altering the natural definition of simultaneity for the primary purpose of resolving theoreti-

7 Bridgman maintains that the clock setting feature that is associated with light is completely unnecessary, arguing that, "[C]locks may be set consistently by transport ... with no direct involvement with the properties of light." (Bridgman, 1983, p. 147)

cal anomalies elsewhere. Einstein's position won out, not only in terms of measurement conventions, but far more importantly in terms of our conception of reality generally.[8]

Dingle talks about the difference between Einstein's approach, and that of Lorentz who argued

> that motion of the clock through the ether changed its rate; but Einstein, discarding the ether, fell back on the fact that it was an *assumption* that a distant event had any unique instant of occurrence at all ... In the absence of any self-evident, necessary way of determining such an instant, Einstein claimed the right to define it in such a way as to save the electromagnetic theory without violating the principle of relativity of motion. (Dingle, 1972, p. 145)

In fact, Lorentz differentiated what he called "local time" from "true time" (or "real time") (Hsu, p. 65; Zapffe, 1977a, p. 17). Poincaré demonstrated (and Einstein restated) that such local time could be established by means of the clock convention, thereby making simultaneity only relative between two distant points, as we have been discussing. For Einstein, "[T]he Lorentz local time is the only possible concept of time" (Hsu, p. 57). Once again in Einstein's own words, "[I]t was only necessary to recognize that one must necessarily define 'time' as the auxiliary quantity 'local time,' introduced by Lorentz" (Einstein, 1907, quoted in Hsu, p. 57).

We are now touching upon what is commonly viewed as Einstein's greatest accomplishment with the Special Theory — that of adopting local time only as the temporal basis of reality. But both Lorentz and Poincaré were more of two minds on the time issue. Thus, Lorentz said that, "I introduced the conception of local time which is different for different systems of reference which are in motion relative to each other. But I never thought that this had anything to do with real time ... real time for me was still represented by the older classical notion of an absolute time" (Lorentz, 1928, quoted in Zapffe, 1977a, p. 17).

Poincaré's position was somewhere between that of Lorentz and Einstein. On the one hand, whereas the local time concept was created by Lorentz purely as a mathematical contrivance, both Poincaré (first) and Einstein (later) treated it as a real phenomenon that would describe moving clocks. On the other hand, Poincaré did not embrace the local time concept exclusively, as did Einstein; and Poincaré is often criticized for not completely getting with the new program in this regard (along with other respects, as we have mentioned). Poincaré never abandoned the true time concept as applying (along with simultaneity) to any system which was at rest in the ether: yet because the ether did not appear detectable, there was no way of knowing whether a system was at rest in relation thereto or not.[9]

8 Notice how Poincaré described the issue, about simultaneity needing to be operationally defined in terms of certain rules: "We therefore choose these rules, not because they are true but because they are the most convenient ... so defined that the enunciation of the natural laws may be as simple as possible. In other words, all these rules, all these definitions are only the fruit of an unconscious opportunism" (Poincaré, quoted in Hsu, p. 58).

Given the complementary orientation which we bring to the issues here, it is not surprising that we turn the tables — against Einstein and for Lorentz and Poincaré. Simply put, we must ask whether Newton's one-sided absolute time is any more of "arbitrary character" (Einstein, 1949, p. 53; see Schaffner, p. 56) than was Einstein's one-sided relative (i.e., local) time; we think not. We maintain that the only coherent way for understanding the issues here is precisely in the form of dual concepts of time — be they called absolute vs. relative, or true vs. local. Once again, Lorentz and Poincaré were on the right track here, whereas Einstein's path has led us the wrong way.

We suggest that one source of the problem we are dealing with here concerns Relativity's joining together of space and time. We have noted that this combined concept is essentially of a spatial character, and Einstein's interpretation of simultaneity is also consistent with this spatial orientation. Thus, the notion that one cannot have a shared temporal moment, which is spatially-distributed, reflects this distorted view of time as simply being a variant of space. By contrast, we maintain that time is different from space, and that the former can be distributed in the latter.

This is clearly the position that Lorentz subscribed to. Thus, in a 1910 lecture, Lorentz propounded the view that, "[S]pace and time are something completely distinct and that a 'true time' exists (simultaneity would then have a meaning independent of position)." (Lorentz, quoted in Hsu, p. 65) Again, "According to Lorentz, true time is something completely distinct from space and allows simultaneity to have a meaning independent from position. This is in sharp contrast with local time, because local time and space are completely interlocked and local time forbids simultaneity to have a meaning independent of position." (Hsu, p. 62)[10]

As we have noted in previous chapters, the present model is not built upon a singular conception of absolute space and time in the classical sense — whereby there is a universal category of true time as corresponding with a (static) ether, in relation to which there would be a variety of local times as corresponding with one or another (moving) object. Rather, these two categories would be seen as corresponding to any two adjacent levels of nature's nesting hierarchy: The true time concept would describe the lower level baseline space and time (normal), while the local time concept would describe the higher level altered space and time (Lorentz transformed).

As far as the simultaneity issue itself goes, we contend that what appear to be spatially-separated events can occur "within time" (i.e., simultaneously) just as much as they can occur "over time" (i.e., successively, by virtue of the time required for something to travel between two points in space). Such a complementary analysis is built upon our

9 Holton also indicates that Poincaré continued to entertain something like a true time concept (Holton, p. 204); and Hsu argues that "Poincaré admitted flexibility in the definitions of time and regarded local time as only a convention" (Hsu, p. 55) — and in at least some of his writing, Poincaré differentiates true time from local time (Hsu, pp. 59 and 61, citing Poincaré from 1900 and 1904). Galison says that Poincaré "*never* equated 'apparent time' (time measured in a moving frame) with 'true time' (time measured at rest in the ether)." (Galison, pp. 241-242)

10 Referring to the ether, Lorentz said that, "One cannot deny to the bearer of these properties a certain substantiality, and if so, then one may, in all modesty, call true time the time measured by clocks which are fixed in this medium, and consider simultaneity as a primary concept."

stasis-motion dimension. As will become apparent in the next chapter, static relationships would be understood as representing the former, in the same way as dynamic relationships would be understood as representing the latter. Indeed, we will discuss a very important experiment — which Einstein first proposed, although it later came back to haunt him — that is directly related to this issue. To put the matter most succinctly, there is no other way to interpret the demonstrated outcome of the Einstein, Podolsky, and Rosen (EPR) scenario except in terms of simultaneity at a distance — absolute simultaneity "within time", essentially in accord with the true time concept.

Unfortunately, the cost of acknowledging such duality (e.g., in terms of both absolute and relative simultaneity) would be far too great for the Special Theory: abandoning the singular and arbitrary simultaneity convention would pull the rug out from under the theory. Christensen makes a similar point: "Yet what would remain of Einstein's epistemology without his critique of simultaneity at a distance and his method for determining it?" (Christensen, p. 53).

That would surely be a pity for Relativists to have to deal with; but it is a far greater pity for the rest of us not to. Thus, consider the serious dilemmas which Einstein's approach would seem to lead to, concerning the whole question of existence. For the fundamental issue is far broader than some bookkeeping problem that is associated with synchronizing separated clocks. If the Theory of Relativity forbids an absolute simultaneity between two distant events — a shared moment of time that is common to them both — then why should we even assume that there can be a shared existence between them at all ... or between any two events that are apparently separated in space? And in the most general terms, if there can be no universal simultaneity, no shared time across the universe, then why should we even assume there there is a common universe at any given moment? On the contrary, we maintain that while it is true that the universe exists over time, it also exists within time ... in a shared and simultaneous sense.

The following complementary chart summarizes the issues and concepts we have been discussing:[11]

11 Unfortunately, there is confusion with the two main category terms here, because they overlap with the standard analysis. Instead of a true time vs. local time distinction for the two entities that are involved, Einstein would describe both entities as having local times — there is no true time in his model — and two different names are required to distinguish between them. In Relativity, these two local time properties are labeled as a)*proper time* which any given system would designate for itself as being static, and b)*relative time* which this system would designate for the other system as moving. By the way, there is also a corresponding spatial pair called "proper space" and "relative space". Here is how Zukav describes both sets: "What we see when we observe our stationary rod and our stationary clock, if we ourselves are stationary, is their *proper* length and *proper* time.... Proper lengths and proper times always appear normal. What we see if we are stationary and observe a rod and a clock traveling very fast relative to us is the *relative* length of the moving rod and the *relative* time of the moving clock. The relative length is always shorter than the proper length, and the relative time is always slower than the proper time." (Zukav, p. 139) In the present 2FT analysis, "proper time" is equivalent to "true time", while "relative time" is equivalent to "local time".

True Time	Local Time
absolute simultaneity	relative simultaneity
simultaneity	succession
within time	over time
instant	duration
static relationship	dynamic relationship

The Space-Time Interval. The primary impetus for the development of the space-time concept came from a mathematician named Minkowski. We will let Dingle set the stage here, by quoting several of his comments on the subject, beginning with his description of Minkowski's 1907 paper which introduced the concept:

> Reduced to its essence, Minkowski's paper is a piece of pure mathematics — as such, extremely elegant and admirable, but, insofar as it purports to contribute to physics, as it does, calamitous.... It is to Minkowski that we owe the idea of a "space-time" as an objective reality — which is perhaps the chief agent in the transformation of the whole subject from the ground of intelligible physics into the heaven (or hell) of metaphysics, where it has become, instead of an object for intelligent inquiry, an idol to be blindly worshipped. (Dingle, 1972, pp. 169-170)

Dingle then describes Minkowski's merging together of space and time into a space-time continuum as being

> utterly unwarranted: it is a conclusion about things not dealt with in physics, drawn from a purely gratuitous interpretation of an arbitrarily adopted mathematical formula.... It is impossible, I think, to form an adequate conception of the harm that this misrepresentation has done.... [P]hysicists ... found themselves faced with the metaphysics of time (eternity) and space fused into a unity more densely obscure than even that of the traditional philosophers.... [T]he speculations — some of them very wild indeed — that have been advanced concerning the nature of "space-time", "the universe", and so on, are meaningless unless they can be translated into terms in which "space" is potentially "place" and "time" is potentially "instant". If that can be done, such a term as "space-time" may afford a convenient means of expression, just as one may speak of "a considerable length of time" without implying that a duration can be measured with a yardstick, but in much modern cosmology, those who speak of "space-time" naively imagine that it refers literally to something existing objectively, and so deceive themselves and others into thinking that the world which is the world of all of us has an esoteric quality that only the specially gifted can understand. (Dingle, 1972, pp. 169-170, 172, 173, and 176)

It may be that, overall, the chief justification for the space-time continuum is in terms of mathematical utility. For, with the help of others, Einstein developed a set of field equations for his General Theory that potentially describes the curvature of the space-time continuum, and thereby all of reality. Indeed, Relativity theorists get so carried away by the mathematical "beauty" of all this that the essential Einstein equation takes on a causal function all of its own: "[O]ne thinks of Einstein's equation as being the fundamental thing all by itself." (Geroch, p. 176) Thus, using the rubber sheet analogy we described in chapter 5, "[T]he curvature of the sheet dictates in detail how the matter (in this case, the steel ball) will move. This is in essence what happens in relativity with Einstein's equation. Matter in the world is forced, via that equation, to react to the curvature of space-time in one way or another." (Geroch, p. 180)

The mathematical crux of Minkowski's unified space-time concept can be found in the notion of a space-time interval. The *space-time interval* is that which joins together two *space-time events* — such an "event" being that which happens at a particular place in space, at a particular moment of time. Minkowski utilized a mathematical expression which describes this space-time interval, and one of its primary values is that it is invariant to the Lorentz transformation.

In that scheme, of course, there are variable space and time metrics which apply to each frame of reference — whereas the space-time interval concept joins together these different space and time metrics into a new kind of space-time combination which applies to both. "Measurements of distance and duration may vary from one frame of reference to another, but the space-time interval between events never changes." (Zukav, p. 158) The analyst attains this end "by converting the time figures into space figures, as the distances that light would travel in the given times; then he treats time, so reckoned, almost as if it were an extra dimension of space." (Calder, 1979, p. 132) "The Lorentz transformation keeps the 'interval' invariant." (Rogers, p. 495) In brief, then, the interval "plays the same role in space-time that the distance does in ordinary geometry." (Clotfelter, p. 361)[12]

Space-time intervals come in two main flavors, depending on whether the basic equation emerges as positive or negative: if it is positive, then the interval is called *time-like*; and if it is negative, then the interval is called *space-like*.[13] While time-like-related events will not appear to be simultaneous, space-like-related events will seem simultaneous (Zukav, p. 146; Geroch, p. 105). Time-like intervals measure the "apparent elapsed time between the events" (Geroch, p. 106), whereas space-like intervals are "a measure of an apparent spatial distance between the events." (Geroch, p. 107)

Look at what we have just stumbled across — an apparent *duality*, even within Relativity! Not only that, but we maintain that this pair can be understood in terms of one particular dimension of our model: that of stasis-motion, or static relationships (which are "space-like") and dynamic relationships (which are "time-like"). We made reference to these two relationship categories in the previous subsection, as corresponding with simultaneity and non- (or relative) simultaneity, respectively; and we shall bring these issues together in the next section. Suffice it to say here that the concepts of a space-like interval and a time-like interval essentially bring us back to the more sensible meanings for the concepts of space and time themselves that we started with!

12 We relay the following terse description from Whitrow: The "invariant space-time interval is defined by the rule that its square is equal to the difference between the squares of the temporal interval and spatial distance between the two world points concerned, units of space and time measurement being chosen so that in terms of them the velocity of light is unity." (p. 123)

13 Technically there is also a third type, "light-like," for cases where the result is zero.

In any event, whereas we shall see that the pair of interval constructs mesh smoothly into the present model, the same cannot be said for the way they fit with other aspects of Relativity Theory. On the contrary, what we shall find in the next chapter is that the construct of a space-like interval creates serious theoretical problems for the Theory of Relativity. For, "One event is space-like separated from another event if there is insufficient time for a light signal to connect them." Thus, "communication between space-like separated events defies one of the most basic assumptions of physics." (Zukav, p. 287) The reason is that such connections are effectively instantaneous (e.g., Zukav, p. 146; Geroch, p. 105): not only would such a phenomenon appear to violate the spirit if not the letter of Relativity — where nothing trumps the speed of light — but even more startling is the fact that such space-like intervals have actually been demonstrated in practice ... a point we noted previously regarding the EPR experiment. Such a result can only be considered calamitous for the Theory of Special Relativity, in spite of attempts by proponents (that we will describe) to sidestep the problem.

Before closing this discussion, it is worth reminding ourselves once again both about the historical background at the time, as well as in how these developments are commonly viewed retrospectively. In particular, this brings us back to the role of Poincaré and the way he is usually treated as being the old-fashioned almost-founder of Relativity who didn't quite get it and make the leap over to the new-fashioned world, as fearless Einstein did — even though it was Poincaré (along with Lorentz) who had assiduously developed all the basic conceptual tools that were involved. Here with the present topic, it turns out once again that it was Poincaré who had earlier (in 1906) put forth the basic idea of, along with the basic mathematics for, a four-dimensional space-time world — including the essential notion of the space-time interval. Minkowski picked up this framework and re-presented it in terms which many physicists found very appealing — including Einstein, after some initial resistance. But by that time, Poincaré himself had backed away from the space-time unity concept, on the grounds that its benefits were not worth the cost — and that our ordinary space and time constructs were a better fit for describing the world around us. Yet once again, although he had pioneered the way forward, Poincaré is faulted for not going down the new four-dimensional path as Einstein did. (Galison, pp. 263-265)

And once more, we dispute the orthodox conclusion as to which was the wisest choice. Thus, we contend that the complementary construct pair of space-like and time-like intervals is an exception to an otherwise bleak conceptual landscape here. And so we concur with Poincaré that the ongoing drawbacks of the space-time continuum construct far exceed the advantages of such a merging. The result is to leave us further removed from a meaningful overall vision of nature.

The Two-Force World of Space and Time

We now want to see how the notions of space and time can be understood in the present model, and what the connection is between them. After that, we shall argue that there is nothing unique about the space-time relationship, and that the same principles which apply to complementary partners in general will apply here, as well. In examining this question, we will return to an issue we raised in the first chapter, concerning a same vs. different interpretation of complementary phenomena.

The Nature of Space and Time. So what is space, and what is time? Our usual mode of thinking treats space as primary: we visualize space essentially in terms of emptiness, with matter being that which fills empty space. By contrast, time serves two functions: one is an essentially static chronicle, which orders our date-time experiences or events; and the other is a dynamic process, which flows from one such temporal event to another. Briefly juxtaposing these with the established physics interpretation, we saw in Chapter 5 that General Relativity explicitly renounces any distinction between space (as an empty framework) and matter (as its contents) — declaring that matter is merely a concentrated form of space(-time). Regarding the other dimension, we have seen how time's dynamic function is eliminated altogether by Relativity; and even the static dating function which remains has been transformed into a mere appendage of space.

In the present model, we have mentioned before that in the most general terms, we would associate space with the *X*-Force of Structure, and time with the *P*-Force of Process. One might also think in terms of the matter-medium dimension, whereby space would correspond with matter and fullness, while time would correspond with non-matter (or medium) and emptiness ... somethingness and nothingness, respectively. Needless to say, this is very different from the usual way of thinking about space and time, as we just noted at the start of this subsection — whereby space would correspond with emptiness, and matter that which fills it ... with time thrown in on the side.

But the most specific way we think about space and time in our model is in terms of the stasis-motion dimension. In fact, there are similarities between this approach and the view of Aristotle, who "defined time as a measure of motion" (Sachs, 1974, p. 7). What we are calling time and space are thus apparent in the way Aristotle "defines nature as an inherent source of motion and rest." (Furley, p. 91) Likewise, "In the Jaina physics.... *Dharma* and *adharma* designate the conditions of movement and of rest respectively." (Jammer, 1957, p. 5) Also, Descartes "made a magnificent attempt to construct a complete theory of the world out of nothing but the concepts of matter and motion." (Holton and Brush, p. 65)

As we indicated in the first chapter when we initially presented our dimensions of analysis, space should be understood primarily in terms of static relationships — in relation to time as dynamic relationships. Of course, both types of relationships are composed of (lower dimension) dominant and recessive components, and both have either matter or medium (higher dimension) characteristics ... along with those from the inside-outside and stable-changing dimensions. In some form or another, therefore, space corresponds with stasis — in relation to time corresponding with motion. Regarding the temporal factor, we all experience the dynamic essence of time as part of our daily lives: even physicists do so, in spite of their theories ... another of those annoying contradictions between established theory and everyday life which never seems to go away.

We should note one area of overlap between the two visions of nature that we are discussing, in that we would concur with the proposition that the static dating function of time can be more properly understood in a spatial context. But that shift hardly justifies the elimination of time as a distinct parameter in its own right, let alone the purging of its dynamic essence.

There is a very significant correspondence between the approach we are advocating and the way in which Minkowski's four-dimensional space-time model is sometimes presented. In particular, we have already noted how the two main types of Minkowski's space-time intervals mesh quite well with the static and dynamic relationship categories. And we attach the notions of simultaneous vs. successive to each.[14] Thus:

Static Relationship	Dynamic Relationship
space-like interval	time-like interval
simultaneous (within time)	successive (over time)

We can develop this analysis a little further by considering how the next lower dimension, of dominant and recessive entities, might fit into this scheme ... although we are merely restating points we have made previously. For example, one way of talking about the components of the static relationship is in terms of a recessive source and a dominant sink, and one way of talking about the corresponding components of the dynamic relationship is in terms of recessive motion backwards (or outwards) and dominant motion forward (or inward). And although we will address the topic more fully in the next chapter, these two dynamic relationship categories can also be described as backward time and forward time.[15]

Complementary Pairs in Nature. We have challenged Relativity's notion of a combined space-time construct, as representing a spurious fabrication — which not only has very little meaning in its own right, but has equally little intelligible relationship with other viable constructs about the natural world. In the previous

14 Note the correspondence of these ideas with that of Leibniz, who construed space as the relationship between coexistent phenomena and time as the relationship between successive phenomena.

15 Quantum Theory sometimes graphs this temporal distinction in terms of the Feynman diagram. In such graphs, forward time would be represented in a downward direction, while backward time would be represented in an upward direction. This approach is related to the idea we discussed in Chapter 3 concerning Feynman's interpretation of the positron as being an electron traveling backwards in time. As we noted in a footnote in that earlier chapter, this same interpretation can also be applied to other matter-antimatter pairs which have been postulated.

subsection, we set forth a simple conceptual framework for how space and time would be understood as separate parameters of nature, which could be jointly applied to nested dimensions of analysis within the present complementary model.

Regarding the main issue in question, we discussed the general pattern in the first chapter, in the section called Same and Different. Briefly, our point there was that complementary partners must be understood in both separate and combined terms — and that this pattern also represents the very essence of complementarity. Two conjoint partners only exist apart from one another by virtue of their existence in relation to one another. The net result is that on the one hand, yes: one can talk about either partner of a complementarity as though it exists by itself, but only with the realization that this partner also exists in relation to its opposite. On the other hand, no: one cannot talk about either partner as being distinct from its partner, and the two only exist as a whole.

Our position is that physics has missed the boat in a variety of ways, regarding this issue, but primarily because it has taken a rigid and singular position on the matter. That is, space and time are treated only in terms of a unified whole in their own right, and not in terms of separate and distinct parts. Of course, the situation is even worse than this — in that the combined space-time construct is understood primarily in terms of only one of the two aspects or components, rather than being shared more or less equitably and inversely between the two. Instead, the spatial mode and its accompanying stasis aspect are determinative ... with time being incorporated almost as an afterthought, and its dynamic essence eliminated altogether. The result has left us seriously hamstrung in our ability to conceptualize the nature of reality — both in terms of these two fundamental parameters, as well as in a more general sense.

The most basic explanation for this state of affairs is simply that physics lacks a complementary framework or template for analyzing these issues, in terms of which the relationship between space and time would be more apparent. Furthermore, the established model is also not based on hierarchical and nesting patterns, as is the present work. In the first chapter, we talked about some important mechanisms, in this regard — using such terms as generalization or synthesis upward in the hierarchy, and discrimination or thesis-antithesis downward in the hierarchy ... among many other descriptions. Thus, two conjoint partners join together into a larger, higher-level whole which they mutually comprise: they are same ... just as the two partners also exist apart as separate, lower-level parts in relation to one another: they are different.

It is in these terms that the Relativity position is unsatisfactory — that "space by itself, and time by itself, are doomed to fade away into mere shadows, and only a kind of union of the two will preserve an independent reality." Consider what the implications would be if we applied this same simplistic and unitary strategy to other basic dualities, such as those we described in Chapter 1.

For example, shall we adopt the same mode of analysis when we speak about male-female, with "male by itself, and female by itself, doomed to fade away into mere shadows, and only a kind of union of the two will preserve an independent reality"? Or shall we speak of stasis-motion, with "stasis by itself, and motion by itself, doomed to fade away into mere shadows, and only a kind of union of the two will preserve an independent reality"? Of course, just such an interpretation of stasis and motion is not far from the Relativity position, concerning the symmetrical interpretation of the Relativity Principle. Finally, we can also mention again what some physicists (including Heisenberg) actually concluded in the realm of micromatter: That is, shall we speak of the combined "particle-wave", with "particle by itself, and wave by itself, doomed to fade into mere shadows, and only a kind of union of the two will preserve an independent reality"?

It is true that just as we have talked about how figure and ground interact with one another — including the fact that the same entity can function as figure in one situation and ground in another — so there is a reciprocal interaction which all complementary partners engage in. In this respect, the idea is perfectly plausible that aspects of time can be converted or transformed into aspects of space, as we have noted previously. There is no objection to such an inversion, in principle — so long as it is mutual and equitable, with the essential attributes of both partners being thereby preserved; and so long as we realize that there is nothing unique about space and time in this respect. Such convertibility merely expresses another "equivalence principle", applicable to any complementary pair within any level of nature's hierarchy, or dimension of analysis in this model.

But in no complementarity of nature is one partner a real independent phenomenon, either explicitly or implicitly, while the other is merely a shadow or extension of the first; and likewise for the whole vs. the parts. Neither partner (nor the whole) is more fundamental than the other (nor the parts), and it will not advance our understanding of nature to pretend or assert otherwise — that the one partner essentially has meaning only in terms of the other, rather than the equivalence being two-way; or that only the combination is real, and the parts mere illusions, if not semantic missteps.

So to say that space does not exist without time (and vice versa) is all very well — but the same is true of every complementarity of nature, including that of their primary counterparts in this model, stasis and motion. Therefore, it is axiomatic that space and time each have just as much independent existence from one another (as well as shared coexistence) as do the two partners of any other complementary pair — and it will not do to simply merge the two together into a larger whole of their combined unity, and obliterate the component parts. Male and female may form a complementary unity: but that unity has meaning both in terms of whatever higher level entity they represent (or produce), but also in the sense that male and female are equally distinct and legitimate concepts or phenomena in relation to one another. In the simplest terms, all complementary partners can be understood both in contrast to one another, as well as in combination with one another.

Nature hasn't changed, then, in spite of Relativity's attempts to transform it: time continues to occupy a category of its own — and to have a dynamic character of its own — as distinct from static space ... with the two parameters being coequal to one another. Space and time form an integral and reciprocal relationship with one another — both in the sense that neither phenomenon or principle exists without its complementary opposite, but also in that neither partner can be treated as simply a mere appendage of the other. Like every other complementarity in nature, the fact that space and time do not exist in total isolation from one another does not mean that they cannot be differentiated from one another. They are same — and they are different.

In the next chapter, we will continue wrestling with the underlying issue — not in terms of space and time, but rather in terms of a host of other pivotal concepts about the way the world works. For the problem is an ongoing theme in established science: the neglect of complementarity in nature. It is the same "no symmetry" category that we keep running up against — whereby one side of a duality is submerged or dismissed or minimized, in deference to the other side as being primary and paramount ... and even exclusive, in a unitary sense. But in addition, we will also focus on the general orientation of this one-sidedness — which is based on the same tendency as we found here, with space co-opting time, and not the reverse. We will address this pro-structure bias, even though that proclivity is not universal.

Diverse Alterations and Transformations

In this section we will look at several categories of space and time changes or alterations — including Doppler effects, the gravitational redshift, and refraction. Our goal is to see how these various phenomena might be brought together within the organizational framework of the present model.

Classical Doppler. Everyone is familiar with the changing pitch of a train's whistle or a car's siren, as it moves closer to or further from us. The phenomenon is caused by alterations in the wave's parameters, which are associated with motion. This is also the principle by which radar operates.

What underlies these effects is the fact that the wave's frequency is different, as it is emitted by the source and as it is detected by the sink (or receiver/listener), in a manner which corresponds with the motion of the sending and receiving objects themselves. If the two objects are moving closer together, then this "pushes" or compresses the wave's frequency closer together — thus higher frequency; whereas if the two objects are moving further apart, then this "pulls" or stretches the wave's frequency further apart — thus lower frequency. (Note that these changes could just as well be described in terms of wavelength — longer and shorter, respectively — since frequency and wavelength are the inverse of one another.) An important point is that the precise details of these wave changes are different, depending upon whether it is the source or the sink which is moving.

Relativistic Doppler. In the modern period, the explanation which we have given so far would be described as only an approximation, which is accurate enough when we are dealing with common wave patterns such as those of sound. But with phenomena that are moving at very high speeds, the classical Doppler formulas fall short — because they do not incorporate the time dilation factor which must be taken into account when an object is moving at velocities that are substantial relative to the speed of light. The clock ticks more slowly for the moving object — which means a lessening in the wave's frequency. And while this relativistic factor is small compared to the classical effect, precision demands that it be taken into account. Note that the main (i.e., classical) factor is called a first-order Doppler shift, whereas the time dilation (i.e., relativistic) factor is called a second-order Doppler shift.

There are actually two additional features which differentiate the classical and relativistic Doppler categories. One is that the classical Doppler effect only comes into play when source and sink are moving closer together or further apart — this being called "radial" or "longitudinal" motion. By contrast, the motion factor which is associated with time dilation makes no direction-based differentiation, and so it would also apply to situations where source and sink are not changing their distance from one another — this being called "transverse" or "tangential" motion, as in one object moving sideways around the other rather than inward or outward. Radial Doppler therefore applies to situations where the motion is *parallel* to the line which joins the source and sink, and thus where the distance between them changes; transverse Doppler entails motion that is *perpendicular* to the line which joins the source and sink, and thus where there is no change of distance between them.[16]

The other distinct feature of relativistic Doppler is the role of the medium, and Einstein's symmetrical interpretation of the Lorentz transformation. Of course, there is no disputing the fact that classical Doppler applies to medium-based phenomena (such as sound waves moving through an air-based medium) — and it is in relation to such a medium that source motion can be differentiated from sink motion. But Special Relativity dispenses with any medium for light transmission, thereby also doing away with motion in relation to such a medium — the result being that there is said to be no way of differentiating source motion from sink motion, and all that we have left is relative motion as between source and sink (regardless of whether we are talking about radial or transverse trajectories).

As far as supporting research goes, the time dilation factor which is associated with high-speed motion has been amply demonstrated, as we noted in the previous chapter. However, virtually all this evidence is provided in the form of source motion, in relation to which the observer can be treated as static — i.e., in relation to a medium. So while the theoretical notion of purely relative motion (as characterizing relativistic Doppler) is constantly touted as representing another important breakthrough for Einstein's model, it has not been convincingly demonstrated.

Universal Redshift. We now want to talk about more generic ways in which the Doppler patterns can be described, and we start by taking a little detour in order to introduce some easy jargon here.

Physics sees the whole range of electromagnetic waves in terms of an electromagnetic spectrum — with high frequency waves at one end of the continuum, and low frequency waves at the other end (short wavelength and long wavelength, respectively). One way of understanding effects like Doppler is in terms of a shift towards one or the other end of this continuum. For descriptive purposes, the two directions which the shift can take are labeled in accord with the color portion of the spectrum, by character-

16 To be clear, the transverse Doppler category is sometimes treated as a pure motion factor, regardless of whether it entails change of distance between source and sink, or not.

izing these two different tendencies in those terms. Since blue has a higher frequency and red a lower frequency, these have become convenient relative labels for indicating the direction which the shift undergoes — *blueshift* for bunching up (as in source and sink moving closer together), and *redshift* for spreading out (as in source and sink moving further apart). Of course, the motion-based relativistic Doppler factor that is associated with the dilation or slowing of time amounts to a small redshift factor in its own right, regardless of the main blueshift or redshift effect.

While the Doppler effect is typically applied to situations where particular entities are moving closer together or further apart, there are two other types of related phenomena which are not construed in those terms — and one concerns the universe as a whole, especially the expansion thereof. What cosmologists find is a generalized redshift from amongst distant objects in the universe — an effect which becomes more accentuated the further away the distant entities are from the earth.[17] The conclusion which is drawn is that most entities in the universe are moving further away from us, as well as from each other: the distance between objects is increasing and the waves are being stretched apart — hence, the universe as a whole is expanding. A related factor is the general principle known as Hubble's Law — which correlates the *velocity* of the universe's expansion to the *distance* between any two components therein. Therefore, the further apart two entities are in the universe, the greater will be the speed at which they continue to move further apart from one another.

Gravitational Redshift. Now let us mention the other type of wave shifting phenomenon which is distinct from the Doppler effect. This change category is not explained in terms of specific motion or universal expansion, but rather in terms of gravitational influences or, equivalently, the curvature of space-time as it is put forward by Einstein's General Relativity.

We discussed the issue in Chapter 5, where we talked about the idea that massive bodies are said to "distort" the space-time fabric all around them, resulting in greater curvature. This would impact objects and waves in that region, including electromagnetic waves. These effects can be seen in the behavior of light as it passes nearby massive bodies, and thus through the altered field. For one thing, the light's pathway would

17 The underlying methodology is based upon the fact that every atom or nucleus in the table of elements is associated with a unique frequency or wavelength, collectively called *atomic spectra*. Such spectra thus serve as a kind of fingerprint, which allows investigators to identify an element solely on the basis of the wave parameters that are emitted. It is by means of this principle that astronomers identify the composition of far-off elements or bodies in the sky. They do this by observing the precise characteristics of the wave patterns which are received, and then by comparing those distant wave patterns with those which are emitted by elements here on earth. When they find an element which emits a similar wave pattern here as that which they are receiving from afar, they then are presumed to have indirectly identified the distant element. For example, hydrogen emits a characteristic set of wavelengths (or frequencies) that is different from the atomic spectra emitted by every other element. When a similar set of spectra is received from a distant source, the presumption is that it is also being emitted by hydrogen. What cosmologists actually find is that the specific set of wave parameters which are emitted from distant elements has been shifted slightly to the blueshift or redshift side of where such elements' emission parameters would be located on the earth's surface.

be affected, by being "bent" or drawn towards the massive body. For another, the light's character would be altered, in the form of changing wave parameters like reduced frequency. Thus the concept of a *gravitational redshift*.

Of course, the effects we are describing would not only apply to waves, but they would also describe temporal phenomena in general. Therefore, the wave would not be alone in oscillating less frequently near a massive body; so would any kind of clock or rhythmic phenomenon.[18] The result is that time slows down, and thus clocks run slower, in a stronger gravitational field.[19]

There are corresponding effects in the spatial dimension. Thus both General Relativity and its Newtonian predecessor presume that gravitational mass increases the closer one gets to a gravitational source — and it is inferred from the General Theory that eventually an object would be crushed by overpowering gravitational forces ... this being one characteristic of a "black hole". Furthermore, because of the equivalence between gravitation and acceleration, both the spatial and temporal distortion effects would also apply to the latter category: thus clocks run slower and masses increase when they are accelerated.

There is one changing wave parameter, in particular, that we need to consider — and this concerns the question of speed. Of course, there is no doubt that the premise of a constant speed of light represents the core of the Special Theory; but its place in the General Theory is far more ambiguous. We discussed the issue in the last part of Chapter 6, where we noted that even Einstein stated that light's speed could change when the limiting conditions of the Special Theory were superseded, such as in a strong gravitational field. On the other hand, the usual consensus — and the public relations myth — is that light's speed (in a vacuum) never changes.

Thus it is that some descriptions talk about a slowing of light's speed of propagation when near massive bodies (cf. Calder, 1979, p. 62; Murchie, p. 581; Whitrow, p. 129). Yet in other cases, we get caught up in the constant speed rhetoric, and dogma has the upper hand: "Because light always travels with the velocity *c*, an increase in energy can occur only by virtue of an increase in the *frequency* of the light wave." (Marion, p. 411);[20] or again, "light could only move at one velocity" (Asimov, p. 123) — thus implying that a change of frequency obviates any change in light's speed. Of course, we reject that position: the actual speed of light will change across different space-time or gravitational force regions of the universe — even if it is measured to be the same in any local region — as we discussed in Chapter 6.

18 Marion argues that a "light wave constitutes a kind of clock" (Marion, p. 41).

19 While the general consensus seems to be that "clocks run slowly in a gravitational field" (Marion, p. 412; also Calder, 1979, p. 52), there are relativists who claim quite the opposite: that the clock in the stronger gravitational field runs faster, more quickly, than the clock in the weaker field (Stewart). Equally bizarre are assertions that frequency *decreases* as it moves from a position near a gravitational source to one further away from it, while conversely frequency *increases* as it moves toward a gravitational object (Marion, p. 411).

20 Why "increase", rather than "decrease"? See the previous footnote.

Refraction. Next we need to introduce one more construct into the phenomenological mix, and this is the conventional category known as refraction. Of course, it may seem very strange to discuss this topic in relation to the other constructs and phenomena we have been discussing. But we are approaching the issues from a different vantage point, in our search for unifying themes.

Indeed, we pick up on the speed of light issue that we just discussed, in this regard. Thus we noted in Chapter 6 that changes of material media are put into a special conceptual box, that of "refraction". Light's speed may vary in these circumstances, whereas it is said to be constant in a vacuum.[21] We question these discrete categories, and the dichotomous treatment of light's speed which is associated therewith. Thus we mentioned in a footnote on page 219 that there is no reason to assume that vacuum refers to something absolute and uniform throughout the universe. Rather, vacuum is the name for one (e.g., idealized) subtype of medium that may exist in a given local (e.g., space-time) region — but which will be of a different character from the vacuum which exists in some other (e.g., space-time) region.

We shall argue in the next chapter that refraction is simply the name we apply to a set of changing-medium phenomena that exists on a small or limited scale — but which has the same general pattern as does the changing-medium phenomena which exists in the universe as a whole on a much larger scale. Variation of medium is variation of medium, on whatever scale it happens and whether it is discrete or not.

And so a light wave which passes through a pane of glass is the same changing medium phenomenon (on a small scale) as a light wave passing through a changing medium in interstellar space (on a large scale). True, an example of refraction like the changing medium of glass is rather more abrupt and hence discontinuous, as compared to what would be envisioned as a more gradual and continuous change of medium in some other situations. But it is the principle of a changing medium — and its corresponding impact upon wave parameters like speed which exist upon that medium — which is the important factor.

Changing Mediums. Now we want to suggest how the constructs and phenomena that we have been discussing in this section can be brought together into a unified conceptual framework that is consistent with the present model. And we emphasize at the outset the exploratory nature of our discussion here.

The most general way to think about a common theme for these issues may be in terms of a changing medium or field, irrespective of the cause. We could talk about a difference in the texture (or even a kind of density) of the regions which are involved — including alterations of space and time themselves, which would apply to both objects and to waves.

One might differentiate the gravitational shift from the Doppler shift in terms of a stasis vs. motion characteristic — with both incorporating dominant and recessive com-

21 Light waves decelerate in denser materials (Hanson, 1962, p. 91).

ponents. Thus, the gravitational shift could be understood in terms of a static relationship between a dominant massive body and a recessive secondary entity or phenomenon; the normal Doppler shift could be understood in terms of a dynamic relationship between a dominant observer and a recessive sender. In the former case, proximity to the large mass would result in contraction-like effects (e.g., strong gravitational field or more curved space-time), whereas distance from the large mass would result in expansion-like effects (e.g., weak gravitational field or more flat space-time). In the latter case, increasing proximity between sender and receiver would result in contraction-like effects (e.g., higher frequency), whereas increasing distance between sender and receiver would result in expansion-like effects (e.g., lower frequency).

Therefore, contraction or expansion would affect not only the field or medium as a whole — but it would alter all characteristics of the contents, including objects and waves: all the components would be contracted or expanded, accordingly. Changes in the field or medium might correspond with a slowing down or spreading out or expansion, in which case it may be referred to as a redshift; or conversely, changes in the field or medium may correspond with a speeding up or bunching in or contraction, in which case it may be referred to as a blueshift. A clock ticks at a different rate in one region of the universe than in another: likewise for differences in speed of light, changes in mass, and dimensions of a meter stick.

Indeed, one might think of a construct like "effective distance" between two entities or regions, as corresponding with the degree of spectrum shift amongst elements, as between them. The greater the effective distance between two regions, the more divergence there will be in an element's spectral lines — in either redshift or blueshift direction. Conversely, two mediums which do not differ from one another would show no shift relative to one another, as this would be tantamount to being the same medium, with no effective distance between them.

Overall, we can talk about a contracting medium or field as opposed to an expanding medium or field, and it is in these terms that we now try to bring together several aspects of the different categories we have been discussing in this subsection, along with refraction, into a complementary chart:

X-Force: Contracting Medium / Field	*P-Force: Expanding Medium / Field*
more curved space-time	more flat space-time
strong gravitational field (higher *g*)	weak gravitational field (lower *g*)
gravitational redshift: near to massive body	gravitational blueshift: far from massive body
increased mass, longer wavelength	decreased mass, shorter wavelength
slower clocks, lower frequency	faster clocks, higher frequency
slower wave speed (including light)	faster wave speed (including light)
radial Doppler "blueshift"	radial Doppler "redshift"
source vs. sink approach	source vs. sink recession
refraction: more dense medium	refraction: less dense medium

Complementary and Hierarchical Patterns

Next we want to further broaden the "unified conceptual framework" that we just proposed, in the form of of summarizing our two main themes of complementarity and hierarchy. We start with the latter, and then turn to the former.

Hierarchical Order.

This is a good point to talk about the hierarchy factor in a number of broad and far-reaching — not to mention speculative — ways.

To start with, of course, the general idea of describing nature in hierarchical terms is not new, although it has rarely been applied in any thoroughgoing way by science ... and especially by physics. As Whyte put it many years ago, "[N]either the term nor the explicit idea is to be found in most physics texts" (Whyte, p. 8), and that is still too often the case today. Whyte himself was one of the pioneers of applying the hierarchy concept in biology, and that field remains the main exception in science. Whyte noted then "that the time has arrived for the gradual development of a comprehensive physical theory of the structural hierarchy of nature." (Whyte, p. 11) Likewise, physicist David Bohm also wrote then that "what is needed in physics as well as in biology, is to perceive the existing facts anew, in the light of the notion of order, and of a hierarchy of orders of order." (Bohm, 1969, p. 39)[22]

As we said initially in the first chapter, the concept of hierarchy that we are working with is based on the idea of an overall pattern of generalization or convergence upwards, and discrimination or divergence downwards. Here is a simple summary that is based upon the hierarchy factor and is intended to incorporate a number of important

22 The concept of hierarchy is a seminal component of Bohm's work (see, e.g., 1980a). He refers to "the existence of level within level of smaller and smaller kinds of entities." (Bohm, 1957, p. 138) He makes the general point that, "[A]ll that is known seems to fit beautifully into the metaphysics of process, assumed to be ordered hierarchically" (Bohm, 1969, p. 49). Bohm goes on to "emphasize the vast order that is implicit in the full concept of hierarchy.... I want to propose that electrons, protons, etc., are merely the names of aspects of a vast, self-regulating, hierarchical process, operating at the level of inanimate matter. Such a hierarchy would in a natural way show many of the properties that are now attributed to quantum mechanical systems." (Bohm, 1969, pp. 51 and 56) There have been others who have applied the idea of hierarchy to physics — like Feynman (Feynman, 1965, p. 125). Sarfatti makes use of "the basic structuring principle of our multilevel hierarchical reality" (Zukav, p. 298). Zapffe talks about "a hierarchical structuring of space." (Zapffe, 1977a, p. 28) And Ruderfer, too, has referred to an "observed hierarchy of matter." (Ruderfer, 1979a, p. 245; cf. Ruderfer, 1975, p. 10.) In cosmology, Wilson cites the concept of hierarchy being used from 1761 onwards, with Charlier proposing in the early 1900's "a universe built up of a hierarchy of 'galaxies.' " (Wilson, quoted in Whyte, p. 114). In 1930, Shapley "pointed to the set of levels into which all matter appears to be organized extending from the sub atomic particles to the 'metagalaxies.' " (Wilson, quoted in Whyte, p. 115) Davies describes the universe as being "a hierarchy of structures — the planets, the solar system, star clusters, galaxies, clusters of galaxies." (Davies, 1974, p. 106) Murchie has proposed a hierarchical ordering of the universe by size (Murchie, p. 232).

characteristics like these. In this case, the two forces can be understood as representing the two end points of the hierarchy:

P-Pole

↑

Size	***Quantity***	***Frequency***	***Speed***	***Density***	***Power***	***Scope***
bigger	fewer	slower	faster	dispersed	weaker	broader
↕	↕	↕	↕	↕	↕	↕
smaller	greater	faster	slower	concentrated	stronger	narrower

↓

X-Pole

The idea is that the higher one goes in the hierarchy, the larger are the components thereof (e.g., in terms of mass and wavelength) — which are fewer in number, slower in their internal frequency or breathing, faster in their overall external movement, more dispersed in their composition, thereby weaker in their local strength ... and likewise the higher the level, the broader and more abstract are the phenomena in question. Conversely, the lower one goes in the hierarchy, the smaller are the components thereof — which are greater in number, faster in their internal frequency or breathing, slower in their overall external movement, more concentrated in their composition, thereby stronger in their local power ... the lower the level, the narrower and more concrete are the phenomena in question.[23]

To be sure, the model of hierarchy that we are striving for is not just flat and singular, but nested and compound. This entails the notion of hierarchies within hierarchies, whereby higher level wholes encompass lower level parts.[24] One could narrow in on any given level, and discern finer and finer levels that are nested within — to the point where every entity would be seen as having subsidiary levels nested within it. In

23 The idea of a relationship between size and frequency has been discussed for many years. Writing about subatomic particles, for example, Capra states that “their lifetime has to be regarded in relation to their size” (Capra, 1975, p. 213). Stevens examines this same theme in the animal world, stating that, “Small creatures breathe faster than larger creatures. Their hearts beat faster, their voices are higher, and their wings or limbs move more rapidly.... [T]he rate of breathing also decreases with increasing size.... Apparently, the small creature lives on a shorter time scale than the large.” (Stevens, p. 23-24) Murchie writes about “the vastly different tempos of existence in the universe” which are “almost inversely proportional to size.” (Murchie, p. 550)

24 It should be obvious that what is a higher level “whole”, in relation to lower level “parts”, is itself lower level “part” (along with some complementary counterpart) in relation to the still next higher level “whole” — just as lower level “part” is itself a “whole” in relation to further lower level “parts”. Indeed, Arthur Koestler coined the term “holon” to describe such dual higher and lower level roles which some hierarchical entity could represent. (Koestler)

this way, the smaller entities that are nested within one larger unit are thereby distinct from the smaller entities that are nested within another larger unit — meaning that we are talking about something that is specific to this or that situation, phenomenon, or realm of nature — yet still exists within one overall hierarchy, which incorporates all phenomena in the cosmos.

It should also be clear that the hierarchy framework we are proposing would encompass both physical and non-physical phenomena, since the latter are also part of nature. In some cases, such as those having to do with products of the human mind, there would presumably be subsidiary hierarchies that are nested within the level of human beings. These would include all our concepts and theories, as well as categories like common sense, metaphysics, and so on (see Chapters 10 and 11).[25] In this respect, one can distinguish between a lower level of the individual human being (and its concepts and theories) as opposed to a higher level of collective human beings (and their concepts and theories). Of course, there are different types or nesting branches of collective theories, such as that represented by the social grouping of scientists.

The essential idea we are talking about — of one all-encompassing hierarchy as representing the entire range of natural phenomena which exist in the universe — is very appealing. Still, Two-Force Theory is a complementary model of nature, and so it should not be surprising that we question whether such a singular conception of hierarchy will suffice. So what is the alternative? We suggest two types of hierarchies, each having different organizational principles; and we shall next present a hazy and exploratory outline of such factors.

Let us first talk about the most obvious kind of hierarchy — which is the kind we usually think of, even if only implicitly. This is the hierarchy we have just been describing — and refer to in most of this book — which is based on the general assumption that we are dealing with different levels of structural phenomena, which are typically arranged in order of size, from the largest at the top to the smallest at the bottom ... albeit entailing subsidiary and nesting forms.

Yet even so, far-reaching as this conception is, it should be apparent what is missing from such an approach: in principle, we would have what is supposed to be a total and complete inclusion of all things in the universe, large and small — but primarily in the sense of structural phenomena. In its essence, this mode does not represent the more dynamic pattern of interactions and energy variations which take place between the entities which exist at those different levels of nature.

The distinction we are talking about completely parallels the theme of the two forces upon which this model is built. That is why we would propose both an overall *structural hierarchy*, encompassing the full range of phenomena in nature — as well as a *process hierarchy*, which would represent the range of dynamic interactions which exist in relation to such structural entities ... again, both types of hierarchy understood in nesting form. The dual hierarchies we are describing here could be construed in terms of a distinction between static and dynamic relationships — which is itself a

25 Popper refers to "hierarchical levels of explanatory hypotheses" (Popper, 1965, p. 173).

prominent way of representing another distinction we have made, between "being" and "seeing".

One way of interpreting the process hierarchy is in terms of certain fundamental energy patterns that we are all familiar with — such as in the notion of input or buildup as describing the upward direction of the process hierarchy, as opposed to output or discharge as describing the downward direction. In terms of the breathing motif, we would be talking about inhalation going upward and exhalation coming back downward. However it manifests itself and whatever names are attached, we are speaking about a bi-directional pattern which could be applied to any entity or local region, as well as to the universe as a whole.

Consider some other ways of describing the two kinds of hierarchy. We suggest that the structural hierarchy can be predominantly characterized in terms of *extensity*, in that it would be based largely upon a *static* range of power, size, and density factors. By contrast, the process hierarchy can be predominantly characterized in terms of *intensity*, in that it would be based largely upon a *dynamic* range of power, control, and energy exchange factors. In many respects, we are talking about *the structural hierarchy as representing space and spatial characteristics*, as opposed to *the process hierarchy as representing time and temporal characteristics*.

Clearly, there would be an interaction between the two types of hierarchies, with process phenomena existing upon a structural phenomena framework: Process phenomena concentrate and build in a narrowing and strengthening/gathering inhale-like input, as they go upward — whereupon they invert, dissipate and discharge in a widening and weakening/spreading exhale-like output, as they go back downwards. By contrast, the structural hierarchy would expand its range going upward, and narrow its range going downward.[26]

The charts on the following page incorporate some of the descriptive labels that can be applied to each of the two types of hierarchy — realizing that each would have an opposite overall polar orientation from the other:[27]

26 Another manner in which the two kinds of hierarchy might interact would concern ways in which process phenomena might be transformed into structural phenomena — either leading to new forms or levels of matter, or being converted from a process to a structural form; and of course, vice versa.

27 Included in this "opposite overall orientation" would be the opposite labeling of the two poles at either end of each hierarchy, something which is only implied in the two charts.

Structural Hierarchy

extensity-predominant static relationship existence (space/being)

X-Force Downward (X-State)	*P-Force Upward (P-State)*
smaller size — greater density	bigger size — lesser density
concentrated / narrow / sharp / precise / specific	dispersed / broad / blurred / vague / general
hard / absolute / real	soft / relative / apparent

Process Hierarchy

intensity-predominant dynamic relationship breathing (time/seeing)

X-Force Upward (X-View)	*P-Force Downward (P-View)*
inhaling / inbound / input	exhaling / outbound / output
narrowing / focusing / specifying	broadening / dissipating / generalizing
strengthening / concentrating	weakening / dispersing
absorbing / buildup / energizing / potential energy	emitting / discharging / activating / kinetic energy
attracting / contracting / binding	repelling / expanding / unbinding
climbing up / ascending	falling down / descending

It is worth noting that the idea of two different kinds of hierarchy is not a new one. Conventional hierarchy theory makes a distinction between a "scalar hierarchy" and a "specification hierarchy", and there is considerable overlap between our thesis and that approach. Thus, the *scalar hierarchy* would be like our structural hierarchy, in that it is characterized as "extensional": its various levels are treated as being quantitatively different, yet synchronous with, one another. The other category of a *specification hierarchy* would be like our process hierarchy, in that it is characterized as "intensional": its various levels are considered as qualitatively different from, and not synchronous with, one another.[28]

As far as the issue of nesting levels of structure goes, we have made that point repeatedly in earlier pages. Some of the most significant applications of this precept concern the classical constructs which have been rejected by modern physics theory. Thus we have argued that every level of the hierarchy entails its own specific versions of these classical constructs — which interlock with levels above and below it. For example, every level will have its own locally-specific absolute space, for the matter domain ... and locally-specific ether, for the medium domain. There is no way of avoiding the crucial logical functions which these classical constructs provide — such as having a medium in which light waves travel — and they remain meaningful precisely because the hierarchical pattern of nesting complementarities provides a coherent framework for the varying (i.e., relative) content which these constructs represent.

The same hierarchical pattern would apply to time concepts, as well. As Wilson puts it, different hierarchical "levels may be characterized by a characteristic time or frequency" (Wilson, quoted in Whyte, p. 55). In Bastin's words, different "levels of control must be distinguished by different time constants." (Bastin, p. 253, italics omitted) This approach is consistent with a point we made above about the relationship between size and frequency. Similarly, Murchie depicts a temporal hierarchy way of ordering the universe (Murchie, p. 241) — and of course we have talked about different levels of time as corresponding with different levels of space. We would suggest that the pattern can also be seen in terms of the notion of resonance that we mentioned in Chapter 3 — the idea that every basic element has its own characteristic or resonance frequency. These are illustrations of the general point that there are different interlocking levels of structural and process phenomena.[29]

Now for a last point about the process hierarchy, in which we bring in the stable-changing dimension. We would then be concerned with the question of whether the going upward in the process hierarchy (which is akin to inhaling) is essentially equivalent to the coming downward in the process hierarchy (which is akin to exhaling). If they are, then we would be dealing with a stable region ... whereas if they are not, then we would be dealing with a changing region. For example, steady-state breathing would be an obvious example of the stable type — whereas unequal breathing or the sudden taking of a deep breath would represent the changing type of process pattern, in which the going up is not equivalent to the coming down. Simple harmonic motion would be of the stable type, whereas its antithesis would be changing. We will pick up on these issues towards the end of Chapter 9.

28 We avoid going off onto a tangent in the main text, but it is worth trying to summarize some of the differences between the two sets of dual hierarchy approaches in this footnote. One of the main problems is that both the scalar and specification hierarchies are based on structural levels — as opposed to that of the present model, whereby one hierarchy is based primarily on structural levels and the other hierarchy is based primarily on process patterns that exist in relation to those structural levels. We would therefore incorporate the structural aspects of the "specification hierarchy" in terms of our structural hierarchy, and the dynamic aspects of both types in terms of our process hierarchy. Another apparent difference is that the various structural levels of the "specification hierarchy" may entail equal size components, whereas the structural components of the "scalar hierarchy" have different sizes. An easy example would be that of different levels of rank in some human-based organization, such as the military. In this situation, a colonel is of the same physical or structural size as a sergeant, but the two exercise greatly varied levels of power — and are thus represented at higher vs. lower levels of the "specification hierarchy". Our interpretation is that this same-size appearance is actually an illusion in hierarchical terms, and that the different-ranked entities can be better understood as representing (e.g., by virtue of being a singular representation or focus of) different cumulative levels of the overall organization, which have different control or scope patterns. Thus, a colonel represents a higher structural level of the overall organization — including in size terms — whereas a sergeant represents a lower structural level of that organization. All of this also explains another difference of approach — whereby the "scalar hierarchy" is understood in nesting terms, but the "specification hierarchy" is not. By contrast, both our hierarchies are of a nesting character, whether we are talking about structural or process constituents. The reason why this would be true in the structural (or "scalar") case is fairly obvious, and both approaches reflect that. Yet we maintain that the nesting factor would also apply to our process hierarchy in the sense that lower level process patterns nest into higher level process patterns. The main reason why the "specification hierarchy" does not appear to incorporate this nesting factor is because of the reasons we just described: it is based on structural components, rather than process ones; and the erroneous same-size assumption does not appear compatible with such a nesting interpretation.

Before closing this subsection, let us mention how different states of matter might be understood in hierarchical terms. Thus, consider the following description:

> What determines whether a particular substance is a solid or a liquid or a gas? In any sample of matter, forces of attraction exist between the atoms that make up the substance. In solids these forces are strong and the atoms are held firmly in place. In gases the interatomic forces are extremely weak and each gas atom moves freely among the other atoms. In liquids the forces are sufficiently strong to maintain the matter in a condensed state but they are not strong enough to prevent the atoms from sliding easily through the surrounding atoms. (Marion, p. 11)[30]

One way of interpreting these patterns is in terms of whether some entity or phenomenon is primarily a manifestation of the *X*-Force and attraction ... as compared to whether it is primarily a manifestation of the *P*-Force and repulsion. In terms of the nesting hierarchy, for example, the same atoms could combine together in a manner that is more oriented towards the *X*-pole of the hierarchy (e.g., solid), or they could combine together in a manner that is more oriented towards the *P*-pole of the hierarchy (e.g., gas) ... or else they could combine in a manner that is somewhere in between the two poles (e.g., liquid).

Of course, in a relative sense, temperature is an important determinant as to which state of matter a substance is in at any given time. If one considers the example of that most ubiquitous substance, water, then we can see how readily it changes from one state to another: at its "normal" or mid-range temperature condition, it might be described as being equi-balanced between the two forces and thereby manifest itself as a liquid. Yet at very cold temperatures, it is more *X*-Force and thus becomes a solid — whereas at very high temperatures, it is more *P*-Force and thereby becomes a gas.

The kinetic theory of heat attributes a body's heat or "thermal energy" to the constant, chaotic motion of molecules within that body (Hecht, p. 179; also Pitt, p. 210). In these terms, we would be talking about the *X*- and *P*-Forces as corresponding respectively with the stasis vs. motion of molecules within a given body. Low temperature

29 For example, it may be an oscillating dominance between two same-level partners which thereby provides (or at least is associated with) the "pump" of interaction with the larger environment — in that one partner would draw "buildup" energy in from without, while the other partner in turn would give "discharge" energy back out. Therefore, it would not just be an isolated exchange of energy or dominance between the main partners that is involved in these situations, for these changes would be enmeshed as part of other interactions and relationships with the larger world. Moreover, not only is the *larger* world involved outside the two entities themselves — that which is *higher*, hierarchically; but the *smaller* world within each of the two separate entities would also be implicated — that which is *lower*, hierarchically. Taken together, for example, there would be (lower) swaps within each entity that would correspond with (same-level) swaps between two entities ... and thence to (higher) interactions beyond. According to this broad approach, then, any changes of state or exchanges of energy must be not only reciprocal (complementary) — but in a larger sense, they must involve a nexus of factors (hierarchical).

30 Murchie provides a similar explanation, talking about "the gently balanced forces of repulsion and attraction [which] produce the fluidity we call the liquid state," while the gaseous state consists of "the repulsing forces overwhelm[ing] the attracting ones." (Murchie, p. 268)

would be associated with the *X*-Force and stasis at one pole, whereas high temperature would be associated with the *P*-Force and motion at the other pole.[31] Indeed, we can also understand Quantum Theory's blackbody concept in the same way — although such a one-sided concept of the perfect absorber must be complemented by a "whitebody" counterpart of the perfect emitter. Summarizing these last points in hierarchical terms might look like this:

P-Pole
↑
motion
hot/burning
more molecular movement
emits/expands/repels
whitebody
bright (visible) / noisy

↕

dark (invisible) / quiet
blackbody
absorbs/contracts/attracts
less molecular movement
cold/frozen
stasis
↓
X-Pole

The hierarchy concept should also be able to incorporate other physical characteristics, like wetness vs. dryness and softness vs. hardness.

Complementary Dimensions. We turn next to a very quick treatment of complementarity itself, in the form of a reprise of the main chart we presented near the beginning of the book (on page 6) which depicted our dimensions of analysis — except that now we will look at an elaborated version (on the facing pages which follow). This incorporates some of the issues we have been discussing thus far in the work, along with a topic that will come up in the next chapter.

There is no point in burdening the reader with the tedious repetition that would be required in order to describe the various items which are listed, since in most cases this is either obvious or has largely been done previously. At the same time, it should be

31 In practice, of course, there is a certain paradox here — in that if the particles in motion get too dispersed, the energy is no longer concentrated enough to create significant amounts of heat. Therefore, the motion must be sufficiently bound within a structure — in the same way that a gas only remains as such, so long as it is bound within a container.

clear that the main point here is the general theme of nesting complementarity — within which we try to illustrate how our dimensions of analysis are general enough to incorporate a diverse array of phenomena, concepts, and theories. Of course, this is only a loose agglomeration, and there is nothing systematic in the way that items are associated in each category.

By the way, note that the complementary chart presented a few pages back, at the end of the Changing Mediums subsection, would be categorized here under the changing outside realm. Also note the point we made when presenting the initial version of this chart in Chapter 1 — that if it were not for page space limitations, the stable and changing regions would be presented side by side ... rather than as here on separate pages; for these two characteristics are complementary with each other in the same way as are the characteristics of the other dimensions.

The one main subject in the chart which we have not yet really addressed concerns the distinction between direct and indirect contact, along with the different aspects of space and time which go with each. This is one of the important issues we will examine in the next chapter.

Nesting Dimensions of Analysis

Changing

Non-uniform / anisotropic / heterogeneous / unbalanced / unbound

Inside

2D / flat / linear / river / horizontal / sideways / charge

Acceleration / electron

Outside

3D / curved / circular / ocean / vertical / up-down / polarity

Gravity / photon

Inside		*Outside*	
Matter	***Medium***	***Matter***	***Medium***
Inner Frame	Closed Field	Outer Frame	Open Field / electromag. - gravita'l. waves
Attract ≠ Repel / Leftward ≠ Rightward	Buildup ≠ Discharge / Inhale ≠ Exhale	Contract ≠ Expand / In ≠ Out	Buildup ≠ Discharge / Inhale ≠ Exhale

Stasis	***Motion***	***Stasis***	***Motion***	***Stasis***	***Motion***	***Stasis***	***Motion***
Static Relationship	Dynamic Relationship	Static Relationship	Dynamic Relationship	Static Relationship	Dynamic Relationship	Static Relationship	Dynamic Relationship
Space (-like interval)	Time (-like interval)	Space (-like interval)	Time (-like interval)	Space (-like interval)	Time (-like interval)	Space (-like interval)	Time (-like interval)
Direct contact	Indirect contact	Static charge / elec. field	Mov.charge/elec.field/AC	Direct contact	Indirect contact	Static polarity/mag.field	Mov'g.polarity/mag.field
Absolute Simultaneity	Relative Simultaneity	Absolute Simultaneity	Relative Simultaneity	Absolute Simultaneity	Relative Simultaneity	Absolute Simultaneity	Relative Simultaneity
Non-inertial mass	Non-uniform motion	Standing wave / ether	Traveling wave	Gravitational mass	Non-uniform motion	Standing wave / ether	Traveling wave

Dominant	***Recessive***	***Dominant***	***Recessive***	***Dominant***	***Recessive***	***Dominant***	***Recessive***	***Dominant***	***Recessive***	***Dominant***	***Recessive***	***Dominant***	***Recessive***	***Dominant***	***Recessive***
sink	source	attraction	repulsion	sink	source	buildup/inhale	dischg./exhale	sink	source	attraction	repulsion	sink	source	attraction	repulsion
left/$mass_i$	right/$mass_{ii}$	leftward	rightward	positive/in	negative/out	leftward	rightward	lower/inner/M_i	upper/outer/M_{ii}	inward/fall	outward/rise	south/in	north/out	inward/fall	outward/rise

Stable

Uniform / isotropic / homogeneous / balanced / bound

Inside

2D / flat / linear / river / horizontal / sideways / charge

Inertia / electron

Outside

3D / curved / circular / ocean / vertical / up-down / polarity

Gravity / photon

Inside: ***Matter***		Inside: ***Medium***		Outside: ***Matter***		Outside: ***Medium***	
Inner Frame		Closed Field		Outer Frame		Open Field / electromag. - gravita'l. waves	
Attract ≈ Repel / Leftward ≈ Rightward		Buildup ≈ Discharge / Inhale ≈ Exhale		Contract ≈ Expand / In ≈ Out		Buildup ≈ Discharge / Inhale ≈ Exhale	
Stasis	***Motion***	***Stasis***	***Motion***	***Stasis***	***Motion***	***Stasis***	***Motion***
Static Relationship	Dynamic Relationship	Static Relationship	Dynamic Relationship	Static Relationship	Dynamic Relationship	Static Relationship	Dynamic Relationship
Space (-like interval)	Time (-like interval)	Space (-like interval)	Time (-like interval)	Space (-like interval)	Time (-like interval)	Space (-like interval)	Time (-like interval)
Direct contact	Indirect contact	Static charge / elec. field	Mov.charge/elec.field/DC	Direct contact	Indirect contact	Static polarity/mag.field	Mov'g.polarity/mag.field
Absolute Simultaneity	Relative Simultaneity	Absolute Simultaneity	Relative Simultaneity	Absolute Simultaneity	Relative Simultaneity	Absolute Simultaneity	Relative Simultaneity
Inertial mass	Uniform linear motion	Standing wave / ether	Traveling wave	Gravitational mass	Uniform circular motion	Standing wave / ether	Traveling wave

Dominant	***Recessive***	***Dominant***	***Recessive***	***Dominant***	***Recessive***	***Dominant***	***Recessive***
sink	source	attraction	repulsion	sink	source	buildup/inhale	dischg./exhale
left/$mass_i$	right/$mass_{ii}$	forward/go	backwd./come	positive/in	negative/out	forward	backward

Dominant	***Recessive***	***Dominant***	***Recessive***	***Dominant***	***Recessive***	***Dominant***	***Recessive***
sink	source	attraction	repulsion	sink	source	attraction	repulsion
lower/inner/M_i	upper/outer/M_{ii}	clockwise	counterclock	south/in	north/out	advanced	retarded

Chapter 9

The Other Side of Nature

Throughout this book, we have been highly critical of physics' tendency to minimize or dismiss one side of nature's duality, in this or that realm, in favor of the other — or to simply merge the two together into some inscrutable melange.

The latter was certainly the case with space and time, as we saw in the last chapter. Not only does such a blending together obscure the vital distinction between the complementary partners, but it is not really a merging — so much as a "hostile takeover". For, rather than the two partners being joined together in some quasi-equal sense, we find instead that the essence of one partner has been converted into that of the other. The result is that while we may be nominally dealing with a combination of the two, we are really dealing with only one ... and the second is thrown in almost as an afterthought. Thus, in appending time to space, Relativity squelches the dynamic essence of time, and deals only with its static aspect. As a result, the spirit of time is lost in Einstein's space-time continuum: it is a quintessentially spatial version of time which is represented in the one-sided combination which results.

It really makes no difference which tactic is used — whether it is one side that is explicitly dismissed in favor of the other, or whether the two are joined together into a lopsided unity. For the end result is still the same, whereby a duality has been collapsed into a unitarity. It is a forced marriage, at best, because it is a distortion (if not a perversion) of nature.

But there is a further problem, which concerns the direction of these reductionist maneuvers[1] ... as typically favoring one overall side of nature at the expense of the other. It is most often the partner which corresponds with the *P*-Force of Process (e.g., time) that is treated as secondary or irrelevant, as compared to the partner which corresponds with the *X*-Force of Structure (e.g., space) that is treated as primary or exclusive.

Let us review a few other examples of this overall trend, which we have previously discussed — here in relation to four specific dimensions of our model. For one thing, we

1 We utilize the term in spite of the supposed fact that, "Physicists have long since abandoned a purely reductionist approach to the physical world." (Davies, 1983, p. 64)

have repeatedly mentioned that (X-like) stasis is treated as primary, in relation to (P-like) motion as a virtual exception to such normalcy — clearly illustrated by the conventional adage that motion only exists in relation to stasis, and not also the reverse specification. Secondly, we find that aspects of the (P-like) medium domain are typically treated either as secondary, or else banished altogether, in favor of the (X-like) matter domain. Thirdly, the outside realm construct called gravity is interpreted almost exclusively in (X-like) attraction terms, as opposed to also entailing — or being equitably juxtaposed with — a (P-like) repulsion counterpart. Finally, recall that the primary distinguishing characteristic between the inertial and non-inertial frame of reference is what we describe as (X-like) stability vs. (P-like) change — with the former being treated as "preferred" by the orthodox model.

We can pick up on what we call the matter-medium disparity by saying that too much attention is given to objects, broadly defined, and not enough to non-objects. Some of these objects are rather small, such as an apple — while other objects are considerably larger, such as the earth; and of course, one can go up and down from there in terms of size, as well. But the main point is that the established paradigm is primarily concerned with *objects* — either on a large or small scale ... all of which are variations on a theme of *particles* (if not a theme of *atomism*). Science makes every attempt to explain nature in terms of objects or particles, and only deals with non-object phenomena to the minimum extent necessary — and in a manner which often has the effect of overshadowing or subverting such non-object phenomena and transforming them into what are essentially object terms.

Furthermore, when this one-sided bias comes into play, explicitly or implicitly, then the essence of the non-object phenomenon is lost — both in terms of the particular entity itself, as well as in terms of the larger non-object portion of nature ... of which the given phenomenon is only one specific manifestation. The upshot is that not only does established physics merely tend to ignore the non-object side of nature: in many respects, it positively rejects it. We have talked about this problem repeatedly — for example, when we mentioned the category of "subatomic particles" in Chapter 3: why is there no corresponding set of equally prominent "subatomic waves"? Similarly, physics' fundamental forces and field-based interactions are all said to operate by means of the exchange of particles (or virtual particles).

Indeed, Karl Popper argues that, "[Q]uantum theory is in a very definite sense a particle theory ... and in a sense which excludes a duality, or analogy, or complementarity, between particles and waves." (Popper, 1982, p. 126) That structural (and not also process) mind-set is even extended more generally, to the point that, "Laws of nature are conceived ... as (conjectural) descriptions of the structural properties of nature — of our world itself." (Popper, 1981, p. 196) It is as though only concrete physical entities are relevant — and only those which can be measured, at that: "There is no place in a physical theory for a quantity that cannot be defined by a measurement." (Marion, p. 438) The result is to define the scope of science in a manner which may exclude less substantial and more subtle phenomena. Thus, biologist George Simpson writes, "This is not to say that science necessarily denies the existence of immaterial or supernatural relationships, but only that, whether or not they exist, they are not the business of science." (Simpson, quoted in Baker and Allen, p. 18)

Yet why should there be *any* phenomena of nature which are "not the business of science"? And how can a narrowing in on "the structural properties of nature" possibly lead to an acceptable representation of a world which is no more inherently structural than it is inherently process? The point is that if nature is not only structure-like, then why should science study only (or even predominantly) those aspects? In sum, why should science insist on defining itself essentially in terms of one side of nature — as in the very name for the field of "particle physics" — and if it does, then what basis have we for concluding that it is a science of *all* of nature, as opposed to only being a half-science of *part* of nature?

We argue that science can have no other fundamental purpose than to describe nature — fully. That was one of the points we discussed in the last part of Chapter 2, where we noted the disparity between what Bohm calls "laws of cause" and "laws of chance". To requote:

> [T]he categories of necessary causal connection and chance contingencies are seen to represent two sides of all processes.... [W]e do not assume, as is generally done in a mechanistic philosophy, that the whole of nature can eventually be treated completely perfectly and unconditionally in terms of just one of these sides, so that the other will be seen to be inessential, a mere shadow, that makes no fundamental contribution to our representation of nature as a whole. (Bohm, 1957, pp. 29 and 143)

Stapp writes about the problem in terms of the Theory of Relativity's treatment of the concept of time:

> [T]here is no reason why something so basic to our intuitive grasp of reality as the notion of process, or the unfolding of Nature, should continue to be banned. For this notion was banished in the first place only by Einstein's cleverly contrived realm of readings.... The empirical fact that some part of our understanding of Nature can be formulated in terms of readings alone does not imply that a full understanding can be expressed in this limited way.... It is unreasonable to impose upon Process relativistic demands drawn from the static realm of reading. (Stapp, 1982, pp. 381 and 383)

Contrast the conventional one-sided approach with that which we advocate in the present model — such as with the matter-medium dimension. On the one hand, the *matter domain* primarily represents objects, based upon a structural or particle predominance, orientation, or perspective. But on the other hand, the *medium domain* primarily represents the relationship between objects, based upon a process or wave predominance, orientation, or perspective ... where the character of the substratum itself represents an essential part of the description. The contrast with one-sided physics could hardly be more sharp — e.g., in Special Relativity's absurd denial of the very concept in question, of a medium for light transmission.

One rejoinder would be to dismiss the general point we are making, by arguing that we are dwelling on trivia here: for example, physics might claim to be able to talk just as well in terms of "subatomic waves" — in the sense that every phenomenon can ultimately be understood to manifest both particle and wave aspects ... in spite of Popper's point above about the particle nature of Quantum Theory. The argument would continue by explaining that it would be too cumbersome to constantly refer to the duo of "subatomic particles and subatomic waves" — rather than simply naming the category with one or the other shorthand ... again with the explanation being that it is arbitrary which of the two terms one cites. A variation on this reasoning would take us back again to the Heisenberg position — that terminology like particle or wave are only linguistic approximations for a natural phenomenon which is neither ... and so it does not matter which of the paired unsatisfactory labels one applies. And as far as the just-mentioned denial of a medium construct in Special Relativity, it would be argued that the theory's longevity provides its own proof of the validity of such a conceptual framework.

We maintain that arguments like these are disingenuous, on several counts. For one thing, the Heisenberg position specifically illustrates the problem — which is a denial of duality. And it is immaterial what form this denial of duality takes: be it in Heisenberg's terms, that the very concept of duality is a false but forced result of our flawed concept system; or the functionally-equivalent position that the two sides of a duality should be merged into one; or that only one side of some duality is real, and the other merely an illusion. Furthermore, if the choice of terminology is truly arbitrary, such as that between the particle and wave appellations, then why is it that physics consistently opts for the former at the expense of the latter — and how can it be claimed that the particle-based description which results does not impact our view of the world? Regarding the point about Special Theory's longevity representing validation of its conceptual framework, it should be clear from our remarks in the last part of Chapter 7 that we reject such a defense as amounting to a dodge that has nothing to do with the core issues.

In fact, the problem we are dealing with in the established paradigm goes deeper than any specific instances, and concerns a fatal bias that represents a far more broadly distorted mind-set, in its own right. One minor example is to be found in Max Planck's approach, which "always regarded the search for the absolute as the loftiest goal of all scientific activity." (Planck, quoted in Jammer, 1974, p. 202). But what about the relative? Likewise, consider Max Jammer's comment that, "It has been claimed that even the most 'progressive' theoretician believes at the bottom of his heart in a strictly deterministic, objective world even if his teachings categorically deny such a view." (Jammer, 1974, p. 253) Yet what about indeterministic aspects — not to mention subjective aspects which are inherent in the role of an observer, as to which perspective is adopted in any given situation ... ?

In any event, even to the extent that modern physics takes such factors like indeterminism into account, what remains is still a mechanistic mind-set. This is the point which Bohm makes, in arguing that

> [W]hat has actually happened has just been a switch from deterministic to indeterministic mechanism..... Thus the usual interpretation of the quantum theory represents, in a certain sense, a rather natural continuation of the mechanistic attitude of classical physicists, suitably adjusted to take into account the fact that the most fundamental theory now available is probabilistic in form, and not deterministic. (Bohm, 1957, pp. 70 and 103)[2]

Then there were people like Einstein, who resisted even the concession to reality which indeterminism represented, given that he resisted notions like probability and uncertainty — "dice-playing", as he called it — and so he was never able to reconcile himself with some of the main findings of Quantum Theory. Actually, we would make the paradoxical argument that Einstein was completely reactionary when it came to a deeper sense of relativity — which would not be based upon a one-sided and misleading principle like "all motion is relative". Far from being truly relativistic, Einstein's model is based upon a deep-seated absolutist and deterministic conception of the world.

Indeed, the argument is often made that Special Relativity would have been better named the Theory of Invariance, both in terms of the Lorentz transformation and in terms of the idea of a fixed and unchanging speed of light — *absolute*, therefore, and not *relative* to the motion of some adjacent object. Furthermore, we would argue that the absolutist style of thinking is also apparent in the theory's commitment to, and foundation upon, the notion of pure symmetry — which we call "false symmetry", and to which we attribute many of the model's illusory conceptions about nature.[3] We contrast this premise with the more nuanced notion of complementarity — which is not based upon a conception of perfectly equal counterparts … and which we contend is inherently relativistic in a way that Einstein's model never can be, beyond its deceptive name.

Of course, complementarity also comes into play when it comes to the other category of symmetry that we keep talking about, which we call "no symmetry". Every instance of one-sidedness — among so many that we keep encountering, everywhere in the established paradigm — represents another violation of that fundamental two-sided essence of symmetry, which precept physics claims to be so committed to. If it were more than skin-deep, such a two-sided symmetry-base would entail an acknowledgement of the two-sides of nature more broadly.

Further exploring this issue is the task we have set for ourselves in the present chapter. Therefore, it is "the other side of nature" that we intend to highlight in the pages ahead — in contrast to the predominantly one-sided orientation of the main theories. We shall do this partly in the overall terms that we have been describing, and partly in terms of some specific other sides of nature that have been neglected or dismissed by the established paradigm. As we shall see, certain of these conflicts represent longstanding controversies, some of which are surely as old as science itself.

Much of the discussion will revolve around the overall issue of connectedness, but we shall also deal with questions about energy changes and exchanges. In our usual fashion, we will first set forth these topics in the standard way that they are described by the established paradigm … followed by the 2FT alternative. After that we will turn to the topic of light propagation and other interactions, to see how these could be reinter-

2 Many observers take a contrary posture and claim that modern physics itself has gone beyond a mechanistic mind-set. For example, Capra says that, "[M]any scientists still hold to the mechanistic paradigm, although physicists themselves have gone beyond it." (Capra, 1983, p. 101) Likewise, Harris makes comments like, "Contemporary physics, then, is thoroughly committed to holism." (Harris, p. 160) He adds that holism "is the metaphysics characteristic of the science of the age." (Harris, p. 161) He goes on to say that, "[C]ontemporary physics … has abandoned mechanism and materialism … and has revealed the essential unity of the physical world." (Harris, p. 172) Were that the case!

3 Likewise, Zapffe states that it is "the original purist position of true symmetry in relativity" which is at the root of "all the objectionable paradoxes of the Special Theory." (Zapffe, 1977a, p. 19)

preted in light [sic] of the connectedness framework that we will present — and we will also discuss some important theory and research in this realm. After that, we will examine the important second law of thermodynamics and its relation to the "arrow of time": we argue that this theory represents another glaring instance of one-sided conceptualizing about nature. Finally, in the last section of the chapter, we will come back to the general issue that we have touched upon repeatedly during the course of the book: this concerns the relationship between symmetry and complementarity, and how the rift between them exposes serious problems with the established paradigm.

All of the discussion will revolve around the overall one-sided, or "unitarity", tendencies of science; and much of it will also provide further support to the notion we have been describing of an overall structuralist or mechanistic mind-set. Of course, this not always the case, in that sometimes the general orientation for the merging or one-sidedness is in the (*P*-like) process direction, rather than the norm of being in the (*X*-like) structural direction. Thus while the bias is not always consistent, the lack of a complementary framework is, as we keep repeating — and this inevitably leads to arbitrary zigs and zags in whichever one-sided conclusion is applied in each case. Our mission here is to explain why both the one side and the other side of nature must be taken into account, if we are to reach a coherent vision of the world in which we live.

Disputing the Essence of Nature

One of the main debates which goes on in science, especially physics, concerns the ultimate nature of the connection that exists between one entity and another — which therefore provides the basis for communication or relatedness amongst elements in nature. In this section, we will first explore the notion of action-at-a-distance, in contrast to intermediary-based interactions, as mechanisms of conveyance in nature. Then we will turn to the related topic of direct vs. indirect contact. After that, we will consider the question of whether the partners of such interactions are to be considered as separate entities and the primary stuff of the interaction, or whether it is the overall relationship between the entities which is most important. Our last topic here concerns the kind of energy changes that an individual entity undergoes, or which is exchanged between two entities, in terms of whether these are discrete or continuous in character.

All of these issues revolve around binary choices, and in each case the conventional models are clear about which is supposedly the correct path. This is true even though the different models do not always agree about the proper choice, and even though the same model may reveal inconsistencies within its overall orientation. Moreover, the debate is actually ongoing — in spite of the definitive conclusions which the established paradigm has reached, regarding the correct outcome. As we shall see, the issues we are discussing here have broad implications, which go well beyond any particular set of natural phenomena or theoretical interpretations: they have a significant impact on our entire view of nature.

Action-at-a-distance vs. Action-via-intermediary. So how are influences conveyed in the world? For example, how does the force of gravity operate in its unitary and pervasive manner, across the universe? As we saw in Chapter 5, classical Newtonian physics was said to be predicated upon the assumption that the force of gravity was distributed by action-at-a-distance, whereby objects interacted directly one with the other, and not by way of some intermediary phenomenon. According to this classical principle, every piece of matter in the universe is somehow connected with every other piece of matter in the universe in a direct, static, and indiscriminate sort of way.

We have seen that such instantaneous action-at-a-distance is explicitly disavowed by the Theory of Relativity, as a model of interaction. Einstein himself puts it this way:

> The success of the Faraday-Maxwell interpretation of electromagnetic action at a distance resulted in physicists becoming convinced that there are no such things as instantaneous actions at a distance (not involving an intermediary medium) of the type of Newton's law of gravitation. According to the theory of relativity, action at a distance with the velocity of light always take the place of instantaneous action at a distance or of action at a distance with an infinite velocity of transmission. This is connected with the fact that the velocity *c* plays a fundamental role in this theory. (Einstein, 1961, p. 48)

In keeping with Einstein's view, physicists say that action-at-a-distance "has long had the after taste of magic" (Hecht, p. 115). By contrast, the Relativity approach is said to provide "a more logically satisfying explanation of interacting matter than Newton's action-at-a-distance concept." (Sachs, 1974, p. 121).

But we must also understand that this choice is not merely one of logical or aesthetic preference. For, in keeping with the view of Einstein just quoted, if entities could interact directly a la action-at-a-distance, then light's central and active communication role would no longer be pivotal. In other words, if such direct interactions really applied — as opposed to Special Relativity's traveling light-mediated interactions — then "the distinction between the Special Theory of Relativity and classical Newtonian theory would all but vanish." (Hecht, p. 144). That is because under those circumstances,"the Lorentz transformations reduce exactly to the Galilean transformations." (Sachs, 1974, p. 54; cf. also Zahar, 1976, p. 258) Obviously, then, the idea of direct action-at-a-distance cannot be contemplated by Relativity, for it would undermine the very heart of that model.

It is true that on its face, the idea of an infinite velocity does make direct action-at-a-distance seem very mysterious! And Einstein's choice is clear, in that he opts solely for the indirect influence approach, with both electromagnetic and gravitational interactions — although we discussed how, by some strange sleight-of-hand, he pretends to avoid the medium which necessarily goes with it ... particularly for light waves in the Special Theory. He accomplishes that trick, of course, by simply rebaptizing the intermediary "medium" as an intermediary "field" — the "electromagnetic field" — and then pretending as though something substantive has accompanied this change of names. As Hecht puts it, "Our modern picture, known as the *field* conception, is a subtle contrivance to make comprehensible the magic of action-at-a-distance — comprehensible if not entirely satisfying." (Hecht, p. 128)

In the case of the General Theory, such indirect influences are conveyed through the intermediary of an abstract gravitational field with the name of "space-time continuum", and by which the terrain or texture all around an object is altered ... with this influence extending outward indefinitely, traveling at the same speed as light. For its part, Quantum Theory provides a whole slew of specific-level fields, as providing the underlying substratum for interaction — each of which might be understood as providing a kind of ersatz-medium, by virtue of the logical necessity for some kind of conceptual linkage between one entity and another.

By the way, not all physicists have a rigid position on the possibility of action-at-a-distance. Again, the wise input of Bridgman:

> To many minds the concept of action at a distance is absolutely abhorrent, not to be tolerated for an instant. Such an intolerable situation is avoided by the invention of a medium filling all space, which transmits a force from one body to the other through the successive action of each other of its contiguous parts.... Now of course it is a matter for experiment to decide whether any physical reality can be ascribed to a medium which makes gravitation possible by the action of its adjacent parts, but I can see no justification whatever for the attitude which refuses on purely *a priori* grounds to accept action at a distance as a possible axiom or ultimate of explanation. (Bridgman, 1927, pp. 46)

Bridgman's cautions notwithstanding, the orthodox position of modern physics is to reject the notion of direct action between objects: only indirect interactions are allowed — although in relation to a "field" or "continuum", and not in relation to a "medium". At the same time, such a field would be understood not so much in terms of a concrete physical phenomenon like the ether, which Einstein rejected — but more in terms of a vague abstraction ... which sometimes seems to amount to no such conceptual function or component at all.

Direct vs. Indirect Contact. Now we shift the topic, so as to get a different perspective on the issue we have been discussing. That is, in talking about action-at-a-distance vs. action-via-intermediary, we have been talking about a mode of interaction between two entities that is either direct or indirect. Either way, the underlying assumption is that the two entities are separated by some distance — and while indirect interactions make sense in those terms, it is the notion of direct interactions at a distance which seems problematic.

But instead of talking about two entities that appear to be separated by some distance, we now want to approach the situation in terms of two entities that appear to be very close to one another. Even then, the mode of interaction issue would still arise: are the two entities interacting with one another directly, such that they are either actually or practically touching one another — or are the two entities only interacting with one another indirectly, and they could never be construed as actually touching?

Let us switch the theoretical framework from Relativity Theory to Quantum Theory, which is in keeping with the close-in perspective we are talking about. Still, the conclusion which emerges would not change: only the indirect kind of interaction is allowed — for both theories. Thus, whether we are talking about action-at-a-distance and Relativity Theory, or direct contact and Quantum Theory, the orthodox position is that the notion of direct interaction is seen as an illusion which does not really exist. Just as Relativity's light-mediated approach does not allow for such direct interaction, so likewise neither does Quantum Theory's atomistic foundation.

According to Quantum Theory, we delude ourselves if we think that individual atoms can ever be in direct contact with one another. Instead, there is the vague assertion that the very idea of actual physical contact is merely "a sudden set of repulsions at very close range.... [T]here is no 'contact' between atoms" (Rogers, p. 145). "[T]he notion of direct contact fails completely at the atomic level." (Davies, 1979, p. 4) Or Hecht:

> [W]hen we begin to think of matter composed of separate, spaced atoms, the idea of *direct contact* becomes meaningless. Atoms do interact, yet they never seem to touch in the familiar way we like to think of macroscopic bodies as touching. Even here, in the very essence of *contact*, there is the mysterious action-at-a-distance. Granted, it is a small distance between atoms, but even though they do push and pull on one another, they do not touch belly to belly, and direct contact in the usual sense is an illusion. (Hecht, p. 28; see also Hecht, p. 128, and Davies, 1979, p. 109)

The same analysis applies to other levels of matter — say, the molecule, which is just a collection of atoms. Thus, "it is not clear what is meant by saying that one molecule 'touches' another molecule ... in general, the effect of a collision between two objects is due to the electrical repulsion and/or gravitational forces that act between them" (Marion, p. 132; see also Murchie, p. 318). Again, we also encountered this same interpretation at the level of the nucleus, where the nuclear force is said to explain what would otherwise appear to be direct contact touching.

We maintain that the issue we are discussing is not merely limited to some obscure question of whether elementary particles can actually touch each other or not. For it also has implications on the very meaning of the physical reality that we all experience. Consider the standard analysis regarding the nature of ordinary matter — for example, our common experience that a physical table seems solid to us. Quantum Theory claims to have "demolished the classical notion of solid objects" (Capra, 1983, p. 80). For one thing, the orthodox interpretation tells us that the table is ultimately understood to be merely an agglomeration of individual atoms. Of course, because the mass of the atom lies in the tiny nucleus, the atom as a whole is taken to be mostly a void. Therefore, the table is composed predominantly of emptiness, and the sense of solidness is merely an illusion.

Not only that, but it is not just a void that we are primarily dealing with, according to this analysis: it is also motion! That is, while the table may appear to be a quiescent object that is static within itself, this is only an illusion — which masks the dynamic patterns of intense activity underneath. "[M]atter is always restless, never quiescent" (Capra, 1983, p. 88) because its constituent elements "are in a state of continual motion" (Capra, 1983, p. 90); a solid object can be seen in a context which reveals an "intrinsically dynamic character ... [as] the very essence of its being" (Capra, 1983, p. 92). What we experience as quiescent solidness should instead be understood in terms of the minuscule high-speed electrons racing around so fast in the emptiness which surrounds the (less) minuscule nucleuses (Capra, 1975, pp. 58 and 62).

That being the case, then, to what extent does a macro object like a table even exist? For, if the constituent phenomena of which these objects are composed can never really interact directly, and even "touch" each other, and if the thing we regard as an object is mostly a void — then why should we assume that the object as a whole is even real? Thus, if direct contact is only an illusion, then does this not mean that the very idea of independently-existing macroscopic objects — all the material content of our everyday experience — must be an illusion as well ... for none of it has the solid aspect we think it does? Of course, this kind of conclusion is also an outgrowth of another aspect of the established model, the notion of fundamental forces of nature. That is, as we first explained in Chapter 1, only the specific levels or regions that are associated with the fundamental forces truly exist: everything else — all of the common phenomena of our experience — is secondary, in this respect ... being just an aggregate of the more basic stuff of reality.

The issue becomes even more bizarre when we think about two such illusory macro objects, and whether they can be in direct contact (e.g., "touching") or not! Since each of the two objects would supposedly be understood in terms of the non-touching atoms of which they are composed — such atoms being themselves mostly comprised of empty space — then in the case of two macro objects, we would have a compounded form of empty space ... and thus a compound denial of direct contact. It is enough to make one wonder where such abstractions will ever end!

Entities vs. Relationships. Now we take an entirely different tack on these issues. Instead of asking about the true character of nature's interactions, whether they are direct or indirect, we now ponder the nature of reality from a very different perspective. Is the essence of nature to be found in the individual entities which interact — or does the essence of nature lie in the interaction itself ... and thus, irrespective of whether the true form of such interactions is direct or indirect? Things, as opposed to the relationship between things: that is the question.

Here, we find a split between the Big Two pillars of modern physics: Quantum Theory is more focused on distinct entities, whereas Relativity is more concerned with relationships. As Sachs puts it, Relativists say that "the relation between things is more fundamental than the things themselves" (Sachs, 1971, p. 93). Thus:

Individual Entities	*Relationships Between Entities*
Quantum Theory	Relativity Theory

Indeed, Sachs argues that there are similarities between Quantum Theory and Newton's classical model, in terms of these issues:

> The Newtonian approach concentrates on things rather than the relationship between things.... [I]t is still the individual, discrete "thing" in Quantum Theory, as it is in Newtonian theory, that forms the basic element of matter (Sachs, 1971, pp. 92 and 144).

But, again, this Newtonian and Quantum view differs from Einstein's:

> [In] contrast with [Newton's] description of matter as an assembly of independent discrete things, it is the continuous relation that is basic in the relativistic approach to physics. Here the relation plays the active role as a fundamental entity, while the "particles of matter" play the passive role of a derived prediction. In contrast, the "particles of matter" play the active role in the description of Quantum Theory, while the relation between the parti-

> cles (i.e., the forces that they exert on each other) play the passive role ... [The Theory of Relativity is,] then, necessarily a *field theory*.... In contrast Quantum Theory is necessarily a *particle theory*. (Sachs, 1973, p. 145)

Sachs' conclusion is that, "Newton's particle theory and Einstein's field theory are clearly logically incompatible. (Sachs, 1973, p. 58) And he draws the same negative conclusion regarding the relationship between Quantum Theory and Relativity, as we shall discuss in Chapter 12.

In the Theory of Relativity, all interactions are indirect and field-mediated, as we have noted previously. Einstein's space-time continuum not only represents the means of interaction, it also represents the stuff of which objects are composed. Therefore, what appear to be individual objects do not actually exist independently in their own right: they are merely concentrations of the space-time continuum. Einstein, then, follows in the tradition of Faraday and Maxwell with his field theory — although Sachs argues that it was the General Theory which brought the "shift from discrete particle to continuous field view of nature." (Sachs, 1973, p. 41)

Unfortunately, even these characterizations are inadequate to account for the shifting ontological foundations which underpin both the Relativity and Quantum models. Consistent with the rather random and scattered — "all over the map" — conceptual underpinnings of modern science, we do a back-flip next when we approach the issue from a different perspective. Here, the Big Two switch sides.

This issue emerged as part of a major philosophical dispute between Einstein and Bohr, which actually concerned the epistemological status of Quantum Theory itself and whether it was a complete theory or not. Indeed, the different positions which the two theorists take on this matter is seen by some as underlying the essence of the important Einstein-Podolsky-Rosen experiment, which we shall discuss later in the chapter. It is in this context that we now return to the entity vs. relationship question.

Bohr argued the relationship position here, and it was in this regard that he established the crucial premise that any experimental situation had to be treated as an entirety — with the observer and observed being considered as one whole relationship, rather than as separate and isolated parts. "[C]omplementarity leads to the conclusion that the world consists not of things, but of interactions. Properties belong to interactions, not to independently-existing things ..." (Zukav, p. 95) Or: "[T]he concept of a distinct physical entity, like a particle, is an idealization which has no fundamental significance." (Capra, 1975, p. 144)[4]

On the other hand, and in contradiction to the interpretation we mentioned just above, particularly for gravity and the space-time continuum, Einstein is also said to have never abandoned the notion of individual material realities. That is,

> [Einstein's] analysis involves in an integral way the implicit assumption ... that the world is actually made up of separately existing and precisely defined "elements of reality." Quantum theory, however, implies a quite dif-

4 It is very curious how Einstein ridicules Bohr's position: Einstein — of all people! — is quoted as saying "that Bohr's thinking was very clear but that when he begins to write he becomes very obscure and he thinks of himself as a prophet." (Einstein, quoted in Jammer, 1974, p. 188)

> ferent picture of the structure of the world at the microscopic level ... [that of] a relational and holistic conception of the state of a physical system. (Jammer, 1974, pp. 235 and 199)

> [Hübner] viewed the Bohr-Einstein debate as merely a reflection of two different or even diametrically opposed principles: one (espoused by Einstein) according to which physical reality consists of substances, and the other (espoused by Bohr) according to which reality is essentially a relation between substances, measurement being a special case of such a relation. (Jammer, 1974, p. 157)

Here is a chart summary which is the opposite of that which we presented just above:[5]

Individual Entities	*Relationships Between Entities*
Relativity Theory	Quantum Theory

Discreteness vs. Continuity. Lastly, we briefly note another longstanding dispute that, at first glance, may seem separate from the ones we have been so far describing; but there are common underlying themes. The main question here is, how shall we characterize the changes which each individual entity undergoes? We can also pose the issue more in the interactional terms we just discussed: How shall we characterize the exchanges which take place as part of the relationship between two entities? Specifically, do these changes or exchanges come in distinct bundles, or do they have a whole range of variability? The issue can be framed in an even more fundamental way, still: is the very texture of nature ultimately discrete or continuous in character?

Here is how Capra frames the theoretical dispute:

> Classical Newtonian physics ... is formulated in terms of solid objects moving in absolute space and absolute time. Relativity theory is formulated in terms of a four-dimensional continuum, space-time, in which all space and time measurements are relative. Quantum mechanics, on the other hand, deals with discrete, that is, discontinuous phenomena ... which are not embedded in space-time." (Capra, 1988, p. 152)

Once again, there is a split between the Big Two theories. Relativity sees nature as being gradual and incremental, symbolized by the notion of the (space-time) continuum. Einstein wrote that, "I do not think the theory [of relativity] can work without the continuum." (Einstein, quoted in Jammer, 1974, p. 122)[6] On the other hand, Quantum Theory sees the essence of nature as being discrete and all-or-none in character, symbolized

5 As far as the disparity between this and the previous chart, it is worth noting that in some respects the two theories are not working on an even playing field here — in the sense that Relativity's position was largely defined by one person, whereas Quantum Theory was a group effort. There have been divergent currents within the latter theory's development and present state. Indeed, one can talk about a general schism in Quantum Theory, on these counts, some of which we discussed in Chapter 2.

6 Note in contrast to this that elsewhere Einstein took the opposite stance — for example, with his photoelectric analysis he tried to particle-ize light, and make it discrete.

by the notion of the quantum jump — this being one of the great insights which that theory supposedly bequeathed to us. Thus:

Discrete and Discontinuous	Gradual and Continuous
Quantum Theory	Relativity Theory

Sachs argues that the basic issue in dispute here goes back well before either of these theories was developed — indeed, way back:

> The most striking point of departure in these two theories is one that is very old in the history of science. This is the difference between discreteness and continuity as basic features for the underlying fundamental description of matter. The question as to the choice between these two descriptions of matter has persisted throughout the entire recorded history of physics. (Sachs, 1971, p. 142)

Not only is this basic dispute not new — but there is no reason to think a resolution can ever be found. For his part, Wolf is not optimistic: "Indeed, the battle of continuity vs. discontinuity may never end." (Wolf, p. 124)

In practice, of course, there really is no controversy — and no balance — regarding the underlying issue: Quantum Theory's position is not merely dominant, it is virtually exclusive. Its "discrete quantum" namesake sets the tone — and not only for physics, but extending outward to the other sciences:

> [I]t was as if these new views in the various branches of science stemmed from a similarly directed change of the mental model used to comprehend fundamental quantities and processes, a change in the guiding idea from a *continuum* to a well-defined particle or *quantum*. (Holton and Brush, p. 445)

Whether this was a move in the right direction is a question we will examine in the next section.

Complementary Connectedness

We argue that contradictions and absurdities abound in the way the conventional theories approach the fundamental issues that we have been discussing. Indeed, it is amazing how desperate contemporary physics' mental gyrations appear, when they are aimed at trying to describe reality with such unsatisfactory, one-sided premises — and how the models flip-flop from one general position to another, based on the particular topic and phenomenon at hand ... not to mention what is in vogue for a given period of our theoretical development.

To be sure, we do not deny the possibility that nature could have been "constructed" in the various one-sided manners that the conventional models suppose. Thus for example, maybe Newton was correct with his direct interaction approach to gravitational influences. Then again, perhaps Einstein was right with his countervailing indirect interaction approach to gravitational influences. Of course, the current vogue is that the former is old-fashioned, has the taste of magic, and is not logically satisfying —

whereas the latter model is newer, represents deep insight, and is more sophisticated, etc.

But as usual, we question the premise that there should necessarily be a winner in these one-sided battles. Simply in terms of principle, and leaving aside other aspects of either model, why would nature be designed in terms of one or the other side of the connectedness issue — as corresponding here to the respective position of either of these esteemed models? It is the unitarity assumption itself, and the arbitrary way (either way!) that it is applied — as opposed to the duality assumption, and the consistent way that it is applied: this is the real choice we must come to terms with.

Organizationally, our alternative model discussion here will not mirror the conventional model topics of the previous section. And although the present treatment is titled "Complementary Connectedness", the issues range more broadly than that. To set the stage, we shall begin by talking about the two basic kinds of relationships that we have been discussing, in both small-scale and large-scale terms. We shall see that there are different ways of thinking about each of the two relationship categories, which go beyond the way we might normally construe them. Then, we will examine some assumptions about whether time duration is always involved with such interactions — no matter the apparent distance we are dealing with. Next we will revisit the issue which concerns the character of energy changes and exchanges. After that we shall find that, to some extent, it is simply a matter of choice and perspective as to which mode of connectedness is involved — which means once again that we are not simply dealing with an objective phenomenon of nature, as opposed to how we approach or view the phenomenon. Finally, we will briefly discuss the general need for two-sidedness in terms of some basic assumptions and predilections about nature, and we will also consider the issue in cognitive and sociocultural terms.

Two Types of Relationships. Obviously, the 2FT answer to the two types of interaction question is *both*. Therefore, we embrace the notion of direct connections as being on a par with indirect ones, regarding the modality for interactions in nature. In fact, most of our discussion in this subsection concerns the need for an expanded view of the former, as providing a more satisfactory counterpart to the latter.

Bridgman says that, "There are, in general, two sorts of influences — interaction between bodies in contact and interaction between bodies at a distance from each other." (Bridgman, 1983, p. 70) That is true — except that we dispute the conventional premise which differentiates these two categories, the distinction between local and distant. We maintain that there is no ipso facto demarcation between these two fundamental categories, of no-distance vs. yes-distance, largely because there is nothing sacrosanct about our human-based judgments thereof. Consequently and in spite of appearances, our position is that there is no clear criterion for determining whether two entities are interacting directly or not — regardless of what seems like objective proximity or distance.

More specifically, there is no absolute spatial criterion as to how close two entities must be, in order to be categorized as interacting directly: in principle, they could appear to be any distance from one another, as gauged by our human-based outlook — so long as in mutual relation they form a unity together. For whether there is "distance" separating them or not is purely a matter of perspective and scale — and the same principle should apply whether we are concerned with two atoms that are part of a solid table, or two similarly-linked objects which may appear to be spread far apart in outer space. Indeed, we maintain that there are problems with the way the conventional categories are even labeled — not only in terms of direct vs. indirect contact, but also in speaking about action-at-a-distance vs. action-via-intermediary.

For example, according to the present model, direct contact *does* exist ... both on the *small-scale* relating to the "touching" interaction of two entities; and — which seems more radical — on the *large-scale* of action-at-a-distance. Our position is that there is no fundamental difference between these small- and large-scale phenomena, except for the factor of scale or perspective. What we typically think of as direct contact "touching" is a microcosm on a small scale, which can be expanded (like a balloon!) into "action-at-a-distance" on a large scale: in principle, the two are essentially the same. In this respect, the solution to understanding action-at-a-distance relates to a factor of scale that we have been overlooking — and a failure of perception or perspective on our part. Thus, again, the appearance of distance separating two entities, according to some human-based perspective, is not decisive as to the character of the relationship between the two entities that are involved.

Recall a different argument of scale we initially made in the last part of Chapter 6, when we compared changes in the speed of light: we maintain that the variability of the fundamental speed of light caused by the "refraction" between one material medium and another (e.g., between air and water) is a microcosm on a small scale of the fundamental speed of light's variability between different large-scale regions or mediums of the universe. Physics acknowledges such changes when they entail material mediums, but typically rejects them in the case of different regions of the universe. We argued there that the two situations are essentially analogous, and differ mainly in terms of scale (i.e., leaving aside any distinct boundary factor). Thus there is no reason to accept changes in light speed for small-scale differences in material mediums, but reject them for large-scale differences in universal regions (or mediums).[7]

Bringing the scale factor back to the present issue, we maintain that if two entities are interacting directly with one another, then there is necessarily no distance separating them — for, together they comprise a combined whole. Then they are no longer separate parts: they are no longer two separate entities *interacting distally*, so much as one combined entity *intraacting* within itself — *locally*, by this definition. In that case, what may seem as a distance separating the two entities is actually part of the space (sometimes mislabeled as "void") which comprises the joint entity ... and which the two (now sub-)entities together provide the boundaries of. The interaction — nay, *intra*action — is then direct and quite local, rather than being indirect and at a distance. There is no gulf between them, because regardless of what we outside observers may deduce as to the supposed "distance" between the two entities, based on some human vantage point, the two entities themselves interact as though this is not the case.

7 We have noted that while such large-scale changes are rejected by Special Relativity, they are acknowledged by General Relativity.

What this also means is that we are questioning the meaning of the very notion of "distance", as in "action-at-a-distance". That is, we propose defining "local" and direct (contact) more in *functional* rather than *a priori* terms. In this sense "action-at-a-distance" is a misnomer, although any other phrase — like "action-in-a-locality" — may seem clumsy. In any event, the terminology is not worth belaboring, so long as we understand that the action-at-a-distance concept we are talking about applies between two entities which by definition are *not* distant in relation to one another, but direct and local ... regardless of how separated they may appear to us. Consequently, we are arguing that the distinction between local and distant, or direct and indirect, should not be based upon any apparent separation between the two interacting entities, since any such determination is perspective-dependent; and that this is an irrelevant factor, as compared to the crucial functional distinction.

Consider: we normally have no difficulty in grasping the idea that one entity exerts a direct action upon another entity if the two entities appear proximal to one another. But such determinations may not always be so obvious, and therefore it is not sufficient to merely assert that, "They are interacting directly only if they are actually touching one another." For, as we have seen from comments about micromatter, it is not always clear what "touching" actually means. Take the case of two entities which appear to be in direct contact: they could be separated a tiny little bit, and still be thought of as in direct contact. After that, we could separate the two entities a little bit further ... are they still in direct contact, or not? Taken overall, our argument is that there is no clear or absolute metric answer as to where one draws the line between direct contact and not, between local and not.

Staying with the micromatter point, how are we to determine whether one nucleus is joined together with another nucleus — and thus whether energy transfer between them is direct? We maintain that there is no simple answer to this question. Of course, there would appear to be little dispute if the one nucleus was immediately and obviously adjacent to the other nucleus, such that the two were bound together into what we would all agree is a molecule. But where is the dividing line between bound and unbound regions — direct and indirect — in this regard ... particularly in light of a discussion to come later in the chapter, whereby Quantum Theory challenges our normal assumptions about localness and distantness?

We are arguing for a very basic and even profound principle here — such that neither the apparent proximity of the two nucleuses, nor the duration for which they stably maintain that proximity, should be the determining factor in whether we conclude that the two nucleuses are bound together into a molecule, or not. Therefore, what constitutes a molecule is not necessarily self-evident. We argue that the concept of a molecule should refer not only to nucleuses which are obviously adjacent to one another, and thus "bonded" in the normal sense — but also to more apparently-detached nucleuses, yet which have a direct interaction relationship with one another ... no matter how far apart they may appear to be, according to our human-based perspective.

The upshot is that, irrespective of whether a concrete molecule can be identified or not, we would go so far as to say that such a perspective is implicit and might be described in terms of a "virtual molecule" or "functional molecule", when no actual molecule is obvious. In that case, the nucleuses would be unbounded, in a conventional sense — i.e., they exist far apart from one another; but within the context of some higher level of structure that they are part of, they are bounded together and convey energy directly between themselves.

Given that there is no precise proximity criterion for such boundedness, then in principle any infinitely distant pair of nucleuses could share between them an electron (or photon) which extends to — and thus gulfs — infinity. Note that the idea of an electron potentially spreading to infinity is something Schrödinger suggested long ago (Cf. Andrews, p. 81; Baker, p. 218; Murchie, p. 408; Cohen, p. 188). We shall return to these issues later in the chapter.

We are equal-opportunity critics — and it is not just Relativity Theory that we are arguing against here, but Quantum Theory as well. In particular, we talked earlier about that model's dismissal of the notion of direct contact, appearances notwithstanding, be it between two nucleuses or two molecules. Related to that is the emphasis on the supposed emptiness or void which characterizes much of the micro world (e.g., the atom), and which lies between two interacting entities. But, of course, we reject that whole approach and line of reasoning: voids have nothing per se to do with separation, distance, and whether an interaction is direct or not ... as we have noted. These are perspective-based determinations which have nothing to do with the issue at hand. Another way of making the point is in terms of hierarchical levels: Thus, we would argue that the region around the nucleus may be empty from the nucleus' point of view; but it is real, solid structure in terms of the next higher level, the molecule.

The hierarchical factor can also be applied to a related issue that we dealt with earlier — namely, whether individual entities represent the essence of nature, or whether it is the relationship between entities that is paramount. Our answer is that both options apply, in that the "relationship" can be understood in terms of the higher level whole, as contrasted with the "entities" which correspond with the lower level component parts. Obviously, neither higher level whole nor lower level parts level are more real or fundamental than the other.

This multi-level approach also explains why we reject another premise of modern physics, which is that only interactions that apply at certain levels or regions are determinative or real, while others are either irrelevant or tantamount to an illusion. By contrast, we maintain that no specific level (such as the nucleus) is any more real than another (such as the table). Every level of matter, at every level of nature's hierarchy, is a real unit of structure in its own right, and not just some epiphenomenal artifact. Every level of structure is just as legitimate, independent, self-determining, and solid as any other level — within that level's frame of reference. Relationships between atoms — or more properly, nucleuses, in the present model — can be either direct or indirect ... and the same goes for tables, and every other level of nature's hierarchy.

Quantum Theory's rejection of direct contact between nucleuses (or any other level of structure) on a small scale parallels Special Relativity's rejection of action-at-a-distance on a large scale. Both cases represent a failure to come to terms with the fact that such direct contact action-at-a-distance relationships exist, and that they occur over a broad range of scale, at every level of nature's hierarchy.

Within Time vs. Over Time. Now let us turn to one preeminent objection that Einstein put forward, regarding the action-at-a-distance concept — and this concerns the speed with which such influence is conveyed between two entities. Again, as Einstein stated it, "[T]here are no such things as instantaneous actions at a distance ..." After all, how could it take no time whatsoever for something to travel from one object to another?

The problem with such an analysis is that it takes time for two entities to interact only if one assumes that the two entities can interact exclusively by means of a third something which moves between them. Yet why can't two entities also interact in a direct manner, and not just indirectly by means of a third entity intermediary?

Of course, lurking beneath the issues we are dealing with here is the question of light, and the conventional model of light that we are accustomed to. In this regard, we must wrench ourselves away from a deeply-ingrained style of thinking about the nature and function of light, and recognize the dual aspect of reality here: there is a difference between direct and indirect relationships. Light in motion may be a perfectly fine example of the latter — of something which travels very fast between two entities. But the two entities can also form a direct and common higher level of unity between themselves ... and in this way also share between them other aspects of the phenomenon called light, as we shall discuss in the next main section.

So, it is true that moving light waves are real, as is the time period necessary to complete the journey: *indirect*, process-mediated interactions between entities. On the other hand, *direct* connections are just as real, and there is no active intermediary or time required to join the two entities: they interact directly — and indeed, *instantaneously*. In the latter case, there is a direct linkage between the two entities themselves: it takes no time to travel or be transmitted between one partner and the other, because the two are effectively the same, in the sense that together they form one larger whole.

Clearly we are not talking about superluminal, infinite, or any other kind of velocities. We are not talking about something traveling faster than light — because we are not talking about something traveling at all! And in that case there would be no motion, no speed ... and by the way, no time duration taken by virtue of which absolute simultaneity must be relativized or denied. There is no speed involved because there is no distance involved: there is no separating medium-terrain to be gulfed. The traveling wave is gone, the moving light is gone, the linking motion and the speed are gone — but the connection remains. Properly speaking we are then concerned with an *intraaction within* a single entity, rather than an *interaction between* two entities. That is why such an influence is instantaneous.

Consider the following comment by Bridgman, which we take exception to: "There is no instrumental method of distinguishing between pure action-at-a-distance and action through a medium." (Bridgman, 1983, p. 157) We would argue that one way of making such a distinction is precisely in terms of whether interactions are instantaneous or over time, as we have been discussing ... and will return to below.

The main patterns we have been examining can be understood primarily in terms of our stasis-motion dimension of analysis. In general, when we speak about two entities that are interacting directly, we are describing a static relationship — whereas if we speak about two entities that are interacting indirectly, we are referring to a dynamic relationship. We will discuss these issues at greater length presently, but the point here is that while the indirect interactions of a dynamic relationship take time — just as Einstein argued — the direct interactions of a static relationship do not: they are instantaneous, just as Newton may have believed. In this regard, note the association we made in the previous chapter, between static relationships and space-like intervals ... and that between dynamic relationships and time-like intervals. Those Relativity constructs represent another viable way of describing the two kinds of connectedness that we are discussing here.

In this regard, consider an embellishment to the summary charts we presented at the end of the previous chapter. In thinking about static relationships and dynamic relationships, one might also describe the dominant and recessive components thereof in quite common and fundamental terms: *here* vs. *there* as characterizing the static relationship or spatial components, and *now* vs. *then* (either future or past) as characterizing the dynamic relationship or temporal component. Thus:

Static Relationship	*Dynamic Relationship*
space	time
simultaneous	successive
direct and static form of contact	indirect and dynamic form of contact

Dominant	*Recessive*	*Dominant*	*Recessive*
here	there	now	then

Of course, in putting forward the above descriptions, one must also be aware of some of the confusing levels of meaning that impinge on such an analysis. In particular, we also depict the components of a static relationship as being "simultaneous" with one another — and one usage of the term is to say that two such events share a common "now". This is indeed a valid way of representing the simultaneous character of a static relationship — but in terms of the relative usage involved with the dual couplets just cited, we are intending to describe two different aspects or components of a dynamic relationship ... as in now vs. then.

Narrow vs. Broad. We have been talking about the two basic kinds of relationships or interactions, which can be called direct vs. indirect — and how a limited and singular world view has blinded physics to this duality of nature. Now we shall see that the same narrow-minded attitude applies to the issue of whether energy changes and exchanges occur continuously over a range, or only in discrete units. And as we shall see, we mean the "narrow-minded" reference quite literally here.

Once again, we are dealing with another one-sided approach to nature, here in its most extreme (yet, illuminating) form. It starts with the unitarity assumption itself — that it is science's mission to identify one or another of the ultimate conceptual singularities of nature. In the case at hand, the choice is between an X-like structural outcome of discrete, as opposed to a P-like process outcome of continuous. Yet there is a subtle bias in making the determination, by virtue of the implicit assumption that the most sharply delineated perspective yields the most desirable result — which leads to the former winning out over the latter. Therefore, underlying the one-sided conclusion concerning the discrete nature of reality is an assumption which goes to the heart of the issue that we are dealing with here — namely, that maximum precision itself is the only proper goal for science, as indisputably representing the best way of revealing the essence of nature.

The principal problem with such a single-minded view is that such sharply delineated focus only represents some one half of nature's pattern — and the other half is its very antithesis, in terms of a more diffuse background. Therefore, what we are encountering is another prime illustration of how the other side of nature is being overlooked — here, the broad or blurred side, in deference to the narrow or sharp side. Our position is that precision and narrowness must be recognized for their limited applicability — and must go hand in hand with generality and breadth. The discrete quantum must yield to the indiscrete non-quantum, as well! Energy is no more merely discrete than it is merely continuous: it is both, as the parallel dual concepts of particle and wave indicate.

We would suggest that much of the problem lies with the very allure of technology itself. Indeed, one could hardly find a better example than this, in terms of revealing how an overdependence upon such a technology-methodology for doing science can all too readily lead us astray. The well-intentioned but flawed premise is that the more precise the measuring technique is, the more revealing of nature's patterns it will necessarily be — thereby favoring the narrower measure as supreme, while treating any broader measure as an irrelevant distraction. In this regard, note that at the end of the previous main section we quoted a passage which essentially talked about how the other sciences were making changes so that they could "get with the program", in terms of the discrete model of nature which Quantum Theory puts forward. The authors go on to state that, "In part, one can explain such changes by noting that the development of better instrumentation and techniques opened up a new level of observation." (Holton and Brush, p. 445) By contrast, our point is that such "better instrumentation and techniques" only serves one side of the duality here, while it ignores and obscures the other.

It could also be argued that it is easier to measure nature's discreteness than its continuity — and that one reason why the wave-like side of nature is subjugated to the particle-like side is because all that we can ever measure directly are emissions and absorptions ... and those are relatively discrete and particle-like. In this sense, once one sets out to prove that nature is granular, then the outcome is already partly determined — because in some respects, *to measure is to structure*, and particle-like structure is (relatively) granular and discrete.

Take the particular example of how the phenomenon of light is understood: "Light propagates from one place to another as if it were a wave ... and yet ... light interacts with matter in the processes of absorption and emission as if it were a stream of particles." (Hecht, p. 260; cf. Capra, 1975, p. 136) Then notice how this dual pattern can be reduced into a unitary result: "The light only appeared as waves if one observed it over fairly long time intervals. If one could just freeze the moment, stop the picture of the steady progression of these light waves long enough, one would see that the waves were made up of tiny granules of light." (Wolf, p. 68) Once again, we have the highly-focused perspective taking precedence over the broader-range perspective — and simply assuming that the former is a more accurate or profound representation of nature than is the latter ... not to mention both together, which is the dualistic position we advocate.

One might also contend that the whole mind-set underlying the Uncertainty Principle is another indication of how the two sides of nature are not approached equitably, as we discussed in Chapter 2. Thus, to the extent that the Uncertainty Principle is understood in terms of a failure or limitation of prediction, then this is more a negation of one side of nature than it is an affirmative acknowledgement of the equally-salient other side. Indeed, as we noted in the earlier discussion, it may be that the essence of the uncertainty phenomenon (as opposed to the Uncertainty Principle) — and the other side of nature that it represents — not only defies and denies measurement and predictability, but is in some sense violated or undermined by such active intrusions. All in all, the process side of nature gets lost as it is forced into a structure side mode and mold, seeking qualities in the former that belong in the latter. This structuring of process effectively distorts half of reality — the other side of nature — if only by circumventing or squelching or denying its essence, and transforming it into what it is not ... in a sense by "pretending" what process would be like if it were structure: such an approach leaves us with a lopsided view of nature and reality.

Two Ways of Seeing. Many of the issues we have been discussing so far in this chapter can be understood in terms of two ways of being. But now, we want to clearly switch to the other side of the being-seeing duality, and talk explicitly about how these issues can be understood just as much in terms of two ways of seeing. The issue was implicit in the previous subsection, and we certainly talked about it at length in the opening chapter: indeed, we use the same title for this subsection as we used there.

According to an *X*-view of some situation, two entities are interacting directly with one another ... and there is no distance separating them. Conversely, according to a *P*-view of that situation, those same two entities are only interacting indirectly by means of some intermediary ... and there is a distance separating them. That duality would apply whether we are talking about action-at-a-distance vs. action-via-intermediary, or direct contact vs. indirect contact.

Applied to the topic of the previous subsection, this means that a narrow and precise *X*-view will reveal discrete changes and exchanges for some given situation — whereas a broad and general *P*-view will reveal continuous changes and exchanges for that situation. Then there is the parts-whole issue: If one looks closely enough at any two entities, by virtue of adopting an *X*-view, then one finds separate entity parts. Conversely, if one backs far enough away and takes a broad enough perspective, by virtue of adopting a *P*-view, then one finds a whole relationship that transcends the component entities.[8]

While circumstances of being may seem to enhance or favor one or the other state, it is also a matter of seeing ... a choice, as to which reality applies. Ultimately, of course, both perspectives must come into play, to one degree or another — which is a lesson we should have learned way back with the wave-particle duality. Nature is never merely one or the other aspect alone.

Here are some ways that we can think about the dual seeing patterns we have been describing, in complementary terms:[9]

X-view	*P-view*
static relationship	dynamic relationship
space-like interval	time-like interval
space / stasis	time / motion

local / narrow	distal / broad
direct contact	indirect contact
intraaction-at-a-distance	interaction-by-intermediary
within time / instant / simultaneous	over time / duration / successive
discreteness	continuity

In one sense, the implications of this complementary viewing approach are startling — for it illustrates anew that no solitary or unitary view will suffice. Furthermore, it describes a degree of relativism which (again) is totally incongruous with its namesake Theory of Relativity. We will develop these issues further, in the pages ahead.

Two-Sided Nature. In closing this discussion of complementary connectedness, we want to reiterate our two-sided approach — in several respects. First we will mention certain other-sided arguments and interpretations to which we might otherwise be sympathetic. Then we will touch upon the one-sidedness factor in cognitive and sociological areas.

We have criticized the conventional models' overall tendency to highlight what we call structural aspects of nature, as opposed to process ones. Now we would like to cite examples of the opposite tendency, both inside and outside of science, which stress process aspects ... that are typically overlooked or minimized. Yet even as we applaud such emphasis on "the other side of nature", that still does not alter our position that both sides must be taken into account.

For example, Capra states that, "The properties of the parts can be understood only from the dynamics of the whole. In fact, ultimately there are no parts at all. What we call a part is merely a pattern in an inseparable web of relationships.... [E]very structure is seen as the manifestation of an underlying process. The entire web of relationships is intrinsically dynamic." (Capra, 1988, p. 147). Harris: "The whole must, therefore, be a dynamic whole and its organizing principle must be self-differentiating, or self-specifying." (Harris, p. 163) Wolf says that, "Unbroken wholeness means just

8 Part of the act of observation would be the association with, or adoption of, a particular perspective (such as a spatial position) — in relation to which there is interaction with the target entity. That might consist of the observer physically moving from one position to another — or it might consist of the observer imagining such an altered position in relation to the target. The observer could even adopt a static *X*-view perspective of (or relationship with) the target (observed) phenomenon, regardless of the fact that the observer and the observed might otherwise appear to be moving vis-a-vis one another; conversely, the observer could establish a moving *P*-view perspective of (or relationship with) the target (observed) phenomenon, regardless of the fact that the observer and the observed might otherwise appear to be static vis-a-vis one another. We are of course talking about a symbolic category at which our own human species is particularly adept at manipulating — though not exclusively so, even in the animal kingdom. Note that "thought experiments" would represent obvious examples of such imaginary perspectives. Also note the correspondence of some of these descriptions with that of Quantum pioneer Niels Bohr: "[C]omplementary descriptions, as conceived by Bohr, seem also to be associated with different vantage points." (Jammer, 1974, p. 102)

9 By the way, one might well argue that in many respects the Theory of Relativity represents a large-scale *P*-view of nature, whereas Quantum Theory is primarily a small-scale *X*-view of nature.

what it meant to the ancient Greeks. You cannot analyze it. You cannot take it apart. For if you do, what you end up with is not contained within the original whole." (Wolf, p. 178) Bohm also emphasizes process and whole aspects, as representing the ultimate reality (see Bohm, 1980a) — arguing, for example, that individual entities "have to be understood as merging with each other ... to make one indivisible whole." (Bohm, 1980a, p. 138[10]) By the way, these one-sided ideas can be contrasted with Bohm's two-sided approach to the laws of nature, as we quoted near the beginning of the chapter.

Outside of science, some philosophers tend to attribute "true reality" to entities of unchanging stasis or being — e.g., Plato's higher forms. Few theories or philosophies are truly balanced in their treatment of both static and structural *being*, as well as dynamic and process *becoming*. In principle, the ancient Chinese philosophy of Taoism is so balanced, to the extent that it incorporates a dualistic yin-yang approach to nature. Sometimes, however, such alternative philosophies or religions seem one-sided in their overemphasis of process, becoming, and whole — as in notions like, "All is change, and stability is only an illusion".

Much as we laud such countervailing sentiments in championing the underrepresented other side of nature, neither side should be treated as paramount or real, while the other side is only secondary or a mere illusion. Unfortunately, the well-intentioned approaches that we just cited are often as one-sided in the opposite way than is much of modern science. Thus, process and whole and change are no more sufficient in themselves than are structure and parts and stability.

Finally, before closing this topic, let us take a very different tack on the one-sidedness issue, by briefly mentioning a couple of cognitive and sociological factors which are relevant to the issues here.

Starting with the sociological, it is noteworthy that physics has long been a strongly male-dominated profession[11] — and this one-sided gender composition could be a significant factor in the kinds of theories which have resulted.[12] Put more pejoratively, the established paradigm's approach to nature may be distorted by a male-dominated mind-set, rather than one which is more equitably balanced between the two sexes.

Gender aside, a similar kind of one-sidedness may be found in the cognitive arena. That is, in the final chapter we will talk about the dualistic roles of the two hemispheres of the brain, each of which represents quite contrary — and complementary — ways of approaching the world. Could it be that the established paradigm corresponds mostly to one hemisphere's cognitive style, at the expense of the other ... leaving us, once again, with a lopsided view of nature? In this regard, it is noteworthy that different world cultures seem to correspond with different cognitive styles; this is another point we shall return to in the final chapter.

10 See also Bohm and Hiley, "On the Intuitive Understanding of Non-locality as Implied by Quantum Theory", 1974, referenced in Zukav, p. 300.

11 Holton's 1988 book mentions "the amazingly low number of women in physics (2½ percent) in the United States ..." (p. 449)

12 Wertheim argues that, "[I]f many women came into physics ... we would start to see different kinds of questions and interests come strongly onto the agenda."

Light Connects

The framework for complementary connectedness provides a very useful template for understanding various phenomena of nature, and now we want to see how this leads to a very different conception about the character of light in particular. Our intention is both to provide a more coherent conceptual foundation for understanding the patterns here, but also to introduce some radically new ways of thinking about the underlying issues ... which go well beyond the conventional interpretation.

This discussion is oriented around light's medium domain role, which is the primary way that we understand this category of phenomena in the present model. We say this despite the fact that light also has matter domain aspects — as is clear from the wave-particle duality, which we treat as medium and matter domain referents, respectively. Nonetheless, we generally emphasize the wave aspect in our analysis.

To set the stage for our medium domain analysis, let us first review how the complementary connectedness issue would be understood in matter domain terms. Consider two objects or particles which are interacting. We can refer to these as dominant and recessive entities, and we have discussed the fact that they can have two kinds of relationships. If they are interacting directly, then they comprise a static relationship, and we can describe the whole area or frame whose boundaries they delineate in various ways — such as in terms of the classical absolute space concept. In static relationship terms, the whole frame would represent the absolute space background, relative to which the dominant and recessive partners would represent the foci. On the other hand, if the dominant and recessive entities are interacting indirectly, then they represent *sink* and *source*, respectively, and they are interacting by means of a dynamic relationship — which latter corresponds with some particle or object that is moving upon the just-mentioned absolute space frame background, between the static relationship's dominant and recessive partners. Such motion would be understood as dominant and recessive aspects of a dynamic relationship — e.g., representing the *coming in* to the sink and the *going out* from the source, respectively.

The medium domain description would be very similar, but we will spell it out so that the parallel is clear: there would also be two entities that are interacting — which we might think of in terms of charge or polarity phenomena (as corresponding with their matter domain object or particle counterparts). Once again, we can refer to these as dominant and recessive partners, and they can have two kinds of relationships. If they are interacting directly, then they comprise a static relationship, and we can describe the whole area or field whose boundaries they delineate in various ways — such as in terms of the classical ether medium concept. In static relationship terms, the whole field would represent the ether medium background, relative to which the dominant and recessive partners would represent the foci. On the other hand, if the dominant and recessive entities are interacting indirectly, then they represent *sink* and *source*, respectively, and they are interacting by means of a dynamic relationship — which corresponds with some entity that is moving upon the ether medium background, between the static relationship's dominant and recessive partners. Again, such motion would be understood as dominant and recessive aspects of a dynamic relationship — e.g., representing the *coming in* to the sink and the *going out* from the source, respectively.

But we need to elucidate this medium domain further, in order to establish it as distinct from its matter domain counterpart. Thus, recall that we normally think of the basic stasis-motion relativity principle — which says that motion only exists in relation to stasis — in terms of one thing (e.g., an object) that is moving, in relation to some other thing (e.g., a reference object) that is static. But that is a matter domain mode of thinking, and now we are concerned with the medium domain. Here we are not dealing with objects, but with waves. So what are the wave counterparts which are either static or moving? Very simply, as we have mentioned before: standing waves and traveling waves, respectively — which, just like their object counterparts, only exist in relation to each other. As a result, the direct interaction static relationship would be understood in terms of a standing wave background medium substratum. The indirect interaction dynamic relationship would be understood in terms of a traveling wave which moves thereupon.

Now we want to consider some variation and elaboration of this basic interpretation, as applied to light. To start with, there is another way of thinking about the two types of wave categories, and their relationship to the kinds of light phenomena that we all experience. This takes us back once again to the pervasive figure-ground theme — and we associate a different kind of light with each of those components. We can speak about a distinct *light as focus*, in contrast to a more general *light as background* — with the former acting as dynamic relationship, in relation to the latter as static relationship. Therefore, the traveling light wave functions as focus, in relation to the standing light wave as medium background.

Such a focus vs. background light distinction is readily apparent in nature — not merely in the laboratory, but experientially, as well. For example, the most basic and evident quality of a visible static relationship background light medium might have to do with the overall pattern and level of brightness-darkness (or simply, "lightness") which surrounds or provides the substratum for any given situation — as opposed (i.e., in relation) to the more specific dynamic relationship focal patterns of contrasting traveling light waves which move thereupon, and are the medium domain component that is associated with particular lighted entities.[13]

The notion of two fundamental kinds of light, focal light and background light, gains support in another sense — and that is from the perceptual apparatus which we and other animals have. It turns out that there are different sensory mechanisms for perceiving the focus of the visual field, as opposed to its background. Thus, there are two types of receptors in the eye, cones and rods. The cones are primarily located in the central region, called the fovea, while the rods are spread around the periphery. The cones operate best in a brightness background, while the rods are oriented towards a darkness background. These two types of receptors also deal with different aspects of the visual field — cones dealing with details of wavelength, sharpness and color, and thus generally with the focus; whereas rods deal with background and the whole image. One interpretation would therefore be that the cones' function is to discriminate features of focus, while the rods' function is to deal with the continuum of background.[14]

Next let us bring these figure-ground issues back to the established paradigm. In particular, we maintain it is vital that we bear in mind the focus-background dichotomy for light phenomena when assessing research and theory. Therefore, not only do we categorically reject the Special Relativity thesis that light waves are exceptional in requiring no medium, we argue that such a medium is indeed a category of light in its own right — just as the waves which exist upon that medium are a different category of light. So when physicists talk about light as the conveyer of information about something, they may be speaking in the typical one-sided sense of light as focus. This is another reason why we should not accept Relativity's picture — the notion that we visually interact with the world solely by means of a distinct light, corresponding with *c*: such a description would represent only the focus of our (visual) perception, and tell us nothing about how we perceive the background, or what its role is ... which role we are trying to explore in this discussion.

In that regard, we can now approach the basic question of light's speed in a new light, as it were. This picks up on the point we made in the subsection above, regarding time factors that would be involved in the two kinds of interaction. Thus, in the case of a direct action-at-a-distance relationship, the two component partners form a static relationship with one another; whereas in the case of an indirect action-via-intermediary relationship, the two component partners interact by means of something that is moving between them — a dynamic relationship. The important point to reiterate here is that direct static relationships are instantaneous or simultaneous, whereas indirect dynamic relationships are over time or successive (as represented in speed, a la *c*).

13 Now for some added complexity to the issues we are discussing, which we footnote here so as to not further confuse the main discussion. To start with, it is worth clarifying that we are talking about background vs. focus aspects of medium domain phenomena here — static vs. dynamic relationships, respectively. That is different from a distinction we made in Chapter 2 between a matter vs. medium domain analysis which we characterized there in terms of lighted vs. light, respectively. The latter "light" term represents a collective (or nested) category for the issues being discussed here, which we are now breaking down into that of background (lightness) vs. focus (light) — so obviously, the multiple meaning wording gets confusing! A second side point is that these focus vs. background patterns might become more complex when the inside-outside dimension is taken into account. That is, one might speculate that in the closed field (i.e., the inside realm), the general standing wave background medium would correspond with "lightness" or "brightness"; whereas in the open field (i.e., the outside realm), the general standing wave background medium would correspond with "darkness" or "dimness". Upon these respective brightness and darkness medium backgrounds would be the countervailing wave type as focus — of darkness traveling waves as focus upon a brightness closed field standing wave background, and of brightness traveling waves as focus upon a darkness open field standing wave background. For example, we could associate daylight and other lighted conditions with the closed field, in which darkness is the focal traveling wave entity which one looks at, as against the brighter standing wave background medium. By contrast, we could associate nighttime or other darkness conditions with the open field, in which brightness is the traveling wave focal thing one looks at, as against the standing wave darkness of the background medium. A final side point here is that one might treat phenomena like a shadow or reflected image as simply amounting to a higher level representation of the main lower level pattern. The apparent vs. real distinction that is inherent in the Plato's cave model might also be understood in these terms.

14 Here is another interpretation, which is consistent with the second point of the previous footnote: cones are best adapted to closed fields that we describe as having a brightness background, whereas rods are best suited to open fields that we associate with a darkness background.

In the direct, standing wave static relationship, we are talking about instantaneous relating between one portion of the standing wave and another — between (same) entities which are bound together into "local" or "short-range" intraactions ... *within* the same larger whole. In this case of a relationship whole, there is no "speed" involved — because there is no distance involved, which would entail a duration of time to be traversed: the standing wave does not "go" anywhere, but rather represents the larger whole medium which the two interacting partners delineate the boundaries of, as we have described.

By contrast, in the indirect traveling wave relationship, we are talking about movement between one static relationship partner and another — between (different) entities which are unbound, and are thus connected only in a distal or long-range sense ... *between* one entity and another. The two entities exist separately as parts in their own right, and there is distance between them — and consequently time duration is required for them to interact, and so speed of interaction applies. In this case, something must move between the source and sink partners, upon the medium-terrain which separates them, in order to convey influences from one entity to the other.

The implications of this distinction are obviously profound, in terms of providing a completely different framework for understanding the character of light: given that there are both static and dynamic relationships for light (as for everything else), then there are both instantaneous and over time light interactions, as well. In this regard, it is also important to reiterate the point we made earlier regarding apparent distance — that there is no *a priori* criterion for differentiating direct contact or short-range local interactions, as opposed to indirect contact or long-range distal interactions.

At this point, we would like to summarize some of the issues we have been discussing. Obviously, there is overlap with the complementary chart we presented earlier, regarding two ways of seeing — which viewing factor we also include here:

Static Relationship	Dynamic Relationship
ether-medium-substratum	wave
stasis / standing wave	motion / traveling wave
light as background	light as focus
space (-like interval)	time (-like interval)
intraaction-at-a-distance	interaction-by-intermediary
direct contact	indirect contact
local	distal
within time / instant / simultaneous	over time / duration / successive
same one entity	different two entities
source-sink whole	source-sink parts
discreteness	continuity
X-view	*P*-view

This completes our discussion of the complementary ways that light connects. Our larger goal has been to eliminate the uniqueness of the light category in these regards, and to subsume it into more general and coherent patterns of nature.

Physics Connects

In this next section, we want to look at two different sets of theoretical conjecture that have emerged from physics, which relate to the issues we have been discussing. One set has been transformed into actual experiments, where it has been validated; the other set remains in the realm of abstraction. Both can be understood in terms of the interaction of complementary partners, and both provide interesting corroboration for — or at least overlap with — the pattern of complementary connectedness that we have been discussing.

Einstein-Podolsky-Rosen Experiment. As we have already suggested, Albert Einstein was never comfortable with Quantum Theory, which he felt "avoids reality and reason." (Jammer, 1974, p. 188) In 1935, Einstein put forward a thought experiment with Podolsky and Rosen which had the intent of discrediting Quantum Theory — in particular, by demonstrating that theory's incompleteness in describing nature. Yet as we shall see, the upshot was that Einstein put his foot in the mouth of his own theory, and the EPR experiment ended up being a thorn in Einstein's side to his dying days. Indeed, we would argue that the impact of this experiment was at least as great as Michelson-Morley — except that the older experiment was taken as confirming Einstein's approach to nature, whereas this newer one undermines it.

Two key issues are raised by the experiment. First, Quantum Theory appeared to allow a phenomenon which occurred in one place to be conveyed instantaneously to a phenomenon in a completely different place. Second, the underlying methodology of such an experiment appeared to allow one to obtain precise quantities for conjugate variables (such as wave-particle) — which would thereby contravene the essence of the Complementarity and Uncertainty Principles. We will consider this second issue at the end of the present subsection.

Starting with the first point, though, how was such a proposition even conceivable, in conventional terms? How could two spatially-separated events be clearly and immediately associated with one another, even over a vast distance, without thereby contravening the Theory of Special Relativity's stipulation that nothing can travel faster than the speed of light? Such instantaneous transmission would appear to violate our usual sense of order in the universe — which can be described in terms of another Einstein premise, called the *principle of local causes*: what happens in one spatial locality cannot depend on what happens in another spatial locality which is physically removed from the first — spatially-separated events cannot directly influence one another.[15] Consequently, the presumption was that, "[T]here is no such thing as immediate, instantaneous action at a distance ..." (Capek, 1988, pp. 99-100; see also Davies, 1977, p. 54)

In order to understand the startling epistemological and theoretical propositions which emerge from this experiment, let us now look more closely at what would be demonstrated in practice. We can do this concretely, because long after it was first proposed as a mere thought experiment, technological sophistication eventually progressed to the point of its actual execution. In fact, there have been a number of ways in which the

EPR thesis has been tested; and it has been verified in numerous cases. Among the best known pioneering studies were those conducted by Alain Aspect and his coworkers in Paris, first published in the early 1980's. The essence of this research can be briefly described as follows.

Light pulses are emitted from a central point and radiate outwards. At the outlying points *A* and *B*, equidistant from the origin, are light sensors. Not all the light which is emitted from the source reaches these sensors, however, because of the experimental manipulations which are involved. The mechanism that is used for this purpose is called a polarizer, which is a device used to block out various angles of light — such as in polarized glass, which is often used in sunglasses (as we mentioned in Chapter 3) to eliminate glare — thereby allowing only certain angles of light to pass through. Polarizers are used both at the source, as well as at the targets (positioned just before the sensors). If the angle of the light which travels outward from the source polarizer is compatible with the angle at which the target polarizer has been set, then the light will pass through to the sensor; otherwise, not. Moreover, since both beams of light have been initially polarized in the same orientation; and because both of the target polarizers are set identically with one another; then if one of the two light beams passes through its target polarizer (e.g., *A*) to the detector beyond, then so will the second light beam pass through its target polarizer (e.g., *B*) to the detector beyond. Here is a diagram of how the two sets of polarizers might be arranged between the source and the detectors:

Target Polarizer *A*	|———————— O ————————|	Target Polarizer *B*
	Source Polarizer	

There is nothing particularly revealing which emerges from the simple layout which we have so far described: the interesting part of the experiment concerns subtle variations of this basic design. For example, the light pulses may leave the source having not the same, but opposite, polarity to one another; or the two target polarizers may be set at different angles from one another. Since there are many combinations amongst the various angles of polarization, there will also be various likelihoods that the polarized light will actually pass through the target polarizer — thus making the available permutations exceedingly complex.

But the most significant outcomes concern those cases where the result is not clearly predictable — i.e., leaving aside cases where the angle of the target polarizer

15 Other descriptions of the EPR experiment and its outcome are framed in terms of "Bell's inequality" — a theorem which has been proposed in different versions since 1964, to mathematically prove "that if the statistical predictions of Quantum Theory are correct, then some of our common-sense ideas about the world are profoundly mistaken." (Zukav, p. 290) These common sense ideas concern our views of reality and of causality — such as we are describing here. In particular, John Bell is said to have proved that any formalism which is consistent with the results of Quantum Theory must entail non-local, or faster-than-light, influences — which is to say that the principle of local causes is false. (d'Espagnat, p. 91; see also Stapp, 1982, p. 379)

will either shut out all the light or let in all the light. It is in precisely those cases where the result is almost chance-like that the underlying logic of the experiment comes into play — for then the crucial question is whether there is any correlation between the near-chance event which occurs at one target polarizer, as opposed to the near-chance event which occurs at the other target polarizer.

The extraordinary conclusion of the experiment is that *there is a significant correlation between these two supposedly-separate chance-like events*. "[T]here is a definite degree of cooperation between these two polarizers ..." (Davies, 1981) Somehow, the polarizer at *A* "knows" what orientation the polarizer at *B* is at — for the two are acting "in collusion," as it were. Perhaps most significantly, Bell concluded that, "the signal involved must propagate instantaneously." (Bell, quoted in Wolf, p. 204)

The question for physics is how such an instantaneous influence is to be explained. How can something which occurs in one place influence something else which occurs in another place, yet by means of a mechanism that physics cannot readily identify — and in an instantaneous manner? Such results seem to clearly conflict with the basic principle that one cannot transmit information faster than the speed of light, and thus violate the principle of local causes. Yet how can this "impossibility" be not only possible — but actual?

The two interacting entities are sometimes described as being separated in a *space-like* manner, which we discussed in the previous chapter. Recall that space-like relationships convey information in an instantaneous fashion — and that for this to happen over a distance contravenes Einstein's model:

> Communication between space-like separated events defies one of the most basic assumptions of physics.... that nothing in the universe can travel faster than the speed of light.... Yet this is exactly what the Einstein-Podolsky-Rosen thought experiment seems to illustrate. (Zukav, p. 287).

Therefore, the problem regarding the entities in this experiment is that, "according to the laws of physics, they *cannot* be connected (if they really are space-like separated), because the only way they can be connected is with signals." (Zukav, p. 294)[16] What looms before physics, then, is "the possibility of a fundamental unity lying deeper than quantum theory and relativity, a fundamental unity which somehow allowed faster-than-light connections between separate 'parts' of physical reality." (Zukav, p. 294) The conundrum which physics is facing here is sometimes dramatically described in terms of "spooky action at a distance".[17]

Now let us consider some of the different explanations which have been put forward to explain the outcome, in a manner which does not threaten the Special Theory.[18] We shall see that, however they are worded, the essence of these responses is to simply downplay the faster-than-light influence which takes place in these experiments, and

16 Healey describes the issue in terms of "the general metaphysical principle that there is no instantaneous action at a distance. In the context of relativistic space-time structure, this principle may be formulated as follows: If two events ... occur at spacelike separation from one another, then there can be no direct causal connection between them." (Healey, p. 51)

to instead propose various convoluted and ad hoc explanations about no "real signal" traveling and no "real information" being exchanged. In this way, the instantaneous, non-"local" influence is said to represent some more subtle phenomenon which does not truly violate the letter of Special Relativity. For example, Sarfatti differentiates speed-of-light "signals" with superluminal "messages" (or transfers of information) — the latter somehow going "beyond space-time" (Zukav, p. 298).

For his part, Davies acknowledges that this EPR result

> seems to conflict with the principle that you can't transmit information faster than light. But in fact when you look at it closely, because everything in Quantum Theory is unpredictable, you can't actually use this device to transmit a message, so that although it seems to violate the spirit of Relativity, it doesn't actually violate the letter. (Davies, 1981)

Jammer writes that:

> It is [the quantum mechanical potential] *U* which transmits instantaneous disturbances from one particle to the other through the medium of the field. Since this disturbance does not increase information it is not a signal and hence its speed, although exceeding that of light, does not violate the principles of relativity. (Jammer, 1974, p. 285)

Coveney and Highfield likewise assert that these experiments do "not violate the theory of relativity since there is no *information* being transferred between the two regions — the non-local correlations in the measurements are only manifest when data are compared *after* the experiment has been performed" (Coveney and Highfield, p. 327).[19]

So much for consistency, coherence, and credibility in our models! There remains to this day no satisfactory explanation by Relativity for this sensational effect — and like the wave-particle duality which was demonstrated in the double-slit experiment, the results of EPR call into question the presumptions of classical physics, as well as our normal expectations of logic and causality in the natural world.

In contrast to Einstein's model, then, "The essence of quantum mechanics seems to demand a nondynamic 'action-at-a-distance' operating faster than light." (Zukav, p.

17 Some aspects of the lab experiments we cited near the end of Chapter 2, sometimes discussed in terms of Schrödinger's cat, provide other examples of such "spooky action at a distance".

18 There are parallels between the kind of bizarre explanations being put forth here in regards to the EPR experiment and those put forth to explain the twins paradox that we discussed in Chapter 7 — anything to salvage Special Relativity.

19 Here is another convoluted phrasing: "A principle characteristic that distinguishes a 'signal' from a general influence is that a signal can be controlled by human choice.... [T]he faster-than-light action associated with the quantum event or act cannot be used to transmit a signal..... This instantaneous change in the knowledge of the human observer evidently does not correspond to any faster-than-light physical action. Therefore, the faster-than-light actions occurring within the Heisenberg ontology are dismissed as spurious or nonphysical.... [T]he basic postulates of relativity theory demand only the nonexistence of faster-than-light *signals*, not the nonexistence of faster-than-light influences.... In a nondeterministic context, such as quantum theory ... it is possible to have faster-than-light influences that cannot be faster-than-light signals." (Stapp, 1988, pp. 44, 45 and 55)

225) And in regards to the above kind of tortured arguments in defense of Einstein's model here, Popper is clear in his rejection: "It is often implied by physicists that only the possibility of *sending signals* whose velocity exceeds that of light would refute special relativity. This, however, is not correct. The moment we say that two distant events are simultaneous in an absolute sense we have abandoned Einstein's relativistic interpretation of the Lorentz-Einstein formalism." (Popper, 1982, p. 20) Popper then goes on to summarize Einstein's conclusion, "that the principles of quantum mechanics, at least in the Copenhagen interpretation, are incompatible with this principle of locality, and that if quantum mechanics is true as interpreted by Bohr there must be action at a distance." (Popper, 1982, p. 21)

Enter the present model, and our humble approach to these issues. We would argue that the confusion which conventional physics encounters with this experiment is based upon a)a failure to comprehend and fully implement a principle of complementarity, and b)a limitation of scale in our thinking. We have discussed both these issues, and can merely restate them now.

To start with, of course, there are complementary modes of connectedness in nature, as we have been discussing in this chapter. Static relationships are direct, local, and instantaneous: it takes no time for influences to be conveyed from one portion to another portion because no traveling is actually involved, inasmuch as they comprise a common whole. By contrast, dynamic relationships are indirect, distal, and over time: something is traveling from one separate entity to the other — e.g., by way of the transmission of an intermediary, whereby traveling takes time and therefore occurs at some speed (e.g., *c*).

In that simple complementary framework, there is no paradox at all which surrounds the EPR experiment: the mysterious aspect of this study can be understood in terms of a static relationship interaction — one that is "space-like", in Relativity nomenclature. Therefore, we can also dispense with explanations regarding a fundamental failure in our notions of causality — or even (perhaps) of the principle of local causes. The core of those concepts remains valid: such functionally-redefined concepts like "local" retain their basic meanings — it is just that their range has been expanded. That brings us to the second point above, concerning a limitation of scale in our thinking: the crux of this problem lies in our implicit definition of concepts like "single physical object", "direct", and "local". As we discussed at length earlier in the chapter, these constructs are based upon arbitrary and presumably human-oriented criteria about the difference between local and distal which are unsatisfactory.

That is because any such factor of apparent scale is irrelevant here: it makes no difference whatsoever how we humans gauge the area which lies between two interacting entities, such as an objective measurement factor would conclude: that proves nothing about which of the two kinds of relationship obtains. Thus, who is to say that two entities are "spatially separated" by a distance — as opposed to being joined by a common whole into one locality? How do we determine whether we have two different physical entities interacting with one another — as opposed to their representing different portions or regions of one and the same single larger entity intraacting within itself? As we stated earlier in the chapter, our position is that the proper criterion is purely a functional one: No matter how close two entities may seem to be to one another, there is distance between them to the extent that they act as separate entities. Conversely, no matter how distant two entities may seem to be from one another, there is no effective distance between them if they act as a single whole unit. Thus, Davies makes the point that two interacting entities may be "millions of miles apart" (Davies, 1980a, p. 119). Or as Coveney and Highfield put it, "Two quantum particles, in widely separated regions of the universe, somehow constitute a single physical reality." (Coveney and Highfield, p. 136)

The non-complementary part of the solution here, then, is to reexamine some of our implicit ideas about separation, distance, localness, and so on — in order that they are not misapplied to situations where they did not belong. What is required is a shifting of our scope and a broadening of our vision — rather than the traumatic wrenching of our fundamental concepts of space and time, in the way that Relativity demands. Thus the term "local" must be understood as having no specific metric attached to it: as we keep repeating, if two entities interact as though they are local to one another, then they *are* effectively local to one another — whether they appear as such according to our usual human perspective, or not.

Of course, there is a temporal difference between the two categories of relationships. Thus, causality presumes chronological sequencing or succession — which we maintain aptly describes functionally distant (and dynamic) relationships, not local (and static) ones. The latter category can be understood in terms of the opposite characteristics to that of the former — therefore, of simultaneity or acausality ... of one larger object intraacting locally within itself, rather than two different objects interacting over some distance. As far as the proposition of superluminal communication, we state again that the local effects we are speaking about do not require traveling faster than light, because they do not entail traveling (and thence, speed) at all.

For example, is there anything strange in the idea that different parts of a single physical object can affect one another in a direct and instantaneous manner? Of course not — this is precisely what a principle of *local* causes must allow; and, indeed, it is in such a given local region that Einstein's Theory of Special Relativity would allow absolute simultaneity to still apply — even when it supposedly cannot apply across distant regions ... in which cases only a relative, light-mediated connectivity is said to exist.

Now let us briefly turn to the other main issue of this experiment, which we mentioned at the outset of this subsection. Given Einstein's resistance to complementarity and the notion of uncertainty in nature, the other purpose of the experiment was to demonstrate that conjugate variables (like wave and particle) do, in fact, exist precisely; and that the restrictions which the Uncertainty Principle quantifies are limitations of measurement, which can be circumvented. The idea was that two correlated entities which appear to be widely-separated in space can actually have their "conjugate variables" — i.e., their complementary aspects — measured precisely ... in a manner which would otherwise be disallowed by Quantum Theory.

Here is why. The part of the EPR experiment that we have so far been describing would demonstrate that the two entities are correlated with one another. By virtue of that correlation, Einstein, et al argued that we should also be able to precisely determine a characteristic of entity A by measuring the same characteristic for entity B — and conversely, determine the complementary characteristic of B by measuring it in A. In this way, we would be directly measuring only one of the two complementary characteristics for each entity — thereby avoiding the simultaneously precise direct measurement of conjugate variables in the same entity, which is what is disallowed by the Quantum Theory principles; but because of their correlation with one another, we will have indirectly measured the complementary characteristic for both. Thus, for example, one could separately measure *A*'s position with precision, and *B*'s momentum with precision. Since the position and momentum of *A* would be correlated with the position and momentum of *B* — then measuring the position of one, and the momentum of the other, would be tantamount to measuring both position and momentum for each. But because one never attempts to simultaneously measure the position and momentum of the same entity, the idea was that the Quantum Theory limitations could be overcome, and absolute precision could be claimed for both conjugate variables.

Given the two main issues of the experiment that we have been discussing, one might have expected that Einstein and his collaborators would succeed either way, in undermining Quantum Theory. Thus regarding the first, their presumption was that a "distant" correlation did not exist, to begin with ... thereby demonstrating Quantum Theory's inadequacy. But if it did exist, then the second interpretation which we have just described would cast doubt on the Complementarity and Uncertainty Principles.

Let us see how Quantum Theory addresses these two aspects of the EPR experiment. As far as the first issue goes, the orthodox Copenhagen position is that the two entities would, indeed, be correlated (as we have described above) — thereby contravening Relativity. Regarding the second issue, the argument was that a measurement performed on one entity would be tantamount to such a measurement being performed on the other entity — the correlation between the two entities would ensure that. Consequently, the Quantum Theory argument continues, such an "indirect" measuring technique would not succeed: it cannot bypass Complementarity and Uncertainty, inasmuch as any "indirect" measurement procedure is indistinguishable (in terms of the disturbance it introduces) from a "direct" measurement. Manipulating either of the two entities would simultaneously influence its distant partner — and Einstein, et al's thesis thus fails.

Quantum Theory clearly won this EPR argument with Relativity Theory, though on epistemological grounds that were hard to grasp. We are attempting to provide a meaningful conceptual framework for understanding the issues, as part of the larger coherent model that we are trying to build in this book.

Absorber Theory. In the previous subsection, we interpreted the mysterious "faster-than-light" communication pattern of the EPR experiment in straightforward terms of a static relationship. Now we look at a different theoretical framework, in which we will find that both static and dynamic relationship patterns might apply. Our discussion here will again entail influences which fall outside the realm of normal information conveyance, as it is conventionally understood.

The main theoretical framework we are examining is actually an important model of light that was put forward in 1945 by Wheeler and Feynman, and is sometimes called the "Absorber Theory". In fact the idea goes back further, to Tetrode (1922) and Lewis (1926). The essence of the proposal is very simple. Lewis wrote: "It is generally assumed that a radiating body emits light in every direction, quite regardless of whether there are near or distant objects which may ultimately absorb that light; in other words that it radiates 'into space.' " Such a proposition corresponds closely to the modern wave theory of light put forward by Hughens. Yet in contrast to that standard assumption, Lewis said,

> I am going to make the contrary assumption that an atom never emits light except to another atom, and to claim that it is as absurd to think of light emitted by one atom regardless of the existence of a receiving atom as it would be to think of an atom absorbing light without the existence of light to be absorbed. I propose to eliminate the idea of mere emission of light and substitute the idea of *transmission*, or a process of exchange of energy between two definite atoms or molecules. Now, if the process be regarded as a mere exchange, the law of entire equilibrium ... requires us to consider the process as a perfectly symmetrical one, so that we can no longer regard one atom as an active agent and the other as an accidental and passive recipient, but both atoms must play coordinate and symmetrical parts in the process of exchange. (Lewis, 1926, pp. 24-25)[20]

The line of thinking we are exploring here has never had much impact upon physics, a fact which Phipps puts down to "the dominance of causal thinking, as promoted by Einstein." (Phipps) On the other hand, these ideas seem never to have been seriously challenged, either. In the later Absorber Theory rendition, Wheeler and Feynman contend that the primary feature is that the absorber (or sink) is "an essential element in the mechanism of radiation." (Wheeler and Feynman, p. 160) Thus light will never be emitted by a source without there being a sink to absorb it: somehow the source must "know" there is a sink in order for the emission to occur.

There are two modalities of interaction in the Absorber Theory model: one is some kind of precursor interaction between or among source and sink; and the second is the subsequent emission of light itself, by the source ... to the sink. In our model, the latter category would be clearly understood in terms of a dynamic relationship — the traveling light wave which moves from source to sink. But what about the former category — how are we to understand that? Let us consider several possibilities, bearing in mind that this discussion is very speculative and abstract, and the ideas are not well developed.

One suggestion is that the prior kind of source-sink interaction is of the same type we saw with EPR, which we describe in terms of instantaneous influences between portions of a larger static relationship whole. A second idea is that the prior kind of source-sink interaction is the conventional kind of dynamic relationship interaction, in this case from the sink to the source. Then there might be a third option, which is some combination of the two. Let us briefly examine each of these, bearing in mind that we are dealing with complementary patterns in all cases.

To start with, it is quite revealing that Wheeler and Feynman speak of "action at a distance" and "field theory" as being "complementary tools in the description of nature" — which, on first glance, might seem equivalent to 2FT's static and dynamic relation-

20 We would interpret the symmetrical patterns that Lewis described more in keeping with our concept of esymmetry, whereby there are dominant and recessive sink and source partners — rather than perfectly equal ones.

ships. However, these authors do not appear to have been thinking of "action at a distance" in terms of an instantaneous transmission — even though they do use the term "simultaneous" to describe the force which the absorbing sink exerts upon the emitting source. Rather, it seems that they are working with a notion of Gauss, based upon a modified "action at a distance propagated with a finite velocity." (Wheeler and Feynman, p. 157). Davies notes as such — that in this Wheeler and Feynman model, the "direct action of particle on particle is not of the instantaneous type which characterizes Newton's theory of gravitation, but is a *delayed* action which propagates at light speed." (Davies, 1977, p. 184)

Rather than fitting with the first (static relationship) option above, it appears that Wheeler and Feynman's approach corresponds with the second (dynamic relationship) one. Basically, the theory is proposing that "advanced waves" would move from the sink to the source as a "stimulus" or instigator for the usual "retarded waves", which are then emitted back out again as regular light waves. Both phases of the Absorber Theory interaction would therefore be understood in the present model as dynamic relationships — one moving from sink to source, as a precursor to the more usual one moving from source to sink. Indeed, one way of interpreting such dynamic relationship interactions of the Absorber Theory would be in terms of reciprocal exchanges: first, the sink signals its energy (or charge) imbalance to the source; and then, the source emits the requisite energy (or charge) rebalance to the sink — a symbiotic, complementary interaction.[21]

But the Wheeler-Feynman dynamic relationship interpretation of action-at-a-distance notwithstanding, we still want to consider the more subtle static relationship interpretation thereof, as represented by our first option. In this case, there would be the EPR-type instantaneous action-at-a-distance static relationship interaction, as precursor to the normal retarded wave dynamic relationship type. In support of this interpretation, we would again cite the prior work by Lewis, in which there is a brief description of how direct contact can be extended to include the notion of "virtual contact", (Lewis, 1926, p. 26) — in a manner not dissimilar from the position we discussed earlier in the chapter, as applying with any kind of static relationship; and therein lies at least an implication of instantaneous action-at-a-distance by the receiving sink upon the emitting source, prior to the normal retarded wave itself.

21 Is the idea of seeking out a complement consistent with the notion of leveling out disparities, or eliminating non-complementarities? Calder says that electromagnetism "guarantees that any discrepancies between one part of the universe and another would be automatically ironed out" (Calder, 1977, p. 61). The radiating waves therefore disperse or "entropize" the discrepant energy. By the way, a variation of this pattern might be found in something like SHM, whereby the two partners are exchanging uniform energy back and forth in the form of a dynamic relationship between them. In the present case, advanced and retarded waves could be understood to simply alternate with one another. Note that some of these ideas bear similarities to the mechanism we discussed in Chapter 3, regarding reciprocal transformations of nucleon charge. There we were talking about oscillations within a single nucleus; but the same pattern might apply to stable interactions between nucleuses, with the two partners reciprocally swapping states with one another as they seesaw back and forth.

Finally, we mentioned a third, combination option. One aspect of this scenario would be an overriding static relationship framework, encompassing the two interacting entities. Within that, there would be the two kinds or directions of dynamic relationship interactions between the two entities, as described with Wheeler-Feynman — of a precursor advanced wave, moving from sink to source, followed by the usual portion that everyone acknowledges — of a retarded wave moving from source to sink.

So much for our discussion of the Absorber Theory itself and the Lewis predecessor. But now we want to use these ideas as a jumping off point for some general and potent conjecture — if not wild speculation — in the last few paragraphs of this subsection. To begin with, it should be clear from our Chapter 4 discussion about stasis and motion, and static and dynamic relationships, that every complementary interaction can be understood as entailing both types of patterns, in some form or degree: instantaneous and direct static relationships (or standing waves), on the one hand ... as opposed to over time and indirect dynamic relationships (or traveling waves), on the other. This would correspond with the conclusion we reached in the last main section — that there are two ways that light can manifest itself, in terms of focus and in terms of background.

In any event, one interpretation here might be very different from the patterns we have discussed, and would not entail advanced waves at all. In this case, the normal retarded wave dynamic relationship interaction would act as a "trailblazer", when it is emitted from the source: that wave eventually arrives at the sink, by virtue of the fact that the sink is the complement of the source. At this point, what was a traveling wave form of light moving from the source to the sink would become transformed into the standing wave whole which joins together the newly formed complementary source-sink whole. In these terms, we might think of radiation outwards as being energy that is in *search* of a complement. Conversely, absorption inwards would be understood in terms of the complement which has been *found*. The former traveling process takes time — but once the linkage has been established and is in place, the subsequent intraactions would be instantaneous. Then there would be nothing traveling anymore, no gulf to be traversed, no separation between partners ... just as is demonstrated in EPR.

Such an interpretation might also be framed in terms of the wave function construct of Quantum Theory that we discussed in Chapter 2. We could think of the evolving wave function ψ as corresponding with a dynamic relationship — similar to how de Broglie described his pilot wave — as the outgoing portion in search of a complement, spreading continuously from the source, outward for unknown distances. Once a complement has been found at the sink, then the interaction would invert into a static relationship, corresponding with the sudden and discrete collapse of the wave function.

Needless to say, all of this raises a whole plethora of questions about reality — which challenge our conventional ways of thinking about these matters. Does it take time for two entities to communicate — for example, over very large distances? In terms of the indirect contact action-via-medium dynamic relationship, it does; but as we have said, by the direct contact action-at-a-distance static relationship modality, it does not.

What could all this mean when we see a star at night? Of course, the established interpretation is that we see traveling light waves which were emitted from the star long ago — and consequently, that we are looking back in time when we see the distant star. Yet is that the only way to characterize such an interaction? In the terms we have been discussing, dynamic relationship traveling interactions would not exist by themselves — and the same entities may also be understood as interacting in terms of a static relationship.

All of this is heresy, of course, according to the conventional view: for it would mean that there are two different modes of seeing or interacting with the star — both in terms of light, as well as in terms of time. Thus, it may be true that it takes time for the light to reach us from the star, in terms of a dynamic relationship: yet could we also see (or otherwise interact with) the star instantaneously, in terms of a static relationship? Would that latter modality also mean that there is a common "now" which exists simultaneously between the star and us, as something which is stretched out between us — such that we are not merely looking backward in time when we see the "distant" star?

We need to have an open mind and broad horizons as we think about such issues, and break out of the narrow template that we have been restricted to for so long.

Order vs. Entropy

Properly speaking, the concept of thermodynamics relates to temperature differences. But such temperature changes really act as a barometer for energy exchanges in general, and as such the category concerns larger issues regarding order and disorder in nature. Our focus here is on the famous second law,[22] which can also be understood as a theory of time — one which can be counterposed to that represented by Relativity.

There are various ways in which the second law can be construed, but probably the standard formulation is attributed to Boltzmann (Prigogine, 1980, p. 9). This states that entropy or disorder always increases over time in any closed system, which category includes the universe as a whole. "Stated roughly, the second law requires that in any change, the universe becomes a little more disordered." (Davies, 1979, p. 157). Or, "[T]he second law of thermodynamics operates inexorably to reduce the organization of the universe to chaos." (Davies, 1983, p. 199) Put another way, "[C]haos is easier to achieve than order and hence is far more likely" (Davies, 1980a, p. 167) Even more simply, "[O]rder tends to give way to disorder." (Davies, 1977, p. 67). Or, "[E]ntropy automatically tends to increase." (Whitrow, p. 160) And in temperature terms, "Heat flows naturally from a higher- to a lower-temperature region." (Hecht, p. 193). Again, "[H]eat flows out of hot bodies and dissipates away in the cooler surroundings ..." (Davies, 1977, p. 64).

This "theory of growing disorder" (Zukav, p. 221) can also be expressed in information terms. That is because information can be understood in terms of order — in contrast to entropy and disorder. Sometimes this concept of information is also expressed

22 While we are primarily concerned with the second law here, we can briefly mention the other two laws, as well. The first is a law of energy conservation, that "in a closed system the total amount of energy of all kinds is constant." (Pitt, p. 380) This is sometimes paraphrased more generally, such that, "Energy can be neither created nor destroyed ..." (Hecht, p. 191) The third law is relatively obscure, and deals with temperature and entropy changes. Miller states it thus: "The entropy of a perfect crystal at 0 K is zero". (G. Miller, p. 66)

as negative entropy, or "negentropy," as opposed to positive entropy ... or, again, simply entropy. Obviously, as information or negentropy decreases, entropy increases.

How could we have a problem with this important law, which has been so well established for such a long period of time? Indeed, Davies says that, "The second law of thermodynamics is so fundamental to all of physics that few physicists would question its validity." (Davies, 1983, p. 199) Our objection is not so much with the law itself, as with the lack of a broader complementary framework in which the law can be understood. In particular, we maintain that the second law represents another stark instance of where one side of nature has been greatly overemphasized, at the expense of the other. Nor is there any pretense otherwise — for this is the second law's raison d'être: to represent a uni-directional trend in the universe.

More power to it in this regards, its supporters would claim, because it is thereby acknowledging — and codifying — the fact that there is a direction to the way things happen in the world: specifically, there is a direction to time. Time moves forward, not backwards — things get blended or made, rather than getting unblended and unmade. Thus:

> We are content to watch a cue ball slam into a wedge of fifteen neatly racked pool balls and see them all fly off, colliding and bouncing around. But if we see moving balls that have been crashing into each other and rebounding all at once converge and pull together into a perfect, motionless wedge with only the cue ball sailing out and away, then something very strange has happened. Disorder has gone into order, and that somehow is not the way of things — not quite the way events seem to naturally occur in time. (Hecht, pp. 197-198)

Examples such as these are used to demonstrate that time moves in one direction only — the movie of life runs forward, as it were, and not backward. Cream is poured into the coffee and then stirred ... whereas we never experience the reverse scenario, in which the cream becomes unstirred and unpoured back out from the coffee again. Davies makes a similar point with wave phenomena, noting that Maxwell's equations have solutions for both the normal outgoing waves (called "retarded") and the never-observed corresponding incoming waves (called "advanced") ... although we encountered a contrary interpretation in the preceding section. Thus if one drops a stone in a pond, the waves travel outward; but one never finds a corresponding case, where the waves converge inward. In other words, while "organised waves get emitted, only disorganised waves get absorbed" (Davies, 1977, p. 180). "But why not? Physicists have never been able to agree on a common answer to this question." (Davies, 1977, p. 84)

In discussing these issues below, we will center our objection to the second law around our twin themes of complementarity and hierarchy. Regarding the former, we dispute the main thesis mentioned above, regarding a uni-directional tendency in nature — both directions must be involved; regarding the latter, there are many levels of nested process in nature — not just one. Taken together, every level of nature's hierarchy has bi-directional aspects associated with it, never just unidirectional ones. Furthermore, we should not let a particular level, even if it be that of the universe as a whole, either confuse us about the existence of other levels or unduly dominate our thinking about those other levels.

No Symmetry. Let us look first at the unidirectionality issue, and the complete lack of symmetry that it represents. Is the second law justified in adopting such an unbalanced approach? On the one hand, the answer is incontrovertibly "Yes!", given the enormous amount of research which supports the proposition. But we would also argue "No!", in the sense that the theory amounts to a one-sided model; and such a stance is not warranted in a two-sided nature.

To start with, one must wonder how all the incredible amount of orderly structure in the universe comes about — which then gets disorderly! How does the body get hot, which then gets cold? Clearly, it is all very well to say that overall entropy increases over time, but there are also situations where entropy decreases over time. That is, for energy to go downhill, as the second law describes; for it to go from an organized to a disorganized state, as in the billiard balls or cream-in-coffee examples above; for radiation to disperse outwards as a retarded wave — all these instances, which tell us how energy is effectively lost or dispersed by an entity, also require some complementary mechanism for how energy is effectively gained by the entity in the first place.

One way of talking about these mechanisms in the present model is in terms of our breathing motif, which may be generally described by saying that energy *comes in* to some entity, on the one hand ... and energy *goes out* from that entity, on the other. In whatever terms, however, there is no disputing the fact that a buildup (or "ordering") must take place ... in order for a discharge (or "entropizing") to occur. We can also frame the pattern in vertical, up-and-down terms: something goes *up* the hill, and it also goes *down* the hill. What we maintain is that it makes no sense to overemphasize the outward or downward aspect — the exhaling, the dispersal, the "entropizing" ... without also giving adequate weight to the inward or upward aspect — the inhaling, the concentration, the "ordering". Growth is just as much a salient feature of nature as is disintegration.

To be sure, the second law does not deny the countervailing tendency towards order in nature that we are describing — only to say that between the two, the tendency towards disorder is paramount. Putting it another way, the second law states that the creation of order in any limited situation can only be obtained at the cost of a larger amount of disorder overall. As Prigogine puts it, "Order and disorder are produced together; entropy is the price of structure." (Prigogine, 1988, p. 129)

Yet there is still the obvious dilemma of how order could emerge or survive, in the face of the greater propensity towards disorder. One way of answering the question is to say that it is the force of gravitation which saves the universe from being dispersed by the entropy described by the second law: "[G]ravitation provides the ultimate and inexhaustible reservoir of negative entropy for the universe." (Davies, 1974, p. 6; see also Coveney and Highfield, p. 154) So were it not for the heroism of the universal entropy-fighter called "gravity", then the universe would already have suffered what may still be its inevitable long-term fate — the famous "heat death" which thermodynamics expects to one day overtake the universe (e.g., Pitt, p. 177). That stage will be reached when the energy in the universe is at maximum entropy, and the temperature is at the same low level (why *heat* death?) everywhere.

In the present model, of course, we do not embrace a unique force of gravity — but instead maintain that every level of nature represents its own raison d'être, its own basis of order and existence. In a corresponding manner, Prigogine talks about "self-organization" principles which effectively work against the onslaught of entropy, as a mechanism for the formation of organization and structure in the universe.

Another way of approaching the two-sidedness issue is to differentiate open and closed systems, and in this regard Davies notes that, "The law of entropy increase refers to closed systems, the law of information increase to open systems." (Davies, 1974, p. 54) Likewise, Miller is referring to an open system when he talks about the "improbable arrangement maintained at a steady state ... by dynamic homeostasis" between the two death-resulting extremes of "excessive order (rigid, no possibility of change)" and "excessive disorder (instability, too much change)" (G. Miller, p. 327). So is d'Espagnat, who wonders how "certain 'open' systems are characterised by the opposite development, namely an increase in their internal complexity and a diversification of their structure." (d'Espagnat, p. 123)

To summarize the issue, it seems clear that the very concept of growth — growth of anything — defies the spirit (though not the letter) of the second law ... for growth is the very epitome of energy being transformed from a disorderly to an orderly state. Moreover, growth and order and structure are not peculiarities: they are not exceptions, and they are not special cases. As Schrödinger puts it, the living organism "feeds on negative entropy" or order (Schrödinger, quoted in Gatlin, p. 22). Indeed, as entropy declines — and thus order increases — the more highly evolved is a living organism (Gatlin, p. 22). Or as Thomas puts it, "To stay alive, you have to be able to hold out against entropy ..." (Thomas, p. 145)

The Arrow of Time. Of course, the main retort to all that we have been saying is that it is just plain wrong: that it is all very well to speak in terms of platitudes like complementarity and two-sided process — but in fact, for the real world there is a singular direction of time. Thus when taking the largest overall view possible, things are actually moving in one direction: whether one likes it or not, there is an arrow of time — and thermodynamics properly represents this uni-dimensional trend. As Hecht puts it, "the game runs downhill, and it is the one-way whine of the universe that we mark off in seconds" (Hecht, p. 197). So let us examine this relationship between the second law and the direction of time.

First, to reiterate: the reason why the second law is primarily a theory of dissolution is because the evidence is said to be clear: the universe is expanding. And since in many respects the second law is first and foremost a theory of the universe, then its one-sided position would seem to be justified; for the evidence is one-sided and overwhelming. The universe as a whole is going down the mountain now: it is going from an orderly to a disorderly state.[23]

Various theorists take such an approach. For example, in contrasting the notion of Newton's absolute time with the more modern theories of relative time, Relativist "E.A. Milne ... argued that, whereas the Newtonian universe *has* a clock, the expanding universe *is* a clock and that time's arrow, to use Eddington's famous expression, is automatically indicated by the recession of the galaxies." (Whitrow, p. 153) Whitrow argues that, "The evidence for the validity of the concept of cosmic time is therefore now impressive." (Whitrow, p. 140) He notes in particular that, "De Sitter's [1917] world is an example of an expanding universe with a cosmic time, its rate of expansion being in fact an exponential function of this time." (Whitrow, p. 141)[24]

Still, we must challenge this standard interpretation, by asking what such a grandiose expression means — the "arrow of time" ... ? And again, is there a singular aspect of nature which this concept refers to, or does it have different meanings in different realms? In the hierarchical terms of the present model, are these theorists justified in uniquely associating time with the topmost level of the universe as a whole?

In Two-Force Theory, we associate dynamic time with process in general: Time is just a way of talking about different aspects of process. This approach has several implications for the current discussion. First, we presume that there are an infinite array of levels in nature's nested hierarchy — and that each one has a separate level of process and time associated with it (which is a point we have made before). And while the concept of hierarchy necessarily also entails the notion of hegemony, this does not deny the fact just stated — that there are separate, autonomous levels of process embedded throughout. Therefore, what happens at any particular level is not necessarily isomorphic with what happens at every other level. Another way to put this is in terms of a point we made in the previous chapter — that there is no single "time", but many different "times" — one for each level of nature's hierarchy.

Unfortunately, the second law introduces or reinforces a one-sided vision of nature, in the hierarchical terms of the present model: by observing uni-directional patterns of process at one level of nature's hierarchy, sweeping conclusions are then drawn as to the general uni-directional character of process at all other levels of nature's hierarchy — to wit, a uni-directional arrow of time alone that applies throughout nature, generally.

We maintain that even if there is a universal time which corresponds to universal expansion, such process at the level of the universe as a whole is only one of an infinite array of time processes — one for every level of nature's nested hierarchy. The fact that it is the highest level has no bearing on the issue, for the highest level characteristics do not diminish the lower levels that are nested within. That is because every level has its own dynamic relationship time metric — and in this sense, Einstein was on the right track, regarding local time. But since every lower level nests into a higher level, this means that at the higher level what were different metrics at lower levels are no longer different at the higher level: they have merged into being one and the same ... common

23 There are interpretations which represent the antithetical point of view, regarding what the expansion is associated with. Davies, for example, says: "The universe began in local equilibrium with the microscopic motions randomly arranged. The beginning was a state of disorder. Organised structure has arisen from disorder quite automatically as a result of the cosmic expansion.... The disequilibrium of the universe is therefore to be sought in its *expansion.*" (Davies, 1977, pp. 216 and 179)

24 It is worth noting that Penrose argues that the direction of time is dictated not by universal expansion, but by the force of gravity, and that even if the universe started contracting again this would not change the direction of time. (Calder, 1979, p. 147) We dispute such an interpretation, as will be made clear in the text ahead.

entity and time metric — and in this sense, Newton was right, regarding universal time. Therefore, at the highest level of the universe as a whole, there will be *one* absolute time metric *everywhere*. Clearly, such a highest level proposition must contradict Einstein's Theory of Relativity, which "showed that there is no "universal" time that permeates the universe." (Zukav, p. 145) "There is no universal time ..." (Bondi, p. 151)

So as far as the notion of a singular universal arrow of time, as would be associated with the second law, it is hard to see any substantial distinction between such a universal time and that of Newton's absolute time. Such a second law interpretation would bring us back to the classical Newtononian view, of a singular absolute time applying throughout the universe — without also having different hierarchical levels of time or process such as we are asserting, akin to distinct levels of Einsteinian local time, as we just mentioned. This brings us to the above conclusion — that it seems like such a singular universal time interpretation would be difficult to mesh with Einstein's model.

There should be nothing mysterious about the notion of a compounding nesting pattern, among different hierarchical levels, as we are advocating. Inside our own bodies, for example, our heart beats at one rate, our respiration cycles at another, our overall sleep and wake rhythms at still another, and so on. There is no end to such temporal patterns in the human body, at all kinds of levels, ranging from, e.g., nerve activity, on upward. When all these different levels (each with its own separate rhythms) are all nested together, they comprise each one of us ... as whole individual people. Likewise with various level groupings of people, and so on.

The same principle applies throughout nature. Go up in the hierarchy and we find a broader and more general unity; go down in the hierarchy and we find a narrower and more specific one — a pattern we first discussed in Chapter 1. Those different levels obviously interact, such that the higher levels are a synthesis of the lower levels — just as the lower levels must be broadly determined by the higher ones.

The conclusion which this leads to is that there is no single "arrow of time": every level has a different "arrow of time" associated with it, as we discussed in the last part of the previous chapter. And since it is axiomatic in the present model that there are also bi-directional processes which exist at every level of the hierarchy, then this in turn means that every level of nature has a regularly changing "arrow of time" associated with it — each with its own oscillating bi-directionality built in, with a frequency that is specific to that particular level ... a point we shall return to shortly.

Note the correspondence of this bi-directional interpretation with that of Feynman which we mentioned in Chapter 3, regarding forward and backward time interpretations (e.g., of electrons and positrons as inward-oriented and outward-oriented phenomena).[25] In addition, there is some additional correspondence, in terms of Wheeler and Feynman's Absorber Theory, which we discussed above. In that case, the

25 Popper makes the general point that, "[T]he direction of time may be reversed in our model world" (Popper, 1982, p. 88); he cites "the famous time-reversibility of the Schrödinger equation." (Popper, 1982, p. 89)

authors propose an incoming ("advanced") light transmission mechanism that is associated with the usual outgoing ("retarded") mechanism for light transmission.

In any event, we need a clearer answer as to what the scope of the second law actually is, given the anomalies that seem to exist. Is it essentially a theory of the universe alone, as we mentioned above — or is it also a broader theory of nature, which is able to differentiate the universe from its contents? They are not the same thing, certainly not in the present hierarchical and nesting model. Nor is the universe the only instance of a "closed system", and not all such closed systems are running "downhill".

Now let us consider an entirely different way of approaching the problem — and we do this by even denying the implicit conclusion above, which cosmologists draw about the universe-wide process. In particular, what is it about the topmost level of the whole universe which makes it appear to be unidirectional? If most other structures in nature appear to go through both phases, then why would the universe be the exception that manifests only one phase — that of the going down, which is what the second law depicts? Recall the hierarchical patterns that we mentioned in Chapter 1. There would be a very short time scale (i.e., faster frequency) at one (e.g., lower) end of the hierarchy, as opposed to a very long time scale (i.e., slower frequency) at the other (e.g., higher) end — which latter is where the level of the universe lies. The time scale for the universe as a whole is obviously incommensurably longer than anything we can relate to, from the perspective of the "lowly" hierarchical level of which we earthly creatures are part.

This must cast doubt on any one-sided conclusions which are drawn about the direction of process or time at the topmost level. Bearing in mind our low-level perspective, what basis would we have for concluding that the one expanding phase of the universe which we happen to be witnessing at present (and in which the second law applies) is the only phase — when there is every reason to think that other levels which we experience, or are aware of, have both phases? Indeed, we argue the contrary — that the universe is on one of its two legs now (outgoing or expanding); but this should not lead us to the false conclusion that there is always and forever only one direction of time for the universe level. Such an interpretation would merely represent the limitations of our own scope or perspective. Therefore, we suggest that our capacity to grasp the bi-directionality of process at any particular (e.g., very high) level may be restricted by our own (relatively) low level vantage point; and that if our vision were large or broad (i.e., high!) enough, then we would grasp the two-sidedness at the level of universe, as well … in spite of the apparent one-sidedness which may emerge from our limited perception at any given time. The universe takes quite big breaths, to be sure — but it does not just exhale: it must also inhale.[26]

We maintain that the general patterns of nature are the same at all levels, including the universe level up around the top of the hierarchy — which means that there are complementary and bi-directional ascending and descending phases throughout …

26 Perhaps the organized structure and disequilibrium which exists in the universe today arose from a previous phase of massive universal contraction — and that previous cycle is now being superseded and steadily eroded by the now-countervailing swing of universal expansion. We will return to these issues in the final chapter.

regardless of which is predominant at any point, despite the limitations of our own perspective and experience. Put elsewise, there are two arrows of time for every level of the hierarchy, each going the opposite way from the other.[27] At any given moment, either one phase or the other phase will be active or dominant — and hence will manifest itself uni-directionally during that swing of the pendulum. How quickly the two modes are seen to alternate with one another is a relative issue, depending both on the target as well as the vantage point from which one is observing. To be sure, lower levels switch more quickly, relative to higher levels switching more slowly. But there is no reason why there should be a persistent pattern of switching for lower levels and not for higher levels, regardless of whether we egotistical little humans are around to witness both modes or not.

False Symmetry. Now we want to turn to another way of talking about the issues here — stemming not so much from one-sidedness or lack of symmetry, as we have mostly been discussing so far; but on false symmetry. Our position is simple: The two modes of process which we have described are not necessarily equivalent. Inhaling and exhaling need not always have the same character — even though both aspects must exist.

One way of understanding the issue is to bring in another dimension of analysis, that of the stable-changing dimension. Our position is that while stable inward and outward processes (or upward vs. downward processes) can be understood as essentially equal or esymmetrical in both directions, this would not be the case for their changing region counterparts. In that event, the incoming (or upward) energy will be different from that of outgoing (or downward) energy. In the changing case, then, the way energy comes in or up (as with inhalation) is different from — asymmetrical to — the way it goes back out or down (as with exhalation). Put simply, breathing can be either stable and uniform, or changing and non-uniform.

We would argue that the assumption behind some of the examples that we mentioned earlier is based on the flawed symmetry conviction: it is presumed that processes should always be identical, in both directions — and the only difference would be the direction in which they (or equivalently, time) operate. This leads to the puzzling sense of dumbfoundness when we are asked to imagine situations whereby certain things should undevelop in essentially the same way that they develop; and the fact that they don't is taken as confirmation that time is uni-directional. But we maintain that such an expectation should only apply to the stable region — whereas for the changing region, buildup (or "ordering") will be different from discharge (or "disordering").

It is in the changing dynamic relationship category where we would put cases of a gradual and continuous buildup, spiraling, or synthesizing upward ... in relation to a sudden and dichotomous discharge, cascading or exploding downward. For example, a volcano builds up, and then erupts; sexual activity builds up, and then there is orgasm;

27 Thus, another intriguing way of understanding the Feynman electron-positron idea is to say that the electron corresponds with the present expanding ("forward time") universe, whereas the positron would correspond with a contracting ("backward time") universe.

a sneeze tendency builds up, and then bursts out; advanced wave radiation comes into an entity, and then there is emission of retarded wave radiation outward; nerve energy builds up, and then the axon fires; a perception or cognition builds up in the background, and then becomes focus. Of course, there is also the grand daddy scenario which we alluded to above — of the universe gradually coming together inward into what is essentially a Big Crunch, followed by a Big Bang explosion outward. Indeed, the hierarchy of nature is a veritable catalogue of such patterns — ranging from universal Big Bangs at one end to more local Little Bangs at the other. In this regard, it should also be apparent that a Big Bang at one level is just a difference of scale from a Little Bang at another level. There is no reason to suppose a different underlying principle for any of these mechanisms.

The general pattern would be that of a gradual build-up of energy leading to a critical point (which in some cases might be called a "singularity") — whereby the whole is greater than the sum of the parts — prior to a sudden inversion, or activation or explosion back downward or outward. It may be that the summing together is the peak or culmination of the binding or coalescing force which holds the partners together, cascading upward (or inward) until the point is reached (or overreached) that the forces are out of balance — and then they bounce back downward (or outward) in the opposite direction, of dispersal, etc.

The stable region would not have this significant difference between inbound and outbound energy flows. For example, what else could SHM, or the swinging pendulum, mean — but an essentially equivalent inward and outward (back and forth) motion or energy? Some oscillating energy fields would also display such reciprocal interaction characteristics.

One-sided Theories. Support for our position that thermodynamics represents a one-sided approach also emerges from arguments that neither it nor some counterpart like Relativity provides a sufficient framework in its own right for understanding nature. Indeed it is sometimes said that thermodynamics picks up where the conventional laws of physics leave off — because the former provides a direction for time, whereas the latter do not. Davies highlights the problem: "Although we are forced to conclude that the laws of physics do not themselves provide a time asymmetry, it is one of the most fundamental aspects of our experience that, as a *matter of fact*, the world is asymmetric in time." (Davies, 1974, p. 27)

By "time asymmetry", of course, Davies is talking about the "arrow of time" issue — meaning that we experience time that goes in one direction (forward), but not the other (backward). The concern he raises is that the usual laws of physics do not distinguish between these two directions, and can in principle apply in a bi-directional manner to either forward time or to backward time. As Whitrow puts it, "[I]t does not seem that the laws governing the fundamental forces and interactions in nature provide any indication of time's arrow." (Whitrow, p. 158). The issue, therefore, is whether this bi-directionality is merely a curious (though ultimately misleading) anomaly of theories like Relativity, or whether these kind of mixed time direction laws are telling us something salient about nature which is missing from our usual uni-directional time experience — as well as from uni-directional propositions like the second law of thermodynamics.

Of course, even to talk about bi-directionality at all, for Relativity, is a dubious proposition. As we saw in the previous chapter, there is no flowing time for the Theory of Relativity, of *any* sort and in *any* direction. Furthermore, of course, that theory treats time as being a mere (static) appendage of space.

Prigogine discusses the relationship between the laws of physics on the one hand, which he calls "dynamics" and in which he includes both classical and quantum theories; and the laws of thermodynamics on the other. He calls the former laws of dynamics a "physics of being", and the latter laws of thermodynamics a "physics of becoming". Thus, "dynamics describes processes in which the direction of time does not matter" (Prigogine, 1980, p. 5), which relates to Davies' point above. Again, Prigogine says that, "In classical and quantum dynamics, the fundamental laws of physics are taken to be symmetrical in time." (Prigogine, 1980, p. 12) Again, we mentioned a prime example of this earlier, with the Feynman-Wheeler Absorber Theory, as well as with the Feynman interpretation of forward and backward time with phenomena like the electron. By contrast, "[T]he second law of thermodynamics ... states that irreversible processes lead to a kind of one-sidedness of time. The positive (i.e., forward) time direction is associated with the increase of entropy." (Prigogine, 1980, p. 6)

The point Prigogine is making is that "we can relate the 'static' dynamic description with *being*; then the thermodynamics description, with its emphasis on directionality, can be related to *becoming*." (Prigogine, 1980, p. 13) He cites Poincaré as concluding that "thermodynamics and dynamics are *incompatible*" (Prigogine, 1980, p. 157), whereas Prigogine speaks of a "complementarity between dynamics and thermodynamics" (p. 196). His conclusion is that "the two formulations of the laws of nature are equally fundamental. We need both." (Prigogine, 1980, p. 212) Here is how the complementarity which Prigogine describes might be represented:

Laws of Dynamics	Laws of Thermodynamics
being	becoming
bi-directional	uni-directional

Whitrow seems to argue that the non-directionality of the "dynamic" laws of physics makes possible the very buildup of orderly structures which the second law treats as secondary: "[O]wing to the symmetry of the laws of motion with respect to both directions of time, it follows that for any arbitrarily chosen state that shows some degree of ordering there is not only a large probability that it will lead to a less orderly state, but equally there is a large probability that it itself arose from a disorderly state" (Whitrow, p. 159).

So what are we to make of all this? On the one hand we have the bi-directional time of ordinary physics, based upon "idealizations" (Prigogine, 1980, p. 215), together with a static perspective of nature — a la the no-directional time of Relativity — which yields only "being" ... as we essentially discussed in the last chapter. In contrast to this, we supposedly have the second law of thermodynamics coming to the rescue, based as it is upon a truly dynamic (and uni-directional) conception of nature, and thereby yielding the counterpart state of "becoming".

That the two approaches contrast with — or, indeed, contradict — one another vis-a-vis the treatment of time, there can be no doubt ... especially in the Relativity case. But it is not clear that the second law offers a satisfactory overall solution to these problems: If the second law of thermodynamics represents a crucial counterpart to Einstein's static model, the former is still one-sided in its overall emphasis — in a nature which is two-sided ... even if that law is accurately describing the one (expanding) side of nature of which we are currently a part. Here is another complementary chart, which is somewhat different from the one above:

Relativity Theory	Laws of Thermodynamics
static model	dynamic model
being	becoming
non-directional	uni-directional

So while we appreciate the nod to complementarity as between the two types of models that Prigogine proposed above, we maintain that neither Relativity nor the second law, either separately or taken together, leaves us with a satisfactory vision of time and process in nature.

Symmetry vs. Complementarity

We want to conclude this chapter on two-sidedness by returning to a topic that we first raised in the opening chapter, regarding the relationship between the constructs of symmetry and complementarity. As we noted in the earlier discussion, there are notable similarities and differences between these important conceptions about nature, making the outcome of a contest between them particularly interesting. That is also the case because these respective concepts so epitomize the differing approaches to nature that we have been discussing in this book — and in the present discussion, we will highlight Einstein's position on these issues.

The established paradigm treats the conception of symmetry with great reverence, as a fundamental principle — an ideal natural pattern, which enlightened theories would be wise to emulate and incorporate wherever possible.[28] Indeed, nature itself seems sometimes to be faulted for failing to reach this perfection: "If nature were all lawfulness, then every phenomenon would share the full symmetry of the universal laws of nature as formulated by the theory of relativity." (Weyl, p. 26) Moreover, Weyl touts the fact that, "[T]he entire theory of relativity ... is but another aspect of symmetry" (Weyl, p. 17). Put differently, Weyl says that, "It is the inherent symmetry of the four-dimensional continuum of space and time that relativity deals with." (Weyl, 1952, quoted in Holton, p. 383) For his part, Holton states that, "From the beginning to the

28 As Zukav puts it, "[T]he ultimate principles ... governing the physical world ... are the laws of symmetry." (Zukav, p. 157) And Clotfelter: "[P]hysicists have found that principles of symmetry seem to express the most profound and most general laws which have been found." (Clotfelter, p. 389) Or Davies: "[S]ymmetry plays a vital role in the organization of the natural world." (Davies, 1983, p. 151; cf. also Feynman, 1965, and Weinberg, in Calder, 1977, p. 185).

end, Einstein's scientific thought was pervaded by questions of symmetry and the closely related concept of invariance." (Holton, p. 388)

Einstein rejected complementarity with the same fervor as he embraced symmetry. Thus, "Einstein was quite contemptuous of the philosophy of complementarity." (Bub, p. 60) Indeed, Jammer refers to "Einstein, the anticomplementarist ..." (Jammer, 1974, p. 90) Likewise, consider Einstein's comment that, "[T]he experimentally assured duality of nature (corpuscular and wave structure) can be realised only by such a weakening of the concept of reality." (Einstein, 1961, p. 157) His conclusion was that, "[T]he disturbing dualism would have to be removed" (Einstein, 1949, quoted in Zahar, 1976, p. 242). Einstein treated the concept of complementarity as being misleading and erroneous, and in this vein, Zahar comments about "a dualism to whose removal Einstein was to devote much of his life" (Zahar, 1976, p. 242).

It is rather curious that Holton discusses Einstein's constructive use of dualistic concepts, including his application of that pattern to understanding "the work and personality of [other] outstanding scientists" (Holton, p. 379). Not only that, but Holton actually credits Einstein's "creative use of apparent opposites" in his work (Holton, p. 380). Holton goes on to praise the very manner of Einstein's treatment of dualities: "It really is the hallmark of Einstein's most famous contributions that he could deal with, use, illuminate, transform the existence of apparent contradictions or opposites, sometimes in concepts that had not yet been widely perceived to have polar character. One need only think of his bridging of mechanics and electrodynamics, energy and mass, space coordinates and time coordinates, inertial mass and gravitational mass." (Holton, p. 380)

Needless to say, we take strong exception to this interpretation — for while Holton may take about "bridging", we talk about "collapsing" or "merging together" ... in a manner which eliminated or obscured the complementary patterns, rather than illuminated them. So what follows are several complementary retorts to the sort of one-sided, hyper-reductionist distortions of nature which were Einstein's trademark — here expressed in terms of the dimensions of analysis of the present model.

There is a difference between stable and changing regions, as represented by both inertial and non-inertial frames. There is a difference between inside and outside realms, as is reflected in the distinction between acceleration and gravitation, along with inertial mass and gravitational mass. There is a difference between matter and medium domains, as described by Newton's mechanics and Maxwell's electromagnetism, or between mass and energy — and waves require mediums (not magic). There is a difference between stasis and motion, and these states cannot simply be randomly flip-flopped with one another ... nor should moving systems be essentially redefined as static systems for the purpose of proclaiming a fundamental speed of light only, thereby denying an effective speed of light which goes with it.

Then there is the profound difference between space and time — and the static essence of the former is no more salient than the dynamic essence of the latter; time is no more merely local than it is merely global; and simultaneous action-at-a-distance interactions are just as real as are those which require some duration to be transmitted over a distance. Finally, Einstein should have learned from Quantum Theory that uncertainty and dice playing exist in nature, not just certainty and determinism — and the necessity for both aspects may also provide another explanation for the fact that while perfect symmetry may correspond with the latter, it must be mitigated by the former.

Clearly, the present model depicts the world in exactly the opposite way from Einstein: complementarity is canon for Two-Force Theory; nothing is more central or essential for understanding nature. By contrast, we treat symmetry as not merely a sideshow — rather, it is a distinctly unsatisfactory depiction of nature ... and in that way a distraction, which has thrown us way off course, in terms of our conceptual evolution. Consequently, 2FT does not see nature as flawed, as implied in the Weyl quote above, by virtue of not mirroring the symmetry fallacy: on the contrary, it is the flawed premise of the pure symmetry precept that is the source of the problem. It is an idealization which does not apply to the real world, and as such actually distorts nature. Therefore, the conventional notion of symmetry is not merely superfluous, it — and not complementarity — is misleading and erroneous.[29]

This concludes our discussion of the other side — and general two-sidedness — of nature. However, most of what we have been talking about in this chapter (as well as previous ones) concerns the unbalanced *content* of our theories — together with their distorted conclusions about what constitutes the ultimate nature of reality — and the fact that they greatly overemphasize structural aspects of nature, as opposed to equally valid process aspects. But our next task is to look at the overall *strategy* that is used, in terms of outlook and methodology — in the sense of how one approaches nature and pursues the task of theory-building. We are talking about the very basis by which one reaches the interpretations and determinations contained in the theories regarding the nature of reality. In this regard, it should be obvious that a certain mind-set would apply just as much to how one does science or studies nature as it does to what one concludes as to the composition of nature ... and the two will be correlated with one another. What we have in science today is just as much a way of thinking, as it is the result of that thinking. And it is the former issue that we shall examine in the following two chapters.

29 It should be clear that while in many cases we compare two constructs (or phenomena) side-by-side in a manner which is intended to reveal their complementary relationship with one another, that is categorically not the case here. We are in no way suggesting that there is a complementary relationship between symmetry and complementarity, such that they could be understood as properly existing in conjoint terms with one another. On the contrary, one must choose between them — between complementarity and symmetry — for they can never coalesce together into any kind of unity.

Chapter 10

The Establishment Rules

As a whole, contemporary science is quite proud of itself: the technological wonders it has spawned are ubiquitous, pervading virtually every facet of modern life. A combination of deep insight by brilliant minds, ingenious mathematical sophistication, and marvelous technological tools — along with the investment of colossal amounts of time, money, energy, and material — have allowed us to expand our horizons ever outward into the distant universe ... and ever deeper into nearby layers of matter. We can predict and manipulate the outcome of events in nature to a degree that was previously unimaginable. Science, in brief, has become the dominant motif of modern existence. Indeed, as Midgley says, "[T]he idea of 'science' is now much the most influential model held up for intellectual enquiry in general." (Midgley, p. 12)

In this chapter, we want to examine some of the main precepts which guide this amazing endeavor, in terms of the ideology and methodology of the established paradigm. We look at the official doctrine in the first part of the chapter — and question these same positions in the second part, in terms of the principles which guide the present work. What results is a pro vs. con parallel discussion of several important topics.

We begin by considering the guiding theses behind these two approaches, which might be summarized in terms of positivism vs. metaphysics. Then we consider the relationship between fact and theory, as to which comes first, and what other elements may also be involved. After that, we consider the role of experiments, in the process of theory-building ... as well as whether science should be restricted to that which can be measured. Aesthetic criteria come next, in our consideration of factors which should guide our theory-building work. Then we look at the role which mathematics should play in this regard. Next up is a discussion of an important feature of modern scientific thought, which is based upon a distinction between truth and falseness, as guiding our theoretical development. We then talk about two opposing overall styles of thinking and theorizing, regarding the best way to build our theories ... followed by a discussion of the role which common sense should have on our work. Then we turn to another important aspect of modern scientific thought, in terms of the struggle between stability and change in our theories. The role of education and information conveyance comes next, followed by the question of who gets to decide on the viability of our models of nature — whether this is to be done by scientists alone, or whether the larger society has a proper input role in the process.

The Scientific Method

The term we use for titling this section can be defined in various ways. For example, one might characterize the "scientific method" as an almost exclusive stress on rigorous analysis and precise argumentation, on a refusal to acknowledge anything but the "objective" facts as they are derived in strictly-controlled observational settings, and on the gradual and laborious construction of a theoretical edifice which stands the test of time and experimental scrutiny. But probably the most generally accepted definition would also be the simplest and most realistic — and it is the one we are working with here: the "scientific method" describes the ideology and methodology with which science actually does its work, and thus is emblematic of all that conventional science represents. In this first part of the chapter, then, we look at the guiding principles and operational practice of what we have called the "established paradigm".

Metaphysics Nowhere: Positivism is All. The predominant ideology of modern science is a very pragmatic one, which goes by various names — like "instrumentalism", "logical empiricism", or "logical positivism". Leclerc refers to "the doctrine that *science* is the *positive* ... that is, the true, genuine, and certain knowledge of nature." (Leclerc, p. 56) The gist of this approach is that the purpose of science is strictly delineated by observation and measurement, and to deny any role for pursuing something deeper than that.

A very different position is represented by the doctrine of essentialism, which is based on the idea that there is a central truth, nature, or reality behind the appearance that we observe; and that it is science's mission — and thus the goal of good scientific theory — to unearth and identify such deeper meaning or essence. Karl Popper describes this deeper essence as that of "ultimate explanation" (Popper, 1968, p. 104).[1]

What we have, then, is a fundamental philosophical dispute: instrumentalism vs. essentialism. Yet in scientific practice, there is no ambiguity: the essentialist approach is given little credence — and, in many respects, it is treated as an affront to the very spirit of modern science, to be relegated to the domain of mere "metaphysics".

Popper describes his own position as representing some kind of middle ground between essentialism and instrumentalism, which he characterizes as a "modified essentialism".[2] For while "reject[ing] the idea of an ultimate explanation" or of "an *ultimate* essence of the world, I do not doubt that we may seek to probe deeper and deeper into the structure of the world or, as we might say, into properties of the world that are more and more essential, or of greater and greater depth." (Popper, 1981, pp. 195-196) Yet he "make[s] it clear that the belief in essences (whether true or false) is liable to create obstacles to thought — to the posing of new and fruitful problems. Moreover, it can-

1 At one point, Popper refers to essentialism as "the reality of *concepts*." (Popper, 1981, p. 323)

2 Popper also says, "I am an indeterminist" (Popper, 1981, p. 215).

not be part of science (for even if we should, by a lucky chance, hit upon a theory describing essences, we could never be sure of it)." (Popper, 1965, p. 107) The more general argument would be...

> that the idea of "pure" knowledge was a mistake; that there could be no knowledge in any other sense but in the sense of *instrumental* knowledge; that knowledge was power, and that truth was usefulness.... [W]hether essences exist or not the belief in them does not help us in any way and indeed is likely to hamper us; so that there is no reason why the scientist should *assume* their existence. (Popper, 1965, pp. 99 and 105)

In fact, according to the instrumentalist doctrine, "a theory is *nothing* but a tool or an instrument for prediction." (Popper, 1968, p. 59) "Instrumentalism can be formulated as the thesis that scientific theories ... are nothing but computation rules...." (Popper, 1965, p. 111) "It is the view that there is nothing there to be understood: that we can do no more than *master the mathematical formalism*, and learn *how to apply it*.... It is the view that all, or some, scientific theories are *nothing but* mathematical formalisms which have useful applications, in particular for the results of experiments." (Popper, 1982, p. 101) "[T]he instrumentalist philosophers of science ... all assert that explanation is not an aim of physical science." (Popper, 1965, p. 104) As Stace puts it, "[S]cientific laws, properly formulated, never 'explain' anything. They simply state, in an abbreviated and generalized form, *what happens*." (Stace, p. 172)

Consequently, to search for anything more meaningful than such narrow function is considered outside the realm of science proper, and for many becomes tantamount to mere metaphysical speculation. "In this conception, *science* stood in contrast to and excluded philosophy and in particular metaphysics." (Leclerc, p. 56) Stapp says that, "[A] principal theme in orthodox quantum thinking is precisely the rejection of metaphysics. The orthodox attitude is to renounce the quest for an understanding of physical phenomena in terms of basic realities and to settle for some computational rules ..." (Stapp, 1988, p. 39) Or as Cassirer puts it, "In opposition to the metaphysical idea of the explanation of nature, there now appears the more modest task of *describing* the real completely and clearly." (Cassirer, p. 114) Specifically, "The value of a scientific theory lies in its ability to predict." (Brillouin, p. 1) The argument is that any "claims about the true nature of reality are ... empirically unverifiable metaphysical hypotheses which scientists should shun, and therefore we must not view a theory as an attempt to explain what we observe." (Brody, p. 118) Davies says that, "By basing itself on utility rather than truth, science distinguishes itself sharply from religion." (Davies, 1983, p. 220)

So the orthodox view is that science's mission is not really a search for deeper understanding and truth, so much as a quest for prediction and control. Accordingly, many orthodox scientists would decry the very notion that science *means* anything at all — for them, science is simply a set of heuristic devices for making specific predictions, and thereby manipulating nature; and there is no valid basis for generalizing beyond the specific phenomenon in question to any larger description of reality. Thus, Kitchener says that the notion of "a *correspondence* between our scientific theories and an independent reality.... is not very popular among contemporary quantum physicists and explicitly disavowed by many adherents of the 'Copenhagen interpretation' " (Kitchener, p. 17).[3] He notes Wigner's view that,

> [Q]uantum physics consists only of mathematical formulas and empirical data, it concerns only how observations are correlated with other observations. Since quantum physics makes no assertions about what is *ultimately real*, it cannot be in conflict with one's metaphysics. (Kitchener, p. 17)

Or as Quantum Theory pioneer Niels Bohr puts it:

> There is no quantum world. There is only an abstract quantum physical description. It is wrong to think that the task of physics is to find out how nature is. Physics concerns what we can say about nature. (Bohr, quoted in Morris, p. 222)

Zukav states it thus:

> The main thrust of the Copenhagen Interpretation, which is the interpretation of quantum theory accepted by the bulk of the scientific community, is that the proper goal of science is to provide a mathematical framework for organizing and expanding our experiences, rather than providing a picture of some reality that could lie behind these experiences.... There is no description of reality.... There is nothing in the rules [of science] which requires that what we propose has to be anything like "reality." Physics is a self-consistent explanation of experience.... All attempts to describe "reality" are relegated to the realm of metaphysical speculation. (Zukav, pp. 307, 79, 80, and 87)

Clearly, the goal of understanding is disparaged as being outside the purview of mainstream science — which must deal only with demonstrable facts, as opposed to mushy meaning systems ... that can never be tested in the way that hard-core science demands. Rather than get involved with such endless abstract semantic controversies, real science stays focused on concrete results — and, in particular, (again) predictions. A theory which merely talks is of little use to mainstream science ... as opposed to the physically meaningful outcomes which really count.

The established paradigm treats such "metaphysical speculation" as representing some discredited system of idle and irrelevant philosophizing — which will actually lead

3 Finkelstein essentially argues that both views are valid. On the one hand he says that there need be no direct correspondence between theory and reality — that, rather, theory can act as an intermediary between the person and the reality. In that sense, the theory is a tool which allows us to perceive the underlying reality. Finkelstein sees a parallel between the way a blind person uses a cane to indirectly perceive the ground he touches. Eventually, he argues, the cane becomes a virtual extension of the person — and the person then feels in direct touch with reality through the cane. Finkelstein's position is that both the cane and the system beyond are real: "A quantum theory is also a cane. At first we perceive only its handle ... but after we grasp it more firmly, we perceive only the system beyond it. The existence of this perceptual shift makes it likely that arguments about the true nature of quantum reality, between those who perceive the ground and those who feel the cane, will go on forever." (Finkelstein, p. 79) According to the position being propounded in the present book, any tool acts as a link between user and object. The issue is how efficient the tool is, how many steps removed it is from the underlying reality object, how difficult it is to use, how many different situations the tool is useful for, and so on. Picturesque analogies like person and cane are all very well, but they should not distract us from the fact that there are better and worse cane tools.

us astray in our attempts to grasp the patterns of nature. Indeed, the orthodox position is that one should have no preconceptions or assumptions or biases about nature: one must not close off any avenues of inquiry, or foreclose any possibilities regarding the outcome in any particular case — but instead, one should have a "blank slate" approach, in which everything is possible. As Feynman puts it, "[I]t is necessary for the existence of science that minds exist which do not allow that nature must satisfy some preconceived conditions ..." (Feynman, 1965, p. 148) All that matters is the bare bones observation of the manifest reality, and the data which emerge therefrom.

Accordingly, the astute scientist must always be on guard not to slip into the murky realm of metaphysics: This is soft, flabby stuff — which represents a virtually pre-scientific stage of intellectual development, and which modern science should not get bogged down with: science is at its best by doing, and not by getting sidetracked with endless abstract speculation about "meaning-of-life" questions which can never be answered. Harris notes that the logical positivist way of thinking treats metaphysics as "neither a respectable nor even a possible pursuit." (Harris, p. 157) And Hanson writes that anyone engaging in such activities might end up being "stigmatized as 'mere philosophers.' This is the ultimate insult amongst physicists." (Hanson, 1962, p. 90)

Facts are Objective. There are two basic ways of construing the role of facts in science. One way is to say that facts are fluid — relative and subjective; and the other says that facts are fixed — absolute and objective.

According to the former, what we see in nature is dependent on the way we approach it, and what its larger framework is — including the larger conceptual-theoretical system in which we are working. Therefore, what is a fact, according to one perspective and in one framework ... is a different fact, according to another.

The instrumentalist doctrine is generally thought to have broken this broadly contextual manner of examining nature. In eschewing qualities like meaning and understanding, the established paradigm necessarily entails the view that facts are fixed, and it is the mission of science to uncover these raw facts ... in the form of precise observation and measurement. Instead of variable interpretation systems, science aims for the most basic and objective proof about what we can all perceive. In this way, a more definitive and rational structure can be built up about the nature of reality, which people can agree upon — precisely because it has been shorn of unnecessary layers of interpretation. As Heisenberg put it, "[P]hysics must confine itself to the description of the relationships between perceptions." (Heisenberg, quoted in Jammer, 1974, p. 76)

Therefore, the phenomenological fact is taken as describing some kind of independently-existing objective reality — a particular kind of "truth", if you will. "Science is said to be characterized by its use of experimentation and its objectivity; it is said to deal with facts, not opinions." (Baker and Allen, p. 2) Such hard facts stand alone — and astute, objective research will illuminate them ... thereby establishing them as cornerstones or proof for theory. The result is that, "[T]ruth and falsity are uniquely and unequivocably determined by the confrontation of statement with fact." (Kuhn, 1970, p. 80) Or as Popper puts it, "[S]cience aims at truth in the sense of correspondence to the facts or to reality ..." (Popper, 1981, p. 59; cf. also Popper, 1981, p. 263) Feynman says that if a proposed law "disagrees with experiment it is wrong. In that simple statement is the key to science." (Feynman, 1965, p. 156) Similarly, Morris refers to, "That arbiter of truth that we call 'experiment'..." (Morris, p. 3) Facts are said to emerge from experiment, and "Physics is built on a fundamental reliance on experiment as the ultimate guide to the universe." (Hecht, p. 11)

Experiments are Key. We just talked about the importance of experimentation for determining facts, and now we want to make three other points regarding this topic. One concerns the notion of testability, and the other two concern particular categories of experimentation which deserve special consideration.

First of all, another outgrowth of the positivist doctrine can be found in the notion of testability: any idea worth its salt in science must be subject to scrutiny. Sagan, for example, says that, "There must be a way of testing any valid idea." (Sagan, p. 73) Elsewhere, he speaks about things which are "essentially untestable and therefore meaningless ..." (Sagan, p. 335) Miller puts it like this: "[A]n alleged description of a possible state of affairs is meaningless in the absence of a means for testing its truth." (R. Miller, p. 531) Similarly, Feynman cites "Heisenberg's idea that you should not talk about what you cannot measure ..." (Feynman, 1965, p. 164) Morris poses the question in this way: "If physicists propose hypotheses that cannot be verified, are they still doing physics? Or are they engaging in a kind of metaphysical speculation?" (Morris, p. 112) Again, "We find no sense in talking about something unless we specify how we measure it." (Bondi, p. 65) Thus, "It seems only reasonable to define *reality* in terms of *measurements* that we can make." (Marion, p. 403). Jammer cites Campbell, in asserting that physics "is the science of measurement." (Jammer, 1974, p. 471) Popper argues, after Landé, "that we call physically real what is 'kickable' (and able to kick back if kicked)." (Popper, 1982, p. 46) Popper insists that theories "which are non-testable are of no interest to empirical scientists. They may be described as metaphysical." (Popper, 1965, p. 257)

Now we want to make reference to two kinds of experiments, about which we will raise questions later in the chapter — collision experiments and thought experiments. As far as the previous topic goes, the first category is obviously testable — whereas the second is not ... at least not in normal, concrete terms.

One particular way that the established paradigm tests its theories is with *particle collision experiments*. The assumption is that, "Using violent collisions, we can unlock new forces of nature buried in the deepest recesses of matter and expose to view for a fleeting moment a bewildering collection of microscopic inhabitants." (Davies, 1979, p. 84) Davies goes on to say that this violent collision "method is crude, but effective." (Davies, 1979, p. 154) He adds that, "[T]here is a respectable history for rendering asunder." (Davies, 1983, p. 161) His position is that,

> [B]y digging deeper into the bowels of matter, unity will emerge. All experience to date has indicated that the smaller the system probed, the broader are the principles discovered.... The belief is that as the energy of the processes of interest soars, so the unity and simplicity become more and more apparent ... [We] smash our way into this regime of simplicity. (Davies, 1983, p. 159)

By bashing together various particles, they also become transformed into other particles — about whose existence we would otherwise be unaware. These are sometimes grouped under the heading of the "subatomic particle zoo", a term that was coined by Kenneth Ford. Many different particles have been produced in this way, including whole nucleuses, although relatively few of these survive for longer than the tiniest fraction of a second.

A very different kind of experiment avoids these sort of techniques altogether: rather than build some elaborate test apparatus, why not simply *imagine* how certain phenomena might interact? Therefore, instead of taking place in the pristine conditions of the laboratory, or even out in the wide-open world, these experiments take place solely within our minds. Such *thought experiments* are seen as a convenient way to examine certain patterns which are either difficult or impossible to reproduce in the laboratory or real life. We already discussed this topic in the last part of Chapter 7, regarding their use in Relativity. But Jammer notes that it has been a widespread practice: "Whenever, in the history of physics, a new fundamental principle has been proposed, it was established on the basis of a standard thought-experiment." (Jammer, 1974, p. 96)

Aesthetic Criteria Considered. One way of talking about what a good theory of science should look like can be described in terms of aesthetic factors. For example, "Scientists talking about their own work and that of other scientists use the terms 'beauty,' 'elegance,' and 'economy' ..." (Wechsler, p. 1)

But what is meant by "aesthetics", or by that which is "beautiful"? One obvious answer is that, "What is beautiful in general and therefore beautiful in science is harmony, order, simplicity, a quality of cleanness." (Gruber, p. 123). Similarly, Davies says that, "Central to the physicist's notion of beauty are *harmony*, *simplicity* and *symmetry*." (Davies, 1983, p. 221) Popper also describes what a good theory should be striving for "in terms of simplicity, coherence with certain other theories, unifying power, intuitive appeal, and above all, fruitfulness." (Popper, 1982, p. 201)

Let us look more closely at some of these factors — namely simplicity, generality, and coherence ... desirable qualities that we have made repeated reference to in this book. We start with the first, although these characteristics are not really distinct from one another.

As with most abstract and subjective features, the quality of *simplicity* is difficult to define clearly. Popper talks about simplicity in terms of depth: "The depth of a scientific theory seems to be most closely related to its simplicity and so to the wealth of its content.... Two ingredients seem to be required: a rich content, and a certain coherence or compactness (or 'originality') of the state of affairs described." (Popper, 1981, p. 197) He also makes reference to "[t]he idea of simplicity [being] ... intuitively connected with the idea of a unified or coherent system or a theory that springs from *one* intuitive picture of the facts ..." (Popper, 1965, p. 241).

For his part, Einstein wrote about science's goals in these terms: "We are seeking for the simplest possible system of thought which will bind together the observed facts." (Einstein, quoted in Laszlo, p. 117) About the laws of nature: "I believe that these laws are logically simple, and that reliance on this logical simplicity is our best guide." (Einstein, from a 1954 letter to Bohm; quoted in Jammer, 1974, p. 219) Elsewhere Einstein spoke of "logical economy. By this last we mean the effort to reduce all concepts and correlations to as few as possible logically independent basic concepts and axioms." (Einstein, 1950, p. 95) He also equated simplicity with beauty (e.g., Rindler, p. 26) — and many scientists consider the Theory of Relativity to be "very beautiful" (Davies, 1977, p. 119).

Feynman associates simplicity with beauty and truth: He argues that ultimately, nature's "laws will turn out to be simple, like the chequer board with all its apparent complexities." (Feynman, 1965, p. 58) Regarding beauty, "[N]ature has a simplicity and therefore a great beauty." (Feynman, 1965, p. 173) And re truth, "You can recognize truth by its beauty and simplicity." (Feynman, 1965, p. 171) "[T]he truth always turns out to be simpler than you thought." (Feynman, 1965, p. 171)

Wheeler writes that, "The beauty in the laws of physics is the fantastic simplicity that they have." (Wheeler, quoted in Davies, 1983, p. 221) Davies says that, "Scientists have long held a belief in the ultimate simplicity of nature." (Davies, 1979, p. 155); and he refers to " ... the tenacious belief that simplicity lies somewhere at the heart of all natural complexity." (Davies, 1983, p. 151) Morris makes the same point: "Physical science has always been based on the idea that natural phenomena can be explained in terms of simple principles. When a theory becomes too complicated, physicists have a tendency to distrust it.... [I]t would not be inaccurate to define science as the attempt to discover the fundamental simplicity that is hidden behind the apparent complexity of natural phenomena." (Morris, pp. 72-73) Conversely: "[S]ometimes so many ad hoc hypotheses are added to a theory that it begins to look too contrived to be a reasonable explanation." (Morris, p. 40)

In the same vein, March says that, "Physicists commonly assume that nature is simple ... and nearly all physicists take it as an unstated article of faith.... If a problem turns out not to have a simple solution, physicists will come to regard it as not fundamental. When pressed to explain what he means by a 'fundamental' law, a physicist is ultimately forced to invoke the essentially aesthetic criterion of *simplicity* or *elegance*." (March, p. 28) Rogers says that, "[T]he test of good theory is not success vs. failure but remains simplicity and economy vs. increasing complexity or clumsiness." (Rogers, p. 349) Weinberg: "That's the most exciting idea I know: that nature is much simpler than it looks." (Weinberg, quoted in Calder, 1977, p. 185)

Glashow says that,

> Those who believe in the simplicity and elegance of nature are usually the ones that discover that simplicity. It's a pragmatic faith — a faith that works: that there is a simple and comprehensible universe underlying the complex and wonderful phenomena of nature. (Glashow)

Many scientists understand both beauty and simplicity in terms of mathematics. Davies: "When physicists talk of beauty and simplicity the language through which these concepts are expressed is mathematics." (Davies, 1983, p. 221) Again, he says that, "We do not understand why the laws of physics are what they are, though we can marvel at their awesome beauty and mathematical simplicity." (Davies, 1980a, p. 180) Speaking about Quantum Theory, Davies writes that its difficulty and abstractness "prevent a lot of people without the necessary mathematical background from understanding its beauty and subtleties." (Davies, 1977, p. 131) Similarly, Heisenberg referred to "the simplicity and beauty of the mathematical schemes which nature presents us." (Heisenberg, quoted in Wechsler, p. 1) Frank writes that, "People who have grown up in a mathematical atmosphere ... will find that Einstein's theory of gravitation is of incredible beauty and simplicity." (Frank, p. 211) Dirac says that, "It is far more important to have beauty in one's equations than to have them fit experiment ..." (Dirac, 1963, quoted in Davies, 1983, pp. 220-221). We will discuss mathematics itself in the following subsection.

Clearly, simplicity is linked to *generality*, in the sense that Einstein and Popper mention above: good, simple theories encompass vast content. Indeed, one of the most basic ways to differentiate goals among theories of nature is simply in terms of scope: How general is the theory? How broad a vision does it have? How much of the whole of nature can it encompass? As Feynman says, "The best characteristic of a physical law is its universality ..." (Feynman, 1965, p. 87) And: "Nature uses only the longest threads to weave her patterns, so each small piece of her fabric reveals the organization of the entire tapestry." (Feynman, 1965, p. 34) In a similar vein, Peat says that, "[T]here are meaningful patterns in nature — or meaningful processes, that are repeated at many levels." (Peat) And Rogers: "The best theory is the one that is most fruitful, economical, comprehensive, and intellectually satisfying ... [W]e aim finally at deducing our whole picture of nature from a few general concepts." (Rogers, p. 349) Likewise, Morris writes:

> Scientists generally demand that a good theory should possess a certain quality of elegance, that it should explain a wide range of phenomena with a minimum of assumptions.... A scientific theory is considered to be useful only if it can account for a wide variety of phenomena.... Significant theories always have a wide-ranging character; it is this that allows us to distinguish them from ad hoc explanations. (Morris, pp. 142 and 40)

Simplicity and generality are also connected with *coherence* — and in many respects these are different ways of describing the same thing. Thus, if we are working with simple patterns and principles, then they are also likely to be general; and if they are simple and general patterns and principles, then they are likely to fit together in coherent ways.

Of course, there are two main meanings of coherence, as we noted in the first chapter. On the one hand, we can ask how *unified* is the theory — how much does it form one whole? Or, how *organic* is the theory — how much do its parts seem to naturally and harmoniously flow with, and evolve out of, each other ... as opposed to merely seeming ad hoc, arbitrary, and fragmented in relation to one another?[4] Then there is the other meaning of coherent, as in *comprehensible*: is the theory itself *understandable*? Bronowski writes that, "[T]here is something more in the human search for knowledge than the wish to get the facts right — basic as that is. We want to feel that the world can be understood as a unity, and that the rational mind can find ways of looking at it that are simple, new, and powerful exactly because they unify it." (Bronowski, quoted in Laszlo, p. 111)

4 Labels like "ad hoc" are often applied to scientific theories in terms of *timing*: a particular explanation is said to be ad hoc if it emerges from the theory (and hence is proposed as a prediction from the theory) *after* an anomaly has arisen, rather than having been fully anticipated by the theory beforehand. However, according to the present approach, timing *per se* is not the crucial factor. Rather, an explanation is considered ad hoc if it was devised mainly to deal with the particular problem in question and if it lacks a more compelling rationale or broader context.

The Central Role of Mathematics. What is to be the function of mathematics in our study of nature? Should mathematics lead, by framing the issues and illuminating the way forward — or should it merely come with us alongside, or even follow? Does mathematics provide a criterion for determining the nature of reality, or merely elaborate the details of the reality which have been identified on other, more meaningful grounds?

The established paradigm is clear: mathematics is primary — it is a high-level system which will yield fruitful concepts and insights into the nature of reality. Indeed, it is nothing less than the key to the universe:

> As Kepler, Galileo, Descartes, Newton, and others had explicitly maintained, the physical world is a mathematical structure, which ensured that the mathematical investigation of nature furnished the full truth about nature. (Leclerc, p. 30)

For his part, Einstein sought "the only trustworthy source of truth in mathematical simplicity" (Einstein, 1938, quoted in Holton, p. 259). He writes glowingly:

> What we call physics comprises that group of natural sciences which base their concepts on measurements; and whose concepts and propositions lend themselves to mathematical formulation. (Einstein, 1950, p. 95)

> [N]ature is the realization of the simplest conceivable mathematical ideas. I am convinced that we can discover by means of purely mathematical constructions the concepts and the laws connecting them with each other, which furnish the key to the understanding of natural phenomena ... The theoretical scientist is compelled in an increasing degree to be guided by purely mathematical, formal considerations in his search for a theory, because the physical experience of the experimenter cannot lead him up to the regions of highest abstraction. (Einstein, 1954, p. 274)

Similarly, Gatlin writes that,

> Theoretical physics consists of the application of mathematical knowledge to the understanding of the physical universe.... Mathematics is simply a way of expressing concepts that anyone can understand in a way that very few can understand. The esoteric language is necessary, however, because it places at our disposal a powerful machinery which in turn gives rise to new concepts that would not have arisen without this manipulative process. (Gatlin, pp. 18 and 51)

Davies says that, "Mathematics is the secret of scientific success." (Davies, 1979, p. 3) And, "Science evolves better and better mathematical descriptions of nature." (Davies, 1977, p. 201) He goes on:

> It cannot be overstressed how important mathematics is to science in general and physics in particular. Leonardo da Vinci once wrote: "No human investigation can be called real science if it cannot be demonstrated mathematically." (Davies, 1983, p. 221)

Siu says that mathematics "is recognized as the acme of certitude among the sciences." (Siu, p. 40) Ziman writes that, "Mathematical thinking is so deeply entangled in the physical sciences that one cannot draw a line between them." (Ziman, p. 37) Laszlo, too, says that, "The more adequate and yet simple a theory is, the further removed it is from everyday concepts, as it makes increasing use of abstract mathematical concepts." (Laszlo, p. 119) And, "[T]he concepts of mathematics are not chosen for their conceptual simplicity ... but for their amenability to clever manipulation and to striking, brilliant arguments." (Wigner, 1967, p. 229)

Feynman also tells us that we really cannot grasp the essence of nature without mathematics:

> [I]t is impossible to explain honestly the beauties of the laws of nature in a way that people can feel, without their having some deep understanding of mathematics.... To those who do not know mathematics it is difficult to get across a real feeling as to the beauty, the deepest beauty, of nature. (Feynman, 1965, pp. 39-40 and 58)
>
> [M]athematics is a deep way of expressing nature.... If you want to learn about nature, to appreciate nature, it is necessary to understand the language that she speaks in. (Feynman, 1965, pp. 57 and 58)

Likewise, Davies writes:

> The neurotic fear of mathematics experienced by most ordinary people is chiefly responsible for their estrangement with physical science. It is a barrier that effectively cuts them off from a full appreciation of scientific discoveries, and prevents them from enjoying vast areas of nature that have been revealed through painstaking research. For, as Roger Bacon appreciated: "Mathematics is the door and the key to the sciences ... For the things of this world cannot be made known without a knowledge of mathematics." Davies, 1983, pp. 220-221)

Greene justifies the lead role of mathematics, because the physical limitations of real life conditions would otherwise bar the testing of theoretical advances:

> Since our theories are so far ahead of experimental capabilities, we are forced to use mathematics as our eyes. That's why we follow it where it takes us even if we can't see where we're going. (Greene, 2003, quoted in Overbye)

We pick up on the issue of whether mathematics should be primary or secondary in the second half of the chapter.

Popper's Falsifiability. One of the most influential methodological principles of modern science was put forth by Karl Popper — and it is based on a fundamental distinction between the character of truth and the character of falseness ... and a further distinction of the former between tentative and actual forms thereof.

Popper starts with the obvious premise that truth is a worthy goal to strive for — but that we may never attain it. "[I]t is not his *possession* of knowledge, of irrefutable truth, that makes the man of science, but his persistent and recklessly critical *quest* for truth." (Popper, 1968, p. 281) "[W]e are seekers for truth ..." (Popper, 1981, p. 47, italics omitted) Yet while science is always seeking after truth, we can never be sure whether *real truth* has ever been found or not. All that a good theory can hope for is that it will be accepted as a *provisional truth*, so long as it holds up to experimental scrutiny.

The distinction between these two varieties of truth is based upon Popper's proposition that there is always the possibility that a theory will ultimately fail experimentally. One important reason why such a negative outcome cannot be precluded is because it is never possible to test a theory under *all* experimental conditions which may eventually materialize. Consequently, we can only judge a theory to be provisionally true — unless and until it is shown to be false; but we can never determine a theory to have fully succeeded or be categorically true, because we can never fully anticipate (or in any case, test) every possible condition which might show it to fail. On the other hand, a theory can be categorically falsified: it need only fail under one set of conditions — put forth assertions which are clearly contravened experimentally, on at least one occasion.

Truth is a singular commodity, in Popper's system: there is no place for notions like local or relative truth — but only for the universal and absolute sort. Thus, he explicitly rejects the idea that "different theories have different ranges of application" (Popper, 1965, p. 113, italics omitted), and can be taken as true within those limited contexts, or experimental conditions.

This, of course, runs counter to the common interpretation — for example, that Newton's theories are valid for a limited set of phenomena. In that respect, Newton's theory had been accepted as true for many years, precisely because it had passed all the experimental tests to which it had been theretofore subjected. But Popper argues that Newton's theory should only have been accepted as a provisional truth, because there surely existed conditions under which it might fail experimentally — and, of course, the consensus is that it did fail eventually, in the context of conditions where Einstein's Relativity Theory was said to have succeeded and thereby superseded Newton's ... first Special, then General. Yet in the same manner, Popper maintains, Einstein's theory should likewise be accepted as only a provisional truth today, inasmuch as we cannot ever exclude the possibility that conditions will be found under which it, too, will fail — indeed, he argues that we must assume this will happen in time. Therefore, as Popper puts it, "[A]ll theories are tentative." (Popper, 1981, p. 287)

Not only that, but it is this very *potential for failure* which is a matter of principle for Popper — *the* matter of principle. For this is precisely how Popper defines a good theory of science — as one which can be falsified. Therefore, Popper's criterion for evaluating a scientific theory is eminently simple: A theory must be specific enough so that it can be explicitly *dis*proved, even though it can never be explicitly verified, or proved. As he states it, in a manner which we have partly discussed previously, "[T]here can be no statements in science which cannot be tested, and therefore none which cannot in principle be refuted ..." (Popper, 1968, p. 47) Popper contends that a methodology based upon the falsification criterion is the only way "that science can hope to learn and to advance." (Popper, 1965, p. 113)

Indeed, Popper calls the falsifiability criterion his "line of demarcation" between science and non-science (or metaphysics): a real theory of science is falsifiable, whereas a theory of non-science cannot be falsified. The former set of scientific theories are precise and yield predictions which can be clearly tested and contradicted — meaning that the theory can be disproved or falsified; by contrast, non-scientific theories are vague and do not yield predictions which can be clearly tested and contravened — meaning that such theories can never be explicitly disproved or falsified.

Overall, Popper's thesis is in general accord with the spirit of modern science, notably in its emphasis on specificity and quantitative detail: but what is different is his treatment of truth vs. falseness. Again, he argues that there are no end of cases whereby a theory can demonstrate its falseness, but there is only one definitive category of truth ... a single universal truth in the world — which we must always strive for, gradually perfecting our theories as we move progressively from one falsified theory to another. In this regard, his theory can be described as a kind of theoretical evolution: as he puts it, "The theory of knowledge which I wish to propose is a largely Darwinian theory of the growth of knowledge." (Popper, 1981, p. 261)

Free Invention. Now we come to a topic where a very particular meaning of the term "scientific method" is sometimes applied. We made reference to this definition in the introduction to this section, which can be paraphrased in this way: science works by means of observation and measurements, in order to identify objective facts — and then to synthesize those results in the form of theoretical conclusions ... which can then be refined and tested in further experiments. Thus, facts are supposed to determine the course of theory, and theory develops in a manner which allows it to incorporate a greater and greater array of facts, more and more precisely.

The essential aspect of this chain of theory-building is the *inductive* step, whereby general rules or hypotheses about nature are drawn from particular experiential events (e.g., experimental data) ... even though it also entails *deductive* aspects, in that specific outcomes are then postulated and examined experientially (e.g., tested experimentally), based upon more general hypotheses (Rogers, p. 343). Induction, then, goes from the specific to the general — generalization; whereas deduction goes from the general to the specific — specification.

Theory-building in science thus incorporates both inductive and deductive aspects. As Spencer put it, "The progress of science is duplex. It is at once from the special to the general and from the general to the special. It is analytical and synthetical at the same time." (Spencer, quoted in Popper, 1981, p. 262) Some approaches to understanding nature are said to put much greater emphasis on one aspect than the other. Pre-scientific development is usually seen as being primarily *deductive*, and is sometimes associated with Descartes ... entailing grand, sweeping hypotheses about nature — but which is not based upon specific experiential (e.g., experimental) events. Classical science, starting with Bacon and continuing through Galileo to Newton and beyond, was said to rely primarily on *inductive* procedures, whereby general hypotheses were drawn from particular observational or experiential datum.

In a crucial dissent from this overall methodology, Einstein sharply rejected the inductive method as a primary mode of theory-building.[5] In fact, not only did he strongly advocate a preponderantly deductive style, he also argued that it was a more mature strategy of theoretical development ... as opposed to representing a reversion to the pre-scientific style, which we just mentioned. Indeed, Einstein argued rather vociferously that induction was a completely wrongheaded way of doing science — the idea that one could draw from and synthesize one's experiences, including those experiences which emerge from experimental observations:

> We now realize, with special clarity, how much in error are those theorists who believe that theory comes inductively from experience. Even the great Newton could not free himself from this error. (Einstein, 1950, pp. 70-71)

> There is no inductive method which could lead to the fundamental concepts of physics. Failure to understand this fact constituted the basic philosophical error of so many investigators of the nineteenth century.... Logical thinking is necessarily deductive; it is based upon hypothetical concepts and axioms. (Einstein, 1950, p. 76)

> The predominantly inductive methods appropriate to the youth of science are giving place to tentative deduction.... [E]very attempt at a logical deduction of the basic concepts and postulates of mechanics from elementary experiences is doomed to failure. (Einstein, 1954, p. 274)

Admittedly, Einstein's wording can be somewhat confusing — such as in the very last sentence, where he seems to mean "induction" rather than "deduction" — but his position is clear.[6] Here is one important way of summarizing it:

> Physics constitutes a logical system of thought which is in a state of evolution, and whose basis cannot be obtained through distillation by any inductive method from the experience lived through, but which can only be attained by free invention. (Einstein, 1950, p. 93)

Indeed, Einstein waxes poetic about how productive this approach is, once the creative ball starts rolling — by allowing one to deduce the essential concepts from the complexity of nature:

> The scientist has to worm these general principles out of nature by perceiving in comprehensive complexes of empirical facts certain general features which permit of precise formulation. Once this formulation is successfully accomplished, inference follows on inference, often revealing unforeseen relations which extend far beyond the province of the reality from which the principles were drawn. (Einstein, 1954, p. 221)

5 Popper also rejects the "myth of a scientific method that starts from observation and experiment and then proceeds to theories" (Popper, 1968, p. 279), as a mechanism for theory-building in science.

6 Other apparent contradictions also emerge — for example, when Einstein uses the terms "logical" and "experience" in confusing ways, which actually conflict with some of the above quoted passages. Thus he says elsewhere that, "Pure logical thinking cannot yield us any knowledge of the empirical world; all knowledge of reality starts from experience and ends in it. Propositions arrived at by purely logical means are completely empty as regards reality." (Einstein, 1954, p. 271). But, again, his point is clear — that our constructs and theories should not emerge from our experiences of the world around us, but from our minds.

On the other hand, Einstein argues that if one fails to identify the essential nuggets, then the experience of nature has no value — including any general inductions which are drawn from it:

> But as long as no principles are found on which to base the deduction, the individual empirical fact is of no use to the theorist; indeed he cannot even do anything with the isolated general laws abstracted from experience. He will remain helpless in the face of separate results of empirical research, until principles which he can make the basis of deductive reasoning have revealed themselves to him. (Einstein, 1954, p. 221)

We need to emphasize the significance of the point we are dealing with here, because it is so crucial to the issues we will be discussing in the pages ahead. What we are talking about is hardly a minor philosophical dispute about the vagaries of the scientific method, or some such peripheral academic topic: rather, the issue is critical for understanding the way in which our theories about nature have developed, especially those of Einstein himself. That is, Einstein must be taken at his word when speaking about "free invention" above, for such a process clearly *did* apply to some of his most radical reconceptualizations about nature — e.g., regarding the speed of light, stasis and motion, or space and time.

More generally, there is little doubt that such "free invention" is a crucial aspect of the way the rest of the established paradigm builds its theories. We can see that now, when we talk about the arbitrary character of the conventional theory-building process, which is perfectly consistent with — and is at least partially an outgrowth from — Einstein's free invention philosophy. Here is a classic rendition of how this methodology would be described:

> [I]f physics is essentially discovery, we are entitled to the hope that in time we shall arrive at the *ultimate truth*.... [But] physics is an arbitrary science.... [based upon] a program in theory building that has become a commonplace of our day in which arbitrary abstract constructs and postulates are freely introduced in the building of physical theories.... [by] the arbitrary use of the imagination in constructing abstract structures.... (Lindsay, pp. 24, 25, and 26)

> [S]uch considerations no longer disconcert the present-day scientist and in particular the physicist, who has developed his own criteria for the success of theories and has found that theories with *arbitrarily* chosen postulates often satisfy those criteria. To him this is sufficient as a pragmatic justification of what he does. (Lindsay, p. 25)

> The clever physicist will always reserve the right to invent in arbitrary fashion the constructs he deems likely to succeed in the theoretical explanation of experience, even if this leads to rather bizarre devices for identifying these constructs with observational data. (Lindsay, p. 26)

In other words, the scientist must have " ... free choice when constructing concepts and hypotheses to serve as the bases of theories." (Lindsay, p. 24) One way of saying this is that, "We do not simply observe the laws of nature ... there is also a sense in which we create them." (Morris, citing Wheeler, p. 230)

Another way of thinking about the issues we are discussing has to do with which comes first — nature or theory. Consider some examples of how phenomena in nature are described by scientists, which concern this preeminence issue. Davies says that the electron "is prevented from falling on to the nucleus solely by the quantum uncertainty energy ... which buoys it up against the electric attraction." (Davies, 1979, p. 72) Similarly, he refers to the electron's " ... outer shells being prevented from collapsing on to the inner ones by the Pauli exclusion principle" (Davies, 1980a, p. 85). Or with the sun: "[S]o strong is the rigidity that when the sun burns out and it shrinks to a size comparable to that of the Earth, even its immense weight will be supported against further gravitational collapse by the Pauli exclusion principle." (Davies, 1979, p. 79) And Davies says that our inability to measure both position and momentum fully "is a manifestation of Heisenberg's famous uncertainty principle" (Davies, 1981).

Morris writes about how, "Heisenberg's uncertainty principle allows high-mass particles to exist for short periods of time anywhere." (Morris, p. 154) In discussing the law of gravitation, Feynman says that, "[O]ur main concentration will not be on how clever we are to have found it all out but on how clever nature is to pay attention to it." (Feynman, 1965, p. 14) Regarding thermodynamics, Hecht raises the question of "whether the second law simply describes the universe as it presently is (i.e., expanding) or whether it actually commands the universe, determining how it will always be." (Hecht, p. 197) And to repeat a quote we cited in the previous chapter, "If nature were all lawfulness, then every phenomenon would share the full symmetry of the universal laws of nature as formulated by the theory of relativity." (Weyl, p. 26)

The issue of preeminence is not minor, in our descriptions of nature; nor is it merely a question of semantics, or of our nit-picking with casual phraseology. For we maintain that something deeper is going on — which goes beyond freely inventing ideas, and then imposing them upon nature: it is also that the theories come to represent the core of reality — which nature, in turn, must conform to or emulate ... rather than the reverse.

Down with Common Sense. One particular way that the positivist philosophy manifests itself is in terms of our usual tools of logical analysis, some of which we discussed in the previous subsection, in the form of a choice between a predominantly inductive vs. deductive approach to theory-building. But now we want to look at another aspect of this ideology, in terms of that body of our individual and collective knowledge which comes under the rubric of "common sense". Science, generally — and physics, especially — rejects this category of insight, as being overly-restrictive in scope and outmoded in relevance.

One of the typical methods used by science to denigrate common sense is to cite mankind's past errors of judgment, which can thus be laid at the feet of our primitive sense of intuition. The general argument is that science continually progresses in a direction which becomes further and further removed from common sense. For example, Heisenberg stated that, "Almost every progress in science has been paid for by a sacrifice, for [with] almost every new intellectual achievement previous positions and conceptions had to be given up. Thus, in a way, the increase of knowledge and insight diminishes continually the scientist's claim on 'understanding' nature." (Heisenberg, 1933, quoted in Jammer, 1989, p. 217).

The idea is that over time, the work of science goes further and further beyond our intuitive grasp — as part of the fresh frontiers which are continually being charted, and the increasingly rarefied atmosphere in which scientists are working. By contrast, characteristics like common sense or meaningfulness are treated as essentially quaint and old-fashioned vestiges of an archaic concept system, that has not kept pace with the stunning breakthroughs which science has achieved ... and which therefore have little relevance in the sophisticated and counterintuitive new world which science claims to be unearthing.

Feynman is particularly scathing, as he writes,

> [A]ll philosophical intuitions about what nature is going to do fail.... [A]ny attempt to express nature in philosophical principles, or in seat-of-the-pants mechanical feelings, is not an efficient way. (Feynman, 1965, pp. 53 and 57)

> In the beginning of the history of experimental observation, or any other kind of observation on scientific things, it is intuition, which is really based on simple experience with everyday objects, that suggests reasonable explanations for things. But as we try to widen and make more consistent our description of what we see, as it gets wider and wider and we see a greater range of phenomena, the explanations become what we call laws instead of simple explanations. One odd characteristic is that they often seem to become more and more unreasonable and more and more intuitively far from obvious.... There is no reason why we should expect things to be otherwise, because the things of everyday experience involve large number of particles or involve things moving very slowly, or involve other conditions that are special and represent in fact a limited experience with nature. (Feynman, 1965, p. 127)

> [N]obody understands quantum mechanics ... I am simply going to tell you what nature behaves like. If you will simply admit that maybe she does behave like this, you will find her a delightful, entrancing thing. Do not keep saying to yourself, if you can possibly avoid it, "But how can it be like that?" because you will get "down the drain", into a blind alley from which nobody has yet escaped. Nobody knows how it can be like that. (Feynman, 1965, p. 129)

As far as our theories becoming "more and more unreasonable", others would agree. Zahar writes that, "Time and again Einstein asserted that progress in physics has to be paid for by an ever widening gap between sense experience and theory." (Zahar, 1981, p. 274; see also Zukav, p. 235) Capek's position is that our normal, commonsensical rules of logic are simply inadequate. He quotes Reichenbach, that "[T]he fundamentals of logic are shown to be the product of an adaptation to the simple environment in which human beings are born." (Reichenbach, quoted in Capek, 1988, p. 96) He also refers to Bergson's view "that our logic ... is a *logic of solid bodies* ..." (Capek, 1988, p. 96) And referring to Finkelstein's position, he says that "[A]fter the profound revision of the foundations of mechanics and geometry, it is now time to broaden the logic itself, which would not be possible without recognizing its empirical, that is, *non a priori* character." (Capek, 1988, p. 96)

In a similar way, Capra says that, "[R]elativity theory has taught us the same lesson as quantum mechanics. It has shown us that our common notions of reality are limited to our ordinary experience of the physical world and have to be abandoned whenever we extend this experience." (Capra, 1983, p. 90) Likewise, Oppenheimer writes that, "Common sense is wrong only if it insists that what is familiar must reappear in what is unfamiliar." (Oppenheimer, 1954, p. 75) Or consider Sagan's comment that, "Our common sense experience and our evolutionary history have prepared us to understand something of the workaday world. When we go into other realms, however, common sense and ordinary intuition turn out to be highly unreliable guides.... [There, o]ur common-sense intuitions do not count." (Sagan, pp. 19-20)

For his part, Davies says that, " ... the internal structure of the atom bears no real resemblance to the world of experience.... Indeed the microcosmos as a whole seems to be ruled by principles, some of which are quite alien to our intuition." (Davies, 1977, p. 131; Davies, 1979, p. 1) And DeWitt wrote many years ago that, "Despite its enormous practical success, Quantum Theory is so contrary to intuition that, even after 45 years, the experts themselves still do not agree what to make of it." (DeWitt, quoted in Hecht, p. 272) Morris: "In spite of its great successes, modern physics has always seemed to contain a residue of theoretical and experimental results that just don't seem to make any sense." (Morris, p. 4)

Finally, Zukav writes about the "totalitarian grasp of 'common sense' " and says that "*common sense must be wrong*" (Zukav, p. 136), noting that, "Commonsense contradictions, in fact, are at the heart of the new physics." (Zukav, p. 27)

Kuhn's Revolutions. We talked above about the views of one of the great modern philosophers of science, Karl Popper, and now we want to examine that of another, Thomas Kuhn. To reiterate, Popper's theory is based upon his falsifiability criterion — both in terms of how one defines a theory of science, as well as in terms of what keeps a theory in favor at any given time. Popper says that a theory of science must be falsifiable — and falsifiable in an absolute manner, such that it can be chucked out and tossed onto the trash heap ... thereby allowing for its replacement with a new theory which is not incompatible with the "facts".

By contrast, Kuhn proposes a very different vision for theoretical stability and change, in his seminal work, *The Structure of Scientific Revolutions*. The essence of Kuhn's philosophy is that there are two basic kinds of science — or two ways of doing science — two vehicles for the advancement of scientific theory. He calls the one *normal science* and the other *extraordinary science* (Kuhn, 1970, p. 82).[7] Each applies to different stages in the course of scientific development. According to his thesis, most science is of the normal sort, the kind that most scientists spend all of their lives doing — eking out this or that relatively humdrum "esoteric detail" (Kuhn, 1970, p. 35) about nature's functioning, accumulating more and more facts, and thereby elaborating upon the current "paradigm":[8]

> Mopping-up operations are what engages most scientists throughout their careers. They constitute what I am here calling normal science. Closely examined, whether historically or in the contemporary laboratory, that enterprise seems an attempt to force nature into the preformed and rela-

> tively inflexible box that the paradigm supplies. No part of the aim of normal science is to call forth new sorts of phenomena; indeed those that will not fit the box are often not seen at all. Nor do scientists normally aim to invent new theories, and they are often intolerant of those invented by others. Instead, normal-scientific research is directed to the articulation of those phenomena and theories that the paradigm already supplies.... (Kuhn, 1970, p. 24)

Inherent in normal science is "an immense restriction of the scientist's vision and ... a considerable resistance to paradigm change." (Kuhn, 1970, p. 64)

> [R]esearch ... [is] a strenuous and devoted attempt to force nature into the conceptual boxes supplied by professional education.... Normal science ... is predicated on the assumption that the scientific community knows what the world is like. Much of the success of the enterprise derives from the community's willingness to defend that assumption, if necessary at considerable cost. (Kuhn, 1970, p. 5)

Not surprisingly, then, one characteristic of normal science which Kuhn stipulates is that of conformity and even rigidity ... which amounts to dogmatism. Such closed-minded clarity is justified, simply in terms of its greater efficiency: When practitioners agree on the basics of the terrain, Kuhn's argument goes, then they can get on with the "normal science" task of elaborating the details of their descriptive map.

> [H]istory strongly suggests that, though one can practice science — as one does philosophy or art or political science — without a firm consensus, this more flexible practice will not produce the pattern of rapid consequential scientific advance to which recent centuries have accustomed us. In that pattern, development occurs from one consensus to another, and alternate approaches are not ordinarily in competition. Except under quite special

7 Kuhn later proposed two different modes of thought, convergent thinking and divergent thinking, which corresponded to normal science and extraordinary science, respectively. (Kuhn, 1977, p. 226)

8 Kuhn uses the term "paradigm" to describe the established intellectual wealth of a scientific community: its language, concepts, theories, visual imagery, tools and instruments, formulas, practical problems and solutions, norms, rules and values, etc. — essentially all those things that are contained in the standard textbooks of the day. However, Kuhn's choice of term here is unfortunate. As he acknowledges in a postscript to his book, he uses the "paradigm" term in an "obscure" manner (Kuhn, 1970, p. 181), and suggests in its stead the term "disciplinary matrix". Other replacements might be even better, such as, e.g., "theoretical milieu" or "conceptual matrix". In any event, the confusion which surrounds Kuhn's use of the "paradigm" concept may underlie other anomalies in his work. One example is the distinction he makes between a paradigm and an interpretation. Another is a distinction between the immature period of "pre-paradigmic" science, and the later stage of a fully-developed "paradigmic" science: once a science reaches its maturity, then one revolutionary paradigm succeeds another, and the pre-paradigmic stage is never returned to. Kuhn acknowledges that this distinction is not substantive: "The members of all scientific communities, including the schools of the 'pre-paradigmic' period, share the sorts of elements which I have collectively labeled 'a paradigm.' What changes with the transition to maturity is not the presence of a paradigm but rather its nature." (Kuhn, 1970, p. 179)

> conditions, the practitioner of a mature science does not pause to examine divergent modes of explanation or experimentation. (Kuhn, 1977, p. 232)

> So long as the tools a paradigm supplies continue to prove capable of solving the problems it defines, science moves fastest and penetrates most deeply through confident employment of those tools. The reason is clear. As in manufacture so in science — retooling is an extravagance to be reserved for the occasion that demands it. (Kuhn, 1970, p. 76)

Nor is Kuhn alone, in these views. Karl Popper likewise praises "The dogmatic attitude of sticking to a theory as long as possible ..." (Popper, 1965, p. 312); like Kuhn, Popper speaks of " ... the value of a *dogmatic* attitude: somebody had to defend a theory against criticism, or it would succumb too easily ..." (Popper, 1981, p. 30) Similarly, Oppenheimer writes that, "A perpetual doubting and a perpetual questioning of the truth of what we have learned is not the temper of science." (Oppenheimer, 1954, p. 24) As Morris puts it, "[S]cientists who have built their professional lives around the elaboration of theoretical ideas are understandably reluctant to give it up." (Morris, p. 40) Rogers says that "in fact most scientists cling to their old theory desperately." (Rogers, p. 348) And as Planck notes, "[A] new scientific truth does not triumph by convincing its opponents and making them see the light, but rather because its opponents eventually die, and a new generation grows up that is familiar with it." (Planck, quoted in Kuhn, 1970, p. 151)

Intransigence and inflexibility are not only seen as desirable attributes for science, generally: they are also seen as essential characteristics for individual theories. So it is that a rigid and brittle theory is considered to be better than a soft and supple one — thus making it easier to discredit and expose the flaws of the reigning conceptual regime, in that they thereby contain the seeds of their own destruction. This quality of conceptual or theoretical brittleness is also strongly advocated by Karl Popper with his refutability criterion, as we have already discussed. Where such narrowly-focused and precisely-defined theories are indeed inadequate, this will be readily revealed as a contradiction between theory and observation.

> The more precise and far-reaching that [theoretical] paradigm is, the more sensitive an indicator it provides of anomaly and hence of an occasion for paradigm change.... By ensuring that the paradigm will not be too easily surrendered, resistance guarantees that scientists will not be lightly distracted and that the anomalies that lead to paradigm change will penetrate existing knowledge to the core. (Kuhn, 1970, p. 65)

The result of these sharp conflicts is that the pattern of normal science is broken at a limited number of high points in the history of science, by a dramatic shift in the established paradigm — not just a relatively minor remodel of the framework, but a complete tear down in deference to something radically different. The contradictions within the old theory become too great, while the alternative solutions proposed by the new theory become too enticing — and the balance then shifts between them: the challenge is successful, the old model is overthrown, and a major reorientation takes place in the group's techniques, language, concepts, theories, values, and belief systems. Kuhn characterizes such breaks in this way:

> Scientific revolutions.... are the tradition-shattering complements to the tradition-bound activity of normal science.... [, necessitating] the community's rejection of one time-honored scientific theory in favor of another incompatible with it.... [S]cientific revolutions are here taken to be those non-cumulative developmental episodes in which an older paradigm is replaced in whole or in part by an incompatible one. (Kuhn, 1970, pp. 6 and 29)

Kuhn's overall thesis, then, "portrays scientific development as a succession of tradition-breaking periods punctuated by non-cumulative breaks." (Kuhn, 1970, p. 208) This contrasts with the more conventional view that, "[S]cience has reached its present state by a series of individual discoveries and inventions that, when gathered together, constitute the modern body of technical knowledge.... [T]he textbook tendency to make the development of science linear hides a process that lies at the heart of the most significant episodes of scientific development." (Kuhn, 1970, p. 140)[9] "Science does not tend toward the ideal that our image of its cumulativeness has suggested." (Kuhn, 1970, p. 96)

Education and Information. Now let us see how these ideas would play out in the areas of education and communication. First consider the way in which Kuhn applies his arguments, in support of a dogmatic approach to the practice of science, to the training of new scientists. His position is that from the beginning of his or her scientific education, the new worker must learn dogma — not how to think freely about nature. Kuhn talks about "educational initiation" (Kuhn, 1970, pp. 165) — noting that it "is a narrow and rigid education, probably more so than any other except perhaps in orthodox theology.... [S]cientific training is not well designed to produce the man who will discover a fresh approach." (Kuhn, 1970, p. 166) Yet he also claims that this educational style is "immensely effective" (Kuhn, 1970, p. 165) — because the professionalism into which the novice is being initiated is not receptive to new concepts, as we just mentioned: "Normal science does not aim at novelties of fact or theory." (Kuhn, 1970, p. 52)

Kuhn sees the tasks of education and communication as being greatly facilitated by having a fundamental body of language, concepts, and theories which are essentially unquestioned. He writes that in the physical sciences,

9 Of course, there are those who dispute the whole overthrow approach. d'Espagnat writes that, "Even the Ptolemaic system is still valid in its limited domain, and in more recent areas equations such as Maxwell's are not only valid but *indispensable* in physics, despite the fact that there is no such thing as an 'ether'. What actually occurs in the history of physics (still speaking from the instrumentalist point of view) is therefore not a sequence of paradigm collapses but rather a steady advance, consisting of successive replacements of a formalism, valid in domain of validity D, by a new formalism which does not necessarily make its predecessor obsolete but which has a domain of validity D' which includes, and is therefore more extensive than, D. From this perspective, all concepts are relative and their objectivisation at a particular time is ultimately of sociological rather than strictly scientific interest. Consequently, their down-grading to merely metaphorical status in no way means that a theory has collapsed." (d'Espagnat, p. 74)

> the various textbooks that the student does encounter display different subject matters, rather than as in many of the social sciences, exemplifying different approaches to a single problem field.... It remains a dogmatic initiation in a pre-established tradition that the student is not equipped to evaluate.... [T]his technique of exclusive exposure to a rigid tradition has been immensely productive of the most consequential sorts of innovations. (Kuhn, 1977, p. 229)

Textbooks, then, are especially important:

> To a very much greater extent than we ordinarily realize, our image of physical science and of measurement is conditioned by science texts. In part that influence is direct: textbooks are the sole source of most people's firsthand acquaintance with the physical sciences. Their indirect influence is, however, undoubtedly larger and more pervasive. Textbooks or their equivalent are the unique repository of the finished achievements of modern physical scientists. It is with the analysis and propagation of these achievements that most writings on the philosophy of science and most interpretations of science for the non-scientist are concerned. As many autobiographies attest, even the research scientist does not always free himself from the textbook image gained during his first exposure to science. (Kuhn, 1977, p. 180)

Kuhn is hardly alone in his approach to these issues: he largely speaks for the established discipline. For example, listen to an expert telling beginners about the indoctrination which they must go through. According to mathematical physicist John Taylor,

> [P]atience [is] required for you to learn the technical language, and the associated ideas, in whatever area of science you wish to become expert. I can't emphasize too strongly the need to acquire this language before trying to solve the problems of the universe. As the old saying goes, "You can't run before you can walk." I — and many scientists, I'm sure — often have letters claiming to have a better explanation of things than given by present science. I call it the "Einstein's wrong syndrome". But these claims usually present half-baked ideas; especially, they show a great lack of understanding of the successes of modern science, and the effort put into it by generations of science. (Taylor, 1980)

A Closed Shop. Another important aspect of Kuhn's ideology is that the work of science is treated as a strictly internal affair — in the sense that it is only the closed circle of scientists who are in a position to determine the fundamental content of the field (i.e., the language, concepts, and theory),[10] as well as to appraise the ongoing research which is conducted. "[M]embers of a given scientific community provide the only audience and the only judges of that community's work." (Kuhn, 1970, p. 209) By contrast, non-specialists are seen as being both incompetent for, as well as spurious (if not detrimental) to, this primary evaluative task. Kuhn speaks about,

10 "Another function of jargon is sociological. It forms its users into an identical clique, which can identify outsiders instantly by their inability to speak the lingo." (Jones)

> The unparalleled insulation of mature scientific communities from the demands of the laity and of everyday life.... Just because he is working only for an audience of colleagues, an audience that shares his own values and beliefs, the scientist can take a single set of standards for granted.... One of the strongest, if still unwritten, rules of scientific life is the prohibition of appeals to heads of state or to the populace at large in matters scientific. Recognition of the existence of a uniquely competent professional group and acceptance of its role as the exclusive arbiter of professional achievements has further implications. *The group's members*, as individuals and by virtue of their shared training and experience, *must be seen as the sole possessors of the rules of the game*.... What better criteria than the decision of the scientific group could there be? (Kuhn, 1970, pp. 164, 168, and 170; italics added)

> For a scientist, the solution of a difficult conceptual or instrumental puzzle is a principal goal. His success in that endeavor is rewarded through recognition by other members of his professional group and by them alone.... Only they look critically at his work, and only their judgment affects the further development of his career.... The practical merit of his solution is at best a secondary value, and the approval of men outside the specialist group is a negative value or none at all.... Scientists who attempt to find a wider audience for professional work are condemned by their peers. (Kuhn, 1977, pp. 290 and 344)

The practice of science, then, is seen to be virtually the exclusive domain of scientists: they set their own standards and goals, and they alone are privileged to judge the results. It is a closed shop, indeed, and all that outsiders may do is look on as spectators.

Indeed, the dogma of the party line is what counts, as opposed to credible explanation — and *conformity* reigns. Any worker who refuses to abide by the established line is effectively excommunicated from the discipline, no longer even considered a scientist — virtually by definition, according to Kuhn. He writes how important it is "that group unanimity be a paramount value, causing the group to minimize the occasions for conflict and to reunite quickly about a single set of rules for puzzle solving even at the price of subdividing the specialty or excluding a formerly productive member." (Kuhn, 1977, p. 291; see also Kuhn, 1970, pp. 19 and 159)[11]

So much for our discussion regarding some of the main mandates that guide science's work — which we will reflect upon in the second half of the chapter.

Question Authority

Given the ideas that we have been discussing in this book, it should come as no surprise that we take issue with all of the ideology and methodology that we have been discussing. We maintain that they represent the wrong standards, which have taken us in the wrong direction. So it is that we shall now reexamine each of the topics presented above, and in the process come to quite different conclusions — and we will not mince words in our evaluations here. As we proceed, note that we will cite many of the same names as

11 We described an instance of this mentality in Chapter 7, concerning the way that physicist Herbert Dingle was treated by his colleagues.

supporting some aspect of our position here as we cited supporting some aspect of the established paradigm in the first part, for reasons that should be apparent in each case.

Metaphysics Everywhere: Positivism is Not Enough. To start with, it should be quite clear that one of the main differences between the established paradigm and the present model is in terms of the topic of metaphysics and positivism. In this subsection, we want to look generally at the role of metaphysics in the development of our theories, and in the following subsection we will narrow in on the topic, in terms of how specific phenomena of nature are conceptualized.

Kitchener makes reference to various uses of the term, *metaphysics*: "as the theory of ultimate reality, the theory of the most general features and principles of all things, or a set of basic assumptions." (Kitchener, p. 6) We use the term in all these ways, as referring to the fundamental conceptions we all have about the essential nature of reality; and we maintain that these are also the principles which guide us in our life experiences, including our explorations of the physical world.

As we have seen, the orthodox approach is to steer clear of these sort of issues — with the result being that scientists are generally unwilling to examine the broad assemblage of assumptions, beliefs, and definitions which underlie their thinking and their theories. For one thing, these kind of nebulous questions cannot be readily tested, and the positivist dogma rejects such matters as not being within the purview of science. And for another thing, the explicit adoption of any such positions would be seen to somehow bias the outcome, concerning whatever phenomenon is being examined.

The arguments might go something like this: "No guiding themes, please — we are open-minded scientists, and we do not wish to be hemmed in by arbitrary and unnecessary preconceptions which may constrain our breadth and stunt our vision. 'Blank slate' is our byword, and we prefer to be free to deal with each situation as we find it, rather than pretend to know any part of the answer (let alone influence the results) beforehand. Moreover, our logical positivism ethos leads us to restrict our study to the particular phenomenon we are dealing with, rather than wasting our energy in pursuit of distracting and unprovable claims about the 'ultimate nature of reality', or some such mirage."

We contend that these sort of explanations are at the very least completely off course, if not actually disingenuous — not only in the sense that good science must help us to understand the real world, as we shall discuss in the next chapter ... and therefore that all theoretical roads in science inevitably *lead to* some general conception of reality. But they are also off course in the sense that all theoretical roads themselves must necessarily *emerge from* some general conception of reality: they do not just fall from the sky, as it were, or even from some clever minds — in spite of Einstein's spirit of free invention. Such a studied detachment from reality is therefore neither accurate nor viable.

The problem with the standard line is that isolated phenomena are treated as though they exist in their own right, whereas we contend that particular entities only exist in relation to larger backgrounds — and that one way of labeling these larger backgrounds is with the term "metaphysics". In this regard, we assume that there are always a whole range of metaphysical factors which represent the background for any situation, in relation to which whatever factors we focus upon exist. Indeed, in the broadest sense, metaphysics is *meaning* and thus *understanding*, in general: it is a template for gauging *truth* by. Therefore, one *must* have a metaphysics, and the very idea that one should shun metaphysics or not have preconceptions or larger concepts when investigating the natural world (or anything else, for that matter) is naive at best.

Clearly, then, we treat metaphysics as referring to the larger reality or meaning system — which provides the broader framework for whatever phenomenon, construct, or theory that we are primarily concerned with in any given case. Put another way, metaphysics is context: it is how we interpret things, and so it cannot be avoided. Consequently, the issue is not whether such a metaphysical background context exists or not in any given situation — as though we were talking about an optional school of thought, some vague and abstract set of speculations about nature. Rather, such background context always comes into play — it is part of all our theories, regardless of whether we choose to label it as such or not. This includes whatever assumptions and methodology are entailed in any particular situation that we are examining. Inevitably, then, such larger context, or metaphysics, is always inherent ... in everything. No field of study is exempt from this pattern — and so science also entails metaphysics, for even science cannot exist in a conceptual vacuum. In these terms, by the way, note the title of a work by Kant: *Metaphysical Foundations of Natural Science* (cited in Popper, 1965, p. 48).

Ignoring such underlying assumptions, or pretending that they do not exist, does not make them go away: it merely adds another layer of obfuscation — of what are essentially uncontrolled variables — into the mix. That is because a "blank slate" is never really blank — but essentially means "hidden slate" or "random slate" or "unknown slate". There is no such thing as a blank slate in our explorations of nature — for the metaphysical preconceptions will always be there in various forms, and the question is whether they are explicitly identified or not.

And whether they are to be more consistent — which leads to the next issue ... of whether there is to be any commonality or not among these background metaphysical factors. For if we don't know which background factors are involved, in some given situation where they are not recognized and identified — then they are likely to be different in some other situation, where they are also not recognized and identified. Instead of a set of factors which we rationally agree upon, we end up with whatever set we happen to stumble upon or implicitly assume. What we are left with, therefore, is subliminal metaphysics, and haphazard metaphysics — which thereby provides justification for an arbitrary approach to theory-building ... by virtue of incorporating diverse sets of background factors. Then instead of meshing together into something coherent, we will have an endless set of ad hoc metaphysics — arbitrary cross sections of nature, based upon randomly-distinct and varied perspectives ... which are ultimately incompatible with one another.

Indeed, this is one of the main grounds on which we criticize mainstream science, in terms of its methodology for theory-building: its refusal to distill and embrace a coherent set of basic principles, or metaphysics, with which to guide its assumptions and explorations about the nature of reality — instead opting to ignore the topic altogether, as being inappropriate for science ... and pretending that it is operating in a more innocent, open-minded, and objective manner by so doing. The result is an overall jumble of contradictory constructs and theories, because there is nothing to unify them: no common theme, no pivotal pattern that is based upon broader and deeper relationships — and relative to which such conceptual components would *not* be ad hoc slices of nature. With this kind of piecemeal strategy — by default and by design — science has defined itself into a conceptual corner.

Here is one way of describing the situation — according to someone who strongly supports the established paradigm, Richard Feynman:

> Discovering the laws of physics is like trying to put together the pieces of a jigsaw puzzle. We have all these different pieces and today they are proliferating rapidly. Many of them are lying about and cannot be fitted with the other ones. How do we know that they belong together? How do we know that they are really part of one as yet incomplete picture? We are not sure, and it worries us to some extent, but we get encouragement from the common characteristics of several pieces. (Feynman, 1965, p. 83)

> Today our theories of physics, the laws of physics, are a multitude of different parts and pieces that do not fit together very well.... If we put all these principles [of contemporary physical theory] together, we discover that there are too many. They are inconsistent with each other. (Feynman, 1965, pp. 30 and 155)

Jauch talks in more extreme terms, about how the process of endless ad hoc theorizing "will lead to a nightmare of general pollution by a random output of theoretical trash.... The proliferation ... is the beginning of the end of our scientific civilization." (Jauch, p. 31)

We suggest that this underlying state of disorder can also be cited as the reason why we are forced to resort to the kind of bizarre conceptual acrobatics which characterize so much of today's scientific reasoning — and the endless mathematical manipulation that goes along with it, and provides cover for it: because there is something inherently flawed with the whole free-invention and situation-specific theory-building strategy of the established paradigm to begin with, which yields such impossible-to-unify components.

There are significant observers who concur with at least some of our position. For example, physicist David Bohm says that, "[W]e cannot thus simply dispense with an overall world view. If we try to do so, we will find that we are left with whatever (generally inadequate) world views may happen to be at hand." (Bohm, 1980a, p. xiii) In the same way, Rudner's comment about "values" can also be understood in terms of "metaphysics": "[S]cientists as scientists *do* make value judgments.... To refuse to pay attention to the value decisions which *must* be made, to make them intuitively, unconsciously, haphazardly, is to leave an essential aspect of scientific method scientifically out of control." (Rudner, pp. 232 and 236)[12]

12 Capra makes a criticism of social science and "values" — that can be equally well applied to physical science and "metaphysics": "By avoiding the issue of values, then, social scientists are not more scientific, but, on the contrary, less scientific, because they neglect to state explicitly the assumptions underlying their theories." (Capra, 1983, p. 190)

Even Karl Popper — who sharply differentiates metaphysics from science — describes himself "as a metaphysical realist." (Popper, 1981, p. 40). Moreover, Popper also assumes that there will always be a context which surrounds our analyses:

> We all have our philosophies, whether or not we are aware of this fact.... [of] our many common sense assumptions — our common sense background knowledge.... Knowledge never begins from nothing, but always from some background knowledge — knowledge which at the moment is taken for granted.... [Such] "background knowledge" does not consist *only* of observation reports ... but also of *theories*: of those conjectural theories that are for the time being ... treated as *unproblematic*. (Popper, 1981, pp. 33, 71, and 364; also cf. Popper, 1981, pp. 71 and 165; and Popper, 1965, p. 238)

In other words, strictly speaking, theories themselves are also "background knowledge" — and in fact are never tested, *per se*. Popper acknowledges this point in the sense that he treats theories as general "universal statements", whereas only "singular statements" derived therefrom are testable: "Every application of science is based upon an inference from scientific hypotheses (which are universal) to singular cases ..." (Popper, 1968, p. 64)

One can apply this argument to both of the two main theories of physical science, the Theory of Relativity and Quantum Theory: the theories themselves are not testable — only predictions derived from those theories are. Even the constant *c* proposition of the Theory of Special Relativity is a definition which cannot itself be tested. And similarly, whereas the Lorentz transformation is confirmed explicitly, the Theory of Special Relativity is only a particular — and peculiar — *metaphysical* interpretation thereof, which also cannot be tested. In the same vein, Einstein's assumptions about a combined world of space-time, the nature of interactions and transformations, and so on — must all be treated as metaphysical interpretations. Likewise in the case of Quantum Theory, concerning the main principles of uncertainty and complementarity ... along with the notion of the discrete quantum itself, as well as the role of observation in manifesting or creating reality.

In the same vein, the radical assertions which Popper makes about testability and falsifiability are themselves examples of general propositions which can only be assumed. Similarly, any assumptions about what are deemed to be desirable characteristics (e.g., aesthetic qualities) of a good theory of science, amount to — guess what? Are all of these inherently untestable theories or concepts to be dismissed as unscientific and metaphysical speculation?

Indeed, the principle can be applied to logical positivism itself: it is a metaphysical contention. That is, one can no more prove that metaphysics should not be taken into account as that it should be taken into account — and therefore, any such decision between yes-metaphysics and no-metaphysics represents a fundamental metaphysical demarcation in its own right. Of course, the typical response might be that no-metaphysics is simpler and more straightforward than yes-metaphysics — and that this is its own justification. Our retort is that no-metaphysics is a mirage — not to mention leading us along a path toward conceptual ruin — a logically bankrupt way of approaching nature.

Returning to Popper's views, he maintains that one cannot avoid the need for an underlying...

> metaphysical faith in the existence of regularities in the world.... [I]n almost every phase of the development of science we are under the sway of metaphysical — that is, untestable — ideas ... [Such a] horizon of expectations plays the part of a frame of reference: only their setting in this frame confers meaning or significance on our experiences, actions, and observations.... [All observations presuppose] a frame of reference: a frame of expectations: a frame of theories.... (Popper, 1968, p. 252; Popper, 1982, p. 161; Popper, 1981, pp. 345; Popper, 1965, p. 47; cf. also 1968, p. 278)

> Science never starts from scratch; it can never be described as free from assumptions; for at every instant it presupposes a horizon of expectations.... [T]he belief that we can start with pure observations alone, without anything in the nature of a theory, is absurd.... Observation is always selective.... [O]bservations ... are always *interpretations* of the facts observed ... *interpretations in the light of theories*. (Popper, 1981, pp. 346; Popper, 1965, p. 46; Popper, 1968, p. 107)

> [T]here can be no pure observational language, since all languages are impregnated with theories and myths.... [T]heories or expectations are built into our very sense organs.... [T]here are no such things as sense data or perceptions which are not built upon theories.... [T]here does not exist anything like pure observational knowledge, untainted by expectations or theories.... [W]e can utter no scientific statement that does not go far beyond what can be known with certainty "on the basis of immediate experience." (Popper, 1981, p. 146; Popper, 1965, p. 181; Popper, 1965, p. 155; and Popper, 1968, p. 94)

In the end, therefore, even Popper cannot avoid the essential principle for which we are arguing here:

> I am inclined to think that scientific discovery is impossible without faith in ideas which are of a purely speculative kind, and sometimes even quite hazy; a faith which is completely unwarranted from the point of view of science, and which, to that extent, is "metaphysical"..... [Such ideas are] indispensable for science, although their character is that of metaphysical or speculative physics rather than scientific physics. Originally they were all metaphysical, in nearly every sense of the word ... They were unifying pictures of the world — the real world. They were highly speculative; and they were, originally, non-testable. (Popper, 1968, p. 38 and Popper, 1982, p. 165)

Not only that, but Popper seems to have mellowed even further, as the years went by. At one point, he presents a set of ideas that might be considered suspect — and acknowledges the predictable criticisms which these might evoke:

> Nothing is easier than to denounce statements of this kind as metaphysical or philosophical or meaningless. "What we want are facts and figures", we hear from all sides, "not philosophical discussions." (Popper, 1982, p. 102)

Popper's response is to ask rhetorically whether a metaphysical idea is, in fact,

> not utterly different from a scientific hypothesis — one in which we are mainly interested *because of its implicit claim to be considered, tentatively, as true*? I no longer think, as I once did, that there is a difference between science and metaphysics regarding this most important point. I look upon a

> metaphysical theory as similar to a scientific one. It is vaguer, no doubt, and inferior in many other respects; and its irrefutability, or lack of testability, is its greatest vice. But, *as long as a metaphysical theory can be rationally criticized*, I should be inclined to take seriously its implicit claim to be considered, tentatively, as true. (Popper, 1982, p. 199)

Other analysts take a similar view, about the importance of background assumptions. For example, Bohm says that:

> Metaphysics is a set of basic assumptions about the general order and structure of existence.... It seems clear that everybody has got some kind of metaphysics, even if he thinks he hasn't got any.... What is called for is therefore that each one of us be aware of his metaphysical assumptions, to the extent that this is possible. (Bohm, 1969, p. 41)

And although he is not talking in the language of metaphysics, Kuhn writes that, "[A] paradigm is a prerequisite to the discovery of laws.... What a man sees depends both upon what he looks at and also upon what his previous visual-conceptual experience has taught him to see." (Kuhn, 1970, pp. 28 and 113) Or as Zukav puts it, "[N]othing can be said about the world 'out there' except via our concepts." (Zukav, p. 71)

Toulmin says that,

> [T]hough Nature must of course be left to answer our interrogations for herself, it is always *we* who frame the questions. And the questions we ask inevitably depend on prior theoretical considerations.... "[P]reconceptions" of this kind are both inevitable and proper ... (Toulmin, 1961, p. 101)

> The scientist.... does not (and should not) approach Nature devoid of all prejudices and prior beliefs.... [T]he scientist's prior expectations are governed by certain rational ideas or conceptions of the regular order of Nature. (Toulmin, 1961, p. 45)

Kitchener refers to Leclerc's position, that, "[P]hysicists cannot avoid doing metaphysics or making metaphysical assumptions" (Kitchener, p. 16). Murchie says simply that: "[P]hysicists are coming to realize that, if they throw metaphysics into the fire, their own subject will soon smolder also." (Murchie, p. 495) Even Einstein said that the "fateful 'fear of metaphysics' ... has come to be a malady of contemporary empiricistic philosophizing" (Einstein, 1944, quoted in Holton, p. 263). Ultimately, as Morris indicates, we have no choice: "If we want to know what physical reality is 'really' like, we must fall back on our own philosophical prejudices and preconceptions." (Morris, p. 230)

Polkinghorne says that, "All scientific observation is a form of 'seeing as'; we have to interrogate the world from a chosen point of view" (Polkinghorne, p. 5). And not just any point of view: "A research programme is defined by its central core of non-negotiable concepts ..." (Polkinghorne, p. 16) — with the end result being that, "It is only when the patterns are interpreted *using current theoretical ideas* that one receives the message ..." (Polkinghorne, p. 5) All in all, it would be putting it mildly to say that, "The need to adopt a point of view introduces a certain precariousness into the scientific enterprise." (Polkinghorne, p. 6)

Capra writes that, "In reality, scientific facts emerge out of an entire constellation of human perceptions, values, and actions ... from which they cannot be separated."

(Capra, 1988, p. 152) Again, "The patterns scientists observe in nature are intimately connected with the patterns of their minds; with their concepts, thoughts, and values." (Capra, 1983, p. 87)

As Prigogine says about today's conceptual impasse:

> [W]e may hope that out of the current situation there will arise a more subtle conception of science in which the traditional conflict between physics and metaphysics would be attenuated, opening the way to a new more unified vision of nature. (Prigogine, 1977, p. 241)

Holton speaks about "the existence of preconceptions that appear to be unavoidable for scientific thought, but are themselves not verifiable or falsifiable." (Holton, p. 13). As he puts it metaphorically, "[N]either research nor a sea voyage is undertaken without some theory that serves as a rough chart." (Holton, p. 437) He acknowledges that this "is an embarrassing conception which did not and does not fit into a purely positivistically oriented philosophy of science." (Holton, p. 35) He even goes so far as to say that, "[T]he scientist, perhaps unknowingly, smuggles the style, motivation, and commitment of his individual system and that of his society into his supposedly neutral, value-indifferent luggage." (Holton, p. 85) Needless to say, "This is indeed far from the conventional idea that the scientist keeps a completely open mind." (Holton, p. 90) In particular, Holton notes that, "It might be said that the foundations of quantum theory were laid under metaphysical auspices." (Holton, p. 408)

Let us look further at some of the particular ideas that Holton put forward, which have specific relevance both to this issue and to our new model. Holton's major theme is precisely the notion of themes, and the influence they play in the theory and practice of science. These "themata" of science can be defined as...

> fundamental presuppositions, notions, terms, methodological judgments and decisions ... which are themselves neither directly evolved from, nor resolvable into, objective observation on the one hand, or logical, mathematical, and other formal analytic ratiocination on the other hand.... [As] guides in the pursuit of scientific work.... themata are not proved or disproved. Rather, they rise and fall and rise again with the tides of contemporaneous usefulness or intellectual fashion.... The themata actually used in science are now largely left implicit rather than explicit.... [But] themata have as legitimate and necessary a place in the pursuit and understanding of science as have observational experience and logical construction ... (Holton, pp. 41, 44, 46, 47, and 206)

> [I]t is precisely such nonverifiable and nonfalsifiable, deep-going thematic hypotheses which are most difficult to advance or accept. They are at the heart of major changes or disputes, and their growth, reign, and decay are indicators of the most significant developments in the history of science. (Holton, p. 206)[13]

13 At the same time, Holton seems to reject any linkage between his themata and that of (presumably Kuhn's) paradigms (Holton, p. 27).

For our part, we treat Holton's set of thematic factors in terms of the general category of metaphysics that we have been discussing — although Holton himself sometimes uses that term in a different way than we do. But leaving this definitional point aside, what is especially intriguing in terms of the present model is that Holton tends to describe these broad thematic tendencies in terms of what we call complementarities. He even talks in terms "of opposing or complementary theme and antitheme." (Holton, p. 83) Examples of this pattern would be various dualities like order vs. disorder, simplicity vs. complexity, parts vs. whole, and reductionism vs. wholism.

In Holton's system, different scientists can be understood as being oriented towards one or the other of such polar tendencies. This is in keeping with the discussion we had in Chapter 2, where we described Heisenberg and Schrödinger as being at opposite ends of a discrete vs. continuous polarity.[14] People like Einstein would, of course, also be guided by the continuum thesis — even though the paradox is that he is also credited with having pioneered the discrete particle orientation of Quantum Theory itself![15] Indeed, in the previous chapter we described the two main theories of contemporary physics in complementary terms that are akin to "themata" and "antithemata", with Quantum Theory being discrete-based and Relativity Theory being continuum-based.

So what about Niels Bohr, and his guiding themata? Obviously, the theme there is complementarity itself — which Holton has essentially subscribed to with his own guiding themata of theme and antitheme. It is noteworthy that Holton pays great professional and personal tribute to Bohr, just as he also seems to have absorbed much of Bohr's thinking. And yet, Einstein also receives lavish praise in Holton's work — even though Einstein is Mr. Anti-complementarity himself! As with Heisenberg and Schrödinger, so Einstein and Bohr represent an amazing pair of oppositional "thema-antithema" orientations.

In closing this subsection, we want to reiterate our main point — which is that it is not the incorporation of metaphysics which represents a primitive mode of doing science: it is, rather, the ignorance of metaphysics which represents such a backward approach. Contemporary science remains in the dark ages when it pretends that such background factors do not exist, or that they represent a diversion which is of little consequence for science's crucial mission. We argue to the contrary — that we need consistent metaphysics and general constructs which can be widely applied. Those simply do not exist in science today — and until they do, we will never have the basis for a coherent model of nature.

Facts are Made. In contrast to the logical positivist view that facts are essentially fixed, absolute, and objective, we maintain that facts are inherently fluid, rela-

14 Note how that conflict can be understood in terms of Holton's point about "the dialectic process ... between a thema and its antithema, and hence between the adherents of two or more theories embodying them respectively ..." (Holton, p. 15)

15 Nonetheless, Einstein wrote to Born in 1920, "that one has to solve the quanta by giving up the continuum, I do not believe." (Einstein, quoted in Holton, p. 116; see also Holton, p. 375)

tive, and subjective ... i.e., they vary with the situation. That "situation" includes both the larger physical context outside in the world, as well as the larger metaphysical context inside our minds.

Obviously, then, we categorically reject the notion that facts are merely some immutable entity in nature, waiting to be plucked as fruit from a tree.[16] "Just the facts", the hard-nosed scientist demands; "no superfluous speculation to distract us, please!" But just where did those facts come from, without the context in which they have meaning and existence? As Siu puts it, " ... the facts and nothing but the facts. But just what is a scientific fact?.... [T]he facts of science imply a theoretical as well as an empirical element." (Siu, p. 46) Here is how Bohm puts it:

> [F]acts are not to be considered as if they were independently existent objects that we might find or pick up in the laboratory. Rather ... the fact is "what has been made" (e.g., as in "manufacture"). Thus, in a certain sense, we "make" the fact.... [I]t has no meaning to separate the observed fact (along with the instruments used to observe it) from the theoretical notions of order that help to give "shape" to this fact. (Bohm, 1980a, pp. 142-143)

Popper says the same thing — the following comments being variations on ones of his that we presented in the prior subsection:

> [W]e have to "*make*" our experiences. It is we who always formulate the questions to be put to nature.... [T]heory — at least some rudimentary theory or explanation — always comes first.... [O]ur discoveries are guided by theory ... rather than that theories are the result of discoveries "due to observation"; for observation itself tends to be guided by theory.... [O]ur scientific theories are not simply the result of a description of nature — of "reading the book of nature" without "prejudice" — but that they are, rather, the products of the human mind.... [A]ll "observables" are *calculated and inferred on theoretical grounds*, rather than observed or directly measured. Thus what is "observable" always depends upon the theory we use.... Facts are something like a common product of language and reality; they are reality pinned down by descriptive statements. (Popper, 1968, p. 280; Popper, 1981, p. 258; Popper, 1965, p. 118; Popper, 1982, p. 3; Popper, 1982, p. 85; and Popper, 1965, p. 214; cf. also Popper, 1965, p. 46)

Toulmin says that, "[T]heories are built on facts, while at the same time giving significance to them and even determining what are 'facts' for us at all." (Toulmin, 1961, p. 95) In Kuhn's words, "[N]o experiment can be conceived without some sort of theory." (Kuhn, 1970, p. 87) Gatlin's remark about the living system applies to our observation of nature generally: "[T]here is no point in worshipping the "hard" experimental fact,

16 A distinction is sometimes made between the *data* which emerge from our observations — data are said to remain objective, as they are merely a "record of our perceptions" (Hecht, p. 9); and the more subjective *fact* — "We make facts out of data by interpreting what we see within the context of what we believe, often within the framework of accepted theory." (Hecht, p. 10) Distinguishing objective data and subjective fact is interesting, but in practice scientists almost never make such a distinction. And in any case even so-called "objective data" is still obtained and perceived in *some* interpretative framework.

because our theoretical perspectives have played no small role in determining which facts we know." (Gatlin, p. 135)

Again, then, there is no such thing as a fact which is not part and parcel of some larger framework of meaning and existence — for, the hard, testable "fact" of focus cannot exist without the soft, untestable background. This means that what we find in nature will at least partly depend on how we look for and at it: There is nothing godlike about the particular perspective we adopt in any given situation, no matter how hard we try to make it objective.

We must always anchor ourselves somewhere upon the conceptual landscape: it is impossible to theorize and observe — or even think — otherwise. Likewise, it is an illusion to suppose that the experimental situation is neutral and objective, in this respect: for what we bring into the experimental situation plays no small part in what we take out of it. Experiments (even so-called "crucial experiments") do not yield hard facts ... shorn of the conceptual and existential framework from which they are derived; and so, the value of any experiment is only as good as the theoretical or metaphysical framework which gives meaning to that exploration. Therefore, *we must spend at least as much time examining the basis of our concepts as we do testing them in the laboratory.*

The idea that a real scientist deals only with observable facts makes no sense — for how could such a mental isolationist deal in any terms at all ... do any sort of research or theorizing whatsoever? What such talk ultimately means is context-free thinking, research, and theorizing — which brings to mind the idealized classical realm of the inertial frame of reference, as we discussed in Chapter 4. There we were talking about a pure force-free realm of the *physical universe*, whereas here we are talking about a pure force-free realm of the *conceptual universe*. Neither corresponds with reality.

Indeed, we turn to Quantum Theory for support of our position here. Thus, Heisenberg said that, "We have to remember that what we observe is not nature herself, but nature exposed to our method of questioning." (Heisenberg, quoted in Murchie, p. 469)[17] We cited this same quote in Chapter 2, where we discussed the point that Quantum Theory formally disclaimed the existence of an independently-existing objective reality — the very proposition of a factual world that is just waiting to be perceived and inventoried by a neutral observer whose observations do not influence that which is observed. Quantum Theory popped that bubble of illusion, and taught us that what we see depends on how we look, on our mode of viewing — and, we claim, on the rest of the context that the observation process is part of ... including the observer's whole conceptual (or metaphysical) frame of reference and mind-set. Consequently, we maintain that there is clearly a contradiction between logical positivism and the Quantum Theory which is built thereupon, and from which is derived much of the theory's epistemological foundation.

In this regard, we would argue that the pre-Quantum science which denied the crucial role of the *process* of examining nature thereby had at least some basis for also

17 "Observation is the interaction between ourselves and the physical entities in which neither is independent of the other. In the whole, there can be no separation between the mind and its object." (Harris, p. 171)

denying the necessity for taking metaphysics into account, in the contextual terms that we are discussing. But a post-Quantum science which then acknowledges the crucial role of *how we see* lacks integrity when it continues to blindly deny the larger metaphysics of which seeing is an important part. In broad Two-Force terms, being and seeing go together — and they cannot be divorced from one another.

The Theory of Relativity also acknowledged that facts were not fixed, but for a different reason than we have just been discussing — bearing in mind that Einstein was never able to break himself away from the notion of an objective and unitary reality. Rather, Einstein's no-fixed-facts position emerged as part of his general strategy for theory-building — which we have seen is based upon a free-invention, deductivist methodology. It is in this respect that he discounted the notion of objective facts waiting to be harvested from nature. For while something (e.g., concrete) was sitting out there in the world, waiting to be identified and catalogued, the issue for Einstein concerned how that phenomenon was to be interpreted and identified in theoretical terms. Thus, Einstein would say that while a fact may fit one interpretative scheme, it may not fit another — and then, "The facts are wrong." (Einstein, quoted in Hecht, p. 12)[18] Again, "It is the theory which decides what we can observe." (Einstein, quoted in Jammer, 1974, p. 57) In his approach, of course, this determination was to be purely arbitrary, in principle — being freely invented in a manner which would mold nature into facts ... that were compatible with the goals toward which the theory was aiming.

So in contrast to the Quantum Theory position that nature changes depending on how one examines it, the Theory of Relativity would say that nature doesn't change — but rather that we can interpret nature in any number of different ways. This latter approach is consistent with Popper's position that, "The laws of nature *are* our invention ... We *try* to impose them upon nature." (Popper, 1981, p. 92). It is also consistent with Kuhn's general description of "normal science" which we cited earlier: "[W]hether historically or in the contemporary laboratory, that enterprise seems an attempt to force nature into the preformed and relatively inflexible box that the paradigm supplies." (Kuhn, 1970, p. 24)

One way of describing this is in terms of the change Kuhn describes, concerning what takes place when one scientific theory overthrows another: "What were ducks in the scientist's world before the revolution are rabbits afterwards." (Kuhn, 1970, p. 111) Nature doesn't change, but a fact does — because a given fact emerges from a given interpretative scheme or theory. Therefore, what is a fact in one theory becomes another fact in a different theory.

At the same time, Kuhn rejects the suggestion that what changes with a change of conceptual framework or "paradigm is only the scientist's interpretation of observations that themselves are fixed once and for all by the nature of the environment and of the perceptual apparatus." (Kuhn, 1970, p. 120) Kuhn argues that this approach is unacceptable, because it is based upon a sharp distinction between the fact and the interpretation: "What occurs during a scientific revolution is not fully reducible to a reinter-

18 Or as Mach put it: "The senses do not lie, only they do not tell the truth." (Mach, quoted in Harrison, p. 580)

pretation of individual and stable data." (Kuhn, 1970, p. 121) Kuhn's point is that the interpretation of an event is not simply an afterthought, as it were, overlaid upon the observation. He says that scientists coming from different conceptual periods or genres would not simply overlay different post-hoc interpretations upon the same data; rather, he says, the data *incorporate* the interpretation in the sense that a different set of data might emerge from another interpretation.

All in all, what we find is that, "The rigid, immutable clarity of the popular concept of fact is, for the most part, an illusion even in the sciences." (Hecht, p. 12) Not only will observations in experiments change the facts which are observed — in the Quantum Theory sense; but the whole prior conceptual framework in which the experiment is conducted is inevitably linked to which facts emerge — in the Relativity Theory sense. So how can we treat "experiment as the ultimate guide to the universe"?

And yet in spite of all this, the notion that, "Facts don't lie" remains a core theme of science — fine-print disclaimers notwithstanding. Immutable, objective, independent facts supposedly emerge from experiments, and a good theory must fit these facts well — even though this description is itself a lie!

Rather than pretend that modern science is firmly rooted upon such pure objective facts, we must realize that this system of thought, too, ultimately rests upon beliefs and assumptions and definitions — and on the nature of the glasses we wear. Thus, for example, our primary conceptions about nature — and all the assumptions and beliefs and definitions which are inherent in them — form the core of our minds and our vision of reality … both inside and outside of science. This is the framework from which theories are generated and facts are identified and amassed. We shall return to these issues in the next chapter.

All is Not Measurable. We title this subsection with the main topic that we shall discuss herein — even though there are two other secondary points we want to cover, as we mentioned in the parallel subsection on experimentation in the first part of the chapter. We begin at the top.

Science faces a real dilemma. On the one hand, and in accordance with its positivist doctrine, it is taken as axiomatic that science can only deal with that which can be clearly observed and explicitly identified and quantified; yet, on the other hand, there are also compelling arguments that the full reach of nature goes beyond such limits. Most established scientists would even concur with such a position — as manifested, for example, in many scientists' personal beliefs in religious and other spiritual pursuits. The usual explanation for this discrepancy would be that certain phenomena are simply beyond the scope of science — precisely because of the measurability criterion that we are discussing. Therefore, the argument would go, there is science and there is non-science — or more strictly, there is that which can be subjected to the methodology of science and that which cannot be.

But is that answer sufficient? We dissent from such a differentiation and argue that the scope of science must include all of nature — no matter its characteristics — precisely because science must be a model of nature … a model of *all* of nature, and not merely a model of a broad subset of nature. This was part of the topic in the previous chapter, where we argued that the "other" side of nature must be fully incorporated into science's theories ... to the same degree as is "this" side of nature. Our focus here is on the testability factor, which we maintain is a false criterion for what the scope of science should be.

There are at least two important reasons for the full-scope requirement. One is rather obvious, in terms of the simple precept that *a science of nature must study all of nature*. But the second reason is equally important — that if science only studies part of nature, it may not be exposing itself to the complete range of patterns which the full range of phenomena exhibit ... but which a partial range does not. The result would not just be a theory of curtailed phenomenological content: it would also be a theory of curtailed phenomenological patterns. That latter outcome would thereby make it more likely that the patterns which science is identifying are only partial and piecemeal, in that they would not reflect the full range of reality.

To be sure, we are not claiming that unless science studies everything and everywhere in nature, then it is thereby only a partial science of nature. Thus, we have only the vaguest inklings of what specific phenomena exist in large parts of the universe, and we still have not completely penetrated the innermost reaches of matter here in our local realm of experience ... nor is it clear that either such goal is ever attainable, in principle. Obviously, then, we are not arguing for the impossible; and these outermost and innermost realms will be increasingly explored by science to the degree that its technological prowess and conceptual range increasingly permits. Rather, our point with the "curtailed phenomena" argument is that science must not specifically exclude whole portions of nature which are within its access, by virtue of treating such phenomena as being either spurious or beyond some self-defined measurability mandate.

Bohm lays out the issue this way:

> [M]easure is a form of insight that has to fit the overall reality in which man lives, as demonstrated by the clarity of perception and harmony of actions to which it leads.... [W]hen measure is identified with the very essence of reality, *this* is illusion.... [T]he primary reality goes beyond anything that can be contained in such fixed forms of measure [as science uses. We must progress to the point where t]he measurable and the immeasurable are then in harmony and indeed one sees that they are but different ways of considering the one and undivided whole. (Bohm, 1980a, pp. 22, 23, 25, and 26)

One can also argue that a restriction of science to things measurable also represents a style of thinking more suitable to a pre-Quantum Theory science than to a later one ... to repeat the kind of objection that we raised in the previous subsection. The point here is that Quantum Theory taught us that we are not living in a world where everything is clockwork, and all that we must do is describe the workings of the clocks. Such a deterministic philosophy will not suffice, because nature cannot be predicted in any complete sense — in principle. As Feynman puts it, "The future ... is unpredictable." (Feynman, 1965, p. 146).

Of course, we noted in Chapter 2 that this acknowledgement was more begrudging than wholehearted: Chance and non-predictability were accepted as an unfortunate necessity — and only integrated into the theory by being essentially turned upside down and made (negatively) predictable, in the form of a specific and isolated factor of "indeterminism" or "uncertainty" — "probability patterns". We noted that in this way, the spirit of non-predictability was effectively squelched, as it was transformed into the mold of (negative) prediction. What remained was a meaningless set of operational constructs, whose sole function was a kind of ad hoc power — but without taking into equal account prediction's equal-and-opposite counterpart of non-predictability.

What we are arguing is that *predictability is tantamount to measurability*, and one way of understanding the connection is in terms of the wave-particle duality — which we know means that one cannot simultaneously measure both conjoint aspects to the same degree of precision. Therefore, nature does not consist merely in that which can be isolated and tested — *or measured* — with precision, and science must not do so either.

Consider another example, which we examined in the previous chapter — that of the discrete quantum. Thus, what if it is true that one can only measure discrete energy changes to the maximum degree, as opposed to continuous ones — what conclusion does that allow one to draw? It would suggest *either* that energy changes in nature are only discrete, and never continuous; *or* it would suggest that we can only measure discrete energy changes, to some maximum degree of accuracy — but that the nature of reality is not so one-sided, that it is both discrete *and* continuous.

Ultimately, of course, one cannot prove either answer: one can either assume that observation and measurement do produce one-sided outcomes, or one can assume that nature is consistently two-sided … measurability notwithstanding. So how does one make this *metaphysical* decision? We suggest that it can either be grounded on ad hoc assumptions, which are largely specific to the particular phenomenon or issue at hand — and we claim that this is the basis for the unitary conclusion of Quantum Theory here; or it can be grounded on general principles, which are very broad and seem to apply to all phenomena — and we claim that this is the basis for the dualistic conclusion of Two-Force Theory.

We strongly contend that science must describe the whole that nature is, and not the parts which scientists want or imagine it to be — or even those which may be the easiest to study. Because an idea is not quantifiable or testable, even in principle, does not make it unworthy of consideration in science. We must learn to accept and understand nature's vagueness, just as much as nature's certainty, if we are to grasp all of its themes.

Now for some brief comments regarding the two types of special experiments that we discussed earlier. In the first case, that of the particle collision experiment, the main question would be whether we are illuminating nature with such techniques — by mimicking its normal processes and witnessing the results; or whether we are, as it were, "short-circuiting" nature, by distorting such phenomena and thereby creating laboratory artifacts. Thus, what basis do we have for assuming that science is actually *discovering* such phenomena, rather than merely *inventing* or *creating* them? Zukav puts the issue this way: "[I]t is no longer evident whether scientists really discover new things or whether they *create* them." (Zukav, p. 9) And Feynman says: "Actually some of these so-called particles last such a short time that there are debates about whether it is in fact possible to define their very existence …" (Feynman, 1965, p. 154) For his part, Roszak wonders,

> Does modern experimentation "discover" or "manufacture" the effects it studies? Eddington raises the issue (in his *Philosophy of Physical Science*), asking if advanced experimental equipment does not tell us how nature can be *made* to behave, rather than how she cares to act.... Nature ... invites and rewards curiosity, but not aggressive pounding. (Roszak, pp. 303-304)

Lastly, let us make a few points about the second type of special experiment — that of the thought experiment. As was discussed in Chapter 7, our position is that such experiments often turn out to be fairly insidious maneuvers, largely because of the obscure and abstract assumptions upon which they are based. These "chess games of the mind" are often used to prove or illustrate especially bizarre points or paradoxes. The peculiar scenarios may concern phenomena of which we have no normal experience — which is, of course, why they are proposed as thought experiments to begin with. But the problem is that by being highly abstract and selective, thought experiments take phenomena out of their normal context — thereby often avoiding natural interactions, and juxtaposing inherently incompatible physical situations. It is too easy to be misled by a logical analysis whose components we cannot readily relate to — precisely because we have no real intuitive grasp of the scenarios which we are supposed to imagine; and there are too many unknown or unstated assumptions which underlie such abstract fantasies for us to really understand what could be happening and what it all means. In this sense, it is hard to know what conceptual wool is being pulled over our eyes as we are subtly led to focus on certain aspects, while neglecting others.[19]

We maintain that one must be very leery of the technique of using wholly imaginary scenarios to demonstrate important phenomena or laws for which we can find *no* analogue in natural circumstances. Therefore, while such bizarre thought experiments may be very entertaining and intriguing, they may easily drift over into what is more in the realm of magic than of science ... a topic that we have touched upon before, and will return to in the next chapter. In most cases, natural phenomena available to our observation are far more dependable and less susceptible to alluring conceptual traps — assuming that we are talking about actual patterns in nature, as opposed to make-believe.

Aesthetic Criteria Reconsidered. Now let us return to the category of aesthetic criteria, which we have specifically talked about in terms of simplicity, generality, and coherence. These are admirable qualities, to which the established paradigm largely subscribes. The problem is that while these kind of characteristics are endorsed in lofty words, they are all too often not implemented in theoretical practice.

To start with, we would argue that the *simplicity* of nature is being obscured by the complexity of existing concepts and theories about nature. Put another way, which is consistent with some of the comments we cited in the parallel subsection earlier, *if some aspect of nature seems complex, it is because we have not yet come up with a sufficiently*

19 Feyerabend's more general comment is quite relevant here: "Methodology has by now become so crowded with empty sophistication that it is extremely difficult to perceive the simple errors at the basis." (Feyerabend, p. 58)

simple conceptual system with which to explain it. Unfortunately, this is what we are facing in science today: Far from being simple, the established paradigm has made complexity into an art form — a kind of modern abstract art. The result is to leave us with the image of an inscrutable nature.

Another important point is that the most simple theory is not the same as the most reductionistic theory — especially to the extent that the latter represents what we have described as a "no symmetry" collapsing which obscures the complementary two-sidedness of nature in all its many forms, such as we discussed in the previous chapter.

As far as the issue of *generality* goes, it is important to point out that terms like "universal" and "general" are sometimes used in very particular ways by the established paradigm. For example, we quoted Popper in the previous chapter as saying that, "[O]ur theories must be *universal* — that is to say, must make assertions about the world — about all spatio-temporal regions of the world." (Popper, 1981, p. 197) But that concept of universality only means that a particular phenomenon (e.g., the speed of light, *c*) would be the same everywhere in the universe. According to such a circumscribed criterion, it matters not how trivial or restricted is the phenomenon with which a theory deals, or how narrowly the theory deals with it — so long as it deals with it in the same way, in all regions of the universe.

By contrast, we maintain that a theory is really general only if it deals with all kinds of phenomena in nature. Thus, theoretical universality for Popper refers to a universe-wide range of *conditions* in which a particular phenomenon manifests itself; whereas for us, theoretical generality refers to a universe-wide range of *phenomena* in which a particular concept or theory manifests itself. We maintain that a principle is general not only because it describes a particular phenomenon in universal terms — but because it describes all phenomena in universal terms ... in the way we interpret the Feynman quote which we cited earlier: "Nature uses only the longest threads to weave her patterns, so each small piece of her fabric reveals the organization of the entire tapestry." (Feynman, 1965, p. 34)

Note that the position we are arguing for runs counter to the established way of thinking in another respect, in that there is a tendency to dismiss such broadly-applicable theories as having little value for science. Horgan makes such a point, in arguing that, "[A]ny theory that is so general that it applies to galaxies and immune systems and cities is pretty much a useless theory. It's so general that it doesn't really say anything useful about any of those particular systems." (Horgan, 1995) Again, we take precisely the opposite approach — and claim that this is the essence of generality.

Then there is the very familiar way in which the established paradigm rejects the kind of pervasive generality for which we are arguing — such as in the notion of a split between the micro world and the macro world, as we discussed earlier in the book. Consider Feynman's position in the following statement — which we maintain contravenes his point that we quoted just above: "[T]he behaviour of matter on a small scale obeys laws very different from things on a large scale." (Feynman, 1965, pp. 32-33)[20]

Morris take the same position — which we claim is also out of the spirit of a remark of his that we quoted earlier, where he said that, "A scientific theory is considered to be useful only if it can account for a wide variety of phenomena." (Morris, p. 142) But then he takes a different position when he says that, "After all, there is no reason to think that the world of the very small (or the very large, for that matter) should look like the world of everyday experience." (Morris, p. 23) Likewise, Bohm says that, "[W]e are not supposing that the same pattern of things is necessarily repeated at all levels ..." (Bohm, 1957, p. 139)

Needless to say, the present model is supposing precisely that: Rather than being grounded at particular levels — or even in terms of a split world of micro vs. macro levels — 2FT is grounded upon the assumption that there are general patterns which exist at all levels. In fact, we would argue that a principle or law is truly general (and thus not ad hoc) if there are examples of the basic principle outside the specific realm (or discipline) in which it was developed.[21] If the principle seems to have no relevance elsewhere, then that alone would make it suspect as a general principle of nature.

This brings us to the matter of *coherence*, with its two aspects — how well the component portions of a theory fit with one another, as well as how understandable the theory is. The present model is based upon an assumption of unity, which means that we live in a single, unified world that fits together; and we also assume that the world can be comprehended. Our models should mirror these characteristics — and so we have the right to expect our theories to present what is largely a single, unified, and comprehensible vision of such a coherent and lucid world.[22]

We should clarify one point in this regard — which is that while our theories must describe characteristics of nature like uncertainty and disorder, such characteristics should be *reflected in* our models of nature and not be *characteristic of* our models of nature. Yet unfortunately, this latter situation is precisely the case — with the result being that there is no coherent and meaningful conception of nature in science today. All one can conclude is that there is no unified nature, but that there are many separate natures — by virtue of conflicting sets of theories and laws coming into play, each yielding a different image of nature's patterns. As we have previously indicated and will discuss at length in the next chapter, the result is a plethora of random conceptual pieces which do not fit together into a coherent whole.

Finally, as we already noted earlier, any consideration of such aesthetic factors, as being desirable characteristics of good theories of nature, should not overlook the fact that all such discussion is pure metaphysics: there is no way to prove even that there are

20 Likewise, Capek speaks in support of this conventional interpretation: "[T]he inadequacy of one of the most fundamental ideas of classical physics: the alleged similarity of the microphysical world and the world of our daily experience. It was generally accepted that molecules, atoms, and electrons did not differ basically from the bodies of our experience.... The microcosmos was thus viewed as a miniature of the macrocosmos ... The failure of all such mechanistic pictorial explanations is one of the most conspicuous features of contemporary physics.... Before the coming of relativity and wave mechanics, the same logic was applied in constructing the models of the megacosmos as that of the microcosmos. Since the space of classical space was Euclidean, similar geometrical figures existed at different scales of magnitude; consequently, it was believed that the only difference between the atomic and the celestial realm was that of dimension." (Capek, 1988, p. 99)

21 One can argue that sometimes the established paradigm also embraces this premise — an outstanding example being that of the conservation laws, as representing general patterns of nature which are not specific to particular fields.

22 Note Feynman's dissent: "What *could* such a statement mean.... 'the underlying unity of nature'?" (Feynman, 1964, p. 12-12)

desirable criteria, to begin with — let alone that any particular set are them ... no matter how widespread is the consensus in this regard. Nor, at the end, is there any way to prove that some theory does or does not satisfy such criteria. None of these are problems for us, with our pro-metaphysics ideology; but they would surely seem to be so for the established paradigm.

The Peripheral Role of Mathematics. We now return to the next topic we discussed earlier — and the issue here is not whether mathematics has its place in our study of nature, but rather what that place should be. In particular, has the mathematization of physical science gone too far?

Not surprisingly, our position is that mathematics should not be primary in our attempts to understand nature: it does not eliminate the need to first have a solid conceptual framework. Thus, we maintain that mathematics is just a tool — a useful technique for manipulating data or describing constructs, and thereby determining the finer details of the phenomena in question. But there is nothing miraculous about these mathematical maneuvers or outcomes: they must be based on meaningful concepts about reality — and it is those constructs which must be primary. If the constructs are not viable, then neither will be the manipulation of those constructs. Here is how Popper puts it:

> [There] are no such things as extra-linguistic mathematical objects; that is to say, thoughts ... [M]athematical talk is *about* these objects; and in this sense, mathematical language is secondary to these objects. But this does not mean that we could construct mathematics without language.... [T]he objects of mathematics owe their existence partly to language ... (Popper, 1981, p. 137)

Popper goes on to paraphrase Berkeley in saying that,

> A mathematical hypothesis may be described as a procedure for calculating certain results. It is a mere formalism, a mathematical tool or instrument, comparable to a calculating machine. It is judged merely by its efficiency. It may not only be admissible, it may be useful and it may be admirable, yet it is *not science*: even if it produces the correct results, it is only a trick.... [T]he question of the *truth* of a mathematical hypothesis does not arise — only that of its *usefulness as a calculating tool*.... Properly understood, a mathematical hypothesis does not claim that anything exists in nature which corresponds to it — neither to the words or terms with which it operates, nor to the functional dependencies which it appears to assert. It erects, as it were, a fictitious mathematical world behind that of appearances, but without the claim that this world exists. (Popper, 1965, p. 169)

Niels Bohr made a similar point, according to Jammer: "Mathematics, Bohr said, could not prove any physical truth, it could only show the consistency of a formalism and its suitability to express relations among physical data." (Jammer, 1974, p. 67) Likewise, and in addition to the pro-mathematics sentiments of his that we cited earlier, Feynman also makes these cautionary statements: "Mathematicians are only dealing with the structure of reasoning, and they do not really care what they are talking about. They do not even need to *know* what they are talking about, or ... whether what they say is true."

(Feynman, 1965, p. 55) He says simply, "Mathematics is a tool for reasoning." (Feynman, 1965, p. 40) Or again, "Mathematics is just organized reasoning." (Feynman, 1965, p. 41)

Bertrand Russell says that, "Mathematics may be defined as the subject in which we never know what we are talking about, nor whether what we are saying is true." (Russell, quoted in Zukav, pp. 98-99) Cartwright makes the same point: "We are misled if we take all mathematics too seriously. Not every significant mathematical distinction marks a physical distinction in the things represented." (Cartwright, p. 200) Likewise, Morris: "It is perfectly possible to carry out calculations without ever wondering what it means." (Morris, p. 220) As Peat puts it, "If you have a beautiful mathematical theory, you still have to interpret it.... Interpreting it means language and concepts." (Peat)

Zukav notes that a mathematical

> proof only proves that we are playing by the rules (We make the rules, anyway).... A scientific "proof" is a mathematical demonstration that the assertion is logically consistent. In the realm of pure math, an assertion may have no relevance at all. Nonetheless, if it is accompanied by a self-consistent "proof" it is accepted. (Zukav, pp. 87 and 272)

The problem is that in practice, what Popper describes as "a fictitious mathematical world" is taken very much as describing reality, even though when transposed to reality it may have little or no — or else a very ambiguous, if not misleading — meaning. As Stapp says, for example, about the Theory of Relativity's treatment of time:

> Einstein's theorizing effectively redefined theoretical physics to be the attempt to create a mathematical structure of relationships between the elements of the static realm of readings, rather than as an attempt to understand or describe the process by which Nature unfolds. (Stapp, 1982, p. 381)

In the case of General Relativity, the mathematics is largely inscrutable and in that respect of dubious value. Some years ago, Davies wrote that, "One of the most notorious features of Einstein's field equations is the extreme difficulty in solving them. Indeed, in the 60 years since their discovery, only a handful of exact solutions are known." (Davies, 1977, p. 104) Actually, there are those who doubt the value of the model altogether: "General relativity is a splendid piece of mathematics built on quicksand and leading to more and more mathematics about cosmology (a typical science-fiction process)." (Brillouin, p. 10)

One argument in support of the view that mathematics plays a crucial role in understanding nature concerns the complexity of the underlying phenomena. This is Popper's position:

> All that we can infer about the nature of the world from the fact that we have to use mathematical language if we want to describe it is that the world has a certain degree of complexity, so that there are certain relationships in it which cannot be described with the help of too primitive instruments of description. (Popper, 1965, p. 330)

If one subscribed to this view, then there might be hope in the fact that we have increasingly sophisticated computer power — so even if the mathematics gets unbelievably complicated, we would have technological help in solving the equations and getting us through the conceptual quagmire.

Yet that misses the point: mathematics do not comprehend the universe,[23] and computers do not really think. Those are secondary tools which cannot substitute for the sound theoretical constructs that should constitute the primary tools of our minds. Mathematics and computers must be subservient to good concepts and theories, not the other way around. As they say with computers, so with mathematics: garbage in, garbage out. In any event, we contend that if our constructs and theories were not so clumsy and intractable to begin with, then we might have less need of such a complex system of mathematics as a means for somehow circumventing the inherent conceptual incompatibilities!

Herbert Dingle writes fervently about the misuse of mathematics in science. He attacks "the false conception of the relation between mathematics and physics (or more generally, experience).... Mathematics in itself ... is independent of experience." (Dingle, 1972, p. 122) Physics overlooks that crucial detail, and what we have instead is the prevailing view, the "unwritten precept in modern physics ... which states that in physics 'anything which is not prohibited is compulsory' " (Bilaniuk and Sudarshan, 1969, quoted in Dingle, 1972, p. 28) — thus, "that *any* (mathematical) hypothesis is *necessarily* true." (Dingle, 1972, p. 28) He describes the result: "The equations worked, so the 'experimenters' became convinced that the theory, whatever it was, must be right." (Dingle, 1972, p. 181) Not surprisingly, Dingle bemoans "the exchange by reason of the cloak of Aristotelian logic for that of mathematics." (Dingle, 1972, p. 29)

Not surprisingly, Dingle disputes Galileo's contention that, "The book of nature ... is written in the mathematical language." (Dingle, 1972, p. 30) In particular, Dingle restates the important point that:

> [A] language is a medium for expressing ideas, and it is just as capable of expressing false ideas as true ones.... Mathematics is an immensely more powerful tool than the Aristotelian syllogism, and its use as a language in which to express the facts of experience has been so successful that the idea has crept unperceived into the minds of physicists that whatever it says must be true. This is openly expressed in the statement already quoted, that everything that is not mathematically forbidden is necessarily observable. Accordingly the habit has developed of assuming that a physical theory is necessarily sound if its mathematics is impeccable: the question whether there is anything in nature corresponding to that impeccable mathematics is not regarded as a question; it is taken for granted.
>
> The fact is, however, that mathematical truths are far more general than physical truths: that is to say, the symbols that compose a mathematical expression may, with equal mathematical correctness, correspond both to what is observable and that which is purely imaginary or even unimaginable. (Dingle, 1972, p. 30)

23 Peat cites Wheeler's references to "magical mathematics", indicating that "If you had the ultimate equation, and you wrote it down, the paper would fly off, it would take off — it would create the universe!" (Peat)

Furthermore, Dingle would not accept the proposition that mathematics represents a higher form of reasoning — and he cites the diminution of our cognitive faculty of common sense (a topic we shall return to below), which would otherwise provide a helpful antidote to the inflated role which mathematics has taken on in our theory-building process:

> [T]he practice of mathematical physics goes hand in hand with lack of elementary reasoning power and of that normal form of human wisdom, somewhat misleadingly called commonsense, that provides its own corrective of premature judgment and never allows the requirements of reason and experience to be overcome by the seductions of attractive speculations.... It is usually taken for granted that the processes of mathematics are identical with the processes of reasoning, whereas they are quite different. (Dingle, 1972, p. 127)

Dingle notes that whereas the creator of *Alice in Wonderland* "recognised the products of his imagination to be wholly fanciful, the modern mathematician imagines, and persuades others, that he is discovering the secrets of nature." (Dingle, 1972, p. 128)

The Theory of Special Relativity is cited by Dingle as a prime case where mathematics has played too great a role in the development of scientific theory. He refers to "the illegitimate assumption that there is some *necessity* for whatever is true in mathematics to impose its inevitability on experience, that is primarily responsible for the error of the Special Theory of Relativity ..." (Dingle, 1972, p. 124) In that context he writes that,

> [A]ll that is necessary to entitle a physical theory, however absurd, to respect is to discover some mathematical process whose symbols can be arbitrarily correlated with the physical entities of the theory, without regard to evidence or probability or commonsense.... [T]he supposed justification of the Special Theory of Relativity ... lies wholly in the impeccability of its mathematical structure: the impossibility of the application to experience of that structure, in the manner postulated by the theory, is left out of consideration altogether ... (Dingle, 1972, p. 127)

Dingle does not dispute the "numerous instances in which mathematics has *suggested* physical possibilities which later have been realised and might not have been thought of without the suggestion; its service in this respect can hardly be over-estimated." (Dingle, 1972, p. 130) But what he objects to is "the *mastery*, instead of the *servitude*, of mathematics in relation to physics ..." (Dingle, 1972, p. 130) — a problem which he considers to be a recent one: starting not with Galileo or Newton or Faraday, but with Maxwell — and increasing up to the present time. Eventually,

> a state of mind was generated in physicists by which, while still formally adhering to the principle that observation was basic and mathematics a useful tool, they were ready to accept mathematical requirements as an adequate substitute for a genuine theory, even though they could see nothing intelligible that corresponded to it physically. (Dingle, 1972, p. 132)

Again, this is the case with Einstein's Special Theory, as we discussed in earlier chapters. We have deplored the fact that physics was willing to embrace an untenable postulate regarding a constant speed of light, at least partly because such a choice may have been more expedient for mathematical purposes. As Zapffe cynically asks, "[A]fter all, is not elegance in mathematics a much more precious thing than the hard facts of life?" — even though "the penalty is intellectual chaos." (Zapffe, 1977a, pp. 17 and 22) This may be one of the more outlandish examples in the modern history of science whereby mathematics has led physics, and in this way dictated reality for the rest of us.

Science has gotten so lost because of its own unworkable constructs, and so mesmerized with its own techniques — most notably mathematics — that it has blinded itself to a deeper conception of nature. In particular, it is the height of intellectual arrogance, and an insult to most of humanity, to state or suggest that nature cannot be understood without a mastery of mathematics. More generally, we vehemently resist any contention that mathematics should lead our pursuit of nature's themes — as opposed to its helping sort out the details of what must be a coherent conceptual framework to begin with.

Truth vs. Falseness. Karl Popper's falsifiability criterion is subject to challenge on various fronts. To begin with, even Popper acknowledges that,

> In point of fact, no conclusive disproof of a theory can ever be produced; for it is always possible to say that the experimental results are not reliable, or that the discrepancies which are asserted to exist between the experimental results and the theory are only apparent and that they will disappear with the advance of our understanding. (Popper, 1968, p. 50)

In a similar vein, Kuhn writes that, "All experiments can be challenged, either for their relevance or their accuracy. All theories can be modified by a variety of ad hoc adjustments without ceasing to be, in their main lines, the same theories." (Kuhn, 1977, p. 281) Kuhn speaks of the "anomaly in the fit between theory and nature.... There are always some discrepancies.... There are always difficulties somewhere in the paradigm-nature fit ..." (Kuhn, 1970, pp. 81-82) A theory "need not, and in fact never does, explain all the facts with which it can be confronted." (Kuhn, 1970, p. 18) "[N]o theory ever solves all the puzzles with which it is confronted at a given time; nor are the solutions already achieved often perfect. On the contrary, it is just the incompleteness and imperfection of the existing data-theory fit that, at any time, defines many of the puzzles that characterize normal science. If any and every failure to fit were ground for theory-rejection, all theories ought to be rejected at all times." (Kuhn, 1970, p. 146) In place of Popper's all-or-none falsification thesis, Kuhn advocates a probabilistic verification approach: "It makes a great deal of sense to ask which of two actual and competing theories fits the facts *better*." (Kuhn, 1970, p. 147)

But more generally, Kuhn criticizes Popper's thesis that the work of science should lead us "closer and closer to the truth" (Kuhn, 1970, p. 170). Kuhn writes:

> We are all deeply accustomed to seeing science as the one enterprise that draws constantly nearer to some goal set by nature in advance. But need there be any such goal? Can we not account for both science's existence and its success in terms of evolution from the community's state of knowledge at any given time? Does it really help to imagine that there is some one full, objective, true account of nature and that the proper measure of scientific

> achievement is the extent to which it brings us closer to that ultimate goal? (Kuhn, 1970, p. 171)
>
> A scientific theory is usually felt to be better than its predecessors not only in the sense that it is a better instrument for discovering and solving puzzles but also because it is somehow a better representation of what nature is really like. One often hears that successive theories grow ever closer to, or approximate more and more closely to, the truth.... Perhaps there is some other way of salvaging the notion of "truth" for application to whole theories, but this one will not do ... [T]he notion of a match between the ontology of a theory and its "real" counterpart in nature now seems to me elusive in principle. (Kuhn, 1970, p. 206)

Frank takes a similar position as Kuhn, in arguing that, "It has never happened that all the conclusions drawn from a theory have agreed with the observable facts.... '[A]greement with observed facts' does not single out one individual theory. We never have one theory that is in full agreement but several theories that are in partial agreement, and we have to determine the final theory by a compromise." (Frank, pp. 210 and 217)

Feyerabend sharply repudiates Popper's falsifiability approach:

> [T]heories shape and order facts and can therefore be retained come what may.... [N]o new and revolutionary scientific theory is ever formulated in a manner that permits us to say under what circumstances we must regard it as endangered: many revolutionary theories are unfalsifiable. Falsifiable versions do exist, but they are hardly ever in agreement with accepted basic statements: every moderately interesting theory is falsified. Moreover, theories have formal flaws, many of them contain contradictions, ad hoc adjustments, and so on and so forth. Applied resolutely, Popperian criteria would eliminate science without replacing it by anything comparable. They are useless as an aid to science. (Feyerabend, pp. 58-59)

Our position is in keeping with many of these comments, and we particularly reject Popper's lopsided treatment of truth vs. falseness. As we discussed earlier, Popper's thesis rests on the proposition that there is only a single truth, which can never be proven; but that there is an infinite array of falseness which can be widely demonstrated. The result is that truth is only provisional, since we can never know for sure about universal positives — whereas falseness is conclusive, since we can readily demonstrate such universal negatives in a variety of circumstances. Thus, the title for one of his books summarizes his differential theory: *Conjectures and Refutations*.

In fact, we maintain that the situation is more complex than this — even in terms of what we mean by truth and falseness, and the notion of a theory-fact fit is only one of several compatibility factors that need to be examined, as we shall discuss in the next chapter. But for our present purposes, let us simply say that our approach here would be based on a distinction between two issues: one concerns scope — i.e., local vs. global (or universal); the other concerns the differentiation itself — i.e., verification vs. falsification. We take exception to Popper's thesis, on both these counts — and we do so in terms of our two main themes of hierarchy and complementarity, respectively.

Starting with the hierarchy issue, it should be clear that Two-Force Theory is a multi-level model of nature, and that one way of describing these different levels is in terms of scope — which range from the very narrow and local to the very broad and global. Our position is that every level of nature's hierarchy is a quasi-independent level of reality in its own right. It is in this respect that we dispute Popper's thesis that a characteristic like truth has only a global or universal referent. For such a definition gives that quality an overly broad meaning which is not warranted: there is *not* only a single universal truth. We argue that there are also local truths, in addition to universal ones: Whether we want local or universal truth depends upon our goal, but the localness of a phenomenon does not diminish its truth in that context. Therefore, Popper is not justified in restricting truth to the singular and universal sort.

Once one abandons Popper's notion that there is only a single universal truth, then one finds that there is no merit to his differentiation between a falsifiability criterion, as opposed to a verifiability criterion. Consequently, our response to the second issue above is that falseness should be treated no differently from truth — and so we reject his one-sided falsification approach, and argue instead for a two-sided and complementary verification-falsification model. This also means that theories are never simply verified *or* falsified: rather, theories are *both* falsified *and* verified — in tandem and never singly.[24]

The crux of the dispute is whether counterinstances are to be taken as (a)falsifying a theory in an absolute and universal sense, thereby disqualifying it as a true description of nature in any respect; *or* (b)delineating the theory's applicability — elaborating the network of factors relative to which it is true on the one hand and false on the other. Popper argues for the former — a single universal truth only, which a theory either is or is not a candidate for ... which is why he rejects the notion, already cited, that "different theories have different ranges of application" (Popper, 1965, p. 113, italics omitted). By contrast, we opt for the latter choice, and in so doing reject Popper's single-truth approach.

Obviously, such a verification-falsification process would be situation-specific, in relation to given sets of referents — as opposed to only being universal and quasi-absolute. A theory which is shown to be false in one context remains true in another context. Thus, any (crucial) experimental decision point substantially elaborates the true-false nexus — by further delineating the proposition's verification referents from its falsification referents ... either in terms of a particular theory alone, or in terms of deciding between two competing propositions that are being tested.[25]

And what about Popper's argument that a theory is only confirmed within the limits of the constraining conditions under which it has been tested — and so cannot represent "confirmed truth" in some still-untested universal sense? That is so — but those same limitations apply also to how well a theory has been falsified. Every theory is *both* true and false, and it is both of these things within the context of certain delineating con-

24 Note that while Kuhn likewise speaks of a "joint verification-falsification process" (Kuhn, 1970, p. 147), he is talking specifically about a competition between two theories; we are not.

25 Kuhn would not accept the former: "In the sciences the testing situation never consists ... simply in the comparison of a single paradigm with nature. Instead, testing occurs as part of the competition between two rival paradigms for the allegiance of the scientific community." (Kuhn, 1970, p. 145)

ditions. Therefore, it is not enough to describe science merely in terms of the refutable. A good theory of science is no more merely the refutable than it is merely the confirmable: it is always *both*.

Neither a theory nor anything else is *merely* falsified; it is also verified — however bizarre the conditions may be which bring that about, and however seemingly unimaginable: for they obviously have been somehow imagined, or we would not even be considering the "unimaginable" theory. All one can do is state the conditions under which the theory is verified, as opposed to those under which it is falsified. Therefore, a superseded truth is not only a truth falsified; it also remains a truth verified, relative to some (e.g., prior) set of conditions. We will pick up on these points when we continue our discussion of truth and falseness in the next chapter.

Arbitrariness in Theory. We give very short shrift to the free invention topic, in our response to the established paradigm's methodology. That is because much of the next chapter is a very broad discussion of the underlying issues, and the short-sighted folly of an arbitrary approach to theory-building.

Simply speaking here, it would not be an overstatement to say that we find such an arbitrary approach to nature to be utterly unacceptable. As Morris puts it, "If the concepts of physics are human inventions, it does not follow that they can be chosen arbitrarily, or be conceived of as interacting with one another in arbitrary ways." (Morris, p. 196).

Needless to say, we also reject the kind of thinking described in the last part of that earlier discussion on free invention — which we talked about in terms of which comes first, nature or theory ... and the very idea with the latter that nature somehow conforms to mankind's laws and theories, rather than those laws and theories merely representing our best attempts to describe the patterns of nature.

In Defense of Common Sense. We have seen that the positivist ideology gives very little credence to the role of meaningfulness, in our theories of nature — and that virtually all that matters is the narrow functional criterion of describing to the fullest extent possible the particular set of phenomenological patterns at hand. Now let us return to that most maligned form of meaningfulness, which is sometimes called common sense or intuition ... or even normal logic.

As we have discussed, this category of knowledge or thinking is treated with considerable derision by the established paradigm. Thus when a conclusion of science appears to conflict with the lay person's common sense, the latter is dismissed by the orthodox regimen — often in a not-so-subtle condescending manner — as being a virtual stone-age vestige that is unworthy of the lofty reality which science strives to represent.

So how are we to resolve this conflict between the knowledge which comes from the established paradigm, and that which comes from common sense? Unfortunately, we must also put off dealing with some of the underlying issues, until we can lay a firmer ground for their discussion in the next chapter ... when we talk more generally about the search for truth and reality in nature. But several points are in order here.

First and foremost, it seems useful to describe the issue in terms of two bodies of knowledge, in these two systems — and we maintain at the outset that neither has some sort of inherent preeminence or what is akin to papal infallibility. Moreover, given that we would argue there are legitimate aspects to both sides of this dispute, then any satisfactory resolution must come in the form of reconciling the differences between them in one manner or another.

Having said that, it should not be surprising where our sympathies lie, in terms of how the imbalance between these two systems needs to be rectified: common sense warrants a far fairer hearing than it normally gets in the temples of scientific power ... whereas the body of knowledge which is associated with the established paradigm is itself grossly overrated. So we opt for defending common sense, and we speak out against the move by science to squelch it as being irrelevant: On no account should people be hoodwinked into thinking that common sense must be discarded or disregarded. For ultimately, to belittle common sense and logic is to belittle meaning and the very notion of meaningfulness; it is to mock the goal of understanding — for these are all the same thing.[26]

Fortunately, there are scientists who realize the importance of common sense. Frank, for example, writes about Bacon's

> rule to keep the principles of science as close to common sense as possible ... [and that] one of the reasons for the acceptance of a theory has frequently been the compatibility of this theory with daily life experience or, in other words, the possibility of expressing the theory in common-sense language. (Frank, pp. 213-214)

Gill notes that,

> While it is now axiomatic that physical theories do not need to conform to "common sense" it is still a fact that a disagreement with "common sense" or with our intuitive ideas of the world, makes a prima facie case for taking a closer and more critical look at a theory. One reason for this is that an important element in scientific common sense is logic, and a logical contradiction in a theory must sooner or later manifest itself; in other words, a conflict with common sense can either be a symptom of a serious defect in a theory, or merely our reaction to the unfamiliar. The first possibility must not be ignored in our overwhelming enthusiasm for the strange and paradoxical, or in a pragmatic disdain for the integrity of our theory. (Gill, pp. 13-14)

Sellars says that,

> [S]cience is continuous with common sense, and the ways in which the scientists seek to explain empirical phenomena are refinements of the ways in which plain men, however crudely and schematically, have attempted to understand their environment and their fellow men since the dawn of intelligence. (Sellars, p. 182)

Popper refers to Kirk and Raven's methodological principle "that gross departures from common sense must only be accepted when the evidence for them is extremely strong."

26 "[T]he laws of logic reflect the structure of being ..." (Picht, cited by Jammer, 1974, p. 376)

(Popper, 1965, p. 148) Popper makes the further points that, "Science, philosophy, rational thought, must all start from common sense," and that, "[A]ll science, and all philosophy, are enlightened common sense." (Popper, 1981, pp. 33-34, italics omitted)

Therefore, common sense is not irrelevant and it is not an antiquated relic — but rather, it is an important clue and a primary criterion for how we are to define reality. Common sense is an invaluable reference system for helping to determine the viability of a scientific construct or theory, and deciding whether it is so much hot air or a viable description of reality. Dingle made a similar point in a segment we quoted a few pages back, in the Mathematics subsection (Dingle, 1972, p. 127). Common sense serves as our mental touchstone — in fact, one might even argue that it provides the gateway to the aesthetic criteria that we discussed earlier.

Indeed, where did our common sense come from? It did not emerge out of nowhere, from a vacuum — it also has roots in "facts". One might suggest that the common sense inside us is a cumulative product not only of our own experience and of the collected history of the ancestors who came before us, but even of all the non-human evolutionary input from the past — the nature we have grown upon. It is our derived, inherited, and synthesized understanding of nature — our "accepted wisdom", our map of the world. Common sense is that body of knowledge we use all the time to guide us in our daily lives.

Part of what "feels right" to us, when we make a judgment according to our common sense — something which is, in Toulmin's terms, "pleasing to the mind" (Toulmin, 1961, p. 115) — is at least partly based upon the degree to which it seems to be harmonious with nature ... both the nature of which we are composed, and the cognitive understanding of nature which we have formed. Therefore, common sense is the sense of nature inside us — it is our "nature sense". The better is our relationship to the larger natural world, and the better order that we can make out of that world, the more valid will be our nature sense judgments about the world.

It is also important that we address one of the main rationales which is put forward (especially by physics) to explain why common sense is not a valid barometer for evaluating good theories — and this concerns the range of experience factor. According to this argument, our normal world experience is quite limited, in terms of the scope of natural patterns that it encompasses: the common sense conclusions which this leads to are said to be similarly narrow, thus leaving us ill-equipped to deal with the deeper patterns of nature. One result is that we are encumbered by obsolete classical concepts, which have subsequently been made outmoded by the discoveries of modern science.

These newer findings are based upon a much more diverse array of natural phenomena, both here on earth and elsewhere in the universe, than any of us encounters in our normal lives. By virtue of this wide-ranging experience, scientists (especially physicists) are credited with having the most intimate and extensive — even exclusive — contact with the physical world. This has allowed them to break free of the kind of inhibiting chains which most people are constrained by ... thereby allowing these privileged individuals to transcend the flawed vision of everyday experience, and lead us into the promised land. The result is that scientists are seen to be in a unique position to objectively describe the full nature of reality — which is why the backward Old Thinking must be superseded by the enlightened New Physics.

One significant illustration of the disparity can be found in the notion that the micro world works in a different manner than the macro world: Our normal common sense experience would be restricted to the macro world of our everyday lives, as distinct from the micro world of the discrete quantum. Then, again, the macro world itself provides another illustration of the limitations: thus, our usual experience is said to be restricted to phenomena having no or slow speeds — as against the Special Relativity world of high speed, which also cannot be fully fathomed in any common sense terms.

The upshot is that contemporary science essentially argues that there need be no particular relationship between the theoretical representation of the world which emerges from science, and the experiential representation of the world which emerges from our minds. For while common experience would yield common sense, superior experience would yield superior sense ... to wit, the extraordinary insights of the established paradigm.

Needless to say, we do not accept these arguments — and in particular, we reject the proposition that our common sense intuition cannot grasp the deeper truths of nature, in either the micro or macro worlds. On the contrary, we contend that just as conventional science is burying its head in the sand by virtue of its refusal to acknowledge that some kind of metaphysics is inherent in everything we do — including our explorations and theories of nature — so likewise it is overlooking an important tool for comprehending nature and evaluating our theories in its rejection of common sense.

Indeed, we would go so far as to suggest that the position of the established paradigm in regards to these issues actually amounts to cover for its own inadequacy — which, of course, is quite a contrary perspective. But in arguing that they are so advanced in their analysis of nature such that they have gone beyond the realm of cogent meaning — and that for anyone else to penetrate deeply into the real secrets of nature, then they too must abandon common sense and ordinary logic — these proponents have put themselves and their theories beyond the reach of the more usual kind of conceptual challenge. Because their ideas would not fare well in that fair domain, it is a matter of necessity to dismiss the domain altogether as being obsolete and irrelevant.

We maintain that the root of the problem is perfectly clear: it is not that our normal cognitive faculties and tools are inadequate, let alone that our range of experience is too limited to discern the real world out there. Quite the contrary is the case, on both counts. Thus, we turn the tables, and argue that far too often it is science which is not dealing with the real world right here, that we are all living in — scientists included — every day of our lives. They have erected schisms that need have no basis in fact: schisms between different levels of nature's realm, and a more fundamental schism between elite science's reality and normal people's reality ... and ultimately between mankind and nature. These are vital issues that we shall return to — with a vengeance — at the end of the next chapter.

The Structure of Scientific Dogmatism. Now we come back to the philosophy of Thomas Kuhn, and his approach to theoretical development. Recall the two types of science which Kuhn proposes: "normal" vs. "extraordinary", which could also be labeled as "dogmatic" vs. "revolutionary". He titles his classic book in terms of the latter type. By contrast, we argue that the most significant aspect of his work deals with the former, and that is the name we give to this subsection.

To start with, it should be clear that there is nothing objectionable for the present model regarding the inherently dualistic thesis which Kuhn proposes — that there are two modes of science (and of thought): Such a state of affairs would represent another manifestation of the complementary theme of the present model. The problem we have is not with the dualism, but with the dogmatism. We completely reject Kuhn's ideological validation of a rigid and entrenched approach to theory-building.

Indeed, it may actually be disconcerting for some people to hear modern science characterized as dogmatic, given that science is typically represented as being a bastion of open-minded, critical, and creative endeavor. Yet, as we have seen, closed-mindedness is inherent in the established paradigm, and Kuhn's influential philosophy actually embraces and validates such a posture for science. To call Kuhn an "apologist" for mainstream science would be too generous, for much of his work can be taken as a justification — which thus contributes to the perpetuation — of that staid mentality and doctrinaire manner. While Kuhn enthusiastically embraces the result, which he describes as "the wonderfully adapted set of instruments we call modern scientific knowledge" (Kuhn, 1970, p. 172), we are less sanguine ... to put it mildly.

For example, take Kuhn's position that a reexamination and potential "retooling" of our concept systems should be treated as an *extravagance*, which a)merely squanders energy that should be invested in the task at hand — that of elucidating the details of the established paradigm; and b)is something akin to a luxury — as though it were something that we might not be able to afford. Shouldn't the issue also depend on the status of the current system with which we are dealing, in terms of how well it meets our theory-building goals? What if the basics of our theories have been agreed upon prematurely — and that we may not have what Kuhn calls a "mature" or "paradigmic science" ... or that even if the dominant theory meets certain criteria, it may not meet others? All in all, why should efficiency be the paramount factor in driving the development of our models about nature and reality — such that consideration of any criticisms or alternative approaches would be treated as not just a distraction ... but virtually an act of frivolity, if not betrayal?

One of the most objectionable aspects of Kuhn's approach concerns the point that there is a consensus at any given time (of "normal science") regarding what the world is really like — and that scientists should strenuously resist any attempts to undermine this picture. The main problem with such a thesis is the implication that there is a coherent world view that is to be rallied around and defended at the given time. By contrast, one of the primary arguments we are making in this book is that there is no coherence in science which is to be upheld — and in this respect, consensus merely gives the illusion of such coherence.

The problem is not merely that there is no overall conceptual framework which brings it all together — no unified model of nature — but that the conceptual pieces we do have actually conflict with one another in significant ways. How does this fit with the Kuhnian mythology that there should be a single and virtually exclusive conceptual framework, as opposed to a variety of different conceptual schemes? He even speaks disparagingly about the diversity of the pre-paradigmic stage as representing "immature science", in contrast to the uniformity and conformity of the paradigmic science as representing "mature science".

Yet the period of modern science that we have known for the past century certainly does not fit with such a description — and Kuhn's own reasoning attests to such a disparate state of affairs. We alluded to this point earlier when we presented his views, where he disputes the notion that science develops in a cumulative fashion — let alone that it develops in a "coherent direction of ontological development." (Kuhn, 1970, p. 206) We will examine both of the issues in the chapter ahead, but the point we want to make here is that Kuhn himself argues that we are dealing with an array of separate constructs and theories in established science, devised at different times to account for distinct problems — and which thus belong to separate epochs. We contend that this also explains why these constructs are unconnected, and why there is no smooth fit between them — and thus why no cohesive unity is to be found. Given that each of these conceptual pieces emerged individually from sharply distinct milieus or contexts (or "disciplinary matrices"), it is virtually inevitable that they would end up being unrelated to one another, in terms of the depictions of nature which they represent. That is precisely the state of science today — and why this reality contradicts the mature science harmony or unity which Kuhn claims.

Therefore, the details which we are so efficiently compiling, as part of "normal science", belong to *different* conceptual frameworks — which do not fit with one another. The end result is that we have all these busy worker bees of science diligently and efficiently moving ahead — at what are essentially cross conceptual purposes, as they attempt to build up incompatible pictures of nature.

Such a conceptual hodgepodge notwithstanding, the issue of rigidity and dogmatism remains — because whether we are speaking about the established paradigm as a whole, or of its melange of component parts, the intransigence which Kuhn advocates is always there. The result is that fundamental constructs and theories are not allowed to evolve but instead are rigidly adhered to — virtually as a matter of faith. Unfortunately, the dogmatism which Kuhn lauds only leads to arrogance: there is no reason to think that it leads to good theories. By contrast, we contend that while greater humility may not be so efficient in the short run, it will be more fruitful in the long run.

Admittedly, there are some in the field who disagree with Kuhn's position, and who argue against such conformity … and stubborn adherence to the status quo. We turn to those voices now, for the fresh air they bring to these important matters.

Karl Popper adopts this anti-dogmatism position (even though it contradicts the pro-dogmatism posture of his that we cited earlier). He thus writes that, "No scientific theory is sacrosanct or beyond criticism." (Popper, 1981, p. 360) He continues:

> [W]e may identify the critical attitude with the scientific attitude, and the dogmatic attitude with the one we have described as pseudoscientific.… [T]heories are passed on, not as dogmas, but rather with the challenge to discuss and improve upon them.… [Science is valued] for its informative content, and for its ability to free our minds from old beliefs, old prejudices, and old certainties, and to offer us in their stead new conjectures and daring hypotheses. Science is valued for its liberalizing influence — as one of the greatest of the forces that make for human freedom. (Popper, 1965, pp. 50 and 102)

> [T]he dogmatic tradition permits only *one* school doctrine ... [whereas the alternative] admits a *plurality* of doctrines which all try to approach the truth by means of critical discussion.... In this rationalist tradition bold changes of doctrine are not forbidden. On the contrary, innovation is encouraged, and is regarded as success, as improvement, if it is based on a critical discussion of its predecessors. (Popper, 1965, p. 151)

Fortunately, Popper is not alone in his ideals. For example, Lindsay "warn[s] all thoughtful scientists to be on guard against the growth of that dogmatism which is the great enemy of scientific advance." (Lindsay, p. 26) Or as Toulmin puts it, "A man unshakeably committed to twentieth-century scientific concepts can display as much 'irrationality' as any Flat-Earther; he simply exercises his obstinacy on a different terrain." (Toulmin, 1972, p. 165)

Hoyle approaches the dogmatism issue not in terms of the goal of efficiency in elaborating the dominant paradigm, but rather in terms of whether such an approach is effective for resolving ongoing questions:

> [W]hen you go for the problems that are unsolved and are interesting, they naturally lead you to the unorthodox; and it's always seemed to me very clear that if the orthodox views were correct about unsolved problems, the problems would be solved. If you have the whole scientific world working on a particular line — on what you would just call the "band wagon" — if the "band wagon" were right, that problem wouldn't exist anymore. Because, there are a lot of very clever people around, don't misunderstand me: technically, the people on the "band wagon" are quite competent. The only thing that stops them from a complete solution of what they're seeking is that they're on the wrong track — that's the only possible explanation. (Hoyle)

Feyerabend expresses a similar skepticism:

> Any ideology that makes man question inherited beliefs is an aid to enlightenment. A truth that reigns without checks and balances is a tyrant who must be overthrown.... There is nothing inherent in science or in any other ideology that makes it *essentially* liberating.... My criticism of modern science is that it inhibits freedom of thought.... Science, of course, must be reformed and must be made less authoritarian. (Feyerabend, pp. 56-58)

> The progress of science, of good science, depends on novel ideas and on intellectual freedom.... Most scientists today are devoid of ideas, full of fear, intent on producing some paltry result so that they can add to the flood of inane papers that now constitutes "scientific progress" in many areas.... [T]he science of today ... is run by slaves, slaves of institutions and slaves of "reason". (Feyerabend, p. 64)

In closing this discussion, let us consider another tack. That is, Kuhn acknowledges that the pattern of evolutionary growth within a paradigm, preceded and succeeded by revolutionary changes from and to other paradigms, is a pattern that is not unique to science — that this same pattern applies to other aspects of society, as well. So let us now consider how Kuhn's reactionary views would be seen when applied to the larger society as a whole.

Translated to that sphere, we would have a rigid society where conformity and uniformity are the norm, and where there is essential unanimity and narrow adherence to strict dogma — in terms of whatever happens to be the reigning "paradigm" or "ideology", as manifested in the established laws, rules, and values of the day. Yet according to Kuhn's reasoning, this would be a very desirable state of affairs, most of the time. We should therefore strive for a society which is rigid and dogmatic, for not only would it function more efficiently, but its very rigidity would more readily and sharply expose its flaws — and ultimately bring about the complete overthrow of any such misguided regimes that were based upon false ideologies.

The contrast to such an authoritarian world would be the society where diversity and pluralism reign and where there is an array of alternative theories, rules, and values to choose from. Yet such an approach would be completely antithetical to the normal science world which Kuhn advocates — for in this more malleable world, governments and societies would reform themselves. Of course, it may well be that such choice and flexibility may reduce society's efficiency, in the sense of distracting its attention away from the main task of implementing some singular ideology or paradigm. Moreover, how else would a false ideology and inept regime be revealed, except by its stubborn and blind adherence to a path that was (later) found to be based on false precepts? According to the Kuhnian model, the regime must discredit itself sharply and significantly in order that it be overthrown.

As to which model is best, that depends upon the goals — and what should be the proper goals of science is one of the main issues we will discuss in the next chapter.

Indoctrination and Propaganda. Next, we look again at the issues of education and information, starting with Kuhn's view that education should amount to a rigid indoctrination. In that earlier discussion, we described Taylor's advice to the novice, and his scathing reference to the "Einstein's wrong syndrome". But we take quite the opposite stance, and argue that such a conformist approach is unacceptable: we turn the tables, and cynically talk about the "Einstein's right syndrome".

What this approach amounts to is a kind of intellectual cloning, whereby the reigning conceptual doctrine is simply carried forward unchallenged — rather than being subjected to any kind of serious questioning or rational critique. That might be all very well if we had reached the pinnacle of our theoretical development in science — assuming such a state can ever be attained — but it hardly provides the climate for new insights and fresh ideas, beyond the sort of minor tinkering that is permitted under Kuhn's "normal science" regimen.

Feyerabend is also critical of the established approach to scientific education:

> Scientific "facts" are taught at a very early age and in the very same manner in which religious "facts" were taught only a century ago. There is no attempt to waken the critical abilities of the pupil so that he may be able to see things in perspective. At the universities the situation is even worse, for indoctrination is here carried out in a much more systematic manner. Criticism is not entirely absent. Society, for example, and its institutions, are criticized most severely and often most unfairly and this already at the elementary school level. But science is exempted from the criticism. (Feyerabend, p. 57)

Bohm is also uncomfortable with this manner of teaching, which basically just conveys the upbeat party line:

> [W]e don't mention problems; it's more or less keeping a stiff upper lip.... I think that we can admit that we have certain problems and talk about them and recognize they're there, without necessarily becoming completely discouraged by them. This is really something which ought to be changed in the teaching of physics. (Bohm et al., 1964, p. 80)

So we ask: Why should scientific education be an initiation rite? Why should scientific textbooks be a source of inculcation only, rather than also being an inspiration for creative thought? And as Bohm says, why can't the outward face of science show more humility, and also acknowledge where theory has failed or is unsatisfactory — rather than pretending to be omniscient? Indeed, this last point applies not just to the scientific education of specialists, but also to the scientific education of non-specialists ... as well as to everyone else outside the classroom.

This brings us to a topic that we did not address in the parallel subsection earlier — which concerns the image of science that is conveyed to the public at large. One particular issue that we want to highlight here is the role of the mainstream media in disseminating scientific ideas to the broader society. We maintain that there are serious problems in this regard, primarily in that the relationship is too symbiotic as between those who do science and those who report science. All too often, science reporters act as the public relations arm of science — sometimes seeming like cheerleaders, in trying to rally support and enthusiasm for science's work from amongst the general populace. To put the matter bluntly, such reporters are simply not tough enough on science, and too often the coverage amounts to pro-science propaganda. Indeed, we would even argue that there is an improper mixing of two different functions here, as between the *reporting* about science on the one hand and the *promotion* of science on the other — and that this amounts to a real conflict of interests.

So while it is right and proper for the media to convey the findings and determinations of science, it is also important to maintain the necessary detachment and perspective which provide some degree of objectivity — and to adopt the same kind of critical and questioning stance that the media applies to most other topics. To be sure, this does happen to some extent in the case of certain technology issues — and far too much of the media takes balance to a ridiculous extreme when it concerns issues of science that have significant political implications. Yet when it comes to the kind of goals, content, and methodology issues that we are examining in this book, any sort of substantial questioning is pretty much taboo.

The reasons are complex and subtle — and they include various psychological, sociological, and cultural factors that we can only touch upon here. The starting point probably concerns a social hierarchy issue, whereby the institution of science is seen as having a lofty and definitive role to play in society; and scientists themselves are commonly thought of as among society's most intelligent individuals — who have special experience with, and keen insights on, the nature of reality. Like most people, science reporters would also feel a psychological sense of deference in this regard — and perhaps they also take pride in being part of an elite group of foot soldiers, who have an almost crusader mission of educating the public (or spreading the gospel) about the established paradigm ... thereby also advocating for science and subtly reinforcing its dominant role in society.

The issues go even deeper, for many of the positions which science has taken regarding the nature of reality (such as we have been discussing in these pages) eventually take on the status of cultural norms, which most people end up absorbing or complying with — making those matters even further immune from sensible scrutiny. Plus, the adoption of (and conformance to) scientific orthodoxy even becomes a measure and display of one's own intellectual enlightenment and sophistication — whereas to question these precepts is to display one's own intellectual inferiority and impoverishment. And for a science reporter to take such a stance is tantamount to giving aid and comfort to the backward forces of ignorance and superstition, which science is trying so hard to conquer and win over — making any such reporter an ideological traitor to the cause. Obviously, such an outcome would make it impossible for that science reporter to survive and flourish — with all the negative personal and social consequences — both in the sense of having a job to begin with, as part of any significant organization, but also in terms of having continued access to science insiders who are the original sources for most of the relevant news stories.

Finally, there is a cognitive factor, in that many of the issues are themselves difficult to comprehend, and therefore be subjected to analysis and critique. Plus, we have seen in these pages that some of the fundamental theoretical ideas simply make no sense at all — and one might argue that it is enough that reporters are even able to regurgitate or convey such inscrutables, let alone be in a position to question them! Of course, that in itself is one of the arguments we are making — that ideas which make no sense *should* be challenged ... although we fully acknowledge that such a position is hardly consistent with the orthodox thinking about these matters. At the very least, however, we contend that far more of the contradictions and limitations of science need to be pointed out, in a manner which is consistent with the Bohm quote above ... and not just the glowing portrait which is so common today.

Elitism and Expertism. For our final topic in this chapter, let us return to the kind of astonishing (if not deplorable) arguments which Kuhn put forward regarding the special (if not saintly) status of scientists — that only scientists can judge science, and anyone else's input represents an unwarranted intrusion ... into what are effectively the private affairs of the scientific community. We claim that this expertism mentality represents another undesirable reactionary posture.

And yet — wait a minute! It's all very well for us to take pot shots at established science for being closed and insular — but one should also raise the flag of fair play here. After all, science is not unique in having sets of insiders, as opposed to outsiders: that is in the character of social interaction, not to mention nature generally. Why should there be a problem with the idea that any association of people would set their own rules and standards — is this not what all groups of specialists do, all the time? Indeed, who but the specialists can know enough about the specialty, such as to be qualified to make these crucial decisions to begin with? And since scientists are supposed to be specialists about nature and the patterns of the natural world, what is so objectionable with the notion that they alone should be putting forth fundamental interpretations about the nature of reality, as best they see fit?

The obvious answer is that science is not just some autonomous institution which is isolated from the rest of society, where the members themselves should be free to determine the rules of their own internal debates. For we are not merely dealing with some peripheral profession here, which primarily impacts itself or some other limited domain. Science lies at the very heart of the larger society — increasingly over time. In many respects, science has become the centerpiece of all of our lives, giving it special powers and special responsibilities. That is because science sits close to the very apex of power in the modern world — both in abstract terms of defining the nature of our common reality, as well as in concrete terms ... and the latter not only regarding *how* we live in a material sense, but perhaps even *whether* we live at all (individually and collectively), by virtue of the fruit of science's endeavors.

Science is special, therefore, in terms of its roles and responsibilities — for we in the larger society must all bear the consequences for what science does: we all gain with its triumphs ... and we all lose when science does not serve our needs. Therefore, we maintain that science must serve society first ... before itself, or any other special interests.[27] As Holton puts it: "[O]ne must conclude that science has a just claim to moral authority when, and only when, it is widely seen as an activity that honors both truth and the public interest." (Holton, p. 467)

Holton also notes that, "The lab remains our workplace, but it must not become our hiding place." (Holton, p. 466) In a similar vein, Bromley cautions other scientists: "[I]f you can't explain what you're doing and why you're doing it to any intelligent layman, that really means that you don't understand it yourself." (Bromley).

For his part, Feyerabend questions the very proposition that science should even occupy such a unique place in society — and he challenges the notion that science operates in special ways which are different from other disciplines:

> [T]he competence, the complications and the successes of science are vastly exaggerated.... [T]here does not exist a single argument that could be used to support the exceptional role which science today plays in society. Science has done many things, but so have other ideologies. Science often proceeds systematically, but so do other ideologies (just consult the records of the many doctrinal debates that took place in the Church) and, besides, there are no overriding rules which are adhered to under any circumstances;

27 Siu pointedly asks, "Should science be regarded as merely amusement for the erudite few?" (Siu, p. 137) He writes eloquently about what this kind of mentality can lead to: "The doctrine 'Science for science's sake' is symptomatic of the scholarly blinkers which are diminishing her potentialities of greater service to man. By adhering to this maxim, science is cutting herself off from her source of significance. Chesterton wrote quite pungently on this score. He found it 'extremely difficult to believe that a man who is obviously uprooting mountains and dividing seas, tearing down temples and stretching out hands to the stars, is really a quiet old gentleman who only asks to indulge his harmless old hobby and follow his harmless old nose.' When an atom is split and a hundred thousand lives are smothered and the world is thrown into terrible anxieties for centuries thereafter, how can the layman believe that the splitting of the atom is a great and commendable accomplishment and the resulting destructiveness a blameless inadvertence." (Siu, p. 137)

> there is no "scientific methodology" that can be used to separate science from the rest. *Science is just one of the many ideologies that propel society and it should be treated as such ...* (Feyerabend, pp. 62 and 61)

As far as the expertism argument goes — that only experts are equipped to understand and criticize science — Feyerabend challenges the very idea that something special is involved here:

> Science is not a closed book that is understood only after years of training. It is an intellectual discipline that can be examined and criticised by anyone who is interested and that looks difficult and profound only because of a systematic campaign of obfuscation carried out by many scientists (though, I am happy to say, not by all). (Feyerabend, p. 62)

Ziman makes a similar point, in arguing that:

> [S]cientific argument, written or spoken, is not very complex or logically precise. The terms and concepts that are used may be extremely subtle and technical, but they are put together in quite simple logical forms ... (Ziman, p. 40)

Popper also takes a hard line against deferring to the experts:

> I am not impressed by the specialist — the expert — as such. We must not allow ourselves to be bowled over by imposing formulae: that is the road to perdition. Admiration for the impenetrable expert puts a premium on sheer difficulty of expression and tempts us to mistake obscurity for profundity. We must insist upon clear thinking, simplicity, and intellectual responsibility in discussing subjects which, after all, ought to be everybody's concern. (Popper, 1982, p. 157)

It is also important to bear in mind that what guides the work of scientist experts is not merely a concern with high-minded issues like theoretical content, elaboration, and exploration — for science workers are subject to the same kind of psychological and social pressures and inducements as everyone else.[28] In this regard, there can be no doubt that institutional influences play a major role in establishing and maintaining science's discipline amongst its workers — including factors like jobs and livelihood (encompassing pay and benefits, as well as promotions), social interaction and support, status and prestige, publications and awards. One result is that most practitioners are properly concerned about advancing their careers — and not risking ostracism and ultimate professional failure by seriously challenging the established line. Yet the influences are not only outward; they are also inward, as well. For to invest much of one's life in the broad field of science is — to a greater or lesser degree — to essentially define a large part of one's existence in those terms[29] ... and that entails an enormous amount of ego involvement, which cannot be dismissed lightly. But not just ego, for much of one's

28 While we do not explicitly deal in this book with arguments about the role of "social construction" in scientific formulations about nature, it should be obvious that such factors represent an inevitable part of how we build our theories, in various respects.

whole concept and belief system, as well as role in society, would also be defined in those terms.

Now let us leave the expertism issue aside and return to the basic argument about science's insulation from the rest of society. We maintain that there must be a better balance between the two extremes of either a completely closed or a completely open science — for dealing with what are, in Popper's words again, "subjects which ... ought to be everybody's concern." Therefore, science must be willing to form more of a two-way relationship with the rest of society — which we contend would benefit science as well as everyone else. In regards to the former, more input from outside that is informed and thoughtful would provide a means for science to gain greater perspective (if not fresh ideas) concerning direction and interpretation — for an institution that is far too inbred and immersed within itself. In regards to the latter, non-scientists have the right — nay, the obligation — to help guide the work of science, and keep it on track ... because of the abstract and concrete impacts upon the rest of society that we mentioned above.

In the abstract realm, it is important to provide a check that those in the larger society are not receiving some false doctrine from on high; and in the concrete realm, it is essential that greater guidance be given for the "valueless" craft of technology which takes no responsibility for the impact of its endeavors (a point we shall discuss in the last part of the next chapter).

Just to clarify our position in regards to these matters, we are not suggesting that science should simply reflect whatever changing fashions and modes are predominant in the larger society. But the fact that, e.g., a dogmatic church attempted to squelch the spirit and the findings of an enlightened Galileo does not justify a dogmatic science closing in upon itself and rejecting influence from outside its domain. It is not enough for a cloistered camaraderie to make decisions which impact us all: we strongly reject Kuhn's position in this regard, and maintain that scientists are answerable to more than themselves.

Of course, changing the balance of the relationship between science and society will not be easy. For any such arguments as we have been putting forth clearly represent a challenge to scientists' vested interests, and threaten their jealously guarded power and privileges. As Siu wrote a half century ago,

> Relinquishing the intellectual throne for the life of a commoner is a hard chore for science after three hundred years of free ranging and a hundred years of lordship. We can fully appreciate her reluctance to make the sacrifice. (Siu, p. 144)

This completes our discussion of some of the ways that science works, and we next turn our attention to the pivotal question which we have largely skirted so far — and this concerns what it is that we primarily want from science ... and whether the institution has succeeded or failed to meet such goals.

29 To requote Morrison: "[S]cientists who have built their professional lives around the elaboration of theoretical ideas are understandably reluctant to give it up." (Morris, p. 40)

Chapter 11

The Myth of Modern Science

> [M]ore and more people are wondering whether reality is not the skeleton in the cupboard of 20th century physics, and whether eventually the crime will have to be cleared up. They are beginning to suspect that the official philosophy of physics ... had gone too far in its eagerness to dispel metaphysical inscrutables. Surely physical hypotheses must be susceptible to experimental test, but why should they not concern the external world and why should they not explain how things work?
>
> After several decades, in sum, several people are starting to ask what became of reality and whether a new philosophy of physics should not be explored, one giving more credit to mother Nature and more freedom to the creative imagination of the theoretician. The new philosophy, it is felt, will have to avoid two extremes that failed long ago: dogmatic subjectivism and dogmatic realism. It will have to be critical and open-minded. (Bunge, p. 5)

Although these words were written many years ago, they are just as relevant today as when they were first stated. So we must ask some very important questions: What do we want science to be, and what do we want scientists to do? Where should science be taking us — what is the purpose of our study of nature? We must ask value questions like these, because it allows us to put into better perspective what science's mission should be — and what science's present role is. The answers represent the premises upon which science functions and scientific theories are built. For until we know what we want from science and its theories, we have no basis for deciding how successful they have or have not been in fulfilling that mission.

In general, established science is not very happy about engaging in discussions of this sort, for the kind of reasons we talked about in the previous chapter. For one thing, there is the argument that such queries amount to unwarranted intrusions into the free-ranging spirit of inquiry which should characterize science — not to mention slowing science down, in its headlong rush forward, by raising doubts about its work and the direction upon which it is embarked. Recall Kuhn's position regarding efficiency, which states that science should not get confused or distracted from what should be its primary task — of investing all its energy directly into further hard-nosed scientific inquiry, and thereby moving ahead with its work in the most expeditious manner possible.

But the main objection is surely a functional one: if a given theory or line of research or technological invention works for what it was intended to do, then that is its own justification and requires no additional "metaphysical" foundation. The reasoning is familiar — that such debates are inevitably vague and general, and merely cloud the air with artful utterances and specious disputation. Of course, such discussions are also said to be inherently unscientific, since those sort of questions cannot be proven or disproven in the way that more basic theoretical propositions can be.

Thus the argument is that we should not get bogged down with idle speculation, when the acid test is whether the thing works or not — and *science works*, there can be no doubt about that ... as one purviews the concrete results of all its inquiries, ranging from the transistor to the H-bomb, from the microchip to television journeys to Saturn and beyond. As Jammer wrote some years ago regarding one of the twin pillars, "[N]ever has a physical theory given a key to the explanation and calculation of such a heterogeneous group of phenomena and reached such a perfect agreement with experience as has quantum mechanics." (Jammer, 1989, p. 386)[1] Therefore, science should be judged on the basis of the enormous technological progress that has been achieved — and it is science's success which has brought us to the amazing point that we are today ... whereas if the theories and inventions had not brought us such progress, then that would have represented testimony enough about the failure of science.

So much for the established line — but it has dodged the issue. For to say that science works still assumes that it works towards certain ends. That is the case even regarding the Jammer comment, because there are different types of "experience" — we shall return to this factor just below — that there can be "perfect agreement with" ... which brings us back to different types of ends or goals. And the fact that science is reluctant to discuss those ends does not make them go away, or that they did not exist to begin with. The question is whether we will consciously and explicitly identify the ends — and judge the relative merit of one set of ends vs. another — or whether we are going to simply run on automatic, and let the ends be implicit and self-justifying. Again, this is the same point we made in the last chapter regarding metaphysics — for that is what we are talking about here.

Call them ends, goals, values, or metaphysics: such factors always operate in any endeavor, whether we choose (or are able) to identify them or not. A theory is only a good theory if it helps us to reach a given goal. This is why the established line above is not satisfactory, for it presumes that science's short-range, pragmatic — but largely unspoken — ends are obviously the only ends which are justifiable; and that is simply not the

1 By contrast, it has also been argued that the theory has not always been so flawless in its predictive power as the standard image suggests. Miller, for example, says that, "Indeed, despite Heisenberg's positivist prejudices, the role of quantum mechanics from the start was to remove paradoxes in causal explanations, not, primarily, to predict. Quantum mechanics for many years added little in the way of predictions, and gave rise to many false predictions." (R. Miller, p. 602)

case. For example, the fact that we are living in what might be called a "wondrous technological society" does not prove the value of science, unless one assumes that the only purpose of science is to help us live in a "wondrous technological society".

> What purpose is the theory to serve? Is it only the purely technical purpose of predicting observable facts? Or is it to obtain a simple and elegant theory that allows us to derive a great many facts from simple principles? We choose the theory according to our purpose. (Frank, p. 218).

As Popper says, "[W]hat is to be called a 'science' ... must always remain a matter of convention or decision ..." (Popper, 1968; p. 52) Or as Siu puts it, "A theory may be useful for a given purpose. It may not be for another." (Siu, p. 22) Once again, then, the crucial question is what we want our theories of nature to do.

Obviously, the issues we are talking about overlap those which we discussed in the prior chapter. The difference is that we addressed these matters there in terms of a list of abstract theory-building criteria, along with arguments for and against each — whereas our task here is to show how these same issues can be framed and construed in terms of the very themes of the present model itself. But before moving down that path, let us first set the stage in the most general of terms as to what should be the main goal for science — understanding or prediction.

Understanding vs. Prediction

There can be little dispute about what the established paradigm sees as its main goal: it is only through *prediction* and control that any theory of science can justify itself. By contrast, any attempts to *understand* nature take us beyond the scope of science — and belong in the realm of philosophy, religion ... or metaphysics.

It should come as no surprise that we contest this conclusion, based on all that we have been saying thus far in the book — and will explain further as we continue to develop our model in the rest of this chapter. But for now, we want to cite a number of observers who are also dissatisfied with the orthodox position on this matter. And while most of these comments are specifically directed at Quantum Theory, the views can be treated as referring more broadly to science in general.

First up is the thinking of David Bohm, who wrote that, "Prediction is by no means the most important function of a physical theory." (Bohm, et al., 1964, p. 64) He says that:

> [W]e are faced with deep and radical fragmentation, as well as thoroughgoing confusion, if we try to think of what could be the reality that is treated by our physical laws. At present physicists tend to avoid this issue by adopting the attitude that our overall views concerning the nature of reality are of little or no importance. All that counts in physical theory is supposed to be the development of mathematical equations that permit us to predict and control the behaviour of large statistical aggregates of particles. Such a goal is not regarded as merely for its pragmatic and technical utility: rather it has become a presupposition of most work in modern physics that prediction and control of this kind is all that human knowledge is about. (Bohm, 1980a, pp. xiii)

Stapp is also concerned about the fact that, "[M]ost orthodox physicists.... will retreat to the position that quantum theory provides testable rules, not an ontology.... [T]he quantum physicist is left with no coherent ontology that is compatible with the validity of the quantum theoretical predictions ..." (Stapp, 1988, pp. 44 and 45)

Karl Popper is very critical of the conventional view:

> Today, physics is in a crisis.... a crisis of understanding. This crisis of understanding is roughly as old as the Copenhagen interpretation of quantum mechanics.... The view that theories are *nothing but* instruments, or calculating devices, has become fashionable among quantum theorists, owing to the Copenhagen doctrine that quantum theory is *intrinsically ununderstandable* ... I think this is a mistaken and even vicious doctrine.... [T]he denial that we can understand quantum theory has had the most appalling repercussions, both on the teaching and on the real understanding of the theory. (Popper, 1982, pp. 1, 42, and 45)

> Not everybody is satisfied with this situation. It is especially unsatisfactory to those who see in physics not mainly an instrument for predictions, or other practical applications, but rather an instrument (if instrument it must be) for understanding the world we live in — for *explaining* this world. (Popper, 1982, p. 102)

> [W]hat we attempt in science is to describe and (so far as possible) explain reality.... [W]e do not only look for biological or instrumental success. In science we *search for truth*.... It is the great task of the natural sciences and of natural philosophy to paint a coherent and understandable picture of the Universe.... [W]hat we are seeking is not so much usefulness as *truth* ... and thus *understanding*. (Popper, 1981, p. 40, italics omitted; Popper, 1981, p. 69; Popper, 1982, p. 1; Popper, 1982, p. 42)[2]

Holton touches upon some of these same points, in arguing that people like Niels Bohr were "motivated by the growing hope that science will lead to a coherent world picture." (Holton, p. 469) The same goes for another Quantum Theory pioneer, Erwin Schrödinger. He was very critical of the fact that:

> [The positivist philosophy] seemed to relieve us from the search for what I should call real understanding; it even rendered the endeavor suspect, as betraying an unphilosophical mind — the mind of a child who regretted the loss of its favourite toy (the picture or model) and would not realize that it was gone for ever.... Discoveries in physics cannot in themselves ... have the

2 It is interesting that Capra reaches a very different conclusion about the coherence factor, for he lauds the quantum model as representing "a coherent dynamic theory" (Capra, 1983, p. 121). Indeed, he chastises other sciences outside of physics for having "no unifying framework." (Capra, 1983, p. 121) He writes that what other disciplines "need to do most urgently is reevaluate their entire conceptual foundation and redesign their basic models and theories accordingly." (Capra, 1983, p. 193) His position is that other disciplines should learn from physics and "be content with a network of interlocking models, using different languages to describe different aspects and levels of reality." (Capra, 1983, p. 369) Needless to say, we would dispute all of these contentions — and in particular, any assertions about Quantum Theory being "coherent" are quite a stretch ... either in terms of comprehensibility or in terms of fitting together into a unified whole.

> authority of forcing us to put an end to the habit of picturing the physical world as a reality. (Schrödinger, 1950, pp. 112 and 111)

Wolf also decries the loss of meaning in Quantum Theory:

> For the first time in the history of science, no one was able to really picture what was going on. The mathematical formula ["Planck's magic formula"] had replaced any visual experience. It worked, but it hardly made any sense. (Wolf, p. 67)

Healey makes a similar point:

> [W]e are bound to find quantum mechanical "explanations" unsatisfying just because they exclude the possibility of forming any "mental picture" of the phenomena. (Healey, p. 18)

> The basic thrust of the [Copenhagen interpretation's proponents] ... is to argue that the goal of science is to develop theories and techniques for successfully predicting our observations, that talk of any "reality" independent of our observations is metaphysical and unscientific, and consequently it is not part of science to describe such a "reality".... (Healey, p. 18)

> I shall simply state my contrary convictions: that it is a primary goal of science to explain phenomena in the natural world; that our observations are only of interest insofar as they give us access to some of these phenomena; and that there are objective criteria for assessing the adequacy of a scientific explanation which have nothing to do with a psychological preference for mentally picturable models.... It would be disappointing if the only way to understand much of contemporary physics were to reject beliefs which motivated one to make the attempt to do so. (Healey, p. 19)

Getting back to the "experience" factor mentioned in the Jammer quote a few pages back, Prigogine laments the fact that, "[W]e may end up with an 'all-too-pragmatic' way of doing physics, taking away one of the most important driving forces of science: The need to connect the various aspects of human experience." (Prigogine, 1988, p. 126) Peat likewise wonders " ... why a theory of the universe never captures our human experience." Laing complains that, "Experience as such is cast out of the realm of scientific discourse." (Laing, quoted in Capra, 1983, p. 55) March says that, "[M]an undertakes the study of nature not merely to master it, but out of a deep-seated need to clothe his life experience in meaning." (March, pp. 239-240) And Margenau: "The purpose of science is to organize the unilluminated facts of our experience into a texture of reason, simplicity, orderliness and multiple perspectives." (Margenau, 1978, p. 336)[3]

There is one valid concern we need to address, which is whether diversion from the instrumentalist path would necessarily rob science of its public and objective character. That is, if specific prediction based upon observation and experimentation were made secondary to general understanding that is in accord with our common experience, then this might take us back to the days before Bacon, Descartes, and the experimental method ... which are typically seen as the turning point for bringing science out of the "dark ages" and into the modern era. Capra writes that, "The nature of medieval science was very different from that of contemporary science. It was based on both reason and faith and its main goal was to understand the meaning and significance of things, rather than prediction and control." (Capra, 1983, p. 53) In its place, what the modern era has brought is a "strong emphasis on control that [is] in keeping with the Baconian ideal that has become characteristic of Western science." (Capra, 1983, p. 174)

The historical point is well taken, and it is perfectly arguable that the exclusive emphasis on prediction may have been a suitable correction to a science not yet grounded on observable phenomena, as opposed to private opinion. The newer perspective was a valid counterweight to overly subjective and speculative approaches to studying nature — and in this respect, the standard positivist or instrumentalist belief that science should deal only with observables may have been valid as a reaction to sheer abstract speculation, which had little or no basis in concrete physical reality.

But all things in nature have their season, and what may have been a valid approach in Galileo's time may not be appropriate today. Indeed, our position is that we have long since gone astray in the opposite direction, in terms of an obsession with prediction and control ... and this is why the pendulum now needs to swing back. Thus it is that the extreme stance of the established paradigm is no longer productive in leading us to a viable overall model of nature.

As Bohm writes, "[M]odern science did in fact grow up, at least in part, in a struggle against the vague, speculative, and untestable features of mediaeval metaphysics. But does it not now tend to overemphasize sharpness of definition of concepts and testability (or refutability), thus tending to the opposite extreme?" (Bohm, 1969, p. 103)

Finally, there is one important issue we need to deal with here — which is that we claim the established paradigm is making a disingenuous argument to begin with, in its contention that the goal of understanding nature is beyond the purview of science itself. We suggest that this is a false stance, which borders on being a problem of integrity. More specifically, we maintain that science is not playing with a straight deck here, in that it wants to have things both ways.

On the one hand, there is a denial of responsibility and a claim of innocence — which comes in the form of the argument that issues which concern understanding and broader meaning are inherently unprovable (or if one prefers, not disprovable) and therefore have no proper place in a good theory of science. The established paradigm would thus disclaim any but the narrowest of interpretations as to the significance of its findings. Instead, one is supposed to be guided by the straightforward principles of logical positivism: what you see is what you get — the facts are the facts, and one must not go beyond them. For example, Zukav says that, "*It does not matter* what quantum mechanics is about! The important thing is that it works in all possible experimental situations." (Zukav, p. 37)

3 So as to not further burden the main text, we here footnote a few more comments on the subject. Thus Kitchener writes that, "[S]cientific theories can be judged to be true or false in some sense larger than 'they allow one to describe, predict and organize the experimental data.' " (Kitchener, p. 17) In Rogers' words, "[T]he full value of a great theory is not just in fruitful predictions but in a deep sense of sureness of knowledge that it gives." (Rogers, p. 345) As Hanson says, "Fundamental physics is primarily a search for intelligibility — it is philosophy of matter.... Theories aim at a deep understanding of phenomena ..." (Hanson, p. 142) Or as Ziman puts it, "[O]ne of the aims of Science is to provide us with a Philosophy of Nature ..." (Ziman, p. 36)

So when pressed to explain what some of its concepts can possibly mean in real life, this face of science would respond by denying such a function: Science can take no position on deeper life issues, and does not pretend to be identifying anything like "truth" or "reality" in any broader sense. Accordingly, there is no general goal of understanding nature, and the very suggestion can be treated as a fabrication — in contrast to the only proper role of science ... which is to predict and make things (or theories) that work, and leave the idle philosophizing to others. This side of the established paradigm would therefore deny any broader ontological significance of its theories whatsoever.

Then there is the other hand, which reveals a contrary attitude — by virtue of how science puts forth new and startling claims about the deeper nature of reality ... and according to which, you get more than what you see. For now we are told that science has identified truly fundamental patterns — of which we were previously unaware, and which are treated as crucial for our more general formulations and conceptions about the world. We have discussed many of these seminal ideas throughout this book, as they have emerged from Quantum Theory and the Theory of Relativity — and which most knowledgeable people consider to be extraordinary insights into the nature of reality itself, which go far beyond any limited matters of prediction. Science's descriptions and theories are treated as describing the essential character of the physical world. The result is that science today is generally considered to be the de facto definer of reality — the standard-bearer for our entire vision of the physical world.[4]

Taking the two faces together, then, the established paradigm wants to have it both ways: on the one hand, the standard framework *is not* meaningful ... on the grounds that it is not science's function to explain nature, but merely to describe (and manipulate) it. Yet on the other hand, the standard framework *is* meaningful, because it has revealed the most essential patterns of nature.

We would advance a more cynical justification for the former no-meaning posture: it is a disingenuous way of avoiding the unfortunate contradictions and logical absurdities which are inherent in and between the individual models themselves, as well as when they are considered in a larger context. By cloaking itself in a detached and abstract scientific aloofness, the no-meaning argument amounts to an excuse (if not a fig leaf) which frees science from the need to commit itself to any sort of comprehensible vision of reality at all — once all the various "insights" are brought together — thereby saving itself from being vulnerable to a good deal of grief and embarrassment. There simply does not exist a coherent conceptual framework which ties all the arbitrary pieces together ... and so there can be no claim of meaning which results.

In this respect, the resistant posture we are dealing with in the established paradigm — regarding such issues as metaphysics, meaning, and common sense — is actually quite sensible, given that so many of the ideas defy comprehension. It is therefore difficult to avoid the conclusion that the narrow-minded pragmatism which characterizes the no-meaning side of the established paradigm is largely born of necessity. Rather

4 "[S]cience ... is the canonical source of what we can know about the nature of reality. People who wish their belief to be based in the solid ground of fact should look to science alone and eschew the treacherous swamps of mere opinion that surround it.... Science is our only reliable guide in the perplexity of human experience." (Polkinghorne, pp. 3-4)

than acknowledge the inadequacy of their constructs and theories, it is much easier for science to dismiss the salience of our common logic and meaning system. It becomes a ready escape to claim that a good theory of science need not trouble itself with such trifling matters, and to deny substantive goals like the attainment of broad truth and meaningful descriptions of nature, etc. — and instead stick to a numbers game.

In the end, though, even the no-meaning argument is a sham: nobody wants to function in that sort of conceptual vacuum, and not many do. For example, few people in the real world seriously deal with Quantum Theory as though it were merely some sort of sophisticated specification set about this or that detail in reality, which is all that a strict approach to the model would permit or tolerate. Instead, people seek answers on a broader scale than that kind of narrow ideology would permit. Scientists, whether they work in the area of quantum physics or not — and non-scientists, as well — do have the goal of understanding their world.

The Concept of Compatibility

Now we want to begin setting forth a broader framework in which issues like understanding and prediction can be seen in relation to one another — based upon the simple notion of *compatibility*. This is a factor we have touched upon at different points recently, sometimes only implicitly, but now we want to explain how the basic concept of compatibility provides a useful touchstone for bringing together many of the issues that we have been discussing. And to be clear at the outset, there is nothing mysterious or exotic in what we are talking about here: compatibility is simply a way of describing whether certain sets of concepts or phenomena fit together or not.

Let us start with the most straightforward and prominent type of compatibility which we are concerned with in the present discussion, that of the vaunted *fact-theory fit* — also referred to as the "empirical fit", of a theory with the facts of the world. Of course, the importance of this type of compatibility is reflected in the standard requirement which is generally put forward — that a good theory must make accurate predictions about phenomena in nature. As Wolf says, "A theory may be beautiful and elegant, but if it doesn't fit the facts, it's just plain wrong." (Wolf, p. 200)

Another kind of compatibility is so inevitable and obvious that we don't really pay any attention to it. In its simplest form, the *fact-fact fit* (or *entity-entity fit*) would refer to the various elements or entities which comprise any given situation in our environment. Clearly, these separate facts or entities must be compatible with one another — for otherwise, there would be no mutually-compatible whole situation. And while the referents we are talking about here would typically be construed as referring to concrete physical phenomena, they need not be.

Indeed, the next category of compatibility we want to mention concerns the abstract counterparts for such physical phenomena, and we can label this grouping as the *construct-construct fit*. The established paradigm also acknowledges the importance of this category, which essentially refers to the factor of internal consistency within a model: a good theory should not contradict itself. Popper says that such internal "consistency is the most general requirement for a system." (Popper, 1968, p. 92) Therefore, one can ask whether the conceptual parts of a theory are compatible with one another ... as component pieces of the whole puzzle which the theory represents. Labeled in terms of this parts-whole relationship, we can talk about the *construct-theory fit*, in the sense that any theory is composed of certain constructs — and there must be compatibility between them.

Corresponding to the construct-construct fit within theories, there is also the *theory-theory fit* between theories. Does the theory fit with other theories or visions of the world? As Feyerabend puts it, "[T]heories cannot be justified and their excellence cannot be shown without reference to other theories." (Feyerabend, p. 58; see also Rindler, p. 26). Therefore, the propositions and descriptions of one good theory should not contradict the propositions and descriptions of another good theory.

In a larger sense, of course, the question of theory-theory compatibility must be applied beyond any two instances, and ultimately encompass all the various theories which are at issue. This leads us to the most general levels of conceptual compatibility, which can be described in a variety of ways, such as a *theory-megatheory fit*. Whatever the label, the essential idea is that the theory and its constructs should be compatible with broader concept systems and modes of relating to the world. One way of referring to this broader concept system is in terms of a general "meaning" framework — which is to say, the way we understand things in the world. This category would also include things like common sense, intuition, metaphysics, aesthetics — or even fundamental belief system, value structure, disciplinary matrix, school of thought, or what have you.

This same general category would also entail something like a *theory-logic fit*, in terms of how well the theory conforms to certain logical rules that we maintain or abide by. One might even extend the notion of logic, in this sense, to incorporate that very particular kind of logic which is called mathematics. In addition, there are a whole host of other compatibility factors which would be involved — including cognitive, social, and cultural ones. The list is pretty much endless, and all such factors are part of the general overall reality, relative to which any fact or theory exists and can be identified.

One way that these factors would apply specifically to science would concern an overall unified theory of nature. In this regard, one can say that, "Science is a unified structure, and in principle it is the structure as a whole, and not its component statements one by one, that experience confirms or shows to be imperfect." (Rudner, p. 235) So the question naturally arises as to whether the various subsidiary theories of science come together into a coherent whole, or not.

It is important to emphasize that our description of these different categories is meant to be illustrative as much as anything else, in that there is nothing comprehensive about this whole analysis — although we will propose a more orderly framework for integrating these topics in the next section. In any event, when we refer to compatibilities relating to facts, for example, we have previously mentioned that we are not concerned in this work with any differentiations between physical entities, data, and facts. The groupings we are using in the present discussion are not rigorously defined, and must be understood in relation to each other ... and as part of the larger argument we are making.

That main argument is a crucial one: our objection to the established paradigm is not that it completely ignores the compatibility factor, but that such compatibility is only invoked selectively. Indeed, the standard line is not usually couched or labeled in these very general compatibility terms — because if that were the case, then it would become immediately apparent how limited the conventional approach is regarding these issues.

The established paradigm emphasizes, almost exclusively, the importance of particular types of compatibility — especially that of the fact-theory fit. But as we are describing, there are various other types of compatibility to consider which go well beyond that between fact and theory. We maintain that a good theory will be consistent or compatible not only with phenomenological facts which we can test experimentally, but it will also be consistent with phenomena which we can test experientially — in ourselves and with nature, from the most specific to the most general categories thereof.

Let us mention three important types of compatibility, which might be grouped under the theory-theory or theory-megatheory categories. The first is that a good theory should also be applicable to (i.e., be compatible with) other domains — even fields of science — outside the realm from which it originates. This is consistent with our position that the fundamental principles are the same everywhere, no matter which realm of nature we are examining — which is part of what we talked about in the previous chapter, in terms of an aesthetic criterion for a good theory of science called "generality". Another way of saying this is in terms of a point we make repeatedly — that a good theory should apply to all of nature, and not just part of nature.

We highlight one of these realms, as representing a second important compatibility factor: cognition. We contend that a good theory of nature must be compatible with (if not entail) a model of cognition — or rephrased, the basic mechanisms and patterns which describe nature in general should also describe (i.e., be compatible with) cognition in particular. Cognition must be embraced, because of the crucial role it plays — both in formulating our theories about nature, as well as in the observation process itself by which we interact with (and also determine the outcome of) nature. Put in more familiar terms of the present model, a good theory of science must not only be a theory of being, but also a theory of seeing. We shall return to cognition in the final chapter.

The third important compatibility factor concerns the theory itself. And by this, we are not talking about the abovementioned requirement that the conceptual components of a good theory should be mutually compatible. Rather, we are saying that the principles which a general theory puts forward as describing nature should also describe the theory itself. That is because the theory is not just *about* nature: it is also *part of* nature — where else did it emerge from? — and in this respect, the theory should not only describe the larger world out there, but it should be self-validating in also describing the character or nature of the theory itself.

Expressed in the terms we just mentioned, all our concepts and theories (which are intended to describe nature) amount to various ways of *seeing* (or "mapping") the world.[5] But what we are now saying is that these concepts and theories (or "maps") — along with the human beings which formulate them and subscribe to them — are also part of the world of nature, and so can also be construed as elements of *being* in that

5 Our usage of the being and seeing concepts here is broader than that which we have utilized elsewhere, where the former was treated more in terms of static relationships and the latter in terms of dynamic relationships; we anticipated this later ("multiple meaning") usage in the first chapter. Also note that the relative distinction we are making between being or existence on the one hand, and seeing or describing that existence on the other hand, may have certain parallels with the philosophical constructs of ontology and epistemology, respectively.

natural order. In this regard, we noted in Chapter 8 how our concepts and theories are products of the human mind — and as such we are dealing with subsidiary hierarchical levels that are nested within the level(s) of human being, either individually or collectively. Thus it is that all of us see-ers are also be-ers ... even when we are looking in a mirror at ourselves.

Truth vs. Falseness Revisited

Next we want to segue from compatibility to truth, and thereby make explicit another important point which may not be apparent: we associate truth with compatibility — just as we associate falseness with incompatibility. That simple relationship can be applied to the various categories of fittingness which we have been discussing.

In these terms, then, the standard criterion regarding the fact-theory fit would be understood as meaning that a theory is true if it is compatible with the facts — or that it is false if it is incompatible with the facts. Likewise, two facts which are compatible with one another would thereby be considered as true, vis-a-vis one another — and the same goes for two constructs that are compatible and thus true, in relation to one another ... or two theories, in terms of one another. Then there is the broader category for compatibility, which we have labeled with various names — such as megatheory: in this respect, a theory which is compatible with some form of megatheory would thereby be considered as true (in relation thereto) — just as if it were incompatible with that broader standard, then it would be considered as false.[6]

Two points should be immediately apparent. One is that we are using the concepts of truth and falseness not just as terms for characterizing the viability of some description of nature. For we are also saying that truth and falseness can be utilized in more general ways — as applying, for example, to aspects of nature themselves ... or to ways of describing nature themselves. Thus again, two phenomena of nature — two facts — which exist in relation to one another are thereby true in relation to one another ... and otherwise false, in those same terms. Likewise, two descriptions of nature are true in relation to one another to the extent that they fit together ... and otherwise false. Briefly expressed in terms of our standard metaphor, truth and falseness do not merely describe the viability of a being vs. seeing relationship (as in fact vs. theory) — but they also describe that of a being vs. being relationship (as in fact vs. fact), or a seeing vs. seeing relationship (as in theory vs. theory).

6 Note how this compatibility approach relates to the three main theories of truth which Popper refers to: the correspondence theory, which says "that truth is correspondence with the facts"; the coherence theory, which says that "a statement is regarded as true if (and only if) it coheres with the rest of our knowledge;" and the last, which sees truth as "pragmatic utility or pragmatic usefulness" — instrumentalism (Popper, 1981, p. 308) Popper himself seems to mostly opt for the first approach, where compatibility with the facts is all that counts. We see that as little different from the third choice, or compatibility with some set of pragmatic facts. But compatibility still comes into play with the second category, in terms of what we have called the theory-megatheory fit — even though we also embrace the broader "coherence" label, as will become apparent later in the section. In any event, all three approaches which Popper cites fit the general model of compatibility and truth (vs. falseness) that we are describing here.

The other point is that truth and falseness are not only absolute — they are also relative, in the same way that compatibility is: one thing can be compatible with a second thing, and not be compatible with a third thing; likewise, something which is true in one sense (in relation to one thing, or set of factors) may be false in another sense (in relation to some other thing, or set of factors). The concepts of truth and falseness, therefore — as with their respective compatibility and incompatibility underpinnings — have a variety of referents. Nothing is only true or false in some transcendent sense, but rather in relation to an appropriate set of referents ... however narrow or broad those may be.

So there are many truths (and falsenesses) — an infinite number — none of which is to be merely embraced outright so much as properly delineated, with their referents specified. For by this definition, *anything can be true or false*, if only we identify a suitable set of referents ... relative to which it is compatible or incompatible.[7] Another way of making this point is to say that there is nothing which one can conceive of that does not exist, in relation to something — even if it is only in terms of the context in which we imagine it.[8] That is the case irrespective of how strange or bizarre the target phenomenon may seem to be, in relation to more conventional circumstances — and therefore no matter how minimal the supporting evidence might appear to be ... even if it is only in relation to the tiny set of contextual factors in which it is little more than a fantasy, as judged in broader and more usual terms.

The upshot is that *any* theory describes reality — *any* phenomenon is reality — if only we define reality in terms which the theory or the phenomenon is compatible with. Put another way, "One can deduce anything if one selects one's premises carefully." (Hospers, p. 95) Or, as Popper quotes Black: "A nice adaptation of conditions will make almost any hypothesis agree with the phenomena." (Black, quoted in Popper, 1968, p. 82) Note Einstein's comment to Schrödinger, as being partially supportive of this point: "Of course, every theory is true, provided you suitably associate its symbols with observed qualities." (Einstein, quoted in Jammer, 1974, p. 24) Truth (and falseness) is relative — to whichever set of referents we choose.

This also means that there are many ways to describe nature — what we can call ways of seeing — an infinite number of ways, in fact, all of which will be found to apply somewhere, according to some perspective. As Feynman says, "One of the amazing characteristics of nature is the variety of interpretational schemes which is possible." (Feynman, 1965, p. 54) Again, he says that, "[E]very theoretical physicist who is any good knows six or seven theoretical representations for exactly the same physics." (Feynman, 1965, p. 168) Kuhn echoes this sentiment: "Philosophers of science have repeatedly demonstrated that more than one theoretical construction can always be placed upon a given collection of data." (Kuhn, 1970, p. 76) Or as Berkeley puts it sim-

7 Notice the similarity between the analysis here concerning truth and falseness, and that which we discussed earlier in the book concerning stasis and motion. Obviously, it is the relativistic factor that we are emphasizing here, and not the respective phenomena themselves.

8 We made essentially this same point at the end of the previous chapter's Truth vs. Falseness subsection.

ply, "[T]he same thing may be explained in different ways." (Berkeley, *De Motu*, p. 67, quoted in Popper, 1965, p. 170)[9]

Now we want to bring these issues back home, in terms of the main themes of the present model. We start with the primary one, that of complementarity — for we suggest that compatibility can be understood in precisely these terms: to speak about whether there is a fit or compatibility between one side and another is a very particular way of talking about whether the two sides are complementary with one another. Expressed in true-false terms, one thing is true if it is complementary to another thing — whereas it is false if it is not complementary thereto.

Yet there is also the question of what scope of relationship is involved, in which such a determination is made — and it is the latter aspect that we explain in terms of the other main theme of our model, that of hierarchy. Such a hierarchy factor was implicit in the references we made in the previous section to different categories of compatibility: we are thereby talking about different hierarchical levels of existential and conceptual breadth.

To be clear about what we are referring to with this last pair of referents, let us go back to the distinction we keep making: by "existential", we mean different levels of being — and by "conceptual", we mean different levels of seeing. When we speak about different levels of existential breadth, we are talking about different levels of phenomena within the overall hierarchy of nature — and when we speak about different levels of conceptual breadth, we are talking about correspondingly different levels of theoretical scope ... of what range (of being) phenomena the theory (of seeing) can be construed to describe.

Putting complementarity and hierarchy together we can say, first, that truth and falseness are absolute within the confines of any given relationship: one thing is absolutely true (or false) in relation to a second thing, to the extent that the two things are mutually compatible or complementary (or not). But then, second, we can say that truth and falseness are relative in terms of which range of relationships are involved, from amongst the hierarchy of possible referents — both in the sense of phenomena that exist in nature, as well as in the sense of how we describe those things. The upshot is that we can contrast an *absolute* complementarity factor with a *relative* hierarchy factor, for ascertaining truth and falseness in nature. Therefore, compatibility and incompatibility — along with truth and falseness — are not only dichotomous and absolute, in relation to particular other referents; they are also continuous and relative, in relation to all other referents.

It should be evident that one of the main issues we are addressing here is not merely an explanation of how truth and falseness can be understood in terms of the framework of the present model — but how we are using that framework to provide a context for understanding the different meanings for this pair of seminal concepts

9 Wigner quotes F. Werner: "How do we know that, if we made a theory which focuses its attention on phenomena we disregard and disregards some of the phenomena now commanding our attention, that we could not build another theory which has little in common with the present one but which nevertheless explains just as many phenomena as the present theory?" (Wigner, 1967, p. 223)

themselves. On the one hand, there is the sense of truth vs. falseness as it applies to any given relationship, which we explain in terms of simple compatibility vs. incompatibility. On the other hand, the truth vs. falseness which applies to any given relationship can also be understood as being higher or lower in relation to all other levels, and thus as representing a greater or lesser scope of compatibility (and truth) vs. incompatibility (and falseness). Thus:

Complementary Truth-falseness	*Hierarchical Truth-falseness*
dichotomous and absolute (simple)	continuous and relative (scope)
complementary truth vs. uncomplementary falseness	higher and broader truth-falseness vs. lower and narrower truth-falseness

With this framework in place, we can now return to the disturbing point we made above — that anything can be true or false, etc. Needless to say, such a contention would be very controversial on its face ... were it not for the hierarchy factor that we are now describing. Indeed, it is precisely the hierarchy factor — ranging from the narrowest set of referents at the bottom to the broadest set of referents at the top — which explains why we are not advocating a free-for-all philosophy of nature that is merely random and chaotic: for there is an orderliness to the multitudinous compatibility referents. Thus it is that there are small facts which are lower, in relation to big facts which are higher — and likewise for theories and megatheories, ultimately leading up to the very highest levels.

Moreover, this whole multi-level framework of compatibility-incompatibility and truth-falseness also explains why we never mean to say that the concepts and theories of the established paradigm are *wrong*, in any definitive and pervasive sense. Rather, it all comes down to how narrowly or broadly compatible (vs. incompatible) a given concept or theory is — and as we have said, not only in relation to particular phenomena of nature, but also in relation to other concepts and theories about nature. In this regards, and to return to a point that we partially made near the end of the previous section, it is not that the standard category of the fact-theory fit is incorrect or insignificant as a criterion for evaluating a theory — but that it is not an exclusive criterion, nor even necessarily the most important criterion ... as providing a referent for truth vs. falseness in nature.

The question of which type or level of compatibility one seeks is to raise anew the topic of science's goal or mission. In this regard, the issue is not whether contemporary scientific theory "works" in the sense that it accurately describes various phenomenological facts, from one point of view or another: the issue is rather that such relatively lower level ad hoc components cannot congeal — nest — into a coherent higher level conception of reality. They cannot yield a broad, consistent, and unified picture of nature — precisely because the perspectives and phenomena upon which they are based are too narrow, random, and specific.

Indeed, we make this claim both on our behalf and on behalf of the established paradigm ... although the latter would not be welcome, as we shall see. For our part, we speak about the highest level concepts and theories in terms of categories like common sense, metaphysics, and coherent meaning — or simply what we call megatheory. Not only does the established paradigm dismiss or trivialize factors such as these, but it is almost a point of pride for science that some of its most crucial constructs and theories not only conflict (i.e., are incompatible) with these categories, but that many of those constructs and theories simply defy comprehension altogether — they are incompatible with a coherent high level meaning system. Yet we also make the same claim on (unwitting) behalf of the established paradigm — simply in terms of the fact that the highest levels which we are talking about can also be understood in terms of a unified model of nature: That is clearly a primary goal of science, although such an outcome has obviously never been attained.

Another way of talking about the highest level theories is in terms of the aesthetic criteria that we discussed in the previous chapter — simplicity, generality, and coherence. A good theory of science is simple in essence because it can be broadly applied — incorporating both "everywhere" and "everything": it is a general and compact big picture of the whole of nature. The greater the scope of a (mega)theory, the higher it is hierarchically and the more it represents all of nature — the longer are the threads of reality which it is trying to weave into a unified whole.[10]

By contrast, lesser theories are relatively lower level: they amount to isolated and independent theoretical fiefdoms and referents that have little conceptual relation to one another, except in a happenstance way — or which may even contradict one another. For although there may be an infinite number of (short thread) ways to describe *particular* aspects of nature, there are not an infinite number of (long thread) ways to describe *all* aspects of nature.

Thus, it is clearly harder to attain higher level unity than what amounts to lower level fragmentation: the latter is a relatively narrow set of hit-or-miss (short thread) requirements, whereas the former is a much more complex nexus of (long thread) interconnections — yet we maintain that this is what science should be striving for. That is how we would interpret the following description by Popper, which overlaps with some of the quotes we included earlier in this section:

> [T]here must be many theoretical systems with a logical structure very similar to the one which at any particular time is the accepted system of empirical science. This situation is sometimes described by saying that there is a great number — presumably an infinite number — of "logically possible worlds." Yet the system called "empirical science" is intended to represent only *one* world: the "real world" or the "world of our experience." (Popper, 1968, p. 39)

And now that we have brought Karl Popper back onto the stage of our discussion, let us return to his argument about a line of demarcation between science and non-science — for we cast the issue very differently than he does.

To start with, we explained in the previous chapter why we reject his differential treatment of truth vs. falseness — on the grounds that every theory is both verifiable and falsifiable, in relation to different sets of referents. In contrast to Popper, we maintain that the fundamental issue — and thus the "line" of demarcation — is not one of

10 Or to requote Feynman: "Nature uses only the longest threads to weave her patterns ..." (Feynman, 1965, p. 87).

falsifiability vs. non-falsifiability ... but rather, of broad verifiability-falsifiability vs. narrow verifiability-falsifiability.

Here is the difference. The conventional view states that the more specific a theory is, the more falsifiable and so desirable it is as a theory of science. By contrast, we would say that the more specific a theory is, the more narrow are the set of referents relative to which that theory is both verifiable and falsifiable — whereas the more all-embracing a theory is, the more broad are the set of referents relative to which that theory is both verifiable and falsifiable. Popper says that once a theory has been falsified anywhere, it remains forever falsified everywhere — whereas we say that while a theory may be falsified in one situation (i.e., incompatible in relation to one set of referents), it remains true in other situations (i.e., compatible in relation to another set of referents).

Hence while Popper delineates science from metaphysics on the basis of falsifiability alone, the distinction we make is between a *general* vs. *specific* verifiability-falsifiability criterion. The truth-falseness which leads to understanding will be of a general sort: it will be more blurred and diffuse, broad and global, fluid and supple. But the truth-falseness which leads to prediction will be of a specific character: it will be more sharp and precise, narrow and focused, fixed and brittle. Not only is such general truth-falseness *not undesirable*, but we maintain that it is the essence of good scientific theorizing.

Therefore, we dissent from Popper's description of science theories as "nets cast to catch what we call the world ... We endeavor to make the mesh ever finer and finer." (Popper, 1968, p. 59) Ah, very true — but we must *also* endeavor to do the opposite, and make the mesh broader and broader. For we want good general theories that are widely applicable, just as much as (if not more than) we want good specific theories that can only be narrowly applied.

We thus reject propositions like Hecht's, that "A theory in which everything is understandable and which runs no risk of being refuted yields nothing understood." (Hecht, p. 5; cf. also the Horgan, 1995 quote cited in the previous chapter) By contrast, Bohm's view is very similar to the position here:

> It seems to me almost as if many scientists would give the highest value to any theory, merely because it happens to be easily refutable, while vague notions of potentially very great scope are frequently brushed aside, as having little or no significance. Here I would like to stress that the mere restrictiveness and refutability of a theory is not always the main consideration determining its value. (Bohm, 1969, p. 103, italics omitted)

It should also be clear that the understanding vs. prediction distinction we are talking about can also be described as a difference of perspective. In this regard, the fact that there are both broad *P*-view perspectives and narrow *X*-view perspectives (about everything) is a central theme of this book. Now we are applying this principle to our theories of nature — resulting in broad and general theories of wide range that are higher hierarchically, as opposed to narrow and specific theories of detail that are lower hierarchically.

Not only are the priorities of the established paradigm skewed, in overemphasizing the narrow at the expense of the broad, but its success in meeting these circumscribed goals tends to overinflate its perceived value as a methodology and as a body of concepts. That is, the fact that the established paradigm defines science primarily in terms of (what we call) relatively low levels of compatibility — coupled with the fact that these theories "hit the jackpot" and thus "work" at those designated levels — gives these theories a mystique or power which they do not deserve: they convey an illusion or promise of broad truth which is not warranted ... for they do not, and never can, fulfill that function. They are chimeras — relatively small islands of narrow truth, which pretend to be more than they really are.

Note that there is another, and more commonplace, meaning of the truth concept which is consistent with the interpretation we are putting forward here, regarding a correlation with scope or breadth. That is, we typically consider something as inherently true if it applies very broadly or has core status in our thinking — whatever that consists of for any individual (or group thereof) — whether it be called fundamental truth, essential beliefs or values, theoretical predilections, metaphysics or meaning system, etc. Regardless of how such characteristics are labeled, they would be understood in this model as very high level referents — and, indeed, many people would regard these highest level referents as representing a kind of absolute truth. Our usage is different, of course, as we have been explaining — but next we shall see that there is a sound basis for the idea that there is an overall direction for truth in the hierarchy.

To start with, let us bring back our old friends of same and different, to see how they can help us demonstrate our twin themes in terms of the present discussion. In simple complementary terms, we can say that one thing is the *same* as another thing to the extent that they mesh with each other — which also means that the two components are mutually compatible, and thus that the one thing is true in relation to the other. Conversely, one thing is *different* from another thing, to the extent that they do not mesh with each other — which also means that the two components are mutually incompatible, and thus that the one thing is false in relation to the other.

Then there is the hierarchy aspect of same and different, as we explained in the first chapter. Thus, we have said that higher level represents *same whole* — synthesis or generalization; whereas lower level represents *different parts* — thesis-antithesis or discrimination. This can be understood here to mean that the upward direction of the hierarchy represents more and more complementarity — and so greater and greater compatibility, coherence, unity (same) ... and thus truth; whereas the downward direction represents more and more uncomplementarity — and so incompatibility, incoherence, fragmentation (different) ... and thus falseness. By the way, one can also apply the two viewing concepts here, *P*-view and *X*-view respectively.

The upshot is that when we talk about hierarchy, there are two different ways or usages in which the relative nature of truth and falseness can be understood. We can either talk about *scope*, as we have been doing in much of this section, thereby referring to lower vs. higher levels of the overall hierarchy we are dealing with:

Lower Levels	Higher Levels
narrower truth-falseness	broader truth-falseness
specific	general
X-view	*P*-view

Or we can say that no matter which relative level is involved for any given relationship or situation, there is still a consistent upward vs. downward orientation or *direction*, as we have just been describing thus:

<u>*Lower Levels*</u>	<u>*Higher Levels*</u>
falseness	truth
thesis vs. antithesis	synthesis
different parts	same whole
discrimination	generalization
X-view	*P*-view

In the former case, we can talk in terms of big truth compatibility and big falseness incompatibility at higher levels, as contrasted with little truth compatibility and little falseness incompatibility at lower ones. But in the latter case, we can simply talk in terms of big truth at higher levels, as contrasted with little truth at lower ones. Which set of truth vs. falseness labeling approaches is most appropriate will vary with the topic, as we shall see in the pages ahead.

Reconciling With Common Sense

In the previous chapter, we talked about how the established paradigm tends to depict common sense as some sort of undeveloped or unsophisticated remnant of an outdated way of seeing the world — equivalent to a base instinct, which is somehow stuck in a primordial stage of thinking, and which can never grasp the deeper insights which modern science has brought us. Whether this antiquated vestige remains in place as a result of our limited range of experience, because of a flawed concept system, out of a devotion to contrary metaphysical assumptions about nature, or simply because of a kind of stodginess of mind ... does not change the character of what we are supposedly dealing with: a rigid and old-fashioned way of thinking about the world, which is not amenable to reform.

Of course we reject this conclusion and the explanations which go with it, as to the value and character of common sense: we see common sense as having just as much merit in our explorations of nature as does the established paradigm. In this respect, we reject the proposition that either side of this "dispute" should be treated as ipso facto superior to the other: both should have equal standing in a discussion of their respective benefits and deficits. As far as which approach is ultimately preferred, we suggest once again that such a question depends upon one's goals.

Comparing the two, we can say that common sense has more general foundations, and the established paradigm more specific ones; the former is more amenable to understanding and the latter to prediction; the former is based on a wide range of long lasting, persistent, albeit mostly conventional experience; the latter rests upon more exotic or transitory realms and modes, including ones which we may not encounter in our daily lives. But we maintain that the crucial difference between them is that while common sense is part of (or a way of describing) a coherent representation of the natural world, the orthodox framework is built upon a scattering of arbitrary premises and ways of representing the natural world.

These factors come into play when we confront an issue which was discussed in the first part of the previous chapter — that as science advances over time, the more its theories contradict common sense. The standard argument is that while science marches forward and explores increasingly unfamiliar terrains of nature, common sense remains rooted within the confines of our normal experience — and this explains why the former is inherently superior to, and increasingly trumping, the latter. But we reject that conclusion, and argue instead that the reason for the gap is the ad hoc character of the theories themselves and the way they interpret the new frontiers which are being encountered. Most important is the established paradigm's failure to embrace or develop a sophisticated and satisfactory (i.e., coherent) metaphysical or conceptual framework ... within which there would be no gap between these two realms of knowledge. We will return to this point shortly.

Then there is a related issue, and the specific charge which the established paradigm is so fond of raising — which concerns particular cases in the past where it would be acknowledged that common sense has been shown to be specifically in error ... i.e., contradicting facts. Thus, it was once mankind's common sense that the sun circled around the earth, rather than the reverse — or that the earth was flat, rather than round. And of course we now know how foolish our ancestors were to have so thought, and how primitive was their common sense in adopting such positions.[11]

But the reality is that it is not just common sense which has had problems in the past — for the orthodox discipline has obviously made errors, as well ... its lofty posture and condescending attitude notwithstanding: clearly, mistakes can be and are made on both sides. So to say that mankind's common sense is fallible does not make it useless, as conventional science suggests; nor does it make the conventional science approach infallible. Why isn't the established paradigm just as eager to deride itself as having little value, because it has made mistakes in the past ... and will continue to do so the future?

The answer would presumably be that the established paradigm has a built-in mechanism for correcting such errors — e.g., in terms of Popper's falsifiability criterion — whereas there is no corresponding correction mechanism for dealing with shortcomings in mankind's common sense. We maintain quite to the contrary — that it is an absurd proposition to claim that our common sense is intrinsically flawed and not self-correcting, by virtue of being fixed and unyielding. That is because our common sense is no more static than anything else in our minds: we can be educated and we can grow — common sense intuition can and should become more enlightened. Just as our detailed understanding of a particular topic may need to develop as part of the process we call "learning", so likewise our more general common sense can evolve, as well — and one can argue that it has evolved over the millennia, along with everything else. There is no reason to assume that we in our minds are incapable or unwilling to cultivate a broader conceptual and phenomenological overview of nature.

11 Actually, a dualistic approach to these matters might yield a different conclusion, issues which we discussed earlier in the book; but we don't want to get sidetracked by such matters here.

How, then, can one conclude that common sense is more rigid in its approach to the world than is the established paradigm — especially given the dogmatic position for which Kuhn advocates and which is so apparent in the incoherent and entrenched models that we are dealing with today?

This brings us to the main point we are trying to make in this section: if there is an incompatibility between the two bodies of knowledge, that of the established paradigm and that of common sense, then there should be some way of reconciling them. And in this process, as we have been saying, neither side should be treated as inherently superior and the other inferior; and likewise, neither side should be taken as fixed and absolute, in relation to which the other must be fluid and relative. In particular, we are not suggesting that our common sense view of the world, as it exists at any given time, is inherently superior to the established paradigm's view of the world, as it exists at any given time: but it is equally important to realize that neither is the reverse the case. Therefore, we are not stipulating which side of the relationship must change in order to fit with the other side.

For example, regarding some of the topics we discussed earlier in the chapter, if there is a conflict between fact and theory, then one or the other (or both) are problematic: either we have misconstrued what the "fact" is, or we have an unsatisfactory theory in relation to that fact — and thus we must interact with or interpret the phenomenon differently. But one way or another, we must attain compatibility; and likewise with construct-construct fits, theory-theory fits ... and especially with the kind of theory-megatheory fit that we are now discussing. For the two sides must ultimately mesh together, in order for the result to be satisfactory.

In terms of the issue at hand, then, both our common sense and the established paradigm must have the capacity to change — it must be a two-way street. Yet Popper describes only one of those directions: "Though I am all for common sense, I am also for enlarging the realm of common sense by learning from science." (Popper, 1981, p. 295) Very well — we would concur: but by the same token, science must likewise be just as flexible and able to learn from common sense ... and not merely reject it as irrelevant. In this regard, we cited various authors, in the second half of the prior chapter, regarding the importance of common sense — including Popper himself.

Again, Toulmin argues that the scientist who encounters phenomena in nature which contradict his pre-existing view of the world "must find some way of applying or extending or modifying his prior ideas about Nature so as to bring the deviant event into the fold." (Toulmin, 1961, p. 46) This is true: but any such alterations which are necessary to meet one set of compatibility requirements must not neglect other compatibility requirements — including that which we have called megatheory. This means that any changes to our highest level representations of nature — called variously common sense, metaphysics, or our general concept system — must happen in a manner which is comprehensible and which does not require us to abandon a meaningful relationship with the world or an overall sense of the ultimate coherence of nature.

Now we come to a salient point in our argument, which concerns how we are to treat significant theories over time, regarding this compatibility factor which we maintain is crucial — and it is not common sense which is in the dock here, so much as the established paradigm.

When a new theory emerges which is (e.g., astonishingly) successful in its ability to account for some given array of previously-unexplained (or even unexpected) facts, but which contradicts other fundamental conceptions (i.e., our common sense or core megatheory) about the nature of reality, then something is out of whack and needs to be altered — something must yield: either the fundamental conceptions about nature, or the new theory. Yet, as we keep saying, somehow we must still end up with a coherent way of representing nature.

Such a process may require that we revise various conceptions about the nature of reality — even to the point that we may have to jettison whole theories that we have previously subscribed to, which are not consistent with the new order. In addition, we might also have to abandon or revise other apparently outmoded personal perceptions, beliefs, models, and values as being out of touch with the new reality — including any conflicting common sense and meaning system. Obviously, such a significant step (cf. Kuhn's retooling) would serve a good purpose, as long as it helps us to attain compatibility and coherence.

This is the case even if we cannot fully understand and wholeheartedly embrace the new theory initially: in time we should be able to comprehend and integrate it — so long as we give it careful and sustained consideration for a sufficiently lengthy period, and so long as we are open-minded and take into account all the arguments and interpretations in its favor. Of course, this reconciliation process task may take one year, it may take five years — it may even take a whole generation: We must give serious attention to every theory which otherwise seems viable or compelling, and we must do so for a suitably extensive period of time. We must be patient, and keep trying to accommodate what many people may claim to be a spectacular new insight into nature — and we must welcome any aids which the theory's proponents can provide in order to help us reach that end.

But this probationary stage cannot go on forever! At some point, eventually, if we still cannot find a comprehensible vision of nature which comports with the new theory, then we must wonder whether we are on the right track, after all. For in time, if the dissonant conceptions continue to make no sense to us, even after many years of mental percolation and in spite of every insight we can bring to bear, then eventually we have the right to make a decision — always provisional, of course — against it. Thus, if we still cannot relate to the new idea or theory, and find some way of meshing it with a satisfactory higher level representation of nature, then there is a prima facie case *either* that the orthodox regime is not presenting its idea well — perhaps because scientists do not understand it themselves — such that we can comprehend it and integrate it into a coherent model of reality; *or* that the idea itself is simply not useful as a general description of nature.

Consequently, whatever particular lower level "micro-facts" seem to underpin the new concept or theory, if far more higher level "macro-facts" continue to contradict it, then the greater weight of the latter makes our decision clear: the new theory must go … and some other way must be found to account for the phenomenon in question. In the end, we must say, "Yep. I know it predicts many things. But after much time and reflection, and very serious consideration, my judgment is that it's simply not acceptable — the theory just won't wash! It makes no sense to me, in terms that I can relate to. I cannot find any way of bringing this idea together with a broader vision that helps me to understand nature. It does not yield a pattern that I find meaningful, relative to the rest of my experience with the world all around me. And that is my ultimate test."

For we cannot have it both ways, if we are going to attain compatibility and coherence: either we reject the upstart concept or theory, as being out of keeping with the vast preponderance of other concepts and theories which we had theretofore subscribed to and were well-served by; or we accept the major new idea and abandon the vast preponderance of other concepts and theories, which we had theretofore utilized to build our conception of nature ... and replace them with something else. One way or the other, however, we must somehow end up with compatibility, coherence — and unity: eventually, it must all fit together and make sense. We cannot have a situation which goes on indefinitely, whereby the two frameworks remain forever unreconciled. Such fundamental incompatibility — between theory and meaning, between theory and common experience — is ultimately unacceptable.

Thus, certain once-novel ideas, like that of the earth moving around the sun, may have made no sense to most people initially: but such ideas obviously made sense to others, and this number gradually increased over time. And yet to whom do the models of the established paradigm make sense today — after many decades, or even a century, of trying? Thus:

> Physicists have now lived with the Theory of Relativity for many years and have become thoroughly familiar with its mathematical formalism. Nevertheless, this has not helped our intuition very much. (Capra, 1983, p. 89)

Likewise, Kitchener wrote nearly two decades ago that,

> Perhaps it is as true today as it was fifty years ago, that we have yet to fully assimilate Einstein's theory into our way of thinking about the world.... Why we have not been able to do so is an interesting question. It is not merely that scientists have not adequately educated the general public, for many scientists themselves do not conceptualize the world in this "relativistic" way. (Kitchener, pp. 8 and 20)

Our core metaphysical position is that the world is characterized by patterns which, though they may not be immediately compatible with much of the way we have learned to think about nature, are — must be — eventually made to fit, in one way or another. Therefore, it is our contention that *any theory of nature which, over a considerable period of time, continues to violate and contradict our evolving common sense is ultimately doomed*. For we have no better guide to the world of nature than our common sense of nature within ... enlightened and developed as far as possible — but never abandoned outright, as modern science so strongly urges us to do.

Indeed, it is difficult to imagine any viable description or model to which our coherent sense of nature cannot, in time, become enlightened and accommodate. Similarly, it is difficult to imagine any other way for understanding of — let alone harmonizing with — nature to take place. And any examples which contemporary science would cite as contradicting this premise are, by definition here, simply evidence of how we have been led astray.

Slipping Backward from Truth

We want to continue discussing some of these issues now, in terms of how science develops over time in relation to factors like common sense and truth. Thus, we have made reference to the conventional view which states that theories of science increasingly diverge from common sense, and there is a similar argument which dissociates the path of science from a realistic conception of nature and a meaningful pursuit of truth.

As a principal contributor to the orthodox thinking in this regard, Thomas Kuhn writes that, "A scientific theory is usually felt to be better than its predecessors not only in the sense that it is a better instrument for discovering and solving puzzles but also because it is somehow a better representation of what nature is really like." (Kuhn, 1970, p. 206) Yet Kuhn rejects such a proposition, saying it "will not do," and arguing for "the implausibility of the view. I do not doubt, for example, that Newton's mechanics improves on Aristotle's and that Einstein's improves on Newton's as instruments for puzzle-solving. But I can see no coherent direction of ontological development." (Kuhn, 1970, p. 206) Kuhn's point is that whereas the later theories are better at solving puzzles (i.e., making predictions), they do not appear to evolve out of a common and coherent thread about nature: there is no clear direction which allows us to say that we are coming "closer and closer to the truth." (Kuhn, 1970, p. 170)

So, the question is: Do characteristics like truth and a meaningful conception of reality potentially exist in a coherent form — towards which our theories of science should be bringing us closer, and in relation to which our common sense can grow to accommodate? Or is any such goal merely a mirage, which can never be identified and will always elude us? In other words, is there a clear ontological (if not conceptually evolutionary) direction before us, or not? These are some of the topics we shall touch upon in the present section.

Let us briefly consider certain aspects of the contrasting paradigms which Kuhn cites, along with a prior example. Starting with that first, we can characterize Aristotle's position as stating that a constant force, between a body at rest and a body in motion, would result in a constant speed of the second body. The idea was that as long as the net force between the two bodies remained constant, then so would the speed of the second body relative to the first. As Toulmin puts it in explaining Aristotle's position, the resulting motion would be constant as long as there was "a more-or-less complete balance between two sets of forces: those tending to maintain the motion and those tending to resist it." (Toulmin, 1961, p. 50)

Galileo's later position represented a challenge to this approach, based upon the concept of natural motion — the maintenance for which required no force at all. In this formulation, "force" was understood as something which brought about a *change* of motion — whereas no such "force" was necessary for the *maintenance* of ongoing "inertial" motion. In fact, Galileo argued that by this interpretation a constant force would lead not to constant speed — but to increasing speed, or acceleration. Needless to say, this later approach is seen today as an extraordinary conceptual breakthrough, and the sensibleness of Aristotle's position was superseded by Galileo's daring new insight.

But consider what has actually happened here. The discrepancy between these new and old positions can be understood primarily in terms of differing conceptions of *force* — both in interactional, as well as what we call stable vs. changing, terms. Regarding the former, Aristotle's approach treated force as a propensity that each object exhibited, and which could be understood in relation to one another — whereas the more modern view treats force as a change phenomenon which is exerted by one entity upon the other.

The stable-changing aspect becomes apparent when we see how the meaning of the force construct has been effectively curtailed by the new approach. That is, Aristotle's position was that there was a relationship between force and motion: if force remained constant, then so would motion — but obviously, the converse would also apply ... that if force was no longer constant, then neither would the motion. But Galileo's use of the inertial motion concept severed this relationship, and thereby narrowed the meaning of force: since the new category of natural motion required no force for its maintenance, the force construct was converted from having both stable and changing aspects to only representing the latter. The former is precisely what inertial or natural motion was understood to mean — no force and therefore stable motion.

What has happened is that the new approach established natural motion as its conceptual centerpiece, rather than what we maintain is a dual force kernel. Thus, Aristotle's model can be understood to mean that stable motion corresponds with stable-force, whereas changing motion corresponds with changing-force. But in Galileo's model, stable motion would correspond with no-force, whereas changing motion would correspond with yes-force. Not surprisingly, we regard the later approach of Galileo not as a step forward here ... but rather as a step backwards. That is because the dual force approach is consistent with (indeed, inherent in) the present model, as representing a persistent pattern of nature — whereas the single force approach represents a typical one-sided orientation. Of course, we have also argued that the no-force concept is an illusion, given that there is nowhere in the universe where forces do not impinge — and the inverse square law is consistent with that position: stable distance means stable (or balanced) force and thus stable motion, whereas changing distance means changing (or unbalanced) force and thus changing motion.

We now move on to a second example, because no matter which of the two force definitions is adopted, these two great thinkers apparently agreed on at least one thing — which is that the trajectory of the constant speed motion would be circular. It is this trajectory conclusion of both Aristotle and Galileo which Newton took issue with in his mechanics, as we discussed in Chapter 4.

Thus according to the official ideology, it is only with Newton's insight that we learn the true character of such natural, or inertial, motion: it is not circular, but linear! Newton said that if no external forces impinge upon a body, then that body will maintain its *uniform linear motion* indefinitely. This, indeed, remains a fundamental tenet of modern physical thought, even though such a notion is a pure idealization — a total abstraction, in the "open field" terms in which it is commonly understood. Again, there are no force-free regions of the universe, " ... and surely the idea of infinite straight-line motion is quite unrealistic." (Hecht, p. 73) Yet in spite of such absurdity, Aristotle's (and Galileo's) constant circular motion thesis was overthrown, with the result being that we have the "replacement of Aristotle's common-sensical paradigm by Newton's new, idealized one." (Toulmin, 1961, p. 56)[12] Even to this day, "Aristotle's paradigm is familiar in a way that Newton's can never be", for the latter "remains rather paradoxical to the commonsense mind." (Toulmin, 1961, pp. 60-61). Needless to say, we certainly do not regard the outcome here as representing a step forward in our understanding of nature, towards a greater truth; nor was it necessary for more narrow puzzle-solving purposes.

Our third example concerns how Newton, in turn, changed from genius to dolt — this time at the hands of his successor, Einstein. Newton had postulated the completely intuitive notion of an "absolute space" in the universe. Einstein's Theory of Relativity is said to have abolished that concept, replacing it with a new four-dimensional *space-time continuum*. Along the way, of course, Einstein also managed to introduce the notion of a *constant speed of light* — another extraordinary and wholly counterintuitive leap. These two crucial conceptions form a significant part of the overturning of Newton's classical universe by the world of Relativity which we still live in today. As Stapp puts it so succinctly: "[T]he basic concepts of classical physics are known to be profoundly and fundamentally incorrect." (Stapp, 1988, p. 54) We have discussed these issues at great length in previous chapters, and we have argued that Einstein's new approach takes us backward along the path towards understanding and greater truth — and that it represents an arbitrary approach which was totally unnecessary for resolving the problem at hand.

Of course, the established paradigm would adopt a very different posture, in that examples like those described above would be commonly cited as demonstrations of how brilliant new theoretical advance flew in the face of common sense — the latter being rightly transcended, and even contradicted, by innovative and radical counterintuitive theorizing. And likewise any notion that we are coming "closer and closer to the truth" (Kuhn, 1970, p. 170) would also be dismissed. Thus, the kind of theoretical developments we have just described — and discussed throughout this book — would be praised *not because they make sense* conceptually, but virtually in spite of the fact that *they do not make sense* conceptually. Indeed, the lesson we are supposed to learn concerns the futility of depending on our normal tools of reasoning and a search for meaningful truth, and how such folly limits us from making bold, new — and here's the key, *productive* — leaps forward. Again, the justification is efficiency, or fruitfulness, or what have you.

So, more productive, more fruitful — how? In terms, for example, of circumventing conceptual anomalies (e.g., eliminating problematic concepts like the ether); by opening the way for the development of new concepts (e.g., the space-time continuum), or by substantiating their supposed necessity (e.g., natural motion, or gravity); by better conforming to mathematical statements, techniques, and limitations — thereby contributing to the hegemony of mathematics in determining our models of nature; or simply by virtue of stimulating new research and theory. But these new approaches are *not* generally hailed for helping us to better understand nature — except in the peculiar sense that they supposedly assist us in casting off false intuitive notions ... and putting nothing meaningful in their place.

12 Admittedly, there is some ambiguity regarding these attributions. In Chapter 4, we already mentioned "the Aristotelian differentiation between terrestrial motions in straight lines and celestial motions in circles ..." (Jammer, 1957, p. 80) Also, we noted Sachs' point that there is not universal consensus about Galileo's opting for the circular trajectory (Sachs, 1988, p. 10).

By contrast, our position is that we are dealing with a catalogue of conceptual missteps — which, although bold, were radical and foolish departures from a more meaningful approach to nature ... that can accommodate prediction, while still leading us to understanding and greater truth. Thus in contrast to Kuhn, we can see a "coherent direction of ontological development", as he puts it — but these steps took us the wrong way.

Obviously, the issue is not whether these bold new "advances" were ingenious and led to short-term breakthroughs — they plainly did; but rather, whether they just buried the fundamental problem ... often in a mountain of mumbo jumbo. For we argue that if these were not just limited solutions, as a means for sidestepping (or otherwise alleviating) certain difficulties, then they should serve a broader purpose. And we maintain that the crucial test of such broader purpose is whether these supposed insights ultimately lead us to a coherent (higher level) vision — not only in general terms of deeper understanding, but also in particular terms of a unified model of nature, as we keep saying. Thus far, they have clearly not — and there is nothing on the horizon to reverse this dismal assessment; so they have failed this key test. Indeed, we contend that the "advances" of each of these kind of counterintuitive "insights" is more than outweighed by the grave long-term deficits that they have left us with. For we have been led badly astray, in terms of what we maintain should be science's primary goal.

The House Called Science

Clearly, established scientific theory did not develop around a common metaphysical theme or perspective about nature. It is clear instead how much of it did develop: essentially as a series of historical accidents. Workers in the field were confronted with challenges in this or that area, and they came up with the best solutions they could find for the period — almost on the fly, in keeping with a kind of "conceptual opportunism" or pragmatism. The result is that most of the components of the established paradigm have been gathered up over time, with the resulting collage forming the body of science we know today.

It should not be surprising, then, that the various constructs and theories have no particular interrelationship with one another — because of the historically-related manner in which they developed, as opposed to a logically-related manner. The assorted constructs and theories emerged at different times, in response to different practical problems and within various conceptual milieus. So why should there be anything coherent about the result, given the haphazard functions and principles upon which the different constructs and theories were based?

This is not to say that we are dealing merely with a process of accumulation, of chronological layering of construct upon construct, and theory upon theory; for there has been elimination, too, in the form of older depictions being superseded by newer ones. But all too often, the older ideas have nonetheless been carried forward — sometimes in a little-changed form, albeit in the shadow of the new theory ... and sometimes in an altered form, in the guise of the new theory. In some cases, the reason why we have the new being superimposed upon the old, rather than merely replacing it, is that many of the new constructs and theories simply make no sense or are very difficult to work with; and it is often easier to revert back to the more meaningful older forms in our everyday thinking and communication. Moreover, the standard argument is that sometimes the older approach continues to be valid in limited, everyday conditions — whereas the new approach goes beyond such constraints and extends to the very small world within and the very large universe without (as well as the very fast) ... and at least in principle eliminates any need for the classical explanations altogether. We have rejected that interpretation in this book — along with the idea that the newer constructs provide a more satisfactory foundation for comprehending nature.

At the end of the day, what we have is still an agglomeration: we have never reached a streamlined new vision. We are dealing with a system that has largely developed in a chronological manner — even if punctuated by occasional and radical innovations in content and tone, as per Kuhn's revolutions — being amassed largely as a series of random increments, with one arbitrary construct after another being adopted over time. We switch between the different renditions and conceptions of nature, which share no common essence, depending on the circumstances.

Now let us consider a very different portrayal of the edifice of science, in the form of a metaphor that was put forward many years ago by one of the fathers of nuclear physics, J. Robert Oppenheimer. Oppenheimer wrote glowingly about the precepts which have guided the growth of modern science, and about the model of nature which it has led to. He speaks about "the house called 'science'," and says that:

> It is a vast house indeed. It does not appear to have been built upon any plan but to have grown as a great city grows. There is no central chamber, no one corridor from which all others debouch.... It is a house so vast that none of us know it.... It is not arranged in a line nor a circle nor a square nor a pyramid, but with a wonderful randomness suggestive of unending growth and improvisation. (Oppenheimer, 1954, p. 84)

Oppenheimer's description is very apt — although we come to a very different judgment as to the value of the result. For it is clearly a rambling and sprawling labyrinth, with its tower of Babel component constructs and theories strewn throughout: too diverse and dispersed to be grasped, conforming to no identifiable pattern, and based upon ad hoc and arbitrary development. In this regard, Oppenheimer's conclusion, about the overall discipline of science, is surprisingly like our own: "[V]iewing all fields together, it seems instead a rather ramshackle structure with little coherence among its various parts." (Oppenheimer, 1954, p. 49)

For his part, Kuhn would not accept such a complacent house of science metaphor, given his dissent from the notion that, "One by one, in a process often compared to the addition of bricks to a building, scientists have added another fact, concept, law, or theory to the body of information supplied in the contemporary science text. But that is not the way a science develops." (Kuhn, 1970, p. 140) Kuhn argues convincingly that we are not dealing in science today with an evolved whole, but with pieces that were devised at separate times to resolve particular problems, and which thus belong to distinct genres or contexts. Needless to say, we would concur with that assessment — although (once again) we are not at all sanguine about the results, as is Kuhn.

One might think of the quasi-random mode of theory-building as a kind of pre-science, where there are no guiding principles — no coherent metaphysics. Plus, not only was there no overall theme to guide the development of science, which would have made it inherently coherent as it progressed — but perhaps more importantly, no such consistent theme has emerged after the fact ... relative to which these concepts and theories could have been modified accordingly, so that they all fit together.

We come back to the point we keep making, that science today amounts to a set of what can be called arbitrary conceptual pivots — based upon arbitrary cross sections of nature — upon which scaffolding is erected an extraordinarily convoluted and meaningless theoretical superstructure. Oppenheimer's house called science is a mass of conceptual contradiction — a paste-up picture of nature, which does not provide us with the guidance we need as to the deeper patterns of nature. It is not something of which we should be proud.

The Ferromagic Physics Museum

Let us narrow in on the ad hoc character and historical fixedness points we have been making about the house of science, in terms of one very specific manifestation — that of terminology. Our concern here is not so much with the arbitrary character of the constructs themselves, which we have discussed, but with the arbitrary labels that are used to represent those constructs.

The task of trying to wade through the nomenclature in most any specialty of science is truly daunting. Instead of a system based upon a logical relationship of meaningful labels, the modus vivendi of established science seems to be that of inventing or erecting an almost impenetrable barrier of obtuse and arcane terminology. The result is the labeling quagmire we have today — which amounts to a terminological garbage dump. Our conventional naming system has not served us well, in various ways.

Of course, both sets of issues go hand in hand: just as the underlying constructs are themselves problematic, the same goes for how they are named. The result is a tangled maze of constructs and construct names, both of which serve to obscure the more general patterns and themes that we should be most concerned with. What results is a completely unnecessary barrier to learning, understanding, and usefulness.

For instance, the whole naming system bound up with electricity and magnetism is weighed down with irrelevant surplus meaning and misleading connotations, as we discussed in Chapter 3 — both the core terms themselves, as well as their combination. Then there are categories like "hyperons", "baryons", and "leptons" — instead of more straightforward labels of "heavy", "medium", and "light" particles. In this regard, Calder notes how subatomic particles have "confusing, mostly Greek, names, so that one of the most eminent of physicists, Fermi, was driven to remark before his death in 1954: 'If I could remember the names of all those particles I would have been a botanist.' " (Calder, 1977, p. 69) And in speaking about the atrocious nomenclature in her field, biologist Margulis says that its purpose "is to keep everyone out of the field." (Margulis, quoted in McDermott, p. 32)

Back to physics, what do we gain by speaking about a "quark" in terms of "truth", "charm", "color", "flavor", and so on? Similarly, in describing patterns in the universe, why not say something more meaningful than "dark energy" — to be differentiated, of course, from "dark matter"? What good is such idiosyncratic and esoteric usage, in which there is little relationship between the common meanings of the terms, and their particular scientific referents? How does it help us to understand anything about nature? Are such labels supposed to be cute or clever — is that their justifying characteristic? It may be the case, as we noted in Chapter 1, that "[P]hysicists delight themselves by using ordinary words for something else." (Feynman, 1965, p. 84) But are they supposed to be entertaining themselves or edifying mankind?

Here is how Jones puts the situation, in a light-hearted way:

> The modern scientific jargon can, in skillful hands, be one of the most opaque and long-winded dialects ever invented.... [O]ne of the great advantages of a developed scientific jargon [is that with its aid] simple and even sunny ideas can be made to appear subtle and portentous and flatteringly complicated.... Indeed, it has been claimed that many students of physics ... never really learn the concepts of their science at all: all they learn is a vocabulary of okay words, with rough rules for stringing them together. (Jones)

Good construct labels would facilitate the task of learning and of exploring nature — rather than hindering it and making it less likely to succeed. That is because the best nomenclature would already be partially organized and predigested, by virtue of the meaning and associations which are inherent in the words themselves — intrinsically conveying the essence of the phenomenon being represented, rather than being completely irrelevant to it. Then our energy could be be better applied to enhancing and expanding our knowledge, rather than in just memorizing convoluted lingo. Needless to say, terminology that is based upon personal names can never fulfill these functions, regardless of how ego-gratifying it may be to those so honored. For example, in terms of the scattering phenomenon there is Tyndal scattering, Rayleigh scattering, and Compton scattering...[13]

Unfortunately, science treats these crucial issues too lightly, and acts as though it does not really matter how we name a notion. But words matter — names count! The name tags are the handles on the conceptual doors: they form part of the constructs — part of what they mean, and how we use them. They represent the tools of our minds — they are what our thinking pivots upon. If we use shoddy (and poorly named) conceptual tools, then we are more likely to end up with shoddy ideas: for we will not be able to readily manipulate and order those constructs — they will clog up our mental pathways, and impair our minds.

It is interesting to approach the issues here in terms of the efficiency factor, which the established paradigm trumpets so loudly as justifying its rigidity — largely on the grounds that such resistance supposedly leads to maximum efficiency and thus progress. This is the basis for Kuhn's argument that it is more important to be productive in the normal science task of filling in details — as opposed to the disruptive science task of challenging (and potentially changing) the framework with which we are working. And yet we maintain that far more long-term progress would be gained by revamping the conceptual framework when appropriate than by sticking to it for short-term effi-

13 "When we reify our concepts we inevitably deify ourselves." (Finkelstein, p. 83)

ciency — both in terms of the constructs themselves, and in terms of how they are named.

We know the historical reasons for why so much of the terminology we use today is named that way. But regardless of why a given term was once used, the name should be changed or eliminated ... if a better way of referring to the phenomenon exists today — both to represent the construct in question, as well as to facilitate that construct's smooth integration with other constructs. As with our theories and constructs, so likewise our terminology must evolve. The constructs and labels we use should be efficient cognitive tools, not antiquated conceptual relics.

In this regard, emphasis should also be placed on curtailing the endless proliferation of new constructs and labels, and eventually allow us to slim down conceptually and reduce the overburden of terminology we are carrying today. The main point is that it is mostly new perspectives that we need, with which to illuminate things — rather than more content, which may only further obfuscate things.[14] What we lack are, in Bateson's terms, *patterns which connect* — which good terminology would help provide, as opposed to more meaningless conceptual clutter.

Science and Religion

We now come to a set of factors that are not generally associated with the institution of science, but which we maintain help to explain both its success and its failure. In brief, we contend that part of science's success can be attributed to some of the very characteristics which it claims to shun and be the antithesis of — whereas part of science's failure can be seen in the continuing need that many people have to look elsewhere than science for a satisfactory understanding of the world in which we live ... to realms like magic, mysticism, and religion.

Starting with the success category, it is indeed paradoxical — and for many, quite ridiculous — to suggest that science is anything but the way it is almost universally portrayed: as the epitome of rationality for approaching nature. Yet it should be quite apparent from our discussion throughout this book that we regard the established paradigm as being based on an arbitrary array of different (typically one-sided) perspectives and assumptions — which have no more inherent justification or validity than would a quite different (e.g., opposite) set of perspectives and assumptions. Moreover, we argue that some of the tools and strategies which science claims to disavow actually play a very prominent — albeit often subliminal or implicit — role in the workings of science. It is reasons like these which underlie our contention that part of science's success is based on factors which are quite different from that which the stereotype conveys.

As we have seen, there is too much about science which is dubiously demonstrated and forever unexplained — characteristics and phenomena that are not based on logic and proof, so much as on conjecture and illusion. So while it is true that science is very

14 Davies has written that, "[I]t is a change of perspective, rather than a new piece of information, which leads to intellectual revolutions." (Davies, 1980a, p. 21)

rigorous and methodical in its approach to a wide array of phenomena in nature, it is only rigorous within the framework of certain assumptions — which are not themselves rigorous. This brings up the topic of metaphysics, which the established paradigm pretends to discount and ignore ... even though such factors are unavoidable and are inherent in all models. Moreover, there is an element of intellectual sorcery in the way science functions, and this is a significant factor in the degree of power which science exercises over our minds.

That is especially the case because of the fertile grounds which the established paradigm provides for such non-logical features. In particular, modern science thrives upon its amorphous theoretical terrain — the incoherent framework and associated conceptual knots and terminological confusion upon which its theories are based, coupled with the total rejection of common sense. What results is a kind of wizardry, whereby scientists can play all kinds of beneath-the-black-cloth tricks, by using theoretical "sleights of the mind" which very few people can follow ... and no one really understands.

Arthur C. Clark once said that magic is science which is not yet explained or understood. It is in precisely this respect that today's science is akin to magic: it manipulates nature, with dazzling technological displays and mathematical acrobatics — but the primary theories (i.e., especially of physics) lack all meaning, in that scientists cannot explain what they are doing or identify the broader patterns which are involved. In this way, conceptual hocus-pocus can easily become a camouflage for the initiated to hide behind — amounting to a secret language in a bizarre game having little apparent relationship to reality, but nonetheless remaining imposing and self-perpetuating. We are confronted with a smokescreen of verbiage, symbolisms, titles, instruments, publications, and awards — not to mention sheer power — which serve more to intimidate than to enlighten us.

Then there is the faith aspect — and here we are talking about science as a house of worship and belief ... to punt with Oppenheimer's metaphor. In this way, one can talk about science as having developed a particular (and peculiar) set of constructs and conventions over time, which are passed on from old to new workers. Whole sects are established, whereby the collective socialization of scientists perpetuates doctrine as much as it does promote reason — which is part of the Kuhnian ideology that we discussed in the prior chapter — and all the apostles recite catechisms that don't mean anything or make any sense. And so we maintain that modern science's combination of myth, magic, mumbo-jumbo — and self-declared central role in defining our common reality — is tantamount to religion. Moreover, the scientist clings to his theories as a priest may cling to his faith: not only is it a source of great strength to him, but it can also provide the basis for considerable power over others.

As an example of the self-declared central role just mentioned, consider how it is often said that the laws of physics govern the behavior of the physical world. What an arrogant and pernicious doctrine — as though the laws of physics represented a kind of holy writ that was handed down from some supreme being! The only thing that the laws of physics govern are the willing minds of devotees, and the submission of many others. Yet the laws of physics are not the same as the laws — or better, patterns — of nature. At most, the laws of physics represent the best current attempt to *describe* nature; but they certainly do not *govern* nature, as such a pompous contention claims. The larger nature does not dance to mankind's tunes.

Neil Postman makes reference to George Bernard Shaw, in commenting that, "In the middle ages, people believed in the authority of their religions, no matter what; today, we believe in the authority of our science, no matter what." (Postman) Capra says that, "[S]cientific knowledge is often considered the only acceptable kind of knowledge.... [T]he high priests of science [are thus considered] guardians of supreme knowledge." (Capra, 1983, p. 39) Capra labels this mentality "scientism".

As a counterpoint, consider Davies contention that:

> Throughout history people have fallen into the trap of attributing special organization to the world where none existed. The gods of our forbearers manipulated the world and maintained its activity. Modern science disposed of the gods and replaced them with laws of nature. (Davies, 1980a, p. 179)

Or did they? Feyerabend says that:

> Science is a religion.... In society at large the judgement of the scientist is received with the same reverence as the judgement of bishops and cardinals was accepted not too long ago. The move towards "demythologization," for example, is largely motivated by the wish to avoid any clash between Christianity and scientific ideas. If such a clash occurs, then science is certainly right and Christianity wrong. Pursue this investigation further and you will see that science has now become as oppressive as the ideologies it had once to fight. (Feyerabend, pp. 55 and 57)

There are even cases where scientists have become not just high priests in society, but virtual gods in society ... and their words our gospel. A compelling example of such deification is that of Einstein and his Theory of Relativity, as we discussed at the beginning of Chapter 6. The very mention of the name "Einstein" evokes great awe and intimidation amongst us all, for we have been taught to deeply revere this greatest scientific hero of the 20th century. And likewise, the very mention of "relativity" is associated with grand truths about the deeper nature of reality — truths so profound that they are beyond the grasp of mere mortals to comprehend. Dingle even speaks of "the evil spell cast by the word 'relativity' " (Dingle, 1972, p. 103).

We should mention one particular source of scientists' power, both over their peers and over the rest of society — and this concerns the specialist and general awards which are bestowed upon them. In some cases, these awards are tantamount to ordination of the individuals involved — and sanctity for their work. This kind of idolatry can be very unfortunate, to the degree that it undermines our capacity to scrutinize the ideas and theories themselves with a proper degree of detachment and objectivity — leading us to treat them as unquestioned truth, rather than as provisional suppositions about nature. And even in the face of showering praise, it is important that we not abandon our criteria for evaluating theories, in the face of such acclaim — the proper goals of science, as we are discussing in this chapter — nor forget the stultifying effect which excessive veneration can have upon our critical faculties. Therefore, it is important to resist the saintly status which very high experts are accorded. Even in the face of universal accolades, we must try to keep these matters in perspective — for even a Nobel Prize does not come down from heaven!

It is one thing to learn from such individuals, and to value them — but they should not be deified. Likewise, science should not be treated or taught as though it were sacrosanct — with its content being treated as divine truth, received through the modality of this or that hallowed figure, with the message being spread by their disciples. Overall, the institution of science needs to be challenged far more than it needs to be consecrated.

Now let us turn to the other side of the issue — not in terms of *science as religion*, to partially explain science's success ... but rather in terms of *science vs. religion*, to partially explain its failure. The main issue here concerns the deep-seated need which people have to make sense out of their world — to understand the patterns of nature and our place in it. For much of mankind's development, it was primarily religion (along with other spiritual pursuits) which fulfilled this function — but then science came along, with its claim to be providing more rational and objective criteria and conclusions about the nature of our common reality.

Unfortunately, this has never been achieved — not in any coherent and satisfactory way whatsoever. What we have from science instead is a vacuous ontology whose contents and consequences we cannot know or understand. Here is how Neil Postman describes the problem, as he contrasts the old world of religion with the modern world of science:

> The medieval world was, to be sure, mysterious and filled with danger; but it was not without a sense of order. Ordinary men and women might not clearly grasp how the harsh realities of their lives fit into the grand and benevolent design; but they had no doubt that there was such a design. And their priests were well able, by deduction from a handful of principles, to make it, if not rational ... at least coherent.
>
> Now, the situation we're in at present is very much different.... The world in which we live is very nearly incomprehensible to most of us. There is almost no fact, whether actual or imagined, that will surprise us for very long — since we have no comprehensive and consistent picture of the world which would make the fact appear as an unacceptable contradiction.... There is no consistent, integrated conception of the world which serves as the scaffolding on which our edifice of belief rests....
>
> [W]e no longer have a coherent conception of ourselves and our universe and our relation to one another and our world.... We have no coherent framework to direct our definition of our problems or our search for their solutions. And therefore we have no criteria for judging what is meaningful, useful, or relevant information.... We live in a world that for the most part makes no sense to us — not even technical sense. (Postman)

In this regard, then, it should come as no surprise that so many people turn to non-objective modalities like "pseudoscience", or to explanations that are based upon some kind of supernatural phenomena or agents ... especially when most people have been brought up steeped in those traditions. People will align themselves with whatever system of explanation they can best relate to, in order to satisfy their vital need to understand and come to terms with the larger world of which they are part — and for most people, science simply does not offer a viable option in this regard. We need wisdom as a species, yet the high priests of science have not delivered this: one can argue that they are not even trying, and that they often ridicule or denigrate the very goal or expectation to begin with.

Of course, the typical counterargument would be that we are failing to differentiate those realms which go beyond the purview of science — realms which mostly belong in the domain of religion. More broadly, it is said that there are two quite opposite ways of approaching nature: science should deal with the material or physical world, while religion should deal with the non-material or spiritual world. Yet such a division-of-territory compromise, which both institutions subscribe to, should be treated with great skepticism. Indeed, we contend that it is primarily a means of avoiding conflict by carving out reality in a manner which does not threaten either institution's legitimacy or existence, or the power of their respective purveyors. By contrast, we contend that the proper solution is not to segment reality into different domains, or pockets of peoples' thinking — which is testimony to the failure of both systems — but to develop a coherent model which will encompass the whole of nature, a point we made in Chapter 9 and will return to just below.

Needless to say, there are times when the conventional border lines break down, and the fundamental conflict between science and religion cannot be avoided. We now turn to one very prominent example, which concerns two very different interpretations regarding the development of our species. On the one side is science and Charles Darwin's evolutionary explanation about the process taking countless millennia (now estimated as several billions of years), and involving random factors of genetic mutation and adaptation. On the other side is religion and the argument that such an explanation is unsatisfactory, primarily because it overlooks the needed orchestration of such events — and the extreme version would be a literal reading of the Bible and its assertion of a few days' duration for the process to occur.

The battle between these two positions came to a dramatic head in 1925 with the famous Scopes "monkey trial" in Tennessee (U.S.), but the struggle between *evolution* and *creationism* has never stopped. In the last few years, the latter has "mutated" into a thesis called *intelligent design* — which basically says that however long it took for our species (and the rest of nature) to develop, it was far too complex a process (and result) to have come about in the blind way that science supposes ... and was instead guided by some higher being. Surveys in the U.S. indicate that there is widespread support among the general populace for this position and its underlying reasoning — and that these ideas should be taught in public schools alongside Darwin's theory of evolution. Of course, there is a larger dynamic going on in the background of all this — which is that the pro-religion forces are fighting back hard in order to reverse what they see as the ascendancy of pro-science "secularism" in modern life.

For their part, many people in the scientific community are quite baffled — indeed, stunned and bewildered (if not thoroughly indignant) — that these issues are even being debated in the modern world, where science's theories and thinking have had such enormous and evident impact. Far from raising any sort of self-doubt as to their overall goals and methodology, the result has been for proponents of the established paradigm to dig in ever deeper and reaffirm even more strongly the power and success of the reigning scientific doctrine. The main question they are willing to entertain is how the general public can continue to be so ignorant about science and all that it has accomplished — and what it will take to bring enlightenment to the masses.

There is no disputing the fact that most lay people are woefully underinformed about science — but why is this the case? The usual way of answering the question is to wonder whether it is due to a failure of one side or the other — the teachers or the learners — in the education process. But we contend that the main culprit is the theoretical framework itself, for reasons that we have stated many times — and this explains why there is so much ignorance about science, and why so many people are considered scientifically illiterate, etc. Why would people be interested in learning a set of convoluted and untenable concepts and theories, which make no sense and mean very little in their real lives? And how could it be a surprising consequence that we would end up with a disempowered and disinterested public? Most people inevitably turn elsewhere for an understanding of nature and their relationship to it, as we have been discussing.

So we maintain that the problem here is largely a result of science's own making. Science's aloofness and smugness — but especially its failure to meet the needs of mankind in the form of providing a coherent model of nature — has left the door open (if not encouraged the necessity) for alternative renditions of reality, such as religion provides, to remain or become entrenched instead. There are too many people for whom the standard theory-building criteria and conclusions are simply unsatisfactory as a basis for ordering their lives.

The fact is that Darwin's approach appears to represent a good theory of science, in terms of the kind of precepts for which we have argued — simplicity, generality, and coherence. But that model is not necessarily evaluated solely on its own terms, and until there is a more satisfactory overall framework which people find credible and compelling — which meets their needs and in which they can feel confidence — then many will look somewhere else for the larger answers they seek ... even at the cost of viable theories like evolution.

Indeed, one might suggest that science should take Darwin's argument itself more seriously, regarding survival of the fittest: science should put forward a more coherent overall model of nature that is a stronger contender to survive in the minds of many — than those which it is seen to be in competition with — in terms of theories that people can relate to and embrace, as an overall conceptual vehicle for carrying their thinking forward and with which they can further evolve. But to succeed in this endeavor, scientists must stop demanding adherence to their authority — and must earn that authority instead, in the terms we have been discussing. In sum, there would be much less need for pursuits like mysticism, mythology, and religion ... if we had a real science of nature.

Now for an important point of clarification, regarding what that "real science of nature" must encompass — and in so doing, we venture into some very hazardous territory ... both because we are breaking the rules, and because we are dealing with subjects that are quite easily misconstrued. The central issue concerns our position that we must have a science of all nature, and not just the "half" of nature which the established paradigm is most comfortable with and decrees to be its domain. Clearly, that whole of nature entails characteristics which are not only measurable, but also non-measurable — and so again, a satisfactory theory of nature must comprise the latter as well as the former, in some manner or other. Indeed, it seems beyond argument that we have already reached this stage with Quantum Theory itself, even though the issue is not necessarily always described in these terms. We are talking about the Uncertainty Principle — and one way or another, that concept must be understood as representing a factor of non-measurability, such as is expressed when we speak about the impossibility of simultaneously measuring conjoint variables.

And so we need to apply that pattern to the present topic. By speaking about "much less need for pursuits like mysticism, mythology, and religion", we do not mean to suggest that all the world is rational and that the overall category of non-material phenomena should be eliminated. On the contrary, there is an essential feature of nature which underlies these pursuits, leaving aside the various belief systems and associated divinities which go with them. It is precisely because there is a category of non-material phenomena that exists in nature — and not just in the minds of man — that this should be reflected in our models of nature. Therefore, the question is how this set of factors is to be construed and integrated into our models.

The essential feature that is involved can be described in broader ways than that of the measurable vs. the non-measurable — or of certainty vs. uncertainty. For, whatever pair of labels one adopts, our position is that we are talking about a larger pattern of complementary characteristics or tendencies, which can in turn be understood in terms of the two forces of nature themselves. Here are some of the associated qualities we are referring to:

X-Force of Structure	*P-Force of Process*
measurable	non-measurable
certain	uncertain
predictable	unpredictable
absolute	relative
explicit	implicit
distinct	subtle
focus	background
rational	irrational
coherent	incoherent
material	spiritual

The consistent posture that we have adopted throughout this work is that nature encompasses both sets of characteristics — not just the one side — and so should science. And while we have not previously identified the "spiritual" factor as part of this pattern before, it now comes into play with the present topic — and this is partly why we find ourselves on risky terrain, as we indicated above. Thus, we have all been thoroughly trained to reach precisely the opposite conclusion — that whatever the "spiritual" category represents, it does not belong in science.[15] Furthermore, how can we even

15 Put in terms of a point we made earlier in the chapter, the fact that it was once necessary for modern science to break away from such non-measurable domains in order to establish itself on a stronger conceptual footing ... should not blind us to the fact that times change, and what was once the optimal approach may no longer be.

be talking about our theories of science as entailing the "incoherent" when we never cease talking about how our theories of science must be "coherent"?

Regarding that latter question, the answer is easy: our theories of nature must be coherent — which means here that they should accommodate both coherent and incoherent aspects of nature, as the above chart indicates. Regarding the former and larger issue, we will explore this complex and delicate topic no deeper in the present work than the little we have to say in these few paragraphs. For one thing, the highly charged term of "spiritual" is fraught with both desirable and undesirable connotations, which makes it very difficult to analyze clearly. Even so, we contend that the associated phenomena are just as valid a topic for science as are the "non-spiritual" ... if only because both types of phenomena exist in nature. Moreover, it is not just that both types exist — but we all experience both types of existence. We make that assertion even though it may be true in different ways and different degrees for different people (if not different cultures) — and we contend that it applies just as much to scientists (even if outside the lab) as it does to non-scientists.

The upshot is that the institution of science must find some way (however constrained or general) of incorporating both aspects of nature into its models. And to be clear, none of what we are saying should be taken to mean that all claims regarding such non-material phenomena are of an equal standing: on the contrary, science should be able to develop an orderly framework that will help us to better sort out and evaluate the crucial features here.

In terms of the present theory, the answers would presumably be found in the same consistent framework we have been describing all along — that there are complementary patterns of both structural and process phenomena, as they manifest themselves at different levels of nature's hierarchy. This would be so regardless of what particular constructs and names are attached in each case — ranging from commonplace references to exotic ones — and therefore whether the phenomenon can be labeled with words like "spiritual" or not. We are talking about general patterns that are not restricted to particular levels — let alone some singular manifestation at the very top of the hierarchy only ... or that is specially related to our own level of the hierarchy, as in human-like or human consciousness terms.

Putting our points together, we contend that the institution of science has two strikes against it concerning the issues we have been discussing: not only are the theories of science generally unacceptable because they are incoherent, meaningless, and irrelevant to most people — thereby necessitating an alternative explanatory framework such as some form of religion or spirituality provides. But science is also unsatisfactory because it shunts aside the form of existence which is contained within these pursuits, nominally on the grounds that such phenomena are outside its domain — yet in practice to the effect that exploring them amounts to a kind of mushy-headed hobby ... as opposed to the hard-headed vocation of studying the real world, which science supposedly represents.

We contend that there is only one real world, and that it includes both the material and the non-material realms of nature. The task of science is to embrace both, in a manner that humanity can relate to.

Science and Technology

Now we want to leave aside science and religion, and turn to the topic of science and technology — and we will bring technology into the picture in two different ways. One is rather unorthodox, in that we will talk about technology as describing a class of theories which we differentiate from that of science. The other is more conventional, in treating technology as the applied form of science, in terms of a set of techniques or a body of products. We begin with the former, and come to the latter near the end of the section.

Throughout this chapter, we have been speaking about two different categories — two different goals, two different ways of approaching nature. And now we would like to apply the above names accordingly. Simply put, we label the broad, higher level type of theories as science — relative to which we label the narrow, lower level type of theories as technology. We contend that they each serve very different functions, and correspond with very different types of knowledge.

Our position is that a good theory of science will pass the higher level compatibility or coherence test (of understanding), whereas a good theory of technology need only pass the lower level compatibility test (of prediction). This means that theories of science should be more concerned with general principles, and those of technology with precise details. One can say that delineating the parameters of reality is the purpose of science and technology — broadly and narrowly, respectively. One might also say that ontology should belong to science, and heuristics to technology; or that science should be considered as metaphysics — background context — to technology.

In proposing the distinction, we are not suggesting that there is no overlap between the two types of theories. Thus, what we call a broad theory of science will also entail more specific aspects, just as what we call a narrow theory of technology will also entail more general aspects.[16] The result is that all theories which incorporate (lower level) prediction also entail some sort of (higher level) conceptual framework which parallels what we call understanding, even if only in operational meaning terms. The issue is one of emphasis and preponderance — as well as how high is the "higher level" portion ... i.e., how compatible that portion is with our megatheory core representation of nature, as we have described.

Let us approach the issue from another tack — which we will also use as a cudgel against the established paradigm. This concerns the standard conclusion about the fact-theory fit — essentially, that a theory which makes correct predictions is thereby considered (provisionally) validated. The flaw here is that we never test whole theories — certainly not general theories; we test aspects of such theories. Even if one tests a representative variety of predictions which emerge from different aspects of a theory, we are still not actually testing the theory as a whole. This is a point we touched upon in

16 Put in other terms, the "line of demarcation" between the two categories would not always be clear — a point which Popper also makes with his distinction between science and non-science: "[W]e must not try to draw the line too sharply," because "the criterion of demarcation cannot be an absolutely sharp one but will itself have degrees." (Popper, 1965, p. 257).

the latter half of the previous chapter, but our emphasis here is different. For what is not being tested with the prediction criterion is whether the larger conceptual or metaphysical framework (of interpretation) which the theory puts forward is viable — and here we are talking about the test of a theory-megatheory fit — and whether there may be better ways of achieving this end that are also able to encompass the predicted phenomena, which may even emerge after the fact.

Needless to say, such an argument is not in keeping with the standard line, which says that a good theory must make predictions in advance — unique predictions, which are subject to being invalidated — and not after the fact. For one thing, this will test the mettle of a theory, to the extent that it is not thereby falsified and dismissed. And for another, it will keep opportunistic theoretical interlopers from jumping in and seizing conceptual territory to which others have already laid claim. Plus, it will discourage the practice of post hoc cherry-picking, whereby proponents of alternative theories select out particular sets of facts as being supportive of their position (while conveniently ignoring contravening facts).

Yet why must a good theory of science crack rather than bend? Why should a theory not be subject to revision — and why can't other interpretations be considered as well? While there may be no way to adjudicate between different theoretical frameworks which make the same lower level predictions, there are viable ways to decide between different theoretical frameworks on the basis of higher level understanding. Once again, we maintain that there has been far too much emphasis on the former kind of compatibility, and not nearly enough on the latter.

The result is to leave us quite vulnerable conceptually, in the sense that claims can be made on our reality far too easily. In this regard, we can talk about something like a lottery mentality — we mentioned the "jackpot" earlier — except that instead of winning a whole bunch of money if one foretells the correct numbers outcome, one gets to claim a whole bunch of reality instead! We object to this methodology — such that if one predicts the correct quantitative outcome of a (lower level) experiment, or even series of experiments, that one then wins the sweepstakes … the result being that one's whole (higher level) conceptual edifice for defining reality wins. We have discussed various examples of the problem in this book, whereby some broad interpretative scheme is embraced … even though another (and perhaps better) one is equally viable.

Given our differentiation between science and technology, several questions are in order regarding several of the issues we discussed in the previous chapter. For example, is such cocky self-confidence as Kuhn essentially advocates really so desirable a characteristic for those contemplating the subtle patterns of nature, which should lay the groundwork for a good theory of science? In other words, does science "move fastest and penetrate most deeply through a confident employment" of the established conceptual tools — or does technology? Which is the best avenue towards the truth and understanding that science should be striving for, as opposed to the pragmatism and prediction of technology — rigid dogmatism or thoughtful flexibility?

Clearly, the whole distinction that we are making between theories of science and theories of technology would be rejected by many people, who would treat technology only in the second way that we described at the outset of this section — as the applied form of science — producing things as opposed to basic theorizing about things. But the very idea that there could be different types of criteria for two categories of theories would be unacceptable. The conventional argument would be that the standard which one uses to judge a theory should not be like jelly — vague and changing to suit the theory — but should be fixed and specific, applicable to all theories. Therefore, the prediction requirement of the orthodox paradigm would be seen as pivotal for determining the viability of a theory, regardless of any distinction between science and technology.

Obviously, we dispute that conclusion: it is in no way self-evident, and instead represents a particular value (i.e., metaphysical) judgment in its own right. Furthermore, we are not alone in the fundamental distinction we are making between two very different mind-sets. Thus Polkinghorne says that, "Technology may be about manipulative power, but science is about the search for understanding.... As scientists they know that prediction, however perfect, is not enough. They want to *understand* ..." (Polkinghorne, p. 13)

Waldron made a similar point many years ago, arguing passionately that...

> [W]hat we have learned in this century is how to do things, and knowledge of how to do things is technological, not scientific.... Understanding, we are told repeatedly, is just around the corner — only when we get round the corner we find that the view is different from what we expected. And so we go on increasing our mastery over nature, but without a proper understanding of what we are doing.... [Long ago,] physics abandoned the attempt to understand nature, and asserted that the business of science is not to understand but merely to learn how to manipulate.... This is a prostitution of science which in the long run would lead to disaster. (Waldron, pp. xi and x)

For his part, Toulmin writes:

> Forecasting, then, is a craft or technology, an application of science rather than the kernel of science itself.... The central aims of science are, rather, concerned with a search for understanding — a desire to make the course of Nature not just predictable, but intelligible.... Prediction is all very well; but we must make sense of what we predict. The mainspring of science is the conviction that by honest, imaginative enquiry we can build up a system of ideas about Nature which has some legitimate claim to "reality". (Toulmin, 1961, pp. 36, 99, and 115)[17]

It should be clear that we would classify the bulk of what we have today as technology — a spectacularly successful collection of details and gadgetry. But more broadly, we lack science — a general overview and ability to understand the bigger picture of nature. The failure is not in the "little truth" mini-patterns and predictions, but rather

17 In a similar vein, Zukav cites "a common misunderstanding when most people say 'scientist,' they mean 'technician.' A technician is a highly trained person whose job is to apply known techniques and principles. He deals with the known. A scientist is a person who seeks to know the true nature of physical reality. He deals with the unknown. In short, scientists discover and technicians apply.... The fact is that most 'scientists' are technicians. They are not interested in the essentially new. Their field of vision is relatively narrow; their energies are directed toward applying what is already known. Because their noses often are buried in the bark of a particular tree, it is difficult to speak meaningfully to them of forests." (Zukav, pp. 9-10)

in the "big truth" mega-patterns and understanding. We have the trees; we miss the forest. By these measures and our definition, the conclusion we come to is that science has failed in its primary mission — indeed, failed abysmally ... both itself, and the larger society. On the other hand, technology has succeeded — spectacularly.

Here is a chart which summarizes various attributes of the two ways of approaching nature that we have been discussing:

Lower Levels	Higher Levels
technology	science
little truth-falseness	big truth-falseness
thesis vs. antithesis	synthesis
different parts	same whole
discrimination	generalization
X-view	*P*-view
prediction / no meaning	understanding / meaning
short, brittle threads	long, supple threads
narrow, shallow, disparate	broad, deep, compact
complex, specific, incoherent	simple, general, coherent
random metaphysics	consistent metaphysics

Now for a few words about the technology which has been so successful — more in terms of the second way that we mentioned at the top of the section, as a means for manipulating nature. This success is all around us and cannot be denied. Thus, the world of technology progresses forward at breakneck speed, and with relatively little debate: fearing no problems and respecting no sanctities — its only apparent mission being to unlock nature's secrets and give mankind greater and greater control, using whatever means are necessary to achieve those ends.

Even so, not everyone is enthralled with the results — and the disparity which has emerged between our ability to control nature and our ability to understand what we are doing in the process. Already a half century ago, Jammer expressed anxiety about the discrepancy between the two very different skill sets:

> In our present age of rapid technological progress the frightening discrepancy between our technical "know-how" and our philosophical incomprehension, in general, of basic scientific conceptions seriously endangers the integrity of our intellectual outlook. (Jammer, 1957, p. vii)

In a related manner, Bohm speaks about the insidious effects of knowledge which is divorced from values, or wisdom:

> Such a division between knowledge and values is indeed most pronounced in science, which is now commonly supposed to yield knowledge that is free of all values (except for the value of objective truth itself). This has helped lead, not only to a dangerously irresponsible use of knowledge, especially scientific, but even more to a general loss of meaning in life as a whole. (Bohm, 1980b, p. 1)

Some years ago, Hans Bethe spoke about the personal evolution from what we call the science mode of trying to understand nature ... to the technology mode of trying to manipulate nature. Here are some of his comments, concerning the "father" of the atomic bomb, whom we spoke about above, J. Robert Oppenheimer:

> It is a very different attitude if you want to find out the deep secrets of nature — which is what you had wanted to do before — and on the other hand if you want to produce something, to produce a mechanism that works. It was a different problem — a different attitude, and he completely changed to fit the new role. (Bethe, quoted in Oppenheimer, 1983)

Freeman Dyson speaks eloquently about the hazards which these imposing technological advances lead to — again, concerning the development of the bomb:

> I have felt it myself, the glitter of nuclear weapons; it is irresistible if you come to learn as a scientist. To feel it's there in your hands to release this energy that fuels the stars, to let it do your bidding, and to perform these miracles, to lift a million tons of rock into the sky — it is something that gives people the illusion of illimitable power, and it is in some ways responsible for all our troubles, I would say — this, what you might call technical arrogance, that overcomes people when they see what they can do with their minds. (Dyson, quoted in Oppenheimer, 1983)

Dyson goes on to add this insight:

> The Faustian bargain is when you sell your soul to the devil in exchange for knowledge and power. And that, of course in a way, is what Oppenheimer did — there's no doubt. He made this alliance with the United States Army, in the person of General Groves, who gave him undreamed of resources — huge armies of people and as much money as he could possibly spend — in order to do physics on the grand scale, in order to create this marvelous weapon. And it was a Faustian bargain, if ever there was one. And of course we are still living with it ever since. When once you sell your soul to the devil, then there's no going back on it. (Dyson, quoted in Oppenheimer, 1983)

Whether all the rest of us have sold our souls to the devil or not, virtually everyone has benefited from the technological advances which permeate our modern world, to a greater or lesser degree. That is obvious from how heartily those in society welcome the convenience of all the utilitarian products, tools, and techniques — not the least of which are medical advances — which have improved our lives in countless ways. The result is that most people typically embrace the material benefits of technology — while they either resist or ignore the teachings of science, for reasons that we have been discussing in this chapter.

So where will our technological prowess lead us? It cannot be only for the good — what about the downside? Davies wrote many years ago that, "Human society can be credited with a few thousand years of civilisation, which might be nearing its own destruction through technology." (Davies, 1977, p. 209)[18] What, then, would be the net value of all of our technological advance? Cosmologist Martin Rees says that unless we can better control the threats which are associated with modern scientific discoveries, then the future of the world might be very dark indeed. His position is that the general public should better understand the implications and consequences of scientific research — not only the potential upside, but the potential downside, as well. He feels that scientists have an obligation to explain what they are doing to the public — and that scientists should have more social concern about how their work is applied.[19]

The negative possibilities we face are endless, of course — concerning such things as the reckless way we treat our environment, the dubious manipulation of the gene pool, and the dangers which loom from weapons of mass destruction. Indeed, here is how one lesson might go, as an outgrowth of the understanding we seek: Selfish fixation upon our own individual selves and species must give way to more collective well-being — of society as a whole and the larger animal kingdom. We humans are soiling and causing the overheating of our nest the earth, and we must learn to live more lightly and more in harmony with nature's rhythms. We are shepherds here, and if we cannot learn that role — if we cannot reduce our reckless sense of false omnipotence — then the future is very bleak ... and the very technology which we have created will surely destroy us. Man must defer, for the broader patterns of nature will ultimately prevail.

Of course, the problem for mankind is not with technology, per se, but with *indiscriminate* technology. The reason we have indiscriminate technology today is that we lack wisdom and values with which to guide its development. That is, in order to choose between good and bad technology, it is first essential that we better understand the larger world of nature and our rightful place in it. And while science has supplied us with the tools of technology, it has not helped us understand how (or indeed whether) to use those tools. Wigner writes: "[O]ur science is more successful in giving us power than in giving us knowledge of truly human interest." (Wigner, 1967, p. 216)

We in society have been given increased power, but we don't know how to use it — because we don't understand the nature of which we and these tools are a part. It is like putting a gun in the hands of a child — and we are like children, with our ignorance about nature. Unfortunately, power is very alluring — and without wisdom, it is also very dangerous.

That lack of wisdom, and why we are in such a state, is one of the main issues we want to consider next.

18 Yet Davies defends the institution of science when he argues that, "The confusion in the minds of many between science and technology has resulted in a backlash against the former on account of the shortcomings of the latter." (Davies, 1977, p. 202; cf. also Cassidy, p. 196)

19 Carl Sagan writes that, "The best way to avoid abuses is for the populace in general to be scientifically literate, to understand the implications of such investigations. In exchange for freedom of inquiry, scientists are obliged to explain their work. If science is considered a closed priesthood, too difficult and arcane for the average person to understand, the dangers of abuse are greater.... Scientific aloofness and opposition to novelty are as much of a problem as public gullibility. (Sagan, pp. 14 and 70)

We Have Been Betrayed

This chapter concludes with a different tone than has been adopted elsewhere in the book — for we let our indignation flow freely here ... in the face of what we claim is a great impropriety that is being inflicted upon mankind, which badly impacts our relationship with the rest of the natural world. For we been left utterly bereft, without any overall guidance from science as to the nature of reality.

We have stated our position repeatedly, regarding what we see as the rightful role of science — and how the orthodox regimen has failed to deliver, in this respect. Indeed, it is as though the mission has been to obfuscate, rather than illuminate, the deepest and most persistent patterns of nature. The conventional model of science is mostly dedicated not to understanding nature, but to manipulating nature. Yet we have argued that such ends as predicting and producing belong not to science, but to technology.

The main counterargument, of course, would be based upon the range-of-experience factor: that science alone has been able to transcend the limited and sheltered realm of our normal lives, and break through to the full panoply of natural phenomena which should be the basis of our theories. Likewise, the very notion of "understanding" would have a different meaning in these two narrow vs. broad contexts — the former corresponding with a quaint and comfortable old-style approach ... and the latter corresponding with the vast new frontier which science is so boldly exploring.

Supposedly, then, we should be grateful not only that science is pursuing new vistas, but that it avoids being stuck in old conceptual ruts as it does so. This is also an argument for science's innocent "blank slate" approach, which ostensibly yields the raw patterns of nature — as opposed to the undesirable "metaphysical" approach, which imposes preconceptions upon nature that are based upon our limited experience and unsatisfactory meaning system. Taken all together, then, science has properly earned its rightful place as the paramount definer of reality ... and even the liberator of our thinking.

Our retort to these counterarguments should be quite familiar. For one thing, science is not innocent and its slate is not blank. In this regard, every scientific inquiry or theory is embedded within a larger framework of conditions and presumptions — which may never be made explicit, but that will apply nonetheless. Nor is it acceptable to claim a new and transcendent set of meaning and logic, when these cannot ever seem to be articulated or reconciled with those that have functioned and evolved over time. Indeed, what basis do we have for believing that we will ever attain any kind of theoretical unity or conceptual coherence if we continue along the present path, given the garbled array of incompatible conceptual fragments that are involved, along with the overly narrow and confining set of theory-building precepts upon which the established paradigm is based ... including the reigning gatekeeping criteria for the admission of new models and interpretations?

More broadly, though, one of the main problems with the "enlightened mind-set" description of science is in the way it subtly reinforces the established paradigm's ideological grip on power — by effectively stifling our initiative and making us subservient to them. It allows science to look down upon the rest of society from its lofty pedestal on high — not to mention upon nature, to which it pretends to dictate with its portentous "laws" about how nature must operate. Indeed, science is aloof not only in what it talks about and how it talks about these things, it is also aloof because it undermines and derides our usual mode of thinking as having no standing or relevance ... which is even more insidious, because of the corrosive and debilitating effect which this attitude has on our own sense of mental competence. Scientists set the agenda — both in terms of defining reality, as well as in manipulating nature: they are free to do what they want, and we have no say in the matter. We are simply told that the established model is valid — and that only crackpots or stupid people would question what all the masters of science have ordained to be true ... and which we are also obliged to accept on faith, as opposed to the outcome of any normal reasoning process. The result is that their reality trumps ours — and we had better get used to it!

The upshot is that many of the most important phenomena which science studies remain shrouded in mystery for the non-specialist: only the relatively tiny coterie of experts, who are determined to slog their way through the conceptual clutter and master the tools of the trade, are admitted to what is supposedly the inner sanctum of nature. Of course, whether it is really the inner sanctum of nature — as opposed to merely being the clubroom of science — is one of the questions we have sharply raised in this book.

Likewise, whether science has such keen insights into the nature of reality is another premise that we have strongly challenged. Indeed, we turn that argument on its head, and claim that such an elitist approach is unacceptable. We contend that science is so enamored with its technological success as to not realize its fundamental ignorance about nature — no matter how many particles it has created, nor how many pages it has published. There is no basis for that kind of intellectual smugness and glib self-satisfaction, given the jumbled conceptual framework of the established paradigm. In fact, one could argue that scientists are no more in touch with nature, in any truly fundamental way, than non-scientists — and the reverse could easily be claimed.

Therefore, we sharply challenge the common conclusion, often accepted out of resignation as much as anything else, that only conventional science is in a position to authoritatively discuss and define the essential character of the natural world. We adopt this stance in spite of the narrow vs. broad range-of-experience ploy, and in spite of the no-sense nonsense which the orthodox discipline touts as being superior to our lowly, logical form of reasoning. So while it is true that the established paradigm may be exposed to a far broader range of phenomena than those to which we commoners are accustomed, any such phenomenological data still need interpretation — and that brings us back to the broader issues of compatibility and coherence that we have been discussing in this chapter. In the end, we maintain that "range of phenomena" does not trump "quality of analysis"; and so we continue to insist that a coherent theoretical framework must always be the goal, and thus the measure, of science's success.

Science, and especially today's conceptually bankrupt science, can have no exclusive or privileged status here, in terms of defining reality. On the contrary, we contend that science has forfeited whatever fiduciary role it may have had, in this regard — by virtue of its manifest failure to grasp the larger patterns of nature, and to share that insight with the rest of society. Instead, what we have today is a science of sheer pragmatism, which takes no responsibility for its actions — neither the technological output, nor the conceptual substance. Therefore, we conclude that it is the established paradigm's perspective upon the natural world, together with its view of science's mission — in relation to nature and in relation to society — which is too narrowly defined. In this light, then, it is not the lay person who suffers from a limitation of vision: it is the established paradigm.

None of what we are saying is meant to deny what has been attained, so much as to put it into perspective ... and into what we think is its proper context. Of course, we cannot hope to change the meaning of our common lexicon overnight, but our analysis differentiates the triumphs of technology from the failings of science. In this regard, we contend that while technology may tend to be ignorant and arrogant — even impulsive and opportunistic — in its short-range quest for things that work, science must be astute and humble ... as it seeks a more profound understanding of nature. Indeed, the goals of science must be lofty ones that are oriented neither towards speed in implementing short-term ends, nor dogmatic in defending long-term ones. For in penetrating deeply the long thread patterns of nature, science must tread softly and slowly, thoughtfully considering — and reconsidering — every step it takes.

Our position is that science has misused its prerogatives and undermined its promise — for true science has an awesome and sacred mission to fulfill, of shepherding society forward ... away from myths which hold us back, and towards a more harmonious and meaningful relationship with the rest of nature. And so in a better world, technology would be the mere handmaiden of a wise science — and this wise science, in turn, would be a humble student of nature, and gracious servant to society ... not its master. Science would be an interlocutor between society and nature — a "wise guide", not a "wise guy" — and surely not a despot, arrogantly trying to whip nature into submission and effectively running roughshod over society. But today's science is aloof — and it is a bully. It makes a mockery out of nature, and tramples upon the sensibilities of mankind that it should be serving.

And so we come back to that single greatest scientific icon of all, Albert Einstein. Clearly, we have reached a very different evaluation from the stereotypical one — for we maintain that Einstein mystified us, rather than enlightened us, as to the nature of reality. Indeed, just as we did not grasp the full meaning of people like Einstein (along with others) in their talk about freely-invented constructs and a deductivist style of theory-building, so society at large has not yet realized how we have been tricked — and how deeply the institution of studying nature has been infiltrated by this deadly utilitarian philosophy. For we in society have been like innocent babes, swallowing science's cant. When we are essentially told, for example, that science has discovered (thanks to Einstein's genius) that the speed of light never changes, we take science at its word there ... even though it makes no sense. Alas, we should not have been so easily beguiled — for no such finding has emerged! It was all a hoax, and this is not hype. As we discussed in Chapter 6, science has not *observed* that light's speed never changes. Science (through Einstein) has simply *decreed* that light's speed never changes, and that is a very different thing.

Here is one of those examples which we've talked about in these pages. Science was confronted with a problem — in particular, about how to reconcile Maxwell's formulas, about light's speed in the ether, with Newton's laws of mechanics. And so Einstein came up with a solution that worked: simply *define* light's speed as never changing — and eliminate the ether at the same time. A bold stroke: it met the immediate requirements, reconciled the formulas with the mechanics — and Einstein was the hero of the revolution!

Except for one thing: it didn't *mean* anything. It was hocus-pocus — a quick fix conceptual patch up, an end run around the problem, which has caused us no end of grief in the long term. Light was said to move in a way that could not possibly be imagined, in any comprehensible manner ... relative to absolutely nothing at all, or relative to everything one cared to identify — take your pick. Either way, it never has and never will lead us to a coherent model of nature.

We need to realize that science is not usually discovering these things that make no sense — and that the "free invention" mind-set is just that: it is about inventing constructs regarding nature ... creating a make-believe abstract fantasy, as opposed to discovering what nature is really like. Look at the argument in support of free invention that we discussed earlier: "[W]e are essentially viewing the purpose of physics as a scientific discipline as *invention* rather than *discovery*." (Lindsay, p. 24) Thus it is that Einstein criticized Mach on the basis that, "[H]e did not recognize the freely constructive element in formation of concepts. In a way he thought that theories arise through *discoveries* and not through inventions" (Einstein, 1948, quoted in Holton, p. 249). Thus, Einstein talked about the "purely fictitious character of the fundamentals of scientific theory" (Einstein, 1933, quoted in Holton, p. 251). Again, "Physical concepts are free creations of the human mind, and are not, however it may seem, uniquely determined by the external world." (Einstein and Infeld, p. 31).

The bold conceptual advances which result from such thinking — deductive leaps off cliffs, as it were — amount to edicts from on high. They represent prime examples of where science has lost its way, and has misled us all. What results are a series of freely-invented definitions, which serve as very convenient solutions to particular problems — aimed as they are at making our fundamental concepts about nature be defined in a manner which makes the equations of physics as simple as possible — rather than the infinitely more valuable goal of leading us to a model which makes sense and helps us to understand the world around us. What we have instead amount to arbitrary definitions ... a series of arbitrary conceptual pivots. And precisely because they are arbitrary, they will never be more than ad hoc solutions to particular problems: they will never have a much broader meaning than that.

Yet it is not enough to solve a problem in one area if, in the long run, it creates new problems in other areas. That is because nature is a unity, and fragmented solutions will not suffice. Arbitrary conceptual pivots are not compatible with very broad themes, with deep and general principles. The arbitrary approach is an idiosyncratic approach, which fosters compartmentalization and cluttered thinking. Such narrowness of vision is the set of definitional blinders that science wears: that is its strength, in focusing its energy and sharply delineating the problems it studies; and that is its weakness, in contenting itself (and bequeathing society) with a blinkered and chaotic vision of nature.

Einstein's deductive strategy, as the way to do science, makes no sense — for it ignores the inductive mechanism which is the proper basis for higher level principles. Thus, even if we label these arbitrary maneuvers with pseudosophisticated covers like *operational definitions*, they still amount to mere ruses, which barely mask the fact that we are running amok in nature's domain — paying no attention to the broad themes, and merely superimposing our own narrow ones ... even if nature goes along with the game, almost as a joke back upon us, thereby deluding ourselves still further. Yet the ultimate (and paradoxical) result is to reinforce our own ignorance, irrespective of whatever short-term advances result, thereby substituting expedience for wisdom.

Given all that we have been discussing, we have basically two choices for dealing with the situation. *On the one hand*, we can continue along the same — what we call dead-end — path, shifting from one meaningless theoretical orientation to another, depending on the subject at hand and the fashion of the era. In this way, even the most ludicrous conceptual tangent is fair game — for it is as though we were passengers on a ship without a rudder (or without a compass) ... lurching from one direction to another, with no overall orientation to guide our way forward. Of course, we are not permitted to question what is happening, or the fact that we are getting seasick in our minds — for the captain-experts know best. After all, our range of experience is so paltry, compared to theirs; they have mastered such potent tools of inquiry and analysis; they have so many facts and concepts and products — so much technological power — at their disposal! In the end, we are completely intimidated and overwhelmed by their way of thinking ... how can we resist?

But thank goodness there is *an other hand*, and another option — and according thereto, we have the right to demand something more from science: in the form of a coherent conceptual framework in which things fit together into a unified whole, and where nature makes sense in terms that we can relate to ... and where there are one or more central organizing principles that encompass both our everyday lives as well as the deeper and vaster world beyond and within. We can say that in the process of studying nature, one must not hide behind a smokescreen of meaningless gibberish — or essentially "patent" domains of our common reality as one's private realm, mauling our most fundamental concepts about nature in the process. For, reality is not a party game — and science must stop playing musical chairs with our minds. They have stolen the integrity of *our* shared meaning system to satisfy the piecemeal and arbitrary requirements of *their* models! Yet whose reality is this world we live in? We are talking about the essential basis of our common existence, and not the private domain of some exclusive clique. Therefore, we can simply reject the established paradigm, as science — just as we must accept it, as technology.

It should be obvious which choice we must make — there can be no contest! Just as the proponents of the established paradigm cannot afford to question the broad foundation of their scientific edifice, so the rest of us cannot afford not to. For if they have too much invested in their dense, ponderous, and impenetrable theories to tolerate any fundamental questioning thereof, so the rest of us have too much invested in life and the future of mankind not to challenge the reigning orthodoxy.

We maintain that it is time to end the tyranny which science has imposed upon our thinking, and to demand common meaning and real logic from science instead. More broadly, we must always remember that it is very unwise to stop thinking and wondering, and asking questions. Likewise, it is very dangerous to accept *anybody's* model of reality, unless we can find some sense in it. Indeed, given its fundamental place in modern civilization, probably no other domain would be more suitable than science for us to apply the credo, *Question Authority* — which is also why we used that as a title for the second and contrary half of the previous chapter.

We *can* understand our world — and we must, if we are to guide its destiny and not be at the utter mercy of experts. As Popper puts it,

> [P]hysicists, and other positivists, try to tell us today: that we cannot, in principle, hope ever to understand anything about the structure of matter: that the theory of matter must forever remain the private affair of the expert, the specialist — a mystery shrouded in technicalities, in mathematical techniques, and in "semantics".... I do not believe a word of this post-rationalist doctrine. (Popper, 1982, pp. 172-173)

All of us can be experts, if we simply make the decision to liberate our minds and become so. We have that power — to break out from the intellectual shackles, the mental strait-jackets, with which we have been burdened. We must seize it: take this power of definition in our hands and in our heads, and not leave it to any masters — no matter what titles or accolades they hold. For in the end, we must keep hammering back with the same kind of nagging refrains: Where have they led us? What are our goals? When do we reach a coherent model of nature? *Who defines reality?*

In conclusion, we maintain that the science of today is not the mecca of the future. Many of its central concepts, most of its style of inquiry and methodology for theory-building, and the general values and goals which guide its work: all have resulted in grave distortions regarding the nature of reality, and all represent a considerable disservice to mankind. Far from being the model of the future, the style and substance of the established paradigm is an ideology which should have no future — at least, not as science, which is that which should be wisely leading our collective way forward in nature.

Chapter 12

The End of the Beginning

We are nearing the end of our journey, in this book — but the end stage here is also intended to be only the beginning of a much longer quest, towards the alternative model of nature that we are striving to reach. Aside from that, however, we are also applying the "end" label to the established paradigm — in terms of two very different evaluations: on the one hand, some advocates of the conventional approach argue that orthodox science has itself reached a conclusion of sorts, in terms of the lofty state of knowledge about nature which has been attained; on the other hand, we apply the "end" label to the conventional discipline, with a quite contrary assessment.

This chapter might also have been called, "A Quest for Unity", because in many ways that is what we shall deal with here — both for the established paradigm, and for the present model. The first section mostly deals with the former, beginning with the two very different assessments of the "end" state of science that we just mentioned. Beyond that, we shall deal with the unity issue in a very particular way: simply put, where is the unified model of nature that we have been continually promised, and which the conventional discipline has desperately sought, for so long? We shall see that there are different ways of answering this question, none of which are satisfactory.

We mostly leave aside the established paradigm for the rest of the chapter, and turn to the very different type of unity which is envisioned for the present model. Having primarily dealt with physics throughout this book, we now want to bring in three other scientific disciplines, in this final chapter. The first one is cosmology — although that topic is not altogether new in this work, nor is it a completely separate discipline; nonetheless, we want to consider several cosmological issues on their own account here, to see how they might be understood and integrated in two-force terms … as opposed to the conventional manner in which they are interpreted. Then we look briefly at an array of issues in the fields of biology and cognition, to illustrate how they might be approached and incorporated as part of a two-force framework.

In the last section of the chapter are some closing comments about the present model itself — and the state of unity which has, and has not, been attained. First we consider a criticism which may be leveled at 2FT, in terms of whether the overriding principle of complementarity can itself be seen as a one-sided interpretation of nature. After that, we look at what stage the present model is at — both in the general terms of how any alternative theory could possibly be in a position to challenge the established paradigm, as well as in the specific terms of what we have tried to achieve in this work ... along with where some of our shortcomings lie.

The End of Science

We have so titled the first section of this final chapter to indicate two very different things. To start with, this is the term which is used by others to describe what they consider to be the extraordinary accomplishments that have been attained by the established paradigm. According to this thesis, science has been so successful in unearthing the patterns of nature that there is not much more to find out, in terms of fundamentals — and it is increasingly just a matter of filling in the blank spots and details, and rounding out the rough theoretical edges — and it is in this respect that we are now nearing "the end of science".[1]

Not surprisingly, we do not share this assessment — for otherwise, this whole book would not have been written. We talked about these issues in the previous two chapters, especially the prior one — and in this respect, the section title is used here in an ironic sense that is meant to cast grave doubt upon such an upbeat assessment. For we maintain that if there is an "end" here, it is the "dead end" of orthodox science.

Of course, the terms of discussion for each of these two judgments are very different — which is largely the point we were dealing with in the previous chapter. For the orthodox discipline, its ability to *predict and manipulate* nature is clearly unprecedented in the history of mankind; and in this regard, science can be said to have reached a high point, if not an end point, theoretically. But for the alternative vision which we advocate, it is the established paradigm's ability to *understand* nature that is utterly inadequate; and it is not even at a high point, let alone an end point, in this regard. One way of talking about "high points" in the present model is in our hierarchy terms — which is consistent with the explanation we have given for differentiating prediction from understanding, as lower vs. higher level criteria, respectively.

Moreover, we speak about an aesthetic factor called *coherence* as having two different aspects — both related to higher levels of synthesis — one being in terms of understanding or comprehensibility, and the other being in terms of a fitting together into a whole unity. And it is in this latter respect that we repeat the question from above: Where is the unified model of nature? In this regard, here is another unflattering "end" usage: time's up! The established paradigm has had its chance, for so many decades now, and it has manifestly failed to come up with the goods. Indeed, we contend that such long-sought unity can only be a mirage, on the conventional path which has been taken. Not only that, but from our perspective it is amazing that one can even talk about having

1 To wit, the name of John Horgan's 1996 work, *The End of Science: Facing the limits of knowledge in the twilight of the scientific age*. Horgan argues that any important questions which remain are essentially unanswerable. Zukav uses the phrase in a different, but related, laudatory sense as the title of the final chapter in his book — to the effect that science has passed a threshold, into a higher plane of awareness.

reached "the end of science" when no such coherence has been achieved or is even foreseeable.

We consider the unity issue next in terms of several topics related to the established paradigm: fundamental forces, Quantum and Relativity Theories, the Standard Model, and String Theory.

Wherefore Physics' Fundamental Forces? In this work, we have repeatedly criticized the fragmented view of nature which emerges from the conventional model, and one of the most significant examples of this problem is that of the fundamental forces. On the one hand, these forces are supposed to explain both the interactions between, and the existence of, all the diverse phenomena of nature; and on the other hand, there must be some way of conceiving how these forces could possibly be related to one another, in the form of some unifying framework. We will examine the latter point shortly, but in this subsection we want to consider the former.

We have made clear our interpretation of what is going on here — that a handful of particular levels or regions of nature have been arbitrarily selected out as being fundamental, in that only interactions which occur at these critical levels or regions of nature are treated as being primary and existing in their own right. All other entities and interactions — i.e., most of the obvious phenomena in the world — are construed as being merely derivative and having no independent physical reality of their own.

There are three or four fundamental forces, depending on how one counts. The nuclear force (or strong nuclear force) has been evoked as a set of rules to explain interactions within and between nucleons. The electromagnetic force is another set of rules for explaining interactions which occur at the level of the atom, including binding into molecules. The weak force is intended to explain particle decay, as well as the behavior of a sub-atomic particle that we mentioned in Chapter 3, called the "neutrino".[2] The force of gravity is an anomaly in that it is commonly ascribed in practice to interactions involving large (heavenly) bodies, although it is said in principle to describe mutual interactions between all objects (or even non-objects, with Einstein's theory) in the universe.

The framework that we are advocating in this book is very different. Instead of a model that is built upon particular levels or regions as representing centerpieces for explaining everything else, a better approach would be a model that is built upon general concepts and principles which could be equally well applied to every level or region. Nor is this to deny that there are specific details which apply to particular levels, in

2 Neutrinos are said to be massless (or virtually so), and without electric charge. They are described as being ubiquitous in nature, in the sense that they are present everywhere, literally filling the universe. At the same time, neutrinos are "undetectable, because they scarcely interact at all with the ordinary matter of the universe" (Calder, 1977, p. 38); "[S]o insubstantial are the neutrinos that they can easily pass through the Earth without stopping." (Davies, 1979, p. 103) On the other hand, neutrinos have long been routinely produced and detected at all large particle accelerators. Ruderfer has gone so far as to characterize the neutrino as the underlying unit for a general medium of nature, arguing that this "neutrino sea model" satisfies all of the requirements for such a function. (See, e.g., Ruderfer, 1975)

terms of how the general principles or patterns of nature manifest themselves there — such as are described by physics' fundamental forces. But that is not enough to accord those specifics unique explanatory precedence, let alone exclusive ontological status.

Here is one way that we would describe these different approaches as to what is fundamental, in hierarchical terms. According to the established paradigm, and leaving ornery gravity aside, it is the *specific lower levels that are construed as being fundamental*, because they provide the basic building blocks for higher levels (a la atomism). But according to Two-Force Theory, and taking gravity into account, it is the *general patterns that are construed as being fundamental*, because they describe all levels of the hierarchy. Putting both in terms of the present model, the conventional approach singles out a few horizontal levels, whereas the new approach looks across vertically at all levels ... up and down the hierarchy of nature. Both are reductionistic, but one is a reduction to primary levels whereas the other is a reduction to primary principles.

One might argue that what we are dealing with here is a variation of the induction vs. deduction conflict that we discussed in Chapter 10, and the higher level vs. lower level theories that we discussed in Chapter 11. In terms of the present topic, the orthodox model could be understood as primarily lower level, specific, and deductive — in the sense that higher levels only exist as compounds of lower levels; whereas the approach we advocate could be understood as primarily higher level, general, and inductive — in the sense that all levels only exist as manifestations of higher level principles.

Nature is a vast hierarchy of nesting, interacting phenomena; and the focus of analysis should be able to move fluidly up and down, between and within these different levels — by virtue of the general patterns which are supreme (even though these may manifest themselves in particular ways at specific levels). Yet so long as existence is reduced to the notion of a handful of specific levels and regions, then we will lose the benefit of such an overall synthesis or generalization ... that is based upon all levels of existence. Specific local orientations alone do not yield global vision: indeed, they can be a concrete impediment by obscuring such global vision, no matter how elementary the hierarchical level upon which they are based.

Joining Quantum Theory with Relativity Theory. In the rest of this section we want to look at the quest towards unity which drives the established paradigm.[3] In this regard, let us first talk about the broad conceptual challenge of bringing together the two main pillars of modern physics.

In Chapter 9, we mentioned the incompatible and arbitrarily one-sided visions of reality which these two models represent — the merging of which is difficult to conceive in any but an arbitrary manner: the spirit of each excludes the essence of the other, the core principles being fundamentally irreconcilable with one another.

3 A small minority of people have actually discounted the validity of such a goal to begin with. As Horgan writes, "[T]he quest for a unified theory is religious rather than scientific.... [T]here is no reason why there should be a single theory that accounts for all the forces of nature." (Horgan, 2003) And to requote Feynman: "What *could* such a statement mean.... 'the underlying unity of nature'?" (Feynman, 1964, p. 12-12)

In broad conceptual terms, the Theory of Relativity is predicated upon an (*X*-like) deterministic conception of nature — an essentially mechanistic universe, where certainty and causality reign ... yet where the dominant characteristics are that of (*P*-like) continuity and the field. By contrast, Quantum Theory is built upon a (*P*-like) indeterminism — where the universe is not mechanistic, where uncertainty and chance are inherent ... yet where the essence of reality lies in its (*X*-like) discrete, particle-like character. Moreover, it is not as though either of these one-sided theories subscribes to a full-fledged thesis of complementarity — according to which each of the two theories would be understood as contributing its part towards such a whole!

Sachs is a physicist who subscribes to the Theory of Relativity, and we cited his views repeatedly in Chapter 9, where we contrasted these two theories in various respects. Sachs believes that the theories are each coherent within themselves, but that there are fundamental incompatibilities between them:

> In view of the logically dichotomous feature of Quantum Theory and the Theory of Relativity (discreteness vs. continuity, non-determinism vs. determinism, intrinsically open vs. closed systems, linearity vs. non-linearity), it would seem necessary to abandon some of the conceptual features of one of these theories in favor of a full acceptance of the conceptual basis of the other, in order to maintain a logically consistent general theory.... [W]hile separately they are logically consistent, together as one general theory they would introduce incompatible ideas as a basis for this theory. Because of this unhappy state of affairs, it is not possible to join the two theories unless the conceptual parts of one that are not compatible with those of the other are removed. (Sachs, 1973, pp. 111 and 142)

Other authors have come to the same conclusion. Capra says that, "[T]he continuous space-time formulation of relativity theory and the discontinuous nature of quantum mechanics has so far stood in the way of unifying the two theories." (Capra, 1988, p. 153) In the same vein, Bohm says that, "[T]he basic concepts of relativity and quantum theory directly contradict each other. It is therefore hardly surprising that these two theories have never been unified in a consistent way. Rather, it seems most likely that such a unification is not actually possible." (Bohm, 1980a, p. 176) Prigogine talks about all the energy which has been invested towards the goal of joining the two theories: "All these efforts, however, whether through the unified theory of Einstein or through the microscopic description of nature with the help of electrons and protons alone, have finally failed." (Prigogine, 1977, p. 241; see also Capra, 1983, p. 92)

Davies seems to be of two minds on the subject. On the one hand, he takes a similar position as the others we just cited, in saying that:

> One of the great unsolved problems of physicists is how to reconcile Quantum Theory, supremely successful on the one hand in atomic physics and subatomic physics, with the Theory of Relativity (a theory of space and time), supremely successful in the subjects of astronomy and large-scale physics.... There seems to be a fundamental antagonism between the two theories that constitute the foundations of the new physics. (Davies, 1981; Davies, 1983, p. 135)

Yet on the other hand, Davies also argues that such discrepancies are acceptable, if not laudable: "Parceling the two theories into separate domains" (micromatter and macromatter) leads him to the conclusion that, "Nature has arranged for quantum theory and relativity to co-exist with wonderful consistency." (Davies, 1979, pp. 86-87) Of course, we reject such a split-realm way of depicting nature, along with the point mentioned in Chapter 10 that nature somehow works to validate mankind's theories in any form whatsoever.

We have also mentioned previously that Einstein was among those who rejected the notion of separate micro and macro laws — and of course he never embraced Quantum Theory itself. Thus, "[W]hile the relativity theory became Einstein's own most enduring life preoccupation, he could not accept quantum physics seriously." (Holton, p. 378) Instead, he spent many years trying to turn the Theory of Relativity into a unified field theory of nature, which incorporated both electromagnetism and gravity; but he ultimately failed in this endeavor. As Weinberg puts it, "Einstein devoted the last 30 years of his life to an unsuccessful search for a 'unified field theory,' which would unite general relativity, his own theory of space-time and gravitation, with Maxwell's theory of electromagnetism." (Weinberg, p. 68)[4]

Following the next subsection, we will look at one particular strategy which has been proposed for merging the point-like particle approach of Quantum Theory with the geometric space-time approach of Relativity, in the form of a combined "quantum geometric" way of construing reality.

The Standard Model. The dominant unifying framework that we have today emerges from particle physics, and is called the Standard Model. This is a quantum-based field theory, which attempts to bring together three of the four fundamental forces (strong, weak, and electromagnetic). Here is how physicist Sheldon Glashow describes the point that has been reached: "We've reduced all the many types of forces — the wind and chemical forces, electrical forces, etc. — to just three. There is the hope of reducing this three to two, and this two to one." (Glashow) Longer term, there is the dream of incorporating the force of gravity, as well — with the eventual goal being a fully unified system, sometimes described as a Theory of Everything.

Yet there is a long way to go, as another of the principals in the field, Steven Weinberg, explains:

> We have ideas about how the theory of strong [i.e., nuclear] interactions can be unified with the theory of weak and electromagnetic interactions (often called Grand Unification), but this may only work if gravity is included, which presents grave difficulties. We suspect that the apparent differences among these forces have been brought about by events in the very early history of the big bang, but we cannot follow the details of cosmic history at those early times without a better theory of gravitation and the other forces. There is a chance the work of unification will be completed by 2050, but about that we cannot be confident. (Weinberg, pp. 69-70)

4 Popper says that Einstein regarded the General Theory itself as " 'ephemeral'; and from the moment of its conception to the end of his life he tried to supersede it." (Popper, 1982, p. 8)

Let us look more closely at some of the main themes that are involved in these unification efforts. The most important one is the assumption that the fundamental forces all emerged from a single force state, which supposedly existed for an infinitesimally tiny fraction of a second around the time of the Big Bang.[5] In pursuit of this agenda, success has already been achieved in the sense that theoretical solutions have been put forward to explain how the electromagnetic force and the weak force can be understood as being based upon a common mathematical description. This combined "electroweak force" was proposed in 1973 by Weinberg, Salam, and Glashow.

One important construct which is utilized in these approaches is the notion of symmetry, and in particular a distinction between circumstances of perfect symmetry and broken symmetry. It is postulated that the former would have applied at the early (high temperature/energy) micromoment of the universe, whereas the latter would apply during the (low temperature/energy) eons of time ever since ... although the broken symmetry case is sometimes described by saying that the perfect symmetry is thereby "hidden".

Take the case of the singular force proposition. This conjecture entails a hypothetical particle which emerges from the Standard Model called the *Higgs boson* — or in a more dispersed form by the Higgs field — which supposedly exists everywhere in the universe. If everything manifested itself in a pure symmetry form, as is projected to have applied at the (high temperature) early micromoment, then the just-mentioned electroweak force would not yet have differentiated. Enter the Higgs boson which is responsible for breaking the symmetry for all the rest of (low temperature) time ever since, thereby causing the electroweak force to separate into the electromagnetic and weak forces.

Not only that, but this Higgs phenomenon is also endowed with another remarkable characteristic: it is also the supposed source of all mass in the universe. The explanation is that without the Higgs, the mathematics of the Standard Model would describe a symmetrical world in which matter had no mass. Obviously, things are not like that — and so something must alter this massless state, without disrupting the rest of the mathematical framework. The Higgs boson is put forward as the agent which performs such a function — and this is the basis for the idea that there is a Higgs field which pervades all of the universe, and which is somehow endowed with a mechanism that can spontaneously break the symmetry of elementary particles, thereby conferring mass upon them. It is only the interaction of matter with this ubiquitous Higgs field which is said to account for the existence of all mass in nature.[6]

If all this seems a bit much, there is more — in the form of a construct called *supersymmetry*. The genesis for this idea had to do with technical difficulties that are related to the projected mass of the Higgs boson which emerges from the standard equations — e.g., the mass is too large for it to perform its function of breaking electroweak symmetry, unless certain undesirable "fine-tuning" is introduced. To get around the

5 Note, by the way, how Weinberg speaks above about "apparent differences", suggesting that the fundamental forces are actually still one and the same as they were way back then — except that they do not seem that way.

problem and "stabilize" the mass of the Higgs boson, a new construct was invented as an extension to the Standard Model, according to which there is a whole new set of elementary particles beyond those which are readily apparent.

More specifically, the supersymmetry construct[7] is built upon the widely-accepted premise that the fundamental forces are conveyed by means of special particles (called "bosons"), as distinct from the more usual elementary particles which comprise matter (called "fermions"). What is being proposed here is that there is a symmetrical relationship between force and matter — called supersymmetry — such that every force particle has a supersymmetrical matter particle counterpart, and likewise that every matter particle has a supersymmetrical force particle counterpart.[8] Furthermore, the symmetry which is involved in each pairing is not perfect but is broken by some degree. By how much is unknown, but if the brokenness is large then the supersymmetrical particle would be quite disparate from (i.e., much larger than) the ordinary particle — whereas if it is small then the supersymmetrical particle would be quite similar to the ordinary particle. What precise mechanism would trigger the symmetry breaking in this scheme is also unknown.

It should be emphasized that all the novel phenomena we are describing here are purely theoretical, in the sense that they have so far never been shown to actually exist — and the question is, can they ever be shown to exist ... and if so what does that prove? One of the problems is that some of these phenomena are said to have existed for only the briefest fragment of time in the primordial, high energy state of the universe in the period of the Big Bang — and there is obviously no way to replay that moment or be able to observe what things were actually like way back then. Even so, physics claims to be able to project what conditions were truly like at that critical instant — and then they try to replicate this hypothetical state in the lab. The argument is that by constructing increasingly powerful particle accelerators, they will be increasingly able to mimic the early moment of the universe — and in those conditions the novel phenomena which they theorize about will become apparent. They will thus be able to demonstrate the viability of their model by discovering the existence of the components thereof, under these extraordinary conditions. Of course, there is always the issue of whether such artificial explorations are a matter of discovery as opposed to creation, which we mentioned in

6 The Higgs concepts are named after the Scottish physicist Peter W. Higgs, who proposed such a mass-endowing mechanism in 1964 (although the idea was not unique to him). Note that the Higgs boson is variously referred to as the "god particle", as the "Holy Grail", or as the "missing link" of particle physics.

7 As an aside, one might naturally assume that the new term "supersymmetry" refers to some novel pattern or mechanism which is different from the way that the "symmetry" concept is already applied by physics (which is itself dubious). But one can argue that this is not the case here, for what is involved is the same principle, merely applied to a different set of referents. What we are dealing with, then, is another instance of the usual disregard for language by physics, as we discussed in the previous chapter ... with all the negative fallout in terms of confusion and intimidation which results.

8 In the case of the Higgs boson — which is not really a force-carrying entity, despite its name, but a new kind of matter — there would not be just one supersymmetrical counterpart but several!

Chapter 10 — along with whether other alternative explanations might be invoked to explain any patterns which emerge in those conditions.

As to where all this astonishing speculation comes from and what it is based upon, much of the answer lies with the mathematical framework itself that underlies the Standard Model, a framework which has been extremely successful in the detail with which it describes many known phenomena. Even so, this framework also entails various anomalies and undetermined parameters, along with ad hoc adjustable constants; and there are a whole plethora of theoretical variations which are in play — not to mention the fact that the model does not incorporate gravity.

We would argue that one of the main problems concerns an issue we discussed in the previous two chapters, regarding the central role which mathematics plays in the established theory-building paradigm — including the assumption that there are always real life counterparts for the abstract mathematical referents. Beyond that, there is the whole arbitrary mind-set which is endemic in the field. For in spite of the claims it makes about being wedded to qualities like aesthetics and concepts like symmetry, physics will adopt whatever conceptual mechanisms it can come up with for resolving the problem at hand. There are no consistent guiding principles beyond the task of making things work. In the case at hand, it is a matter of somehow constructing a scenario for attaining unity — regardless of how contrived and implausible the result seems.

Look at how the time factor changes, as part of the framework of explanation which physics puts forward in its quest towards unity. On the one hand, there are the attempts to reduce various ordinary phenomena to the three micro forces — this could be called a "symmetry", in physics-speak — which process involves a common point in time, such as the present. On the other hand, there are the strategies we have described which are used to explain and thus reduce the three micro forces to two (and ultimately to one) — another "symmetry" — which process involves two different points in time, with one being said to exist for only the tiniest fraction of a second around the beginning of the universe.

This in itself also entails the dubious assumption that a unique set of extraordinary circumstances must be invoked as applying to that single early micromoment — entailing extremely precarious speculations about hypothetical states and convoluted mechanisms — for explaining totally different patterns of nature which have existed throughout all the billions of years ever since. How much of this is based upon reality and how much are we talking about abstractions, such as we discussed with the inertial frame construct — unreal idealizations that may exist in the imagination, but nowhere else … none of which can ever be proven, one way or the other?

Let us wrap up this discussion about unity by looking next at one particular conceptual approach that has been put forward in recent years, which is a further outgrowth of the ideas we have been discussing.

Strings Galore. The basic challenge which has been looming before physics for many years now is how to combine some version of the Standard Model, which deals with the very small … with some version of a model which deals with the very large. What this has normally been understood to mean is how to merge the very successful twin pillars of modern physics: Quantum Theory and its treatment of the three micro forces (electromagnetic, weak, and strong nuclear), together with Relativity Theory and its treatment of the one macro force (gravity). If such a quest were successful, it would finally bring us to the ever-elusive unified model of nature ... that Theory of Everything. For many contemporary physicists, the most promising candidate for fulfilling this function is one that has emerged in recent years, called String Theory.

As we noted earlier, there is an inherent conflict between the point-like particle approach of Quantum Theory and the extended space-like approach of Relativity. One of the main features of String Theory is the way it converts the former conception into something closer to the latter. At the same time, gravity itself is treated as a "quantum field".

The essential idea is that all the universe is composed of tiny string-like phenomena, and that it is the different vibration patterns of these strings which provide the basis for the difference between one elementary particle and another. Moreover, since the whole fundamental force framework presumes that all macrophenomena are mere aggregates of elementary particles operating upon a space-time tableau, then the whole world of nature is like a symphony that is ultimately based upon the individual notes of music which such vibrating strings produce. In the process, matter becomes a geometric phenomenon, and space-time becomes an exotic form of matter.

If this peculiar picture weren't fantastic enough, add this one in: all such vibrating strings have many dimensions. Initially, the theory proposed that there were 26 dimensions for the vibrating string surfaces, although this was later reduced to a total of 10 — but more recently expanded to 11 ... 10 for space and 1 for time. The fact that nature is commonly conceived in terms of 3 dimensions for space and 1 for time leaves the theory with some explaining to do! But no problem: the 7 extra spatial dimensions are simply treated as being curled up or compacted into incredibly tiny balls which exist, unobserved, throughout the universe. (By the way, more recent renditions of the theory do not require that these dimensions be inconceivably tiny — just plain small.)

These changing number of dimensions corresponded with a changing theory, including its name. What was still called String Theory in the 1970's became Superstring Theory in the next decade, by incorporating supersymmetry — and in more recent years the model has become M-Theory. What the *M* stands for is unclear, but it has been variously described as representing " 'magic,' 'mystery,' 'mother,' (as in mother of all theories) or, more prosaically, 'meta,' 'matrix,' or 'membrane.' " (Johnson) The "membrane" reference is to another construct in the recent versions: called "branes", for short, these correspond with various other dimensions — anywhere from 0 to 9, which exist in relation to the 10-dimensional string-based (spatial) framework. One prominent subset are the "D-branes", which represent the multi-dimensional surfaces in relation to which strings interact — and such D-branes may even be more fundamental than strings themselves, being that of which strings are composed!

This whole approach arose in the first place as a way to describe the nuclear force, but that initial venture was unsuccessful. On the other hand, the theory did appear to smoothly account for gravity, again by treating it as a quantum field — thereby making it the first consistent theory of quantum gravity. Over the years, attempts have been made to incorporate the other fundamental forces, as well. Success would mean a satisfactory theory of quantum gravity, finally uniting Quantum Theory and General Relativity — in spite of the fundamental differences of character between them that we have discussed.

As for as the compacting or "curling up" goes, the fuller explanation goes something like this: Once upon a very long time ago, in the tiniest fraction of a moment after the universe was created, such as we referred to in the previous subsection — 10^{-43} second here, to be precise — all the many dimensions which String Theory proposes actually manifested themselves in full-fledged form. That original state of the universe was very dense and very hot. But those initial high energy conditions collapsed (less than) an instant later — which is when the Big Bang happened, and the universe exploded outward ... thereby switching to the relatively cold, low energy state of the universe that we know today. It was then that nature was transformed, with the extra dimensions suddenly and inscrutably curling up or compacting, becoming embedded in the remaining 3+1 dimensions of space-time ... which expanded into the existing world.

It should be emphasized that there is no single theory of strings, but a highly flexible framework encompassing many potential candidates — entailing an assortment of associated principles and calculational techniques, which can fit all sorts of data. And yet regardless of any such uncertainties and apparent shortcomings, supporters of the evolving theory are enthralled with the elegance and consistency of the highly sophisticated mathematical structure upon which this approach is built. The model's proponents, who currently dominate the field of particle physics, are very enthusiastic about String Theory's prospects for ultimately uniting the fundamental forces of nature — indeed, many advocates see it as the only viable option in this respect — however long a time it takes for this end state to be attained.

Needless to say, we take a very different stance on all of these matters, and we discussed some of the underlying issues in the last part of the previous subsection. In particular, we have made clear our position that mathematics should play a limited role in the conceptual development of our theories — regardless of how elegant and impressive such mathematics may be. In the case at hand, there can be no question about the lead role which mathematics has played in the unfolding of String Theory. In fact, the mathematics is not merely primary here — it is exclusive, in the sense that there is no other corroborating support whatsoever: there is no actual phenomenological basis of any form which substantiates the conjectures of this theory, be they in real life terms or even in the artificial conditions of the laboratory — and that is the case both now as well as what is expected in the foreseeable future: we are dealing with pure mathematical speculation which is thus far completely lacking in plausible demonstrability, even after many years of trying. Indeed, in Horgan's view, "[P]hysicists have gone off the deep end, postulating particles and energies and dimensions whose existence can never be experimentally verified." (Horgan, 2003)

Then there is the whole issue of meaningfulness or comprehensibility. Thus, it is difficult to see how this theory makes any sense, in real life terms — or how it is to be differentiated from science fiction. The theory's compactedness conundrum alone is enough to boggle the mind! This mysterious feature has been described as "the most vexing problem of all such theories: explaining exactly how the extra dimensions are "compactified" so as to make them unobservable in our four-dimensional world." (Garisto, p. 2996) In this regard, there is the issue we discussed previously — the idea that there was a one-time momentary state of affairs which supposedly existed for the most unbelievably tiny fraction of a second at the start of the universe ... as opposed to all the eons of time ever since. Such far-fetched and ad hoc conjecture is certainly not consistent with the precepts of the present model, according to which there are persistent general patterns that manifest themselves throughout nature and over time.

A final point of objection concerns an issue that we have made repeated reference to about the fundamental force framework, and which also applies to String Theory — which is that we are still dealing with what is in essence an atomistic model, in the sense that the whole approach presumes that there are basic building blocks of nature which explain everything else. It is the same old idea that all (higher) phenomena in the universe can be understood simply in terms of these fundamental (lower) levels of nature — be they called particles, strings, D-branes, or whatever the next iteration turns out to be.

By contrast, of course, the model which we are proposing is based upon the precept that all phenomena of nature can be understood in terms of certain fundamental patterns — which can be understood not in terms of lower level building blocks, but rather in terms of general attributes which apply to all levels. Whether they be called dimensions of analysis or not, these general features are not "compacted" or hidden at all — nor were they only manifest at a singular and unique micromoment at the birth of the universe. On the contrary, they are pervasive, ever-present — and complementary[9] — aspects of nature which are completely accessible to the full range of our normal experience. Whether the parameters of String Theory's dimensions bear any resemblance to these properties is not a matter that we are pursuing in this book.

Cosmology

With that, we mostly shift our focus away from the attempts at unity by the established paradigm and to that of Two-Force Theory. We start, in the next three sections, by considering how the two-force ideas that we have been developing might be applied to some other areas of science, which we have not focused on so far in this book. We begin with cosmology, and then move on to biology and cognition. Note that much of the treatment in these three sections is very general and speculative — especially this one.

9 In mentioning complementarity, we should make note of a rather bizarre usage of the term *duality* that comes into play with the topic we are discussing. This comes about when different versions of String Theory yield identical outcomes for particular problems; and in such cases, the two overlapping mathematical frameworks are given the name of "dualities" — which simply means that they are interchangeable ways of representing the same result. Needless to say, this usage of the duality term bears no relationship to its traditional usage in Quantum Theory (or its newer usage in the present model). As to why the corresponding solutions would be offhandedly described in terms of a "duality", rather than more straightforward names like "alternatives" or "equivalents", is hard to say — beyond the general tendency of physics to use ordinary words in capricious and esoteric (if not pretentious and misleading) ways, as we noted in the previous chapter ... along with Feynman's point again about how "physicists delight themselves by using ordinary words for something else." (Feynman, 1965, p. 84). On the other hand, given the manner in which the symmetry label is often applied by physics, why don't they simply refer to the alternative mathematical frameworks (of which there can be more than two) as all being "symmetrical" to one another?!

It seems clear that many of the precepts we discussed earlier in the book obviously apply to cosmology, or astrophysics. Indeed, it may well be that the guiding themes of complementarity and hierarchy have considerable light to shed on issues here. This might also help us to deal with the fact that even today, the point that Wigner made long ago still stands — that "[W]e simply do not have a cosmological picture 'in the large' which does not contradict some simple and general law of physics." (Wigner, 1976, p. 7)

Let us first mention a few general descriptions of cosmological phenomena, which are consistent with earlier topics that we have discussed. For example, most (if not all) bodies in outer space can be construed as figure, in relation to an enveloping background — which background can in some cases be characterized in terms of an atmosphere and/or a magnetic field. Even galaxies may be surrounded by very faint shells of material. Of course, breathing-like patterns are also apparent: Thus, the sun is breathing in and out, alternately expanding and contracting over a 76-year period. There may also be such a cycle on a much smaller scale, over a 2-hour period. Similarly, sunspots vary in number and size on an 11-year cycle, and in each cycle the sun's magnetic field reverses.

The planets in our solar system are sometimes grouped into two complementary categories — consisting of the inner planets ... as opposed to the outer planets of Jupiter, Saturn, Uranus, and Neptune. One set of planets are smaller, with higher density, less atmosphere, and slower rotation; by contrast, the other set of planets are larger, with lower density, more atmosphere, and faster rotation.

Likewise at the galaxy-wide level, cosmologists may distinguish between so-called Population I objects and Population II objects. One type lies in the very thin center, or focal, layer of the galaxy; the other type lies around the periphery, or background, of the galaxy. The inner type of objects are younger, metal-rich, and move more slowly — while the outer type are older, metal-poor, and move more rapidly (Clotfelter, pp. 238-239). Complementarity is also apparent in terms of our galaxy's shape, which has been described as being like a double convex lens — two fixed eggs packed together back-to-back.

Now let us discuss a few specific examples in cosmology that are in dire need of a complementary analysis.

Black Holes and White Holes. The black hole construct has become so widely known that it is often used in common discourse, as referring to a place into which large amounts of energy and resources continually disappear. Here is how its primary usage can be described, in terms of physical objects in the heavens: "A black hole is an area of space which appears absolutely black because the gravitation there is so intense that not even light can escape into the surrounding areas." (Zukav, p. 184)

Although black hole phenomena are generally thought of in connection with the Theory of Relativity, they need not be. As far back as 1795, they were being considered from the standpoint of Newtonian physics (Zukav, p. 184) Sometimes the term is thought of in connection with a very massive collapsed (or collapsing) star, but more recent models seem to associate it with the centers of (active) galaxies. On the other hand, many binary star systems are understood to consist of a visible star rotating about an invisible star — and it is the latter which some sources identify as a black hole (e.g., Geroch, p. 218; Zukav, p. 185). It is also said that if there is sufficient material in a small enough volume, then by definition this is a black hole.

The mathematics of General Relativity indicate that at some point inside a black hole there exists a phenomenon that can be described in terms of a "singularity". Davies says that in the case of a collapsed star, the singularity "is the prediction that all the matter in the star is squeezed into the same mathematical point. At this point the density and the space curvature must be infinite. This absurdity is called a *singularity.* A singularity is not an object, it is a place where known physics has come to an end." (Davies, 1977, p. 120) Again, "Occasionally it is conjectured that any matter which encounters a singularity leaves space-time altogether." (Davies, 1977, p. 120) For his part, Penrose says that, "[I]n the center of the black holes you have these singularities, where the geometry ceases to make sense: you find that curvatures become larger and larger, and apparently go to infinity — and the whole body of the star seems to be concentrated at a singular point, if you like, at the center. And that makes very little practical sense." (Penrose)

Not surprisingly, there are doubts about whether singularities actually exist in any pure sense, in the real world: "Many physicists now suspect that the infinite densities predicted by the general theory of relativity are not reached in nature, and the most likely reason is that quantum effects, which are ignored in the general theory, must come into play when space-time curvatures become very great." (Clotfelter, p. 381) And once again, the problem may go back to the flawed perfect symmetry construct, as underlying the singularity construct: "Is it possible that the prediction of a spacetime singularity is purely a consequence of this mathematical assumption of exact symmetry; that if we were only to assume approximate symmetry we would no longer predict singularities?" (Wald, p. 54)

So how shall we make sense out of these black hole conjectures? To start with, one must question the notion that black holes exist in the idealized, one-sided form in which they are usually described, along with the pure singularity phenomenon that goes with them. Regarding the latter, our position is that this construct is similar to other idealizations we have dealt with — like Quantum Theory's blackbody, or the general concepts of a complete vacuum, totally empty space, force-free zones, and so on. None of these constructs exist in the pristine form in which they are described, any more than does symmetry itself — which is why we call it "false symmetry". Moreover, to whatever degree these phenomena approach their ideal form, they do not exist by themselves, but rather in relation to some complementary counterpart: this concerns the "no symmetry" category — the notion that real nature is always a complementarity, and one never finds such a one-sided phenomenon without some manifestation of its opposite.

It seems reasonable to presume, therefore, that we are not dealing with something distinct and idiosyncratic here; and that whatever the black hole phenomenon refers to, the patterns at issue are not unique … and they are not one-sided. Consequently, if one discounts the notion that a black hole has an infinitely high density, then we are left with the simpler characterization of a black hole as having a (very) high density. But if we are then dealing with the real world of density, we are obviously talking about a range or continuum from the very low to the very high. The result is that a less dense object will display similar properties as an idealized "black hole", but to a proportionately lesser degree. In that respect, one could argue that any object can be seen as possessing properties akin to a "black hole", relative to some lesser entity that is locked into its realm. For instance, the nucleus' intense attraction might be characterized as a black hole, relative to smaller phenomena nearby. Or the earth's gravitational attraction might seem like a black hole, from the perspective of something falling towards its center.

The obvious retort to these analogies is that while some entity can be pulled within or towards the nucleus or the earth, energy (or matter) can also be emitted by such "normal" phenomena — whereas in the case of the black hole it is a uni-directional affair: this is also called a "one-way membrane" by Finkelstein, one of the originators of modern black hole theorizing (Zukav, pp. 183-184).

But that is a restatement of the one-sided thesis, which is supposedly what makes black holes unique — an interpretation which we reject, as being completely antithetical to the complementarity premise at the core of our model. Such an in-only proposition of a black hole singularity makes no sense whatever, and as such it is akin to the one-sided essence of the second law of thermodynamics which we discussed in Chapter 9; only, in that case the one-way was *out*, whereas with the black hole the one-way is *in*. In both instances, however, there must be an opposite side to the puzzle. In the thermodynamics case, we suggested ways in which the unidirectional time or process portions broaden into a bi-directional whole. Likewise, we contend that all black hole "singularities" must broaden into "dualities" when one looks at the whole instead of the part.

So what could be the complementary "out" counterpart for the energy and matter which goes "in" to a black hole? There are several ways of approaching this question. For one thing, even the basic assumption that black holes are one-way affairs is only "a first approximation. Some physicists theorize that black holes actually shine due to photons and other particles quantum-tunneling out of the one-way membranes." (Zukav, p. 184) More specifically, "In 1974, Hawking showed that black holes are not really black at all, but emit heat radiation at a characteristic temperature that depends on their mass." (Davies, 1980a, p. 166) This phenomenon is referred to as "Hawking radiation".

In fact, the mathematical equations of General Relativity yield two outcomes of relevance here — one corresponding to the notion of the black hole, and the singularity which goes with it; and the other corresponding to the opposite solution of a "white hole", where matter and energy pour outward. One way of understanding this pattern is in terms of issues we discussed in relation to the second law, which concerns the direction of time: the black and white hole outcomes could correspond with forward and backward times, respectively. Continuing with that analysis, one might argue that black holes manifest themselves in the present expanding universe, while white holes would manifest themselves in a contracting universe.

A related idea is to think of the Big Bang itself in terms of a white hole, as opposed to the black holes that have existed in the expanding universe since then.

Another approach is to assume that black holes and white holes might co-exist in some sense. Not so long ago, the thesis was put forward that quasars should be understood as representing white hole counterparts to black holes,[10] although this idea is no longer in favor. Even so, one might still say that just as an (in-oriented) black hole is said to exist at the center (i.e., focus) of each galaxy, so the rest (i.e., background) of the galaxy of (out-oriented) stars could be understood to function like a white hole.

On the other hand, the current assumption is that quasars can be understood as massive black holes. Yet rather than coexisting, is it possible that the in-oriented black hole builds up to a point where it explodes as an out-oriented white hole (quasar)? There would be a buildup of energy or matter coming in to the black hole, until the balance shifts — perhaps there is an implosion, followed by an inversion ... and then the energy or matter would go back out again. In this case, when the pendulum swings too far over to one side, the scales "tip" and the pendulum swings back in the reverse direction — it doesn't keep going in the same one-sided "singularity" direction forever, but reverts back in the other direction.

Then there is the theoretical idea that black holes and white holes actually do coexist, and are connected by "worm holes" — with the black hole existing in our universe, and the white hole existing in another parallel universe ... perhaps amounting to a Big Bang there. A variation of this is the idea that what went into a black hole emerges at "another time and another place in this universe (through 'Einstein-Rosen bridges')." (Zukav, p. 185)

One way or another, we would argue that complementarity must apply to the black hole construct — and that the general pattern is the same as applies at other levels of nature's hierarchy.

Sun and Nus. As we just indicated, we take exception to the notion that, "A black hole cannot be approximated by anything from familiar experience." (Clotfelter, p. 378) Not only do we argue that the in-oriented aspects of a black hole are just one side of some in vs. out complementary whole phenomenon, we also maintain that (as we just stated) the same pattern applies to every level of nature's hierarchy. In fact, a far-fetched conjecture is that there is something akin to this pattern in the heavens right near us, at the core of our own solar system.

We can think of the sun is an out-entity which primarily emits energy, but is there any complementary in-entity for the solar system? There could well be: The earth and other planets orbit the sun, right? Well, only partly — or the orbit would be a proper circle. In fact the earth's motion is not precisely circular: it is elliptical; and while a circle has one focus, an ellipse has two. The sun is only one of the two foci which the planets orbit.

What do we know about the other focus of the solar system? Very little, aside from occasional remarks to the effect that this other focus supposedly has "no physical significance." (Marion, p. 102) Or put elsewise, that the one focus is occupied by the sun and that the other focus is unoccupied.

Could it be that this second focus has an inverse function to that of the sun? For want of a better name let us call this second entity the *nus*. Therefore, just as the sun has the function of giving out or emitting energy, so the nus would have the complemen-

10 "There are objects in our universe that appear to be the reverse of black holes. They are called white holes (of course). These objects are quasi-stellar radio sources, or quasars for short. Quasars are extraordinarily intense energy sources.... The 'output' side of a black hole, according to this hypothesis, is a quasar." (Zukav, p. 186)

tary function of drawing in or absorbing energy from outside the solar system. One might suppose that the energy which the nus draws in to the solar system is passed on directly to the sun, which then gives it out. Note that even the orbiting pattern of the planets around the solar system appears to be consistent with this premise. Thus, the earth moves faster in the direction of the sun than in the direction of the nus — this being compatible with the proposition that overall energy flow within the system goes directly from nus to sun.[11]

Furthermore, consider the phenomenon known as "aberration of starlight". This refers to the fact that an earth-bound astronomer viewing stars must change the direction or angle of his telescope, depending on the season — which is to say depending on where the earth is in its orbit about the solar system. "The telescope must be angled in a continually changing direction as the earth moves about its curved orbit." (Asimov, p. 73) One explanation would be that the starlight comes not directly to the earth but via the solar system as a whole — or more specifically, it comes in via the nus and then passes towards the sun. Since the earth's position relative to the nus is continually changing, so must the angle at which the earth views starlight coming in to the nus.

The overall complementary pattern, therefore, is that of an in-partner called "nus" and an out-partner called "sun". The former would be dark and invisible, just like a black hole. Of course, we do not label this phenomenon as such, but how would it appear to a distant observer?

In summary, we assume that there is an in- and out- complementarity at every level of nature, and there is no reason to think there is any fundamental departure from this pattern anywhere in the universe. In some form or other, black hole and white hole would be in- and out-partners at a galaxy level, just as nus and sun would be in- and out-partners at our solar system level.

In-partner	Out-partner
black hole	white hole
nus	sun

Cosmic Background Radiation. The phenomenon known as cosmic background radiation provides a conceptual locus for discussing a number of issues that relate both to complementarity and hierarchy, which we have examined earlier in this work. Some of the topics we will touch upon in this brief subsection are the Big Bang, absolute space, and universal lumpiness vs. smoothness.

Cosmic background radiation, or CBR, is a faint phenomenon that is said to exist in a largely uniform manner throughout the universe.[12] It was discovered in 1965, and

11 Does this raise questions about theories of the sun's life and energy development, which are not consistent with the kind of ongoing energy input suggested here?

12 In fact, the actual topic that we are discussing is more properly known as *cosmic microwave background* (or CMB), because it is only a subset of the more general category of cosmic background radiation. As its name indicates, we are talking about the microwave portion of the electromagnetic spectrum — whereas other background radiation phenomena exist in the infrared, x-ray, and other portions of the spectrum. Those other cases are said to have different (usually specific or local) causes, as opposed to the universe-wide phenomenon under consideration here.

is generally understood in relation to the Big Bang, as representing a "feeble, dwindling glow from the primeval fireball.... a fossil of the fiery birth of the universe." (Davies, 1977, p. 168). Indeed, cosmic background radiation is usually cited as providing compelling evidence for the Big Bang itself, and for theories that relate thereto.

The standard interpretation is that initially, immediately after the Big Bang start of the universe, there was a smooth and evenly distributed fireball — a generalized radiation that existed everywhere, with a temperature of some 3,000 degrees Kelvin. Immediately thereafter, the universe began expanding; and in the process, the high temperature dissipated. Throughout this period, however, there was only a kind of "matter soup" amidst all the radiation, with protons and electrons moving around and interacting but unable to bind together into whole atoms. It was only after some 300,000 years that the universe became sufficiently cool for the phase called "recombination" to take place. It should actually be called simply "combination", since this was the first time that sustained proton-electron binding could occur. The result was the emergence of full-fledged material, whereby matter actually existed independently of the generalized background radiation itself.

During the millions of years since then, the universe has continued to expand and the temperature has continued to drop — so that today it is something just below 3 degrees Kelvin. Throughout this time, the universe is said to have been more or less evenly distributed, a situation that is encapsulated in doctrine: "The assumption that the universe is uniform on the large scale is now accepted by most (though not all) cosmologists, and is known as the cosmological principle." (Davies, 1977, p. 146)

Decreasing temperature aside, there remains a theoretical question as to what triggered the emergence and continued development of matter into the kind of complicated structure ... from amidst the "nothingness" that existed prior thereto. So how could the "small scale lumpiness" of matter have emerged spontaneously from the "large scale smoothness" of non-matter ... whereby matter and radiation became "decoupled"? This question of how lumpy structure emerged from smooth radiation represents a serious theoretical challenge for our models of cosmology.

In fact, what more recent research has indicated is that CBR is not completely uniform, or isotropic, in the universe. The primary exception is referred to as *dipole anisotropy* — sometimes described in terms of a "yin-yang pattern"! — and the assumption is that all the local galaxies are moving in relation to the background radiation at some 600 kilometers per second. The result is a Doppler-like pattern, with the background waves that we are moving towards being compressed or blueshifted (and appearing hotter), as opposed to the background waves that we are moving away from being expanded or redshifted (and appearing colder).

Still, this anisotropy would not explain the emergence of matter, but only our perception of wavelength or temperature variations based upon motion. The other kind of anisotropy is on an even smaller scale, amounting to small ripples of variation in the sea of smoothness. Whatever their explanation, such fluctuations from uniformity are seen as essential for the formation of matter — thereby resolving what is called the "galaxy formation problem".

Now let us raise some questions about this interpretative scheme — for it could be that this whole conceptual framework is way off the mark, in various respects. To start with, there is the standard interpretation that CBR should be understood in terms of the Big Bang. In this regard, everything that exists (or existed) in the universe somehow emerged from the Big Bang, not just CBR. Moreover, what we would emphasize is the overall pattern here — and the general notion of a background phenomenon that exists in relation to figure counterparts: there is nothing unique to CBR in this respect, given our contention that this is the most ubiquitous pattern of nature there is.

The persistence of this figure-ground pattern is also why we would question the notion that the early universe only (or primarily) existed as radiation alone, apart from matter. Yet why would such a pattern manifest itself, whereby background alone (radiation) would exist apart from figure (matter)?

Indeed, we would suggest that such an arbitrary interpretation represents another instance of the "no symmetry" problem. This is why reject the notion that the universe was once a unitarity, consisting of a smooth sea of non-matter — which at some point became "lumpy", with the "decoupling" of matter and medium ... which requires some deep theoretical explanation which is specific to that issue. We assume, by contrast, that there has always been matter and "lumpiness", as well as medium and "smoothness" — from the start. By this complementary interpretation, the Big Bang would be understood as an explosion outward of both matter and medium, which would distribute these two domains more or less evenly in all directions as it went.

One might also question the conclusion that CBR has the same precise character everywhere in the universe — leaving aside the anisotropies that we discussed above, which are attributable to either motion or microfluctuations. All we know is how the CBR phenomenon manifests itself in some given locale, but why should we conclude that this background phenomenon manifests itself in the same way everywhere else? Just as Relativity sees its space-time universal continuum as varying in flatness vs. curvedness, why would a similarly varying pattern not apply to CBR? In this respect, it might be better to speak about *local background radiation*.

By the way, regardless of whether one treats CBR as local or universal, it should also be apparent that the very notion of a uniform background referent would clearly provide "an absolute standard of rest", which can be understood in terms of standing gravity waves (which we talked about near the end of Chapter 5). (Rucker, p. 382;[13] also, Prokhovnik, 1979, p. 227). Needless to say, such a notion completely contradicts a fundamental tenet of the Theory of Relativity, according to which there is no such absolute standard of rest in the universe ... no matter whether such a reference is called absolute space or an ether medium.

Missing Matter. Now let us mention how the notion of nesting hierarchy may provide an answer to theoretical problems concerning the overall mass of the universe. There have long been significant discrepancies between the amount of mass which has been conjectured to exist — or which is required for explaining aspects of other phenomena (such as that relating to universal gravitation and expansion) — as opposed to that which has been deduced by calculating various estimates of the quantity of mass

13 "This faint electromagnetic hiss fills the whole universe with something like a pattern of standing waves." (Rucker, p. 382)

that actually exists. The problem is that there is not enough of it — the reality does not match the requirements of the theory: the issue is encapsulated in terms of what is called *missing matter*, or *dark matter*, and physics has long sought an explanation for it.[14]

We would like to raise the possibility that the source of this problem could be very simple: it may well be a result of the faulty assumption that we are merely dealing with an array of isolated low-level particles (i.e., what are called normal atoms, or "baryons"), which do not collectively add up to the required total mass. Such a piecemeal approach would be consistent with physics' fundamental force model — that only certain levels of nature's hierarchy actually exist, relative to which all the usual forms of matter are effectively illusions. By contrast, the present model is predicated upon the precept of nesting wholes, and the associated idea that every level of nature's hierarchy represents its own legitimate structural existence … which would also include having its own mass. In these terms, the principle that the whole is greater than the sum of its component parts would apply in a cumulative fashion — thereby building up to a much higher total quantity of mass than would be assumed by merely adding together the lower level constituent units by themselves.

The same approach might be applied to what is called the "hierarchy problem". This is the name which is given to the fact that there is an inexplicable disparity in proportionality between the different fundamental forces of nature — and especially that between gravity and the others. One explanation would again be based on the notion of nesting levels of nature's hierarchy — and that if these many other levels were taken into account, between the very low levels of the nucleus, etc. and the much higher levels associated with gravity, then a more orderly relationship of scale might emerge.

One-Time or Recurring Universe? One of the issues we dealt with above concerned variations in the universe over time, and we encounter that factor again here under a different topic: are we living in a one-time universe, or one that comes and goes?

There are also two sub-options for the one-time universe question. The first is called the 'steady state' model, and it says that the universe is static — and although it is expanding, there would also be 'continuous creation' of matter to fill the expanding void (Kauffman, p. 116). The other is more of an evolutionary model, and it says that the universe is expanding: this is the usual choice. Based on evidence like the universal Doppler redshift and on Hubble's Law correlating distance with speed (which we discussed in Chapter 8) there has long been a very broad consensus in favor of this second alternative — although the judgment is not unanimous.

We noted in Chapter 5 that Einstein added a "cosmological constant" to his General Relativity field equations in order to stabilize their outcome and thereby predict a static or steady state universe, i.e., the first approach above. But when Hubble later

14 One mainstream explanation that has been put forward in recent years is that the missing matter exists in the form of the lightest supersymmetric particle (the "neutralino"), which is not readily detected.

amassed data to show that the universe was in fact expanding, Einstein abandoned the constant. For the fact is that without the constant, the equations lead to the conclusion that "the universe must always be expanding or contracting. The notion of an ever-constant, static universe is incompatible with general relativity; the universe must be dynamic." (Wald, p. 49, italics omitted) Consequently, the General Theory and the Big Bang model go hand in hand.

So the universe is expanding and not static. But what if we are not living in such a one-time expanding universe? Put another way, did something come before the Big Bang? Unfortunately, when we ask what came before this cataclysmic explosion, we only get the equivalent of a blank stare from theories like General Relativity. In fact such theories deny the validity of the question: such a query is ruled out of order, as being based upon ignorant pre-Relativity notions of space and time:

> What happened before the big bang? This is a natural question to ask within the context of pre-general relativistic notions of space and time. But within the context of general relativity, this question does not make sense since there was no such thing as space and time "before" the big bang.... [L]iterally speaking, nothing happened before the big bang. Note that it is difficult even to express this idea, since our language is attuned to prerelativistic notions of space and time. In particular, the word *before* presupposes the notion of time. (Wald, p. 50)

Indeed it does, and with good reason. Because Einstein's General Theory must remain mute on the question of what came before the Big Bang is no reason to assume that such a question, and its "before" connotations, are necessarily unfounded.

The most obvious alternative answer is that we are not living in a one-time universe — which brings us to the other main possibility that we started this topic with: before the Big Bang and the expanding universe, there was the opposite ... of a contracting universe, into what is sometimes called the Big Crunch. This leads to the reasonable conclusion of an oscillating universe, which expands outward and then contracts inward, cyclically (Capra, 1975, p. 183; Calder, 1979, p. 152; Davies, 1980a, p. 180). Unfortunately, there is a problem, again, in terms of General Relativity: for Relativists like Penrose and Hawking have supposedly "proven" that such oscillating universe hypotheses are "not in accord with general relativity." (Wald, p. 54-55)

In fact, more recent theory and observation are said to support a conclusion that the current state of universal expansion is occurring at an ever-increasing rate — a phenomenon that is described in terms of some mysterious *dark energy*. Some of the explanations for this pattern involve a kind of negative or repulsive gravitational force, that is once again equivalent to Einstein's cosmological constant. The meandering logic of this interpretation states that whereas attractive gravity was predominant in the first half of the universe's development, since then repulsive gravity has become predominant. This repulsive gravity is supposedly weaker than attractive gravity, and only kicked in when the universe had expanded far enough that the space between galaxies was sufficiently large. Furthermore, the reasoning continues, because there has been a change from the attractive to the repulsive stage of gravity, then the universe will continue expanding forever — and will not turn back inward again: it will simply expand faster and faster, indefinitely. Note that such a one-way expansion interpretation would be consistent with the unidirectional time thesis of thermodynamics.

By the way, as to why this repulsive form of gravity is not generally observed and incorporated into common theories, the answer would supposedly be that normal attractive gravity still predominates on smaller distance scales, such as in our solar system. The repulsive form would only apply for the universe as a whole; and even then, only for the last 10 billion years of the "second half".

Summarizing the points we have discussed in this subsection, it should be clear that, once again, we are faced with a variety of theories and interpretations which seem utterly ad hoc and implausible, from the vantage point of the present two-force model. Of course, there is one point about which we do concur with most observers — that the steady state proposition is not viable: nothing is only stable in nature, and there is no reason to think that the universe is an exception (which is also the point we made about the cosmic background radiation of the universe). However, we also reject the notion of a one-time universe that will continue expanding, faster and faster ... forever. Such a model would seem to be inconsistent with more general patterns: everything waxes and wanes, or breathes; and there is no reason to think that the universe would be an exception to these cycles. Consequently, the idea of an oscillating universe is in complete accord with the present model — while neither a steady state model, nor a one-time Big Bang model, appears to be.

Complementary Universes. In our discussion so far, and in virtually all common discourse, the universe is treated as a singular whole. But what would be its complementary counterpart at that topmost hierarchical level? Needless to say, we reach into very abstract territory here.

One answer would be to say that if we are talking about the universe on the one hand, then the counterpart might be labeled as non-universe on the other hand ... whatever that might be. But another answer would be to turn the question back, and ask why we should assume that there is only *one* universe. This is not to argue along the lines of certain subcurrents in Quantum Theory, which assume an infinite array of ever-splitting universes — as in the Many Worlds hypothesis which we discussed in Chapter 2 — but something more coherent and meaningful.

Of course, one could argue that this highest level expands and contracts only in relation to itself. But why make such a specific-level assumption? No other level interacts solely with itself — on the contrary, the general pattern of nature is that things come in pairs: two by two. Why should the universe be any different? Isn't it more plausible and consistent to assume that there are two complementary universes interacting with one another in the same way as do so many other complementary partners? Just because we have reached the highest level of the hierarchy, there is no reason why we should drop the "two" and suddenly think only in terms of the "one". There may be no compelling reason for such a conclusion, other than our tradition of always thinking in terms of "the" universe. So are there really *two* universes, rather than the singular one we generally assume?

Here is another way of posing the issue. When we say that the universe is presently expanding, where is it expanding *to*? We assume that the energy is going somewhere, as it spreads out and becomes "dispersed" as part of the overall expansion — only, where is it spreading and giving out energy *to*? The energy must be going from somewhere to somewhere, and one approach is to assume that something (a sink) is receiving the energy from that which emits it (a source) — or it wouldn't be going anywhere. This interpretation would presume that if one thing is expanding, then it is doing so in relation to another thing which is contracting: that there could be no expansion without some corresponding contraction.[15] The two go together: complementarity tends to work like that. One never has simple repulsion (or expansion) alone, or simple attraction (or contraction) alone — the two must go hand in hand, inversely to one another. For example, on the earth surface the tides go out and the oceans recede; but they go somewhere, for in other parts of the globe, the tides are coming in and the oceans approach.

Thus, could it be that if our universe is expanding, then perhaps it is doing so in relation to a complementary universe that is contracting? So is our universe spreading out and giving forth energy ... to another universe? Such an interpretation may also give more satisfactory answers to questions about what is outside our universe, in the sense that there would be two universes which are interacting with one another, perhaps one enveloping the other — or each being enveloped by the other, as seen from within either one.

Consider that there is an "event horizon" at the edge of the visible universe, beyond which we cannot see. True, this event horizon seems to move further out as our technological efficiency improves; but no matter, in principle. For might it be that on the other side of this "event horizon", out and beyond that point — and perhaps past the yonder "nothingness" — that there is a whole other universe out there, much like our own? Only *it* is contracting, in contrast to ours which is expanding? And while we see a "universal" Doppler redshift, they would see a complementary "universal" Doppler blueshift? Would there always be one universe which is expanding, at the same time as the other universe is contracting? Furthermore, might this scenario also shed light on the so-called dark energy factor, in the sense that what is causing the speeding up of the expansion of our universe could be the increasingly contracting force of the complementary universe?

Note how the dual universe thesis can be understood in both spatial and temporal terms: By assuming a figure-ground reciprocal pair of complementary universes, the spatial aspect is incorporated, as just mentioned with "enveloping"; and the temporal factor is also incorporated, by virtue of each partner alternately and inversely expanding and contracting over time. In this way, we would have a complementary pair of universes — ours which is now expanding, relative to the other which is contracting. This also means that in the present phase, the other universe is dominant and ours is recessive. At some point, the pendulum will reach its apex and then swing back — with the previously-expanding universe then going into a contraction phase, and the previously-contracting universe going into an expansion phase. The two universes would

15 Note the correspondence here with the reasoning that Lewis put forward in the context of Absorber Theory, which we discussed in Chapter 9.

oscillate in dominance-recessiveness, just as each one would separately oscillate its cycle of contraction-expansion. Complementarity in space, and complementarity in time.

This ends our brief foray into cosmology, where we saw some examples of how our themes of complementarity and hierarchy might be applied. Now we shift over to a very different field of study, right here on earth, and to some extent inside ourselves.

Biology

The field of biology is a veritable catalogue of complementarity and hierarchy, far more so than any of the other sciences. We will briefly scan through an array of phenomena in this section that illustrates the range of these themes, and then summarize those patterns at the end.

To start with, consider how our two basic forces of nature would be perfectly compatible with a biology description like this: "Living organisms represent a constant interaction of two opposing forces: the constructive process, *anabolism*, and the destructive process, *catabolism*." (Baker and Allen, p. 20) Or as Wolf says generally, "[W]e are composed of complementary properties." (Wolf, p. 205)

Then there is the longstanding philosophical or metaphysical struggle between two opposing currents of thought in biology — that between *mechanism* and *vitalism*.[16] Mechanism is basically a fixed and deterministic view of nature, in the style of the classical Newtonian world view: "Essentially, mechanists view organisms as very complicated machines." (Baker and Allen, p. 15). Vitalism, on the other hand, is more "spiritual" and assumes that every organism has a "vital force",[17] which cannot be measured and isolated. "This force is assumed to be wholly different from other known physical forces and ultimately to be unknowable — not subject to physical and chemical analysis." (Baker and Allen, p. 15)

These two approaches to nature would be readily compatible with the complementary and two-force basis of the present model. Even the great Quantum pioneer Niels Bohr cited the fundamental pattern when he said that, "An example is offered in biology where mechanistic and vitalistic arguments are used in a typically complementary manner." (Bohr, p. 318)[18] And Jammer notes that the issue came to physics even earlier: "In 1932 Pascual Jordan published similar ideas in an essay on quantum mechanics and biology, in which he presented vitalism and physicalism as complementary aspects in the study of living nature." (Jammer, 1974, p. 88)

16 This controversy parallels, in turn, a larger philosophical dispute — *materialism* vs. *idealism*: materialism assumes that all phenomena in nature can be understood in terms of matter, while idealism opts for the contrary view that ultimate reality must be understood on a level which transcends matter.

17 This concept has reappeared in modern garb in terms of a "self-organizing force". Also, the question of how biological entities differentiate themselves into coherent structures is sometimes discussed in terms of "formative causation". See, e.g., an interesting theory about these matters by Sheldrake.

Now let us do a quick survey of some highlights in the field, which illustrate the themes of 2FT: in the rest of this subsection, we will emphasize (i.e., italicize) the complementary partners which are involved.

First up is a very obvious example, in the form of the famous polynucleotides or giant molecules known as *DNA* and *RNA*. Chemically, DNA and RNA molecules are very similar, although they are not identical — the chief difference being that whereas DNA molecules are double-stranded, RNA molecules are single-stranded. In terms of the present model's dimensions of analysis, DNA and RNA might be understood in terms of static and dynamic relationships, respectively.[19] Of course, DNA itself is typically described as being a "double helix", composed of *two complementary chains*; furthermore, "the two DNA strands have opposite polarity." (Gatlin, p. 2)

Each DNA molecule or polynucleotide is itself composed of two complementary nucleotide bases, *purine* and *pyridimidine*. Both of those bases, in turn, have complementary variations — *adenine* and *guanine* for the former, and *thymine* and *cytosine* for the latter. The result is that each DNA molecule consists of one purine nucleotide base (either adenine or guanine) together with one complementary pyridimidine base (either thymine or cytosine) — in either one or the other order (i.e., the purine base being primary and the pyridimidine base being secondary, or vice versa. In fact, only two complementary combinations exist: adenine can be joined together with thymine, and guanine can be joined together with cytosine.

Another obvious example concerns the complementary processes of cell transformation known as *meiosis* and *mitosis*. Meiosis is the process whereby the focal germ (gamete) cells complement with one another to produce a higher-level synthesis. Mitosis, on the other hand, is the process whereby normal background soma cells uncomplement within themselves to produce same-level separation or replication. Beyond this, the details are too complex to warrant our attention here.[20] The important point here is that overall and relative to one another, meiosis is an "upward" synthesizing, whereby entities complement one another and thereby produce a higher level unity; whereas mitosis is a "downward" differentiation, whereby the cell divides or uncomplements and thereby produces replicated parts.[21]

The subject of both of these transformations is the chromosome, which consists of a DNA molecule that resides in the *nucleus*, which is embedded in a protein coat of the

18 More broadly, Bohr "hoped he had found (in the complementarity point of view) a new platform from which to evaluate and solve the basic problems in a variety of fields ... [and he] was able to point to situations in psychology and biology that also present complementary aspects..." (Holton, pp. 138 and 134)

19 Thus, RNA appears to link together the polynucleotides with proteins, and it may be that one type of RNA ("messenger RNA") is the outgoing component to the protein (by way of the "ribosome"); whereas another type ("transfer RNA") comes from the other direction and so acts as the incoming component, by bringing individual amino acids to the ribosome for assembly into proteins. There is also said to be a third type of relatively-static RNA ("ribosomal RNA") at the ribosome, which may also be double-stranded.

20 In any case, the arbitrariness of the stage categorizations that are involved here is dubious, and the terminology horrendous — in keeping with the criticism on this topic that we put forward in the previous chapter, including words from biologist Lynn Margulis.

cell. This is a familiar focus vs. background pattern, which can also be seen operating at a subsidiary level — between *chromosome* and *nucleus*. In this regard, there are two fundamental stages of a cell's life, which may be simply described in terms of periods when it is *undergoing transformations*, as opposed to periods when it is *not undergoing transformations*.[22] In the "cell transformation" period, the chromosome is predominant and visible, whereas the nucleus itself recedes and becomes invisible. But this relationship inverts in the "interphase" period, when the cell is not undergoing transformations: then, the nucleus of the cell becomes predominant, whereas the chromosomal material is dispersed and invisible therein (even though it is said to be actively preparing or reproducing itself).

There are two complementary modes of cell reproduction, which parallel that of meiosis and mitosis — *sporulation* vs. *replication*: The cells can either differentiate into separate sexes and then sporulate, or they can proliferate by self-dividing and replicating. Yeast cells which sporulate have even been found to interchange rapidly between the two sex roles, by jumping from one chromosomal position to another (a "jumping-gene" factor). Indeed the complementarity of *male* and *female*, of *sperm* and *egg*, is a *sexual relationship* which itself can be seen as complementary to what has been termed a *parasexual relationship*.

Considering an individual's genetic makeup more generally now, every organism possesses a certain number of complementary ("homologous") chromosomes, which at a secondary level are called *diploid* if the chromosomes are complemented and *haploid* if the chromosomes are not complemented. In humans, for example, the male's sperm cells carry 23 uncomplemented (haploid) chromosomes, and the female's egg cells also carry 23 uncomplemented (haploid) chromosomes. When the two sets of 23 uncomplemented chromosomes come together and complement one another at conception, one partner of each pair becomes *dominant*, while its counterpart is *recessive*.[23] Each chromosome contains 500-5,000 specific character genes — and that in each case the dominant gene is the active and determining one, producing the outward characteristics which can be seen (the *phenotype*) ... while the recessive gene is the passive and retiring one, producing no outward characteristic but remaining in the organism's gene pool (the *genotype*).

We can also consider examples of *continuous vs. dichotomous* genetic changes. Thus, there are two fundamental types of changes in flu virus patterns, which are labeled *antigenic drift* and *antigenic shift*, respectively: one is the small-scale gradual type of slow drifts in the antigenicity of the virus; and the other is the large-scale abrupt

21 The details include such things as "spindle fibers" and "astral rays" which are akin to lines of force and poles in an electric field.

22 "[T]he cell is a kind of resonating system which cycles constantly through a set of states." (Goodwin, p. 155)}

23 Here we finally come to the terms "dominant" and "recessive" as they are used in established science. Of course, in 2FT we use the concepts to analyze nature generally and not just in the isolated field of genetics, which is what we are discussing here. Note the consistent usage: "Dominant and recessive are relative terms; that is, one factor is always dominant with respect to some other factor." (Baker and Allen, p. 273)

type, which are associated with pandemic changes, that occur periodically (typically between ten and thirty years) — a shift in the antigenicity of the virus. More generally, evolution theory differentiates between *gradual changes* which occur over time — and sudden large-scale genetic revolutionary changes in a population, called *punctuated equilibria* (which are correlated with mutations).

Aside from these continuous vs. dichotomous points, we can also briefly mention some other conceptual parallels between Darwin's theory and the present two-force model. One is that the notion of biological evolution can be understood as a kind of growth upwards, in the hierarchy of nature. A second is that the randomness which is intrinsic to evolution theory corresponds with the factor of uncertainty or unpredictability that is built into the present model. Finally, the survival of the fittest notion can be understood in terms of the best adapted species, or the most compatible species ... the one having the most *complementary* relationship with its environment.

In the first chapter, we talked about cyclic patterns of alternating dominance in nature, which we have sometimes discussed in terms of our breathing motif. The fact that every structure, at every level of nature's hierarchy, oscillates between (e.g.) active and passive phases is very evident in the cycles of living organisms, and of their component (nested) parts. These phases or stages are often described in terms of *biological clocks*. The different biological clocks of an organism typically include phenomena like cell cycles, heart beat, sleep-wake, and respiration — all nesting together into the same larger system. And of course the particular system (e.g., the human) is in turn part of higher physical systems (earth, moon, solar system, etc.), whose rhythms must interact with the lower system. In these terms, for example, it can be no coincidence that the human female's menstrual cycle has the same periodicity as do phases of the moon. Even the annual ring pattern of a tree is not something continuous alone, but oscillates between phases of growth and dormancy — which, of course, are linked to phases in the environment. In this regard, note that it makes sense to think of the earth's seasons as two: spring-summer (*growth*) and fall-winter (*dormancy*).

Now let us summarize the issues we have been talking about for biology, in terms of our two main themes. To start with, hierarchy plays an important role in biology. Gatlin talks about, "a structural hierarchy of control levels. The hierarchy concept is so fundamental to the description of the living system ..." (Gatlin, p. 11) Bastin says, "[T]here is direct evidence for hierarchical organization in the very existence of cells." (Bastin, p. 257). These comments are consistent with one from the pioneering work of biologist Whyte that we mentioned in the first chapter, regarding a "structural hierarchy of nature." (Whyte, p. 11)

As far as complementarity itself goes, we cited various instances above as to how this theme applies to the field of biology. Here is a much more sweeping list of examples, not at all organized, with which we end this topic:

<u>X-Force of Structure</u>	<u>P-Force of Process</u>
mechanism	vitalism
meiosis	mitosis
conjugation	fission
concentration / convergence / constriction	diffusion / divergence / dilation
tight / tense / squeeze / inhibition	loose / relaxed / release / excitation

wax / ebb	wane / flow
anabolism and buildup	catabolism and breakdown
differentiation	de-differentiation[24] - integration[25]
heterogeneous	homogeneous
speciation	variation
sperm / chicken / male / penis / androgen	egg / egg / female / vagina / estrogen
animal pole	vegetable pole
left side body and parts	right side body and parts
phenotype / active gene	genotype / inactive gene
activity / excited state	rest / ground state
constructive enzymes and catalysts	destructive enzymes and catalysts
development	decay
sex / life / creation	aggression / death / destruction
sexual relationships	parasexual (symbiotic) relationships
competition / attack / advance	cooperation / defend / retreat
function	malfunction
ATP	ADP - AMP
DNA / e.g., DNA strand *A* / + / tRNA	RNA / e.g., DNA strand *B* / - / mRNA
purine	pyridimidine
eukaryotes	prokaryotes
mitochondria	chloroplast
neuron	synapse
axon / all-or-none	dendrite / graded potential
afferent / parasympathetic n.s.	efferent / sympathetic n.s.
gray matter	white matter
nucleus	vacuole
acid: excess positive ion	base radical: excess negative ion
IPSP (inhibition)	EPSI (excitation)
speciation	variation
active transport concentrates → disequilibrium	passive transport diffusion → equilibrium
phagocytosis (solid) / pinocytosis (liquid)	reverse phagocytosis (solid) / pinocytosis (liquid)
endergonic	exergonic
photosynthesis	respiration
eye cones	eye rods
microorganism	macroorganism
pandemic change / punctuated equilibria	slow drift / gradual change
evolution	devolution (dissolution?)
natural selection	random mutation

24 Sherman and Vilia, p. 136 (re cells).

25 Spencer (see Popper, 1981, p. 267).

Cognition

One of the points we have made is that a good theory of science should be applicable to realms beyond that in which it originates, including entire other disciplines in science; and the previous section on biology was intended to provide some general ideas as to how that requirement might be addressed by the present model. The same is true of the present section on cognition, but with an additional twist. That is because we have also stated that a good theory of science must encompass aspects of both what we have called *being*, as well as *seeing* — which, of course, is a radical departure from the conventional paradigm of physics. We make that assertion in spite of the fact that Quantum Theory broke important new ground with its inclusion of the observational mechanism and the role of the observer, as part of its analysis of the micro world. For we contend that the issue is far broader than such limited factors would suggest.

To be sure, the new physics does talk about consciousness — even to the point where the dominant Copenhagen interpretation ascribes an actual function thereof in producing or triggering the underlying reality.[26] But we have rejected any special role for human (or any other) consciousness in this regard, as opposed to more general interaction patterns which are involved. On the other hand, we would indeed emphasize the role of human consciousness, as being a vital part of the process by which we produce our models of nature. That is because any theory is a conceptual device which is created by the human mind — and certainly any deep theory about reality cannot be completely divorced from the mind mechanism which generated that theory. Of course, there is also a larger context in which that mind mechanism operates, as we discussed in the previous two theory-building chapters; for those contextual elements inevitably impact the resulting theory, and thus color the way the world is depicted. We talked about a broad array of such factors which are involved in a theory's development, ranging from areas of philosophy, metaphysics, and values ... to those of psychology and sociology.

But it is the central role of the mind itself in the production of our theories which explains why we maintain that a good theory of science must ultimately incorporate (or at least be compatible with) a viable theory of cognition ... which can also be construed as a theory of seeing on the broadest scale. Our task here, then, is to briefly suggest how Two-Force Theory might be applied to aspects of cognition.

To start with, we can speak about a cognitive hierarchy (or hierarchies), as representing the whole gamut of cognitive entities and patterns that an individual has been endowed with at birth (thus encompassing evolution), or which have emerged since then. The resulting whole would represent the "universe" of the mind. The lower levels of the cognitive hierarchy would be more concrete, as opposed to the upper levels being more abstract. The former would be primarily sensory, and include representations from all parts of the body, especially the nervous system which connects them. Signifi-

26 In this regard, Wolf wrote some time ago that, "So far all attempts to modify quantum mechanics to include consciousness have fallen short of explanation." (Wolf, p. 220)

cant in this realm would be the conventional mechanisms connected with touch, taste, smell, seeing, and hearing.

In this system, what we call *perception* would be understood not merely in terms of outside phenomena that we experience — something "out there", which then comes into our brain (via our eyes and visual pathways, etc.) and is "seen". Rather, there will be an interplay between lower level sensory levels (akin to figure) and higher level conceptual ones (akin to background) — such that what we perceive in lower level sensory terms is partly determined by whatever higher level conceptual associations have been triggered or stimulated by the situation.

One would normally expect to find a good fit, or compatibility, between the internal representations and the external reality ... which is akin to the standard theory-fact fit criterion of the established paradigm. At the same time, there would not always be a one-to-one correspondence between such internal and external phenomena — meaning that something may exist in the external world that does not exist internally ... or vice versa. One way that the former case might be described is in terms of "ignorance", and one way that the latter case might be described is in terms of "imagination".

The above approach also provides a framework for understanding why our cognitive preconceptions are such vital determinants of what we perceive in our observations of the world. For the higher levels of the cognitive hierarchy that are inevitably involved with our mental activities would not be limited to formal *theories* (of any sort), but would also include all sorts of other values, goals, or *metaphysics*. We could also talk about *common sense* in terms of the collective set of cognitive entities, or maps or rules, for how the world works and how one interacts with it. Likewise, *experience* (including that derived from evolution) is just a general way of describing how we got to where we are, as reflected in our resulting cognitive hierarchies. In all of this, of course, we are also talking about the overwhelming importance of context, in terms of various hierarchical levels — no matter what content we are dealing with. As we have said previously, context is background to whatever we are focusing on.

The cognitive hierarchy is also the locus for overall memory, which would presumably be subject to being further broken down into our dimensions of analysis, in terms of reflecting the character of the world — stable vs. changing, inside vs. outside, matter vs. medium, static vs. moving, and dominant vs. recessive. Also, the difference between *short-term memory* and *long-term memory* may be related to the realm of the overall cognitive hierarchy we are concerned with, at any given time: The region of the nested hierarchy which is more active would be associated with short-term memory ... as opposed to the rest of the (non-active) hierarchy that would be associated with long-term memory.

At the same time, we could talk about a generalized state of awareness or *background of attention*, as opposed to a more distinct *focus of attention*. We can all sense what something feels like which is in the periphery of one's consciousness, as opposed to something that is central thereto — which is the pattern we have repeatedly cited in the case of the Rubens faces-vase image. But the dual pattern we are talking about would apply not only in spatial terms (e.g., foreground vs. background), but also in temporal terms (e.g., gradual buildup vs. sudden discharge).

One could approach the two established types of conditioning in terms that are familiar in our model. *Classical conditioning* would represent a spatial-like (simultaneous) static relationship, whereas *operant conditioning* would represent a temporal-like (successive) dynamic relationship. In either case, the *reinforcement* mechanism could be either *positive* or *negative*.

More generally, *learning* would be understood in terms of the usual hierarchy mechanisms that we have discussed: *generalization* upwards, whereby some larger whole (or pattern or entity) is synthesized from subsidiary components or examples thereof; and *discrimination* downwards, whereby some smaller subpatterns or components are specified within that larger whole (or pattern or entity). It should also be clear that learning doesn't come out of nothing: we are always starting somewhere, in the cognitive hierarchy — and once again, the notion of a blank slate (here, as mind) has very little meaning, for there is always an associative context or background from (or at least, in relation to) which new entities emerge and old entities exist.

Note that the dual mode of consciousness theme that we are working with — such as focus vs. background — is nothing new, with the present model; for such a conception has emerged repeatedly, in theories of the mind. Indeed, the overlap is complete, as between the 2FT approach and others in this regard — with the one mode representing that which is sharp, in relation to the other mode representing that which is diffuse.

This kind of dual mode model may have influenced Niels Bohr's thinking, as he developed his complementarity principle. Here is an excerpt from Max Jammer's book, in which he cites the writing of psychologist William James as being "one of the major factors which influences, wittingly or unwittingly, Bohr's formation of new conceptions in physics ..." (Jammer, 1989, p. 183) Jammer quotes from James:

> Like a bird's life, [the stream of consciousness] seems to be made of an alternation of flights and perchings.... The resting-places are usually occupied by sensorial imaginations of some sort, whose peculiarity is that they can be held before the mind for an indefinite time, and contemplated without changing; the places of flight are filled with thoughts of relations, static or dynamic, that for the most part obtain between the matters contemplated in the periods of comparative rest. Let us call the resting-places the "substantive parts," and the places of flight the "transitive parts," of the stream of thought. (James, *Principles of Psychology*, quoted in Jammer, 1989, p. 183)

In other writings, James uses a very interesting metaphor for visualizing the flow of consciousness, in terms of stones in a creek bed, with the water rushing past: The flow or motion of the water becomes analogous to the flow of consciousness (or attention), in relation to the creek bed stones (along with the banks of the creek, and the terrain beyond).

Now let us highlight a particular parallel with physics, in terms of the standard interpretation of Quantum Theory — according to which, "[T]he world changes in two fundamentally different ways: it pops and it flows." (Wolf, p. 219) Leaving aside whatever catalyst brings about the change from the one mode to the other, we are speaking about the *evolution* (*flow*) vs. *collapse* (*pop*) of the wave function. In Chapter 2, we talked about these in terms of a *buildup and potential* stage vs. a *discharge and actual* stage. Now we are citing the correspondence between the two modes of mental awareness and the two modes of the wave function, macro cognitive world and micro physical world respectively.

Consider next a very prominent set of cognitive patterns, which fit quite neatly into the overall two-force framework of the present model. This concerns the two cerebral hemispheres, which are the highly developed parts of the brain of advanced species where the most sophisticated workings of the mind are anchored physiologically. Research indicates that the two hemispheres each act, to some degree, as separate mirror entities. However, the manner in which they function is not at all identical: the two appear to play different roles in cognitive behavior, with each hemisphere performing aspects of certain tasks better than the other.

For most people, the *left hemisphere* is typically described as being the primary locus for cognitive activity like rational or logical thought, language processing, linear perspectives, and relating to the external world more in an objective or mathematical manner. By contrast, the *right hemisphere* is commonly described as being the primary locus for cognitive activity that is less analytic and more intuitive, that entails relationships and relating to the world more in a subjective or aesthetic manner. Not surprisingly, then, the left hemisphere has sometimes been depicted, in overly general ways, as "the scientist" and the right hemisphere as "the artist".

Part of the problem with these kind of descriptions is the implication that we are dealing with completely independent structures and functions; but any such conclusion is too simplistic. For one thing, the two hemispheres constantly interact by means of connecting brain tissue (especially the corpus callosum). In any case, there is no reason to suppose that each of the two hemispheres is engaged in unrelated activities from one another, but rather that they work together in tandem by approaching the same tasks from different vantage points or styles. Thus it is that the functional patterns we have mentioned above are sometimes fittingly described in terms which are compatible with the main theme of our model, as *complementary hemispheric specialization*.

Indeed, there is every reason to think that all the general issues which we have discussed regarding complementarity and the two forces of nature would come into play here: both hemispheres would always be involved and never just one; the two hemispheres would never be perfectly equal, but one would be dominant and the other recessive at any given time; such disparity would not be permanent, but would alternate between the two hemispheres;[27] each hemisphere would entail aspects of the other, yet the two are opposite to one another — plus, each hemisphere controls the opposite side of the body. Then there is the related same vs. different pattern that we keep coming

27 As far as the dominant-recessive issue goes, we would remind the reader how the pattern fits with our concept of *esymmetry* — according to which the two partners are not quite perfectly symmetrical to one another in any given instance. Beyond that, one might posit a distinction here between *structural esymmetries*, where there is more of a fixed physical dominance between bilateral partners; and *process esymmetries*, where there is more of a fluid functional dominance between bilateral partners — such as in the hemispheric specialization terms that we have been describing. The former category is admittedly reaching outside the cognitive realm, but it is relevant to the issues here. Thus, while there are typically fixed physical differences between various bilateral portions of the human body — e.g., feet, hands, eyes, hemispheres — such that they are not perfectly equal to one another, these physical or structural esymmetries would be tempered by the fact that the two partners still work in tandem, as they alternate dominance in functional or process esymmetry terms.

back to: each partner and the functions it performs are *different* from the other at a lower level — whereas they join together into the *same* overall single-organism unity at a higher level.

Let us summarize the issues here in terms that are familiar in the present model. Thus the left hemisphere would typically provide what we call a narrow and local *X*-view focus of the world, whereas the right hemisphere would typically provide what we call a broad and global *P*-view of context and background. Likewise, it may be that the left hemisphere better represents temporal relationships which exist sequentially over time, whereas the right hemisphere would better represent spatial relationships which exist simultaneously within time.

There are other ways in which "dual process" patterns are thought to manifest themselves in the realm of cognition — expressed both in terms of contrasting sets of personality characteristics (such as *introvert* vs. *extrovert*, or *sharpener* vs. *leveler*), as well as in terms of contrasting stages of maturational development; but also in terms of the human species' evolutionary development (sometimes described in terms of a progression from "implicit" to "explicit"). Of course, some of the contrasting patterns of cognition that we have discussed may also be culturally (and even specifically hereditarily) determined, as well as being gender-based. Nisbett and colleagues have researched differences between Eastern and Western thought, revealing patterns which are compatible with the distinction in cognitive style and modality that we are describing here (Nisbett; Chua, Boland, and Nisbett); and we also mentioned the gender issue in Chapter 9.[28]

It also seems reasonable to suppose that various types of consciousness can be at least partly understood in dual process terms. For example, certain phases of sleep (i.e., non-REM), meditation, and other passive types of cognitive activity could be described in terms of less focus and more background. On the other hand, dreaming (i.e., especially REM), high states of alertness, or certain drug-induced states, would correspond with more focus and less background. Of course, we already mentioned in the first chapter that sleep itself is the complement to wake, and that the two alternate in function.

In concluding this brief treatment of cognition, here is a chart that encompasses many of the complementary patterns that we have been discussing. Once again, this is a loose array of relative factors as between each pair of partners, and we are certainly not suggesting that all the items in either column go together.[29]

28 At least some of the style and modality differences mentioned in this and the following paragraph may be related to, if not explicable by, corresponding variations in hemispheric specialization patterns that we have been discussing.

29 As we have noted at various points in the book, which fundamental force a given attribute is associated with would be a relative determination, depending upon the particular meaning and context. Regarding some of the cases in the chart, we would propose a distinction between the overall character of the two forces themselves, and the character of the choices which some entity encounters in a particular situation. One example would be that how a given interaction is labeled may vary, depending upon perspective — with one partner seeing it as positive and attractive, and the other as negative and repelling. The important point we again emphasize here is the complementary nature of the analysis, regardless of how it is applied in any given instance.

X-Force of Structure	*P-Force of Process*
X-view	*P*-view
focus of attention	background of attention
sharp / precise / narrow figure	diffuse / hazy / broad ground
discrimination / pop / actual	generalization / flow / potential
sudden collapse	gradual evolution
left hemisphere	right hemisphere
rational / logical / linguistic	emotional / intuitive / pictorial
objective / deductive / divergence	subjective / inductive / convergence
temporal / sequential / parts / local	spatial / simultaneous / whole / global
operant conditioning	classical conditioning
negative reinforcement	positive reinforcement
avoidance / repel / dislike / hate	approach / attract / like / love
sharpener	leveler
extrovert	introvert
Western thought	Eastern thought
normal consciousness	alpha - meditation
wake / REM / dream	sleep / NREM / non-dream
consciousness	unconsciousness

The Beginning of Two-Force Theory

At long last now, we want to take stock of where we are with the new theory that we have been presenting — turning the tables from scrutinizing the established paradigm to scrutinizing ourselves. We start by asking what it is that would be required for a theory like this to be even considered as a serious model of nature. Then we talk about what our main task has been — and in what ways we have succeeded ... and failed. Related to these measures is the question of what it is reasonable to expect from an initial venture like this. Lastly, there are some comments from a more individual perspective.

But before turning to those topics, there is one small piece of conceptual housekeeping that we want to deal with first, in order to clarify a point that may be in the back of some readers' minds. This concerns what might be seen as an ultimate paradox regarding the nature of complementarity itself.

Complementarity and Not. The issue at hand is fairly simple: One of the most important aspects of the principle of complementarity that we have been developing is the notion that nothing exists all alone, completely apart from a relationship whereby its opposite counterpart also exists. But that definition must also be applicable to the very concept from which it derives — meaning that complementarity itself would not be immune from this pattern. Therefore, the principle must be turned back upon itself: complementarity only exists in relation to some form of non-complementarity. Yet what can this mean?

There are different ways of answering this question, depending on the particular context that is involved — including the general point that there can be various meanings for the same construct, which is an issue we talked about in the Multiple Meaning section of the first chapter. But in the most basic terms here, if complementarity is treated as the topmost principle of nature, then what does it exist in relation to?

We answer the question in hierarchy terms, in a manner that we have touched upon at different points in the book — such that complementarity would be understood in the upward direction of order and unity, in relation to its opposite in the downward direction of disorder and chaos. Another way that we have described these two tendencies is an upward direction towards same whole, as opposed to a downward direction towards different parts. These various characteristics might be charted as follows — bearing in mind the point we always make that it is the complementary categories which are primary, regardless of how the pair of factors are allocated in any particular instance:

Upward Direction	*Downward Direction*
complementarity	non-complementarity
order / unity / certainty	disorder / chaos / uncertainty
synthesis	thesis vs. antithesis
same whole	different parts
generalization	discrimination
X-view	*P*-view

Now having addressed the basic question about how complementarity itself can also be understood in an overall sense as existing in relation to its opposite, let us briefly consider the issue in terms of the main topic of the present chapter.

One of the arguments we have made repeatedly, at different points of the book, is that a good theory of nature must incorporate what we have called both sides of nature, in various respects — e.g., both order/unity/certainty and disorder/chaos/uncertainty. However, we do not want both sets to describe the character or composition of the models themselves, either individually or collectively — because we explicitly seek coherence and unity in our theories, not fragmentation and confusion. Yet this latter set is precisely what we are encountering with the established models, both individually and collectively. Thus we maintain that conventional science as a whole, with its hodgepodge of incompatible theories and constructs, corresponds very well with a non-complementary and chaotic world view. Such a non-complementary world is many forces, as we discussed at the beginning of this chapter — or Many Worlds, even, to mention a school of thought from Quantum Theory that we dismissed in Chapter 2 ... among so many other haphazard aspects of the conventional paradigm that we have rejected.

Of course, the problem is also reflected in the mainstream treatment of the complementarity concept itself. Thus, we have seen that its primary theoretical patron, Quantum Theory, applies complementarity in an inconsistent and almost spastic manner — to the point where it does not have the deep significance that it should, but rather is merely a kind of flag of convenience, to be run up the mast when the seas get rough ... such as to account for the wave-particle duality. All in all, there is a fundamental incompatibility — a "disconnect" — between the overall character of existing scientific models on the one hand, and a thoroughgoing understanding and integration of complementarity on the other. This is why a pervasive implementation of the principle of complementarity is impossible within the confines of orthodox theory, to begin with.

It is complementarity which has been neglected for far too long by science, in terms of the fundamental pattern that it represents. By contrast, complementarity is our star in this work — the unsung hero of nature — and so we do our best to highlight that role and importance.

The Challenge. Before we can take stock of what we have accomplished in this regard, we must first confront a sobering reality: How is it even remotely conceivable that any alternative theoretical structure, such as that which we have been presenting herein, could possibly be taken seriously — in an intellectual and psycho-socio-cultural milieu in which the deck is stacked mightily against it, and thus where the odds of success are virtually nil? Seen in these terms, of course, it will probably never happen — at least, it is hard to foresee how that could come about, given how massive are the entrenched institutional and theoretical structures, and how deep is the commitment thereto ... not only personally and professionally, but (again) socially and culturally.

Indeed, the reality is likely to be quite the contrary, in that this work will no doubt be pilloried in various ways — to the extent that it is even considered at all — as metaphysical drivel, as pretentious new age nonsense ... and as both unwarranted and impudent in the way it challenges well-established principles and icons, etc. The broad complementarity template which we are proposing will surely be dismissed as an abstract conceptual scheme that is far more fluff than substance — which can have no standing as the basis for a theory of science. The most that we may end up with, at least in the near future, is something that is disparaged as a minor but curious philosophical commentary — while having no impact whatsoever on the orthodox paradigm, leaving it completely unscathed and intact in its exclusive position at the apex of modern civilization.

Still, we can dream, can't we? Nature dreams! And, anyway, that is what we have been doing throughout this book, to some extent: the ideas that we have been presenting represent such a very different mode of thinking about the world, than that which conventional science normally presents to us, that they amount to a kind of dreaming — about the deeper patterns of nature ... and about what a viable model of science might look like that fits the criteria for which we have been arguing.[30] So while our realistic chances are not very promising at present, we can only hope that over time enough people will alter the way they think about these issues, in the direction that we have been advocating, such as to gradually bring about a change of outlook in the established system.

The main thing that must happen, of course, is for people to be able to look afresh at the goals of science, and determine whether the established paradigm is meeting those ends or not. If that cannot happen — if the existing rules are treated as beyond question and any reexamination is dismissed as being out of order — then there is nothing to talk about, and the conversation is over; and all the ideas that we have been pro-

30 Siu asks: "After all, are not scientific concepts but dreams petrified? Inability to dream strangles the source of the raw stuff of science." (Siu, p. 150)

posing in this book can be dispensed with ... and perhaps we are approaching the "end of science" which we described earlier in the chapter. On the other hand, if there is room to discuss whether we are on the right path towards the proper goals of science, then we can do two things: talk about what a better path might look like, as well as consider other labels for the path that we have been on. In other words, if science should be doing something else, then what is that something else — and what do we call that which we have been doing, if it is not "science"? Of course, we have tried to answer both of those questions in the previous chapter, and there is no point in repeating that discussion here.

As an aside, it should be emphasized that the model being proposed in this book is not the only possible candidate worth considering for the new path of science which we advocate: it is merely the best one we have been able to come up with — and even that in fairly primitive form, as we shall discuss shortly.

The Mission. We have been quite explicit regarding the goals of this work, from the very first pages: to develop a model that is based upon a pervasive theme — what we treat as the central organizing principle of nature — and to explain why the result would be consistent with a desirable set of criteria for doing science. The central principle, of course, is complementarity; and the main criteria are simplicity, generality, and coherence ... with the last point encompassing both compatibility and comprehensibility. Put in other words, we have been working to develop a theory of complementarity which is both simple in essence and general in scope; which is coherent within itself; and which makes sense in terms of the patterns of nature we see all around us, as well as those which we know (or can learn) about inside ourselves.

There are several distinct ways that we have attempted to implement or extend these features, in the sense of fulfilling other important functions and representing other desirable characteristics. For one thing, we have said that a good theory of science should describe realms beyond that at which it may be primarily targeted — in other words, it should describe all kinds of phenomena, and not just a limited or narrow domain. One particular application in this regard is that a good theory should entail, or at least be compatible with, a viable model of cognition. This, in turn, overlaps with another function, in that a good theory should not only describe the world outside of ourselves, but also the way we interact with and how we construe that world — meaning that it should be a theory of seeing as much as it is a theory of being. In addition, we have said that a good theory should also describe itself — the main themes should apply not only apply to nature as a whole, but should also be descriptive of the self-same model which is built upon them.

Finally, the model we are striving for must be able to fully represent competing aspects of nature, in a manner which is alive and dynamic in spirit as much as it is structured and consistent in form. In terms of the most nebulous factor, we want a picture of nature and the universe which is aesthetically pleasing in the broadest sense, including being both harmonious and fulfilling.

Putting all these characteristics together, what we have attempted to do throughout this book is to apply the complementarity thesis in as all-encompassing a manner as possible — both in the *being* sense that reality cannot be reduced to anything further than two essential force principles, in whatever form and combination they manifest themselves for any given situation; and in the *seeing* sense that complementarity acts as the binary set of glasses with which we view nature and the world.

Clearly, then, complementarity has been our guiding conceptual light, as we have attempted to tease apart the crucial characteristics which are common to all natural phenomena. We have used complementarity both as a prism and as a cudgel, in order to scrutinize (as well as to critique) many conventional constructs and assumptions ... thereby exposing the limited (i.e., arbitrary and one-sided) vision upon which they are based, as well as elucidating how they can be incorporated into the orderly dual-force model that we are proposing. In case after case, we have worked to show how existing theory has erred in minimizing, ignoring, or sweeping aside valid instances of the basic principle — at our collective conceptual peril.

In the end, it should be clear that complementarity has become the core criterion in its own right. In other words, the overall criteria we have spoken of (e.g., simplicity, generality, coherence) have devolved into complementarity itself. Clearly, we are talking about something that is far more fundamental than a quaint theoretical relic which is left over from the world of Niels Bohr.

So much for what we have tried to achieve. Next comes the downside — and there is plenty.

Uncomplementary Thinking. To start with, we are not there yet! In spite of all that we have tried to achieve in demonstrating the salience of complementarity in so many ways, we can still only claim to comprehend it in limited form — and therefore we have been able to implement the two-force theme with only partial success. For although it is superficially simple, the principle of complementarity that we have been trying to elucidate is not only startlingly profound in its meaning and scope, but in many cases it is also exceedingly subtle and elusive in its essence. All in all, the notion of an ever-present duality of nature is a very difficult conception to truly embrace, let alone to fully assimilate into our thinking.

For one thing, we must contend with the shortcomings of our conventional mind-set — and the issue here goes well beyond the content of scientific theories themselves, as well as ways of doing science. For we are talking about a pervasive non-complementary mode of thinking about the world, which we first noted in the opening chapter, that includes our deepest common conceptions about ourselves and the nature around us.[31] Clearly, it will take some time for us to develop, both intellectually and culturally, to the point where we can fully apprehend the two-force theme of the universe ... and the new mode of thought which goes with it.

In this regard, it may be that what we need most is new logic — a new kind of complementary logic. We noted in Chapter 2 that Bohr adopted a similar position — that, in Jammer's words, complementarity represented "a new logical instrument." (Jammer, 1989, p. 363)[32] One way of describing the result is to say that complementarity also

31 As we indicated earlier, there may be geo-cultural differences concerning such cognitive style issues.

functions as a kind of conceptual wedge — the "prism" and "cudgel" we mentioned above — which gives us leverage in perceiving or describing things ... because it is a consistent and persistent pattern that applies to all phenomena. In this regard, we maintain that complementarity represents an indispensable tool of analysis.

Speaking of Niels Bohr, it is fitting to pay homage one last time to this great pioneer's lofty vision in regards to complementarity. To this end, we present some extended comments from Holton's work:

> [O]ver the years Bohr came to regard the complementarity principle as more and more important, extending far beyond the original context in which it had been announced.... [For] the situation in quantum physics is only one reflection of an all-pervasive principle.... [I]t was the *universal* significance of the role of complementarity which Bohr came to emphasize.... During the last thirty years of his life, Bohr took many opportunities to consider the application of the complementarity concept in fields outside of physics.... Bohr's attempt to understand the unity of knowledge ... on the basis of complementarity could be seen as precisely the fulfillment of the desire to discover the "great inter-relationships among all areas of knowledge." (Holton, pp. 105, 134-135, and 138)

> Bohr had great expectations about the future role of complementarity. He upheld them with unshakable optimism, never discouraged by the scant response he got from our unphilosophical age.... Bohr declared, with intense animation, that he saw the day when complementarity would be taught in the schools and become part of general education. (Rosenfeld, 1963, quoted in Holton, p. 145)

We lament the unfortunate lack of progress which has been made in fulfilling the mission and vision that Niels Bohr laid out before us. For things have not progressed very far since Bohr first proposed the basic thesis much of a century ago. As Holton puts it:

> Bohr's aim has a grandeur which one must admire. But while his point of view is accepted by the large majority in physics itself, it would not be accurate to say that it is being widely understood and used in other fields.... Even those who in their professional work in physics have experienced the success of the complementarity point of view at first hand find it hard or uncongenial to transfer to other areas of thought and action, as a fundamental thematic attitude, the habit of accepting basic dualities without straining for their mutual dissolution or reduction. Indeed, we tend to be first of all reductionists ... Perhaps, also, it is just a matter of time — more time needed to assimilate a new themata widely enough ... (Holton, p. 138)

Time is not enough, of course — for complementarity is a tool that must be developed and honed in its own right. In the process, we must also identify and sweep away contrary premises and themes which only serve to impede and undermine the further

32 Also note Harris' comment that, "The metaphysics required by the new physics should, then, be first holistic and second, as a necessary consequence, dialectical.... It will be the guiding principle of the dialectical process ... [I]t will be nothing less than dialectical reason itself." (Harris, p. 171)

development of our thinking along the path of complementarity. Perhaps the chief culprit is the one Holton just mentioned — which we have referred to as an excessive (or hyper-) reductionism — which only has the effect of continuing to obscure the essential pattern of complementarity itself, by reducing it to a unitarity ... as we have described many times this book.

So it is that we continue to think and talk in terms of absolutes and one-sidedness, even though nature is also relative ... and dual. It seems like a dilemma that will always nag at us: there are two ways of seeing, and when we slip into one view we neglect the other — and feel like we can legitimately deny the other. But the one-sidedness which results is a limited vision, at best — a distorted one, that we must be constantly on guard against ... assuming our goal is to comprehend the whole. For whenever we fail to properly incorporate complementarity, there will be an inevitable gap in our analysis.

And failed we have, throughout this work — in who knows how many ways! The kind of limitations we have been discussing in this section have no doubt impacted a great many topics that we have touched upon in the book, whereby our own concepts and thinking must evolve more before we can progress further ... in pursuit of Bohr's dream. All in all, we cannot pretend to have fully grasped the relativisms which should be involved in all our interpretations, or to have correctly identified complements in different situations ... as well as having confused together different ways of viewing the same situation. The present work is also surely full of one-sided assertions, which must give way over time to a broader and more comprehensive two-sided approach to nature. In addition, there are no doubt various ways that we have engaged in flip-flopping between different aspects and perspectives — such as in some of our charts — in the same way for which we have faulted others. There is a long road ahead!

Compatibility Problems. There can be no doubt that there are various places in this book, and therefore in the current state of the new model, where our explanations may seem forced or dubious — a stretch or a strain, to put it mildly — and where the conceptual components do not fit together as harmoniously and coherently as they ultimately must. For example, the dimensions of analysis at the model's core are still not completely satisfactory — either in their conception, or in their presentation. Taken all together, there are many issues to deal with in terms of the present model's internal problems, contradictions, and failures of interpretation.

One way to talk about these issues is in terms of the compatibility factor, and our broader shortcomings in this regard. Thus, we have said that a good theory of science must be able to incorporate or be consistent with subsidiary theories, constructs, and data points — and of course the most notorious category of compatibility is that of the fact-theory fit. Regrettably, there are surely weaknesses and errors in the present work, in this regard. We can only seek forbearance: and so we offer a plea for leniency, partly on the basis of mitigating circumstances which must be taken into account.

To understand why, we can start with the obvious point that there is a virtually infinite array of fact and interpretation that can be drawn from and reconciled with, in the modern world of science and technology. Moreover, because of ever-increasing specialization, it is impossible for anyone to have anything approaching comprehensive knowledge of such facts and interpretations within any given realm ... let alone across many (not to mention all) such realms. The glut of information, which is exploding exponentially everywhere and every day, makes that patently impossible. In this regard, we have brought upon ourselves a plague of detail in science today — which may have the effect of hindering, rather than facilitating, our progress. As Wigner wrote long ago, "[S]o many phenomena are being discovered and investigated, so many new ideas are advanced, that no one can fully keep track of them.... In short, science begins to form such a large body that it threatens to fall to pieces." (Wigner, 1976, p. 4) Holton refers to "the growing flood of changing statistical data." (Holton, p. 416)

Kuhn also notes the difficulties which increasing specialization in science engenders, and he wonders where it will lead. "Does the number of incompatible viewpoints employed by the increasing number of communities of specialists grow with time? Unity of the sciences is clearly a value for scientists, but for what will they give it up?" (Kuhn, 1977, p. 289)

The quandary is that only the specialist can hope to encompass the range of factual details, which have emerged throughout the many areas of science and technology — and that can only be done within the narrow realm in which the specialist is immersed. On the other hand, only a generalist can try to bring all these details together, if the goal is a broad model of nature. Yet no generalist can know all the details: The scope required is too vast for all of these particulars to be subsumed in one mind. Consequently, any broad new theory must be imperfect in its initial rendering, for it will necessarily entail errors of detail — how could it be otherwise?

Actually, there is an even more important issue that needs to be highlighted, which makes the situation much more complex than this — for we are not merely talking about an endless array of isolated factual details which must be identified and incorporated by a general model of nature. That is because all of those factual details only exist as part of various conceptual frameworks. And the really big hurdle in developing an alternative vision is that to get through to the former, one must contend with the latter — including all the arbitrary constructs, arcane terminology, and theoretical incompatibilities which are rampant in the mainstream models. The essence of the problem is that in pursuing the individual targeted phenomenon or pattern, it is extremely difficult (if not impossible) to avoid getting drawn or sucked into the established theoretical mind-set which goes with it. Black hole, anybody?

The obvious consequence of such immersion is the likelihood that one will be swept into the wrong direction — and become absorbed by, as well as invested in, the established framework ... rather than remaining outside that framework and developing an alternative thereto. The effect of such paradigm immersion is clear — and the conflict should be apparent: the more fully one is absorbed inside the dominant conceptual box (i.e., orthodox science), in order to comprehend all the relevant constructs and data points which exist or are framed therein, the more difficult it becomes to think outside that box. Indeed, the more ensconced one is inside the box, the greater the conceptual leap that is necessary to get outside the box — and the less likely it is to happen ... in the sense of how high the box's walls are, which must be transcended. Nor is this disparity problem a minor one, if we are talking in terms of a major paradigm shift. For then we are not merely concerned with being able to recognize some certain anomalies within the existing framework, but rather of essentially being able to challenge the entire set of primary assumptions therein … as part of a virtual epiphany about its overall inadequacy.

All in all, there is a trap here, in terms of this conflict between the inside and outside of the theoretical box — even if it is only an implicit one. On the one hand, we are weighted down inside the box by an infinite mass of detail and interpretation, which any new model is required to account for; and yet on the other hand, we must be light enough to spring over the walls and spirit together an entirely new conception about how this mass of reality is to be construed. It is an almost impossible mission for a new model — the immense breadth and depth of constructs and phenomena that must be mastered, together with the creation of a radically new and powerful alternative conceptual framework that is able to encompass it … not to mention meeting the other broad requirements we have described for building a good theory of science.

Indeed, what we are talking about with this trap is not merely a methodological constraint — but a methodological stranglehold, which the dominant paradigm exerts … upon any conceivable challengers to the status quo.[33] The result is to effectively foreclose the possibility of developing any broad, new, unified theory about nature. For as long as science demands that such a theory must have all the answers — meet all the criteria, solve all the problems, integrate all the data and facts and concepts that we know about today (as well as make new predictions going forward) — and that this must be done at the very beginning, immediately upon the first entrance of the new theory, then it *must* fail!

Growing Theory. We maintain that such instant theoretical coming-of-age as we have been discussing has never happened in the past, and it never will in the future. There is no viable way that any serious alternative can ever emerge, which will meet all the conditions we have been discussing … in one fell swoop.

And why should it? Fully mature beings do not emerge from the womb: they must first be born, and then must develop. Likewise, a full-grown, comprehensive model of nature does not fall from heaven. A good theory must evolve — it must be given the space and time to grow, so that it is not effectively stillborn. Therefore, it is important to have a realistic set of expectations … as to what it is reasonable for the first rendition of a new theory to accomplish, right from the start. Some degree of tolerance is essential, if the new theory is to really see the light of day — and have any chance of ultimately developing to the point of meeting the high expectations which we have been discussing.

In a similar vein, Bohm says that, "[I]n science it is generally necessary to begin with vague general notions at an early stage, and to articulate these notions later as the

33 Of course, we are not even talking about a whole other set of psychological and social factors which could intervene and add to the problem. We discussed these in Chapter 10, concerning the pressures to embrace the theoretical status quo which would apply if one's professional life activity forms an organizational part of the reigning paradigm … not to mention if one's very livelihood and status are completely dependent thereupon. At the very least, one could call this "biting from the hand which one is fed by" — resulting in a terrible conflict of motivation and goals!

subject develops." (Bohm, 1969, p. 104) Or as Dyson puts it, "We should have learned by now that ideas that appear at first sight to be vague and metaphysical sometimes turn out to be true." (Dyson, quoted in Morris, p. 230) Even more, the new ideas may initially seem very improbable: "For any speculation that does not at first glance look crazy, there is no hope." (Dyson, quoted in Hecht, p. 268)

As to the conventional argument that any new theory of science should make better predictions than its predecessor, there is clear acknowledgement in the field that this may be unrealistic. Thus, Sachs talks about "criticism which could be applied to any newly instituted approach that the theory has thus far given very little that it can claim over other theories." (Sachs, 1973, p. 107)

Kuhn writes about the "clumsiness inevitable in a first venture ..." (Kuhn, 1970, p. 33) He also says that, "[I]f a new candidate for paradigm had to be judged from the start by hard-headed people who examined only relative problem-solving ability, the sciences would experience very few major revolutions." (Kuhn, 1970, p. 157) For his part, d'Espagnat writes:

> Is it not the case that genuinely new ideas, at the moment when they take shape in the human mind, may well have no consequences that can be verified or falsified by technical means? It is plausible that to make major conceptual leaps and establish real connections calls for a less rigid methodology. (d'Espagnat, pp. 67-68)

Regarding the present endeavor, there should be no confusion: what is presented in this book is a sketch — not a blueprint — for a new model of nature. It is intended to be almost an illustration, rather than anything definitive. So this is very much a work in progress, a theory in evolution ... a model in making. Our hope is that the rudimentary framework which we have put forward will eventually mature into a complete theory. Therefore, if we have merely succeeded in demonstrating the power and potential of the complementarity paradigm, then we will have accomplished our mission — in spite of all the obvious shortcomings herein.

A Personal Adieu. And now for a change of tone, in these last moments of the book — or a change of pronoun, I should say — moving from the first person plural in which all of this work has been written ... to the first person singular here at the very end. That latter is a posture which I would have been loathe to adopt in the body of this book, for various reasons that I would now like to explain.

Some of it is a matter of style and personality, and a question of taste and comfort, in terms of how one presents oneself and one's ideas to other people. It has been easier for me to discuss the issues here as though I was speaking from the perspective of something larger than myself — rather than as a lone individual, trying to sell or promote his own private views. Plus, I have not wanted the factor of ego to unduly intrude on the presentation of content, and I saw no way to avoid that with the first person singular form. In this regard, it would have felt very arrogant, and quite distracting, to be constantly referring to *I* or *me* throughout this book. I believe that the issues are important enough to be considered on their merit — and strong enough to stand on their own —

without being continually obscured by repeated references to an individual personage as part of the discussion.

Aside from that, such a collective presentational approach represents a perfectly legitimate mode of classical argumentation, which seems completely appropriate for this kind of material. In some ways, I have tried to carry on this discussion as though there were a group of us who had been conversing about these matters — and I happened to be the one reporting on the outcome. In the same vein, I have tried to present the material as representing a new school of thought, of which I happened to be the interlocutor — and in this sense I was speaking for a larger group of persons than myself alone ... which, of course, I hope is and becomes increasingly the case. Thus, while I may have been the person to bring many of these ideas together here, and am clearly responsible for their current form, I also feel confident that there are many other people who (do or will) share some or all of these views, in various respects.

Moreover, while the model itself is new, in terms of the wide scope and particular manner in which the main precepts are presented and applied here, there is also much that is old, as well — and, again, it would have felt presumptuous of me to speak as though they were mine alone. For example, as was mentioned in the very first chapter, the essential concept of two opposing forces in nature goes way back to ancient China. Plus, there have been repeated references in this book (including just above) to the invaluable contributions which Niels Bohr made in developing and applying the idea of complementarity, especially in the early days of Quantum Theory. So there have obviously been various predecessors who developed some of these ideas before me, and it is impossible to imagine how my own thinking could have ever progressed as far as it has without these trailblazers ... in whose footsteps I have been following.

Finally, it's very lonely out here ... when I think of myself as only one person having these ideas, and challenging the massive edifice of science! As an individual, I represent the tiniest speck of both power and information, as compared to the gargantuan structure which is science today. Science speaks with a commanding and largely unified voice — certainly in terms of its ideology and its mission, even as it argues about details here and there, and while different people vie for territory and supremacy in particular realms. By contrast, again, I am a very small voice in the wilderness. So perhaps I've needed the feeling of support which the others of "we" provide.

Nonetheless, I apologize to anyone who has been offended by my manner, with this usage: I have certainly not intended it in any kind of pretentious or gratuitous way. For that matter, I also regret any other offense which I have caused in this book — e.g., by what is perceived to be a haughty or overly-critical (if not sarcastic) style of presentation, regarding the issues themselves or the personalities that are associated with them.

For my part, I hope to pass the baton off to others, who can help in making corrections and clarifications — and in carrying the model forward and further in ways that are well beyond my own comprehension or vision or abilities. I am only one individual, whose knowledge and experience are far too limited in so many ways.

Alas, however, it is now time to go, and so I bid you adieu — and send thanks again for your company on this journey!

Bibliography

________ (1977) "Binary Experiments Confirm that Light Velocity is Independent of Source", *New Scientist*, 76:422, 17 November.

________ (1978) "New Limit on Constancy of Velocity of Light", *Physics Today*, 31:19, March.

Achuthan, P., Shankara, T.S., and Venkatesan, K. (1979) "Ether — as Advocated by Einstein and Others", *Speculations in Science and Technology* 2:277-284, 1979.

Alvåger, T., Farley, F.J.M., Kjellman, J., and Wallin, I. (1964) "Test of the Second Postulate of Special Relativity in the GeV Region", *Physics Letters*, 12:3, pp. 260-262.

Andrews, Donald H. (1974) *Chemistry: A Humanistic View*, New York: McGraw-Hill.

Asimov, Isaac (1966) *Understanding Physics: Light, Magnetism, and Electricity*, New York: Mentor, 1969.

Bailey, J. et al. (1977) "Measurements of Relativistic Time Dilation for Positive and Negative Muons in a Circular Orbit", *Nature*, 268:301-305, 28 July.

Baker, Adolph (1970) *Modern Physics and Antiphysics*, Reading, Mass.: Addison-Wesley, 1972.

Baker, Jeffrey and Allen, Garland (1979) *A Course in Biology*, 3rd edition, Reading, Mass.: Addison-Wesley.

Bastin, Ted (1969) "A General Property of Hierarchies", in Waddington, C.H., ed.

Bateson, Gregory, *Mind and Nature*, 1979, New York: E.P. Dutton.

Bergstrand, Erik (1966) "Velocity of Light", in Besancon, Robert M., ed., *The Encyclopedia of Physics*, New York: Van Nostrand Reinhold, 987-992.

Bohm, David (1957) *Causality and Chance in Modern Physics*, London: Routledge & Kegan Paul, Philadelphia: University of Pennsylvania Press, 1980.

Bohm, David (1969) remarks in Waddington, C.H., ed. (1969).

Bohm, David (1980a) *Wholeness and the Implicate Order*, London: Routledge & Kegan Paul, 1982.

Bohm, David (1980b) "On Insight and its Significance, for Science, Education, and Values", in Sloan, Douglas, ed., *Education and Values*, New York: Teachers College Press, Columbia University.

Bohm, David and Hiley, B.J. (1993) *The Undivided Universe*, New York: Routledge.

Bohr, Niels (1948) "On the Notions of Causality and Complementarity", *Dialectica*, 2:312-319.

Bondi, Hermann (1964) *Relativity and Common Sense*, New York: Doubleday, New York: Dover edition, 1980.

Brecher, Kenneth (1977) "Is the Speed of Light Independent of the Velocity of the Source?", *Physical Review Letters*, 39:1051-1054.

Bridgman, P.W. (1927) *The Logic of Modern Physics*, New York: Macmillan, 1960.

Bridgman, P.W. (1983) *A Sophisticate's Primer of Relativity*, 2nd edition, Middletown, Conn.: Wesleyan University Press.

Brillouin, Léon (1970) *Relativity Reexamined*, New York: Academic Press.

Brody, B.A. (1980) "Towards an Aristotelian Theory of Scientific Explanation", in Klemke, E.D. et al., eds. (1980).

Bromley, Alan (1997) NPR *Science Friday* program, 14 March.

Brown, G. Spencer (1972) *Laws of Form*, New York: Julian Press.

Bub, Jeffrey (1988) "How to Kill Schrödinger's Cat", in Kitchener, Richard F., ed. (1988).

Buckley, Paul and Peat, F. David (1979) *A Question of Physics: Conversations in Physics and Biology*, Routledge & Kegan Paul.

Builder, Geoffrey (1957) "The Resolution of the Clock Paradox", *Australian Journal of Physics*, 10:246-262.

Builder, Geoffrey (1958a) "Ether and Relativity", *Australian Journal of Science*, 11:279-297.
Builder, Geoffrey (1958b) "The Constancy of the Velocity of Light", *Australian Journal of Physics*, 11:457-480.
Builder, Geoffrey (1959) "The Fundamental Concepts of Relativity Theory", *Australian Journal of Science*, 22:87-97.
Builder, Geoffrey (1960) "Relative Velocity", draft manuscript.
Bunge, Mario (1967) *Quantum Theory and Reality*, New York: Springer.
Calder, Nigel (1977) *The Key to the Universe*, London: British Broadcasting Corporation.
Calder, Nigel (1979) *Einstein's Universe*, London: British Broadcasting Corporation.
Capek, Milic (1975) "Relativity and the Status of Becoming", *Foundations of Physics*, 5:607-617, December.
Capek, Milic (1988) "Do the New Concepts of Space and Time Require a New Metaphysics?", in Kitchener, Richard F., ed. (1988).
Capra, Fritjof (1975) *The Tao of Physics*, Boulder, Colo.: Shambala, New York: Bantam edition, 1977.
Capra, Fritjof (1983) *The Turning Point*, New York: Simon and Schuster, Toronto: Bantam Books.
Capra, Fritjof (1988) "The Role of Physics in the Current Change of Paradigms", in Kitchener, Richard F., ed. (1988).
Cartwright, Nancy (1983) *How the Laws of Physics Lie*, Oxford: Clarendon Press, 1989.
Cassidy, Harold G. (1976) "Henry Margenau's Philosophy as Frame of Reference in Teaching Science", in Laszlo, ed. (1976).
Cassirer, Ernst (1923) *Substance and Function and Einstein's Theory of Relativity*, Chicago: Open Court.
Choi, Charles Q. (2007) "Not So Neutral Neutron", *Scientific American*, 297:37, December.
Christensen, Ferrel (1981) "Special Relativity and Space-like Time", *British Journal for the Philosophy of Science*, 32:37-53, March.
Chua, H.F., Boland, J.E., and Nisbett, R.E. (2005) "Cultural Variation in Eye Movements During Scene Perception", *Proceedings of the National Academy of Sciences of the U.S.A.*, 102:12629-12633, 30 August.
Close, Frank (1981) BBC World Service *Discovery* program, 22 September.
Clotfelter, Beryl E. (1976) *The Universe and Its Structure*, New York: McGraw Hill.
Cohen, M.H. et al. (1977) "Radio Sources with Superluminal Velocities", *Nature*, 268:405-409, 4 August.
Coveney, Peter and Highfield, Roger (1991) *The Arrow of Time*, New York: Fawcett Columbine.
Davies, Paul (1977) *Space and Time in the Modern Universe*, Cambridge: Cambridge University Press, 1978.
Davies, Paul (1979) *The Forces of Nature*, Cambridge: Cambridge University Press.
Davies, Paul (1980a) *Other Worlds*, London: Dent.
Davies, Paul (1980b) "Why Pick on Einstein?", *New Scientist*, 87:463-465, 7 August.
Davies, Paul (1981) BBC World Service *Discovery* program, 22 October.
Davies, Paul (1983) *God and the New Physics*, New York: Simon and Schuster, Touchstone edition, 1984.
de Beauregard, Olivier C. (1988) "Space-Time and Probability", in Kitchener, Richard F., ed. (1988).
d'Espagnat, Bernard (1989) *Reality and the Physicist*, New York: Cambridge University Press.
Dingle, Herbert (1957) "The 'Clock Paradox' of Relativity", *Nature*, 179:865-866, April 27.
Dingle, Herbert (1960a) "The Doppler Effect and the Foundations of Physics (I)", *British Journal for the Philosophy of Science*, 11:11-31, May.
Dingle, Herbert (1960b) "The Doppler Effect and the Foundations of Physics (II)", *British Journal for the Philosophy of Science*, 11:113-129, August.
Dingle, Herbert (1972) *Science at the Crossroads*, London: Martin Brian & O'Keefe.

Einstein, Albert (1905) "On the Electrodynamics of Moving Bodies", reprinted in Lorentz, H.A., Einstein, A., Minkowski, H., and Weyl, H., *The Principle of Relativity*, New York: Dodd, Mead and Company, 1923.

Einstein, Albert (1920) *Ether and the Theory of Relativity*, London: Butler & Tanner.

Einstein, Albert (1950) *Out of My Later Years*, Totowa, New Jersey: Littlefield, Adams & Co., 1967.

Einstein, Albert (1954) *Ideas and Opinions*, New York: Crown, 1963.

Einstein, Albert (1956) *The Meaning of Relativity*, Princeton, New Jersey: Princeton University Press, 1974.

Einstein, Albert (1961) *Relativity: The Special and the General Theory*, New York: Crown.

Einstein, Albert and Infeld, Leopold (1938) *The Evolution of Physics* Simon and Schuster, 1966 Touchstone edition.

Evans, Melbourne G. (1962) "The Relativity of Simultaneity: A Critical Analysis", *Dialectica*, 16:61-82.

Feyerabend, Paul (1980) "How to Defend Society Against Science", in Klemke, E.D. et al., eds. (1980).

Feynman, Richard (1949) "The Theory of Positrons", *Physical Review*, 76:749-759, 15 September.

Feynman, Richard (1964) *The Feynman Lectures on Physics*, Vol. II, New York: Addison-Wesley, 1966 edition.

Feynman, Richard (1965) *The Character of Physical Law*, London, British Broadcasting Corporation, Cambridge, Mass.: MIT edition, 1967.

Finkelstein, David (1988) "The Universal Quantum", in Kitchener, Richard F., ed. (1988).

Frank, Philipp G. (1980) "The Variety of Reasons for the Acceptance of Scientific Theories", in Klemke et al., eds. (1980).

French, A.P. and Taylor, E.F. (1978) *Introduction to Quantum Physics*, New York: Norton.

Friedman, Jonathan R., Patel, Vijay, Chen, W., Tolpygo, S.K., and Lukens, J.E. (2000) "Quantum Superposition of Distinct Macroscopic States", *Nature*, 406:43-46, 6 July.

Frisch, Otto (1972) *The Nature of Man*, London: Thames & Hudson.

Froome, K.D. and Essen, L. (1969) *The Velocity of Light and Radio Waves*, New York: Academic Press.

Fujiwara, Kunio (1980) "Is the Light Velocity in Vacuum Really a Constant?", *Foundations of Physics*, 10:309-331, April.

Furley, David J. (1976) "Aristotle and the Atomists on Motion in a Void", in Machamer, Peter K. and Turnbull, Robert G., eds., *Motion and Time, Space and Matter*, Ohio State University Press.

Galison, Peter (2003) *Einstein's Clocks, Poincaré's Maps*, New York: Norton.

Gardner, Martin (1981) *Science: Good, Bad, and Bogus*, Buffalo, N.Y. : Prometheus Books.

Garisto, Robert (1998) "Curling Up Extra Dimensions in String Theory", *Physical Review Focus*, 9 April.

Gatlin, Lila L. (1972) *Information Theory and the Living System*, New York: Columbia University Press.

Geroch, Robert (1978) *General Relativity from A to B*, Chicago: University of Chicago Press.

Gill, T.P. (1965) *The Doppler Effect* London: Logos Press.

Glashow, Sheldon (1994) BBC World Service report by David Whitehouse on *Newshour* program, 23 June.

Goodwin, Brian C. (1969) "A Statistical Mechanics of Temporal Organization in Cells", in Waddington, C.H., ed. (1969).

Grieder, Alfons (1977) "Relativity, Causality and the 'Substratum' ", *British Journal for the Philosophy of Science*, 28:35-48, March.

Gruber, Howard E. (1978) "Darwin's 'Tree of Nature' and Other Images of Wide Scope", in Wechsler, ed. (1978).

Hanson, N.R. (1962) "Postscript", in *Quanta and Reality*, American Research Council.

Hanson, N.R. (1980) "Observation", in Klemke, E.D. et al., eds. (1980).

Harris, Errol E. (1988) "Contemporary Physics and Dialectical Holism", in Kitchener, Richard F., ed. (1988).

Harrison, David (1979) "What You See is What You Get!", *American Journal of Physics* 47:576-582, July.
Healey, Richard (1989) *The Philosophy of Quantum Mechanics*, New York: Cambridge University Press.
Hecht, Eugene (1980) *Physics in Perspective*, Reading, Mass.: Addison-Wesley.
Heisenberg, W. (1925) "Quantum-Theoretical Re-interpretation of Kinematic and Mechanical Relations", in van der Waerden, ed., *Sources of Quantum Mechanics*, Amsterdam: North-Holland Publishing, 1967.
Heisenberg, W. (1955) "The Development of the Interpretation of the Quantum Theory", in Pauli, Wolfgang, ed., *Niels Bohr and the Development of Physics*, New York: McGraw-Hill.
Hewitt, Paul G. (1981) *Conceptual Physics*, 4th edition, Boston: Little Brown.
Holton, Gerald (1988) *Thematic Origins of Scientific Thought*, revised edition, Cambridge: Harvard University Press.
Holton, Gerald and Brush, Stephen G. (1973) *Introduction to Concepts and Theories in Physical Science*, 2nd ed., Reading, Mass.: Addison-Wesley.
Horgan, John (1995) NPR *Science Friday* program, 3 November.
Horgan, John (1996) *The End of Science*, Reading, Mass.: Addison-Wesley.
Horgan, John (2003) from internet discussion on whether a unified theory will be attained by 2020 [http://www.longbets.org/12].
Hospers, John (1980) "Theory", in Klemke, E.D. et al., eds. (1980).
Hoyle, Fred (1985) BBC World Service *Behind the Universe* program, 8 April.
Hsu, J.P. (1979) "The Analysis of Time: Is the Relativistic Time Unique?", *Foundations of Physics*, 9:55-69, February.
Ives, Herbert E., in Hazelett, Richard, and Turner, Dean, eds. (1979) *The Einstein Myth and the Ives Papers*, Old Greenwich, Conn.: Devin-Adair.
Jaeger, F.M. (1917) *Lectures on the Principle of Symmetry*, Amsterdam: Elsevier.
Jammer, Max (1957) *Concepts of Force*, Cambridge, Mass.: Harvard University Press, New York: Harper Torchbook edition, 1962.
Jammer, Max (1974) *The Philosophy of Quantum Mechanics*, New York: Wiley.
Jammer, Max (1989) *The Conceptual Development of Quantum Mechanics*, 2nd edition.
Jauch, J. M. (1973) *Are Quanta Real?*, London: Indiana University Press.
Johnson, George (1998) "Almost in Awe, Physicists Ponder 'Ultimate' Theory", *New York Times*, 22 September.
Jones, David (1983) BBC World Service *Science Through the Looking Glass* program, 22 August.
Jung, C.G. (1955) *Synchronicity: An Acausal Connecting Principle*, London: Routledge & Kegan Paul.
Kaufmann, William J. (1973) *Relativity and Cosmology*, New York: Harper and Row.
Keating, Richard E. (1977) "Clock Paradox", *McGraw-Hill Encyclopedia of Science and Technology*, 4th edition, 3:219-219D.
Kitchener, Richard F. (1988) "Introduction: The World View of Contemporary Physics", in Kitchener, Richard F., ed. (1988).
Kitchener, Richard F., ed. (1988) *The World View of Contemporary Physics*, Albany: State University of New York Press.
Klemke, E.D., Hollinger, Robert, Kline, A. David (1980) *Introductory Readings in the Philosophy of Science*, Buffalo New York: Prometheus Books.
Koestler, Arthur (1967) *Ghost in the Machine* New York, Macmillan.
Korenman, Victor (1972) "Letters", *Physics Today*, January 1972, p. 15.
Kuhn, Thomas (1970) *The Structure of Scientific Revolutions*, 2nd enlarged edition, Chicago: University of Chicago Press.
Kuhn, Thomas (1977) *The Essential Tension*, Chicago: University of Chicago Press.
Laszlo, Ervin (1976) "Nonempirical Criteria in the Development of Science", in Laszlo, ed. (1976).
Laszlo, Ervin and Sellon, Emily B., eds. (1976) *Vistas in Physical Reality*, New York: Plenum.

Leclerc, Ivor (1988) "The Relation Between Natural Science and Metaphysics", in Kitchener, Richard F., ed. (1988).

Leibfried, D., Knill, E., Seidelin, S., Britton, J., Blakestad, R. B., Chiaverini, J., Hume, D. B., Itano, W. M., Jost, J. D., Langer, C., Ozeri, R., Reichle R., and Wineland, D.J. (2005) "Creation of a Six-atom 'Schrödinger Cat' State", *Nature*, 438:639-642, 1 December.

Lewis, Gilbert N. (1908) "A Revision of the Fundamental Laws of Matter and Energy", *The London, Edinburgh and Dublin Philosophical Magazine and Journal of Science*, 16:705-717.

Lewis, Gilbert N. (1926) "The Nature of Light", *Proc. N. A. S.*, 12:22-29.

Lewis, Gilbert N. and Tolman, Richard C. (1909) "The Principle of Relativity, and Non-Newtonian Mechanics", *The London, Edinburgh and Dublin Philosophical Magazine and Journal of Science*, 28:510-523.

Lindsay, R. Bruce (1967) "Arbitrariness in Physics", *Physics Today*, 20:23-26, December.

Ljung, Donovan A. (1972) "Letters", *Physics Today*, p. 13, January.

Lockwood, Michael (1989) *Mind, Brain and the Quantum*, Cambridge: Basil Blackwell.

MacRoberts, D.T (1980) "Clocks and Special Relativity", *Speculations in Science and Technology*, 3:365-378.

March, Robert H. (1970) *Physics for Poets*, New York: McGraw-Hill.

Marder, Leslie (1971) *Time and the Space-traveller*, London: Allen and Unwin.

Margenau, Henry (1964) *Ethics and Science*, Princeton, N.J.: Van Nostrand.

Margenau, Henry (1978) *Physics and Philosophy*, Dordrecht, Holland: D. Reidel Publishing Co.

Marion, Jerry B. (1979) *General Physics with Bioscience Essays*, New York: Wiley.

Maxwell, Nicholas (1984) *From Knowledge to Wisdom: a Revolution in the Aims and Methods of Science*, Oxford: Basil Blackwell.

Mazrui, Ali (1979) BBC Reith Lecture, November.

McCaig, Malcolm (1977) *Permanent Magnets in Theory and Practice*, London: Plymouth.

McDermott, Jeanne (1980) "Lynn Margulis — Unlike Most Microbiologists", *CoEvolution Quarterly*, 25:31-38.

Mercier, A., Tredder, H.J., and Yourgrau, W. (1979) *On General Relativity*, Berlin: Akademie-Verlag.

Midgley, Mary (1989) *Wisdom, Information, and Wonder*, New York: Routledge.

Miller, Arthur (1983) "Introduction", in Bridgman, P.W. (1983).

Miller, G. Tyler, Jr. (1971) *Energetics, Kinetics, and Life: An Ecological Approach*, Belmont, Calif: Wadsworth Publishing Co.

Miller, Richard W. (1987) *Fact and Method*, Princeton: Princeton University Press.

Morris, Richard (1987) *The Nature of Reality*, New York: Noonday Press.

Morrison, Philip (1978) "On Broken Symmetries", in Wechsler, ed. (1978).

Mortimer, Charles E. (1979) *Chemistry: A Conceptual Approach*, New York: Van Nostrand.

Murchie, Guy (1967) *Music of the Spheres*, New York: Dover.

Neilburger, Morris, Edinger, James, and Bonner, William (1973) *Understanding Our Atmospheric Environment*, San Francisco: W.H. Freeman.

Nisbett, Richard E. (2003) *The Geography of Thought*, New York: Free Press.

Noonan, Thomas W. (1972) "Letters", *Physics Today*, p. 15.

Nordenson, Harald (1969) *Relativity, Time and Reality*, London: George Allen & Unwin.

Oppenheimer, J. Robert (1954) *Science and the Common Understanding*, New York: Simon and Schuster, 1966.

Oppenheimer, J. Robert (1983), segment on Oppenheimer from NPR *All Things Considered* program, 5 September.

O'Rahilly, Alfred (1965) *Electromagnetic Theory*, Vol. I, New York: Dover.

Overbye, Dennis (2003) "Pure Math, Pure Joy", *New York Times*, p. 7, July 29.

Palacios, Juan (1963) "Inner Inconsistency of Einstein's Theory", *Rev. de la Academia de Madrid*, 57:586.

Parish, Leonard (1977) *The Logical Flaws of Einstein's Relativity*, London: Cortney Publications Luton.

Peat, F. David (1989) "Quantum Theory, Creativity and Consciousness", New Dimensions Radio interview with Michael Toms (Host), 22 June.

Penrose, Roger (1980) BBC World Service *Discovery* program, 2 October.

Phipps, Thomas E. (1979) Letter to the Editor, *Speculations in Science and Technology*, 2:326.

Pierce, John R. (1974) *Almost All About Waves*, Cambridge: MIT Press.

Pitt, Valerie H., ed. (1977) *The Penguin Dectionary of Physics*, New York: Penguin, 1978.

Podlaha, M.F. (1978) "Some Suggestions in Relativity", *Indian Journal of Theoretical Physics*, 26:181-184, September.

Podlaha, M.F. (1979) "Light Signal Synchronisation and Clock Transport Synchronisation in the Theory of Relativity", *British Journal for the Philosophy of Science*, 30:376-380, December.

Polkinghorne, John (1996) *Beyond Science*, New York: Cambridge University Press, Canto edition, 1998.

Popper, Karl (1965) *Conjectures and Refutations*, New York: Basic Books, Harper Torchbooks edition, 1968.

Popper, Karl (1968) *The Logic of Scientific Discovery*, revised edition, New York: Basic Books, Harper Torchbooks edition.

Popper, Karl (1981) *Objective Knowledge*, revised edition, Oxford: Oxford University Press, Clarendon Press edition.

Popper, Karl (1982) *Quantum Theory and the Schism in Physics*, Totowa, NJ: Rowman and Littlefield.

Post, E.J. (1967) "Sagnac Effect", *Review of Modern Physics*, 39:475-493.

Postman, Neil (1994) "Technology in Culture", Talk at Portland City Club, 8 April.

Press, Frank and Siever, Raymond (1974) *Earth*, San Francisco: Freeman.

Preston, M.A. and Bhaduri, R.K. (1975) *Structure of the Nucleus*, Reading, Mass.: Addison-Wesley.

Prigogine, Ilya (1977) "Physics and Metaphysics", *Advances in Biological and Medical Physics*, Vol. 16, New York: Academic Press.

Prigogine, Ilya (1980) *From Being to Becoming*, San Francisco: Freeman.

Prigogine, Ilya (1988) "The Rediscovery of Time", in Kitchener, Richard F., ed. (1988).

Prokhovnik, S.J. (1963) "The Case for an Aether", *British Journal for the Philosophy of Science*, 14:195-207, November.

Prokhovnik, S.J. (1979) "An Introduction to the Neo-Lorentzian Relativity of Builder", *Speculations in Science and Technology*, 2:225-229.

Redhead, M. L. G. (1977) "Wave-Particle Duality", *British Journal for the Philosophy of Science*, 28:65-74, March.

Rees, Martin J. (2003) *Our Final Hour*, New York: Basic Books.

Rindler, Wolfgang (1977) *Essential Relativity*, New York: Springer.

Rogers, Eric M. (1960) *Physics for the Inquiring Mind*, Princeton: Princeton University Press, 1977.

Roos, C.E. et al. (1980) "Lifetimes and Longitudinal Acceleration", *Nature*, 286:244-245, 17 July.

Rosser, W.G.V. (1978) "The Clock Hypothesis and the Lorentz Transformations", *British Journal for the Philosophy of Science*, 29:349-353, December.

Roszak, Theodore (1972) *Where the Wasteland Ends*, Garden City, New Jersey: Doubleday & Co., Anchor Books edition, 1973.

Ruch, E.A. (1958) *Space and Time: A Comparative Study of the Theories of Aristotle and Albert Einstein*, Pretoria.

Rucker, Rudy von (1981) "Faster than Light, Slower than Time", *Speculations in Science and Technology*, 4:375-383.

Ruderfer, M. (1962) "Re-Evaluation of the Existence of an Ether", *Proceedings of the IRE*, 50:325-326, March.

Ruderfer, M. (1975) "Neutrino Structure of the Ether", *Lettere Al Nuovo Cimento*, 13:9-13.

Ruderfer, M. (1979a) "Introduction to Ives' 'Derivation of the Lorentz Transformations' ", *Speculations in Science and Technology*, 2:243-257.

Ruderfer, M. (1979b) "Comments on the Twin Paradox and Allen's Experiments", *Speculations in Science and Technology*, 2:327-330.

Ruderfer, M. (1979c) "Further Remarks on the Twin Paradox and Allen's Experiments", *Speculations in Science and Technology*, 2:333-337.

Rudner, Richard (1980) "The Scientist *Qua* Scientist Makes Value Judgments", in Klemke, E.D. et al., eds. (1980).

Sachs, Mendel (1971) "A Resolution of the Clock Paradox", *Physics Today*, 24:23-29, September.

Sachs, Mendel (1973) *The Field Concept in Contemporary Science*, Springfield, Ill: Thames.

Sachs, Mendel (1974) *Ideas of the Theory of Relativity*, New York: Wiley.

Sachs, Mendel (1988) *Einstein vs. Bohr*, La Salle, Illinois: Open Court.

Sadeh, D. et al. (1968) "The Effect of Mass on Frequency", *Science*, 161:567-569, 9 August.

Sagan, Carl (1979) *Broca's Brain*, New York: Random House, Ballantine Books edition, 1981.

Schaffner, Kenneth F. (1974) "Einstein vs. Lorentz: Research Programmes and the Logic of Comparative Theory Evaluation", *British Journal for the Philosophy of Science* 25:45-78, March.

Schlegel, Richard (1973a) "An Interaction Interpretation of Special Relativity Theory, Part II", *Foundations of Physics*, 3:277-295, September.

Schlegel, Richard (1973b) "Flying Clocks and the Sagnac Effect", *Nature* 242:180, March.

Schlegel, Richard (1974) "Comments on the Hafele-Keating Experiment", *American Journal of Physics*, 42:183-187, March.

Schlegel, Richard (1980) "The Light Clock: Error and Implications", *Foundations of Physics*, 10:345-351, April.

Schoffeniels, E. (1976) *Anti-Chance*, New York: Pergamon.

Schrödinger, Erwin (1950) "What is an Elementary Particle?", *Nature*, 9:109-116.

Schrödinger, Erwin (1952a) "Are there quantum jumps? Part I", *British Journal for the Philosophy of Science*, 3:109-123, August.

Schrödinger, Erwin (1952b) "Are there quantum jumps? Part II", *British Journal for the Philosophy of Science*, 3:233-242, November.

Schrödinger, Erwin (1960) *Space-Time Structure*, Cambridge: Cambridge University Press.

Sellars, Wilfred (1963) *Science, Perception and Reality,*, London: Routledge.

Sheldrake, Rupert (1981) *A New Science of Life*, Los Angeles: J.P. Tarcher.

Siu, R.G.H. (1957) *The Tao of Science*, Cambridge: MIT Press, 1979.

Sherman, Irwin W. and Vilia G. (1979) *Biology: A Human Approach*, Oxford University Press (New York) 2nd ed.

Smith, Cyril S. (1978) "Structural Hierarchy in Science, Art, and History", in Wechsler, ed. (1978).

Stace, W.T. (1980) "Science and the Physical World", in Klemke, E.D. et al., eds. (1980).

Stapp, Henry P. (1982) "Mind, Matter, and Quantum Mechanics", *Foundations of Physics*, 12:363-399, April.

Stapp, Henry P. (1988) "Quantum Theory and the Physicist's Conception of Nature", in Kitchener, Richard F., ed. (1988).

Stevens, Peter S. (1974) *Patterns in Nature*, Boston: Little, Brown.

Stewart, John (1981) BBC World Service *Science* program, 13 November.

Stroke, George W. (1977) "Light", *McGraw-Hill Encyclopedia of Science and Technology*, 4th edition, 7:563-571.

Taylor, John G. (1975) *Special Relativity*, Oxford: Clarendon.

Taylor, John G. (1980) BBC World Service *Advice to Beginners* program, 6 October.

Terrell, James (1972) "Letters", *Physics Today*, p. 9, January.

Thomas, Lewis (1974) *The Lives of a Cell*, New York:Macmillan.

Toulmin, Stephen (1961) *Foresight and Understanding: An Enquiry into the Aims of Science* , Indiana University Press, New York: Harper Torchbook edition, 1963.

Toulmin, Stephen (1972) *Human Understanding*, Vol. I, Princeton, New Jersey: Princeton University Press.

Trefil, James (1978) *Physics as a Liberal Art*, New York: Pergamon.

van der Wal, Caspar H., ter Haar, A.C.J., Wilhelm, F.K., Schouten, R.N., Harmans, C.J.P.M., Orlando, T.P., Lloyd, Seth, and Mooij, J.E. (2000) "Quantum Superposition of Macroscopic Persistent-Current States", *Science*, 290:773-777, 27 October.

Waddington, C.H., ed. (1969) *Towards a Theoretical Biology*, Vol. 2, Edinburgh: Edinburgh University Press.
Wald, Robert M. (1992) *Space, Time, and Gravity: The Theory of the Big Bang and Black Holes*, Chicago: University of Chicago Press
Waldron, R.A. (1977) *The Wave and Ballistic Theories of Light — A Critical Review*, London: Frederick Muller Limited.
Wechsler, Judith (1978) "Introduction", in Wechsler, ed. (1978).
Wechsler, Judith, ed. (1978) *On Aesthetics in Science* Cambridge: MIT Press.
Weinberg, Steven (1999) "A Unified Physics by 2050?", *Scientific American*, 281:68-75, December.
Wertheim, Margaret (1995) Interview with Daniel Zwerldling, All Things Considered Weekend, 10/14/95, re 1995 book, *Pythagoras' Trousers: God, Physics, and the Gender Wars*, New York: Times Books/Random House.
Weyl, Hermann (1952) *Symmetry*, Princeton: Princeton University Press.
Wheeler, J.A. and Feynman, R.P. (1945) "Interaction with the Absorber as the Mechanism of Radiation", *Reviews of Modern Physics*, 17:157-181.
Whitrow, G.J. (1972) *What is Time?*, London: Thames & Hudson.
Whittaker, Edmund T. (1953) *A History of the Theories of Aether and Electricity*, Vol. I, New York: Nelson.
Whyte, L.L., Wilson, A.G., and Wilson, D., eds. (1969) *Hierarchical Structures*, New York: American Elsevier.
Wigner, Eugene P. (1967) *Symmetries and Reflections*, Bloomington: Indiana University Press.
Wigner, Eugene P. (1976) "On Some of Physics' Problems", in Laszlo, ed. (1976).
Williams, R.W. (1972) "Letters", *Physics Today*, p. 11, January.
Wolf, Fred Alan (1981) *Taking the Quantum Leap*, New York: Harper and Row, Perennial Library edition, 1989.
Zahar, Elie (1976) "Why Did Einstein's Programme Supersede Lorentz's?", in *Method and Appraisal in the Physical Sciences*, Howson, C., ed., Cambridge, Cambridge University Press.
Zahar, Elie (1977) "Mach, Einstein, and the Rise of Modern Science", *British Journal for the Philosophy of Science*, 28:195-213, September.
Zahar, Elie (1981) "Second Thoughts about Machian Positivism: A Reply to Feyerabend", *British Journal for the Philosophy of Science* 32:267-276, September.
Zapffe, Carl A. (1977a) *Seven Short Essays*, Brainerd, Minn.: Lakeland Color Press.
Zapffe, Carl A. (1977b) "Einstein and the Aurora Borealis — the World's First Gedankenexperiment of the Second Order", *American Journal of Physics* 45:775-777, August.
Zapffe, Carl A. (1980) "Reply to Comments", *Speculations in Science and Technology* 3:483-485.
Zapffe, Carl A. (1982) *A Reminder on* $E=mc^2$, Brainerd, Minn.: Lakeland Color Press.
Ziman, John (1980) "What is Science?", in Klemke, E.D. et al., eds. (1980).
Zukav, Gary (1979) *The Dancing Wu Li Masters*, New York: William Morrow, Bantam edition, 1980.

Index of Names

Alvåger, T., 188
Aristotle, 140, 141n, 142n, 287, 400, 440-441
Aspect, Alain, 336
Bacon, Roger, 368, 370, 406, 422
Bell, John, 336n, 337
Bergson, Henri, 278, 374
Bergstrand, Erik, 190n
Berkeley, George, 398, 429
Bethe, Hans, 458
Bohm, David: chance, 81, 310; context, 72; determinism/mechanism, 311; education, 413; facts, 389; hierarchy, 297; light in mirror, 243; light, 226n; mechanism, 81; metaphysics, 386, 423; new theories, 508; prediction, 420; quantum vs. relativity theories, 471; refutability, 433; relativity, 250n; testability, 393; universal patterns, 397; values, 458; wave function, 47n; whole, 330; world view, 383
Böhme, Jakob, 9
Bohr, Niels: biology, 490; classical concepts, 56, 63; complementarity, 37, 56, 62, 80, 82, 388, 504; complementary views, 329n; consciousness, 69, 497; correspondence principle, 45; description, 361; dispute with Einstein, 318-319, 335, 388 (*see also* Einstein-Podolsky-Rosen experiment); dispute with Heisenberg, 55-57, 80; indeterminism, 56; mathematics, 398; probability, 41; psychology, 497; quantum postulate, 34, 63; testability, 43
Boltzmann, Ludwig, 345
Bondi, Hermann, 228, 235, 350, 363
Born, Max, 32, 39, 42, 49, 64, 69, 78, 252, 255, 388n
Bragg, William, 46
Brecher, Kenneth, 188, 189
Bridgman, Percy: action at a distance, 315, 321, 325; empty space, 127; event-frame, 271; experience-theory, 133n; explanation, 227; laws of nature, 122; Lorentz transformation, 253n; Maxwell's equations, 210; relative velocity, 131; simultaneity, 280; speed of light, 187, 195
Brillouin, Léon, 190, 196, 360, 399
Builder, Geoffrey: acceleration, 239n; clock-acceleration, 239; ether, 208n, 214; Galilean vs. Lorentz transformation, 266, 268; length-clock changes, 248; special relativity, 266n; time - light speed, 195; unidirectional speed of light, 192n, 193n, 194
Bunge, Mario, 418
Capek, Milic, 39, 153, 277n, 335, 374, 397n
Cartwright, Nancy, 51, 68, 70
Cassirer, Ernst, 253, 360
Christensen, Ferrel, 196, 283
Compton, Arthur, 33
Copernicus, Nikolaus, 141n
d'Espagnat, Bernard, 44, 46, 266n, 336n, 348, 378n, 509
Darwin, Charles, 370, 451-452, 493
De Broglie, Louis, 34, 34n, 48-49, 82, 344
De Sitter, Willem, 188, 349
Descartes, René, 41, 140, 287, 367, 370, 422
DeWitt, Bryce, 53, 375
Dingle, Herbert: common sense, 401; Einstein vs. Lorentz models, 210-211, 236, 254, 256n, 266, 281; equivalence principle, 156n; mathematics, 400-401; Minkowski, 284; relative motion, 120; renegade, 236n, 380n; space-time continuum, 284; spell of relativity, 449; twin paradox, 236
Dirac, Paul, 90, 91, 366
Dyson, Freeman, 459, 509
Eddington, Arthur, 121, 186n, 279n, 348, 395
Einstein, Albert: absolute space, 230; at speed of light, 241-244; complementarity, 40, 356, 388; continuum, 319; dispute with Bohr, *see under* Bohr; duality, 135; electromagnetic field, 213; empty space, 127; ether, 199, 209, 210-215, 233; experiment with Podolsky & Rosen, 283, 286, 318, 335-341, 344; free invention, 371-372, 391; indeterminism, 42; induction-deduction, 370; inertial frame, 127; law of nature, 121-123, 364; lightning, 244; Lorentz transformation, 122, 229; mass-energy, 262-263; mathematics, 367; Michelson-Morley experiment, 208; micro vs. macro laws, 46; playing dice, 42, 311, 356; principle of relativity, 118-135; quanta, 33, 34; relative motion, 120; simplicity, 364; space-time continuum, 274; symmetry, 356; train, 234, 244-245, 269-271; unified theory, 472; Einstein vs. Lorentz models, 199, 202, 210-211, 254, 281; Einstein vs. Poincaré models, 256-258, 278, 281
Escher, M.C., 272
Everett, Hugh, 53, 53n, 66n
Faraday, Michael, 314, 318, 401
Fermi, Enrico, 113, 445

Feyerabend, Paul, 395n, 403, 411, 412, 415, 426, 449
Feynman, Richard: absorber theory, 341, 350; beauty, 365; clever nature, 373; double slit experiment, 35; experiments, 362; forward-backward time, 111, 350, 352n, 354; intuition, 374; mathematics, 368, 398; multiple interpretations, 429; positron, 111; preconceptions, 362; predictability, 393; puzzle, 383; simplicity, 365; symmetry, 18; terminology, 446; testability, 363; truth, 365; understanding, 374; unity of nature, 397n, 470n; universality, 366, 396
Finkelstein, David, 361n, 374, 446n, 481
Fitzgerald, George, 199n
Gale, Henry, 218n
Galileo, 118, 140, 154, 367, 370, 400, 401, 417, 423, 440-441
Galison, Peter, 256, 257, 274, 282n, 286
Gauss, Karl, 343
Glashow, Sheldon, 365, 472
Graham, Neil, 53
Grieder, Alfons, 211
Hafele, J.C. *See* Hafele-Keating experiment
Hawking, Stephen, 481, 487
Healey, Richard, 43, 43n, 49n, 65, 66n, 67, 68, 70, 123n, 337n
Hegel, Georg, 4, 13, 22
Heisenberg, Werner: causality, 41n; classical concepts, 63, 275, 311; complementarity, 39, 56, 63, 80; description, 362; dispute with Schrödinger, 57-59; observation, 51, 390; dispute with Bohr, *see under* Bohr; simplicity, 365; testability, 363; uncertainty principle, 40, 44, 57; understanding, 373; wave function, 68, 69
Higgs, Peter, 474n
Holton, Gerald: Bohr's complementarity, 491n, 505; complementarity, 388; correspondence principle, 46; duality, 356; information overload, 507; Lorentz, 254n; observation, 83; Poincaré, 256, 282n; preconceptions, 387; scientific responsibility, 415; special relativity postulates, 225; symmetry, 355; themata, 387; understanding, 421; women in science, 330n
Horgan, John, 396, 468n, 470n, 477
Hoyle, Fred, 411
Hsu, Jong-ping, 214, 221n, 225, 281, 282, 282n
Hubble, Edwin, 163, 293, 486
Hughens, Christiaan, 342
Ives, Herbert: Lorentz transformation, 266, 266n; mass-energy, 262; special relativity, 267; speed of light experiments, 218n; speed of light ritual, 195
James, William, 497
Jammer, Max: Bohr's complementarity, 37, 504; Bohr's thinking, 497; Bohr-Einstein dispute, 319; Bohr-Heisenberg dispute, 55; complementarity, 37; complementarity vs. uncertainty, 57; determinism, 311; faster than light, 338; Heisenberg's goal, 44; indeterminism, 77; Newton's gravitational mechanism, 151n; observation, 52; quantum theory accuracy, 419; technology vs. science, 458; thought experiments, 364; uncertainty, 43
Jauch, J.M., 383
Kant, Emmanuel, 9, 21, 25, 382
Kaufmann, William, 163, 258
Keating, Richard, 238 (*see also* Hafele-Keating experiment)
Kepler, Johannes, 367
Kitchener, Richard, 360, 361, 381, 386, 423n, 439
Koestler, Arthur, 298n
Kuhn, Thomas: building science, 444; dogmatism, 376-378; education, 376, 378-379, 412-413; facts, 391; metaphysics, 386; multiple interpretations, 429; new theory, 509; paradigm, 376n; revolutions, 375, 408-412; specialists, 379-380, 507; truth-falseness, 362, 402-403, 440-443
Langevin, Paul, 234
Larmor, Joseph, 253n
Laszlo, Ervin, 368
Leibniz, Gottfried, 288n
Lewis, Gilbert, 341
Lindsay, R. Bruce, 372, 411, 464
Lockwood, Michael, 43, 51n, 53, 67, 68, 69, 70
Lorentz, Hendrik: ether, 210, 225, 233; mass-energy, 262; transformation, 253
Mach, Ernst, 258n, 391n, 464
MacRoberts, D.T., 190, 194, 195, 235, 236, 238n
Margenau, Henry, 195n, 422
Margulis, Lynn, 445, 491n
Maxwell, James, 123, 201, 206, 253n, 264, 280, 314, 401
Mendeleyev, 86
Michelson, Albert, 204
Miller, A., 267n
Miller, G., 345n, 348
Miller, R., 36, 39, 43, 47n, 51, 58, 363, 419n
Milne, E.A., 348
Minkowski, Hermann, 274, 284-286, 288
Morley. *See* Michelson-Morley experiment
Newton, Isaac: gravity, 150, 314; inductive thinking, 370; laws of mechanics, 116-118, 123; principle of relativity, 118
Nisbett, Richard, 499
Oppenheimer, J. Robert, 37, 375, 377, 444-445, 448, 459
Parish, Leonard, 213

Pauli, Wolfgang, 49n, 69, 187, 373
Peat, F. David, 54, 366, 399, 400n, 422
Penrose, Roger, 349n, 480, 487
Planck, Max, 32, 33, 34, 252, 255, 262, 311, 377, 422
Plato, 274, 330, 333n
Poincaré, Jules-Henri: dynamics vs. thermodynamics, 354; Lorentz transformation, 253; mass-energy, 262; simultaneity, 281n; space-time, 286
Polkinghorne, John, 115, 386, 424n, 457
Popper, Karl: action at a distance, 339; common sense, 407, 437; complementarity, 38; context, 385; defining science, 420, 455n; dogmatism, 377, 410; entropy, 69; essentialism/instrumentalism, 359; experience, 432; expertism, 416, 466; facts, 389; faster then light, 339; forward-backward time, 350n; general relativity, 472n; indeterminism, 77; induction-deduction, 371n; internal consistency, 425; inventing laws, 391; mathematics, 398, 399; metaphysics, 363, 384-386; observation, 68; prediction-understanding, 421; propensity, 72n; quanta, 34n; quantum theory, 309; realism, 69; simplicity, 364; subjectivism, 69; testability, 363; theories as nets, 433; truth-falseness, 362, 368-370, 377, 402-405, 428n, 432; universality, 396; wave function, 47n
Postman, Neil, 449, 450
Prigogine, Ilya: being vs. becoming, 354; dynamics vs. thermodynamics, 354; experience, 422; metaphysics, 387; order-disorder, 347; quantum vs. relativity theories, 471; self-organization, 348; time dimension, 278
Rees, Martin, 459
Roszak, Theodore, 395
Russell, Bertrand, 399
Sachs, Mendel: atomism, 108n; Einstein vs. Newton gravity, 314; ether, 209; new theory, 509; quantum vs. relativity theories, 317-318, 320, 471; space-time, 274; vacuum, 219n; young Einstein, 241
Sagan, Carl, 363, 375, 460n
Sagnac, Geroges, 218n
Schlegel, Richard, 240n
Schrödinger, Erwin: atomism, 64; cat experiment, 66-67, 74; dispute with Heisenberg, *see under* Heisenberg; infinite electron, 324; order-disorder, 348; quantum jumps, 58; uncertainty principle, 41n; understanding, 422; wave function, 47n, 48, 57
Sheldrake, Rupert, 490n
Siu, R.G.H., 368, 389, 415n, 417, 420, 502n
Stapp, Henry: classical physics, 442; faster than light, 338n; mathematics, 399; metaphysics, 360; prediction, 421; time, 277, 310
Suppes, Patrick, 43n
Taylor, John, 236n, 379, 412
Tetrode, Hugo, 341
Toulmin, Stephen: facts, 389; forces, 440; modifying ideas, 411, 437; preconceptions, 386; prediction-understanding, 457; scientific concepts, 84; theoretical appeal, 407
Voigt, Woldemar, 253n
Von Neumann, John, 51-52, 68, 69, 82
Weinberg, Steven, 365, 472, 473n
Werner, F., 430n
Wertheim, Margaret, 330n
Weyl, Hermann, 355, 357, 373
Wheeler, John: absorber theory, 341-344, 350; jiffyland, 54n; many worlds, 53; mathematics, 400n; simplicity, 365
Whitrow, Gerald, 152n, 248n, 277, 285n, 345, 348, 353, 354
Whyte, L.L., 21, 297, 493
Wigner, Eugene, 360, 368, 460, 479, 507
Young, Thomas, 35
Yukawa, Hideki, 90, 104, 108, 112
Zahar, Elie, 211n, 256n, 265, 356, 374
Zapffe, Carl, 214n, 253n, 281, 297n, 312n, 402
Ziman, John, 368, 416, 423n

Index of Subjects

2FT. *See* Two-Force Theory
absolute motion. *See* motion, absolute and relative
absolute space. *See* space, absolute and relative
absorber theory, 341-344, 350, 354, 489n
acceleration, 117
acceleration and gravitation, 154, 156, 167-170
action at a distance, 174, 314-315, 321-324, 333, 342
ad hoc, 84, 264n, 366n, 424, 435, 442, 443, 444, 445
advanced waves. *See* waves, retarded and advanced
aesthetic criteria, 1, 364-366, 384, 395-398, 427, 432
Alice in wonderland, 200, 401
alive and dead, 12, 14, 67, 494
anabolism and catabolism, 490
anti-gravity. *See* gravity, negative
antimatter, 91, 102-105, 288n
anti-semitism, 260n
arbitrary conceptual pivots, 445, 464, 507
arrow of time, 348-352
atom, 85-115
atomic spectra, 86
atomism, 64, 85, 108-115, 309, 316, 470, 478
attraction and repulsion, 25
awake and asleep, 11, 14, 23, 350, 493, 499, 500
awards, science, 416, 448, 449
background knowledge. *See* context; metaphysics
backward time. *See* time, forward and backward
beauty, 153, 264n, 284, 364-366, 368, 399, 425
being, 10-15, 94, 330, 354, 427n, 428
being and becoming, 354
being and seeing, 300, 495
Bell's inequality, 336n
big bang, 91, 102, 103, 162, 353, 473, 474, 477, 478, 481, 482, 484, 487
big crunch, 353
bimatter, 97n
biological clocks, 493
biology, 490-493
blackbody and whitebody, 32, 304, 480
black hole and white hole, 173, 294, 479-482, 507
blank slate, 362, 381, 382, 461
boat, 185, 196-197
bonding, molecular. *See* molecular bonding
breathing, 17, 28, 99, 111, 298, 300, 301, 302, 352, 479, 493
brittleness of theory, 377
broken symmetry, 19, 473
buildup and discharge, 303n, 347, 352
cat state, 74
Cavendish experiment, 176
centripetal and centrifugal forces, 117, 138-139, 143, 147
chance, 67, 76, 81, 251, 310, 337, 393, 471
changing mediums, 295
changing region. *See* stable-changing dimension
charge, 88, 92, 95
Cheshire cat, 210
chicken and egg, 23
circular motion. *See* motion, uniform and non-uniform
classical vs. operant conditioning, 496
clock convention. *See* simultaneity
codes and rods of eyes, 332
cognition, 18, 70, 260, 330, 353, 408, 414, 427, 447, 495-499
coherence, 2, 84, 102, 163, 250, 263, 366, 397, 409, 432, 437, 444, 450, 455, 461, 468, 503
common sense, 186, 336n, 373-375, 401, 405-408, 424, 426, 435-439, 462, 496. *See also* experience
compatibility, 425-428, 506
complementarity, 1, 3, 20, 22-24, 37, 59, 60, 62, 80, 82, 355, 500-502
complementarity, Bohr vs. expanded, 62
complementarity as mumbo jumbo, 39
complementarity as sideshow, 27-30
complementarity vs. symmetry, 18, 20, 355-357
complementary connectedness, 320-330
complementary hemispheric specialization, 498
complementary stages, 73
complementary states, 71
complementary universes, 488-490
compressions and rarefactions, 101
Compton effect, 33, 58, 446
concepts. *See* tools of the mind
conductors and insulators, 95
conformity, 60, 236n, 259, 260, 376, 380, 410, 412, 414

consciousness, 17, 18, 50, 51, 53, 65, 66, 68, 70, 76, 237, 274, 277, 454, 495-499. *See also* observation and observer
conservation of momentum, 117n
contact, direct and indirect, 146, 315-317, 321-324, 325
context, 19, 72, 102, 112, 170, 269, 270, 382, 384, 389-391, 395, 404, 410, 429, 444, 455, 495, 496, 499
continuous. *See* dichotomous and continuous
contracting and expanding universe, 293, 348, 351, 481, 486-490
Copenhagen interpretation, 49-52, 54, 65-66, 68, 70, 71
correspondence principle, 45-46, 81, 82
cosmic background radiation, 483-485
cosmic rays, 240, 240n
cosmological constant, 163, 486
cosmological principle, 484
cosmology, 478-490
Coulomb experiment, 179
creation and destruction, 105
creationism, 451
current, 94, 95
dark energy, 163, 445, 487, 489
dark matter, 445, 486
data, 389n
DC and AC, 94
deduction and induction, 370-372, 470n, 500
dialectics, 4, 13, 22, 388n, 505n
diamagnetism, 95
dice playing, 42, 311, 356
dichotomous and continuous, 7, 64n, 319-320, 326, 388, 492
dimensions of analysis, 4-7
direct contact. *See* contact, direct and indirect
discovery vs. invention, 221, 252, 264, 370-373, 394, 463-464, 474
discrete quantum, 32-34, 59, 63, 64, 64n, 394
discrimination and generalization, 22, 497
DNA and RNA, 491
dogmatism, 184, 250, 258, 294, 376, 378, 380, 408-412, 417, 418, 437, 448, 456, 463
dominant and recessive, 492
dominant entity. *See* dominant-recessive dimension
dominant-recessive dimension, 5
Doppler effect, 291-293, 296
double-slit experiment, 35-36
double stars, 188
dreaming and not dreaming, 499, 500, 502
duality, 79, 311, 478n, 481. *See also* complementarity
duality within relativity, 285
dynamic relationship. *See* stasis-motion dimension
dynamics vs. thermodynamics, 353-355
dynamic time, 276-278, 349
earth, flat and round, 143, 411, 436
Eastern. *See* Western and Eastern
education, science, 376, 378-379, 412-413, 452
Einstein's right syndrome, 412
Einstein's wrong syndrome, 379, 412
Einstein-Podolsky-Rosen experiment, 283, 286, 318, 335-341, 344
electric and magnetic fields, 85, 93
electric and magnetic materials, 95-97
electricity and magnetism, 93
electromagnetic force, 85, 90, 93, 105-106
electromagnetic wave, 86, 106
electron, 83, 85-92, 110, 484
electron and photon, 33, 35, 37, 84n, 86, 105, 106, 111
electron and positron, 90, 105, 110, 111-112, 350, 352n
electroweak force, 473
elevator, 154-155, 167, 170-171, 261
elitism, 379, 414-417
elsitron, 111
enlightenment, 417, 422, 463
entropy and order, 14, 25, 69, 345-348, 354
EPR. *See* Einstein-Podolsky-Rosen experiment
equivalence principle, 154-157, 167, 169, 170-172, 240, 261, 290
essentialism, 359
esymmetry, 19, 342n
ether drag hypothesis, 217n
ether medium, 121, 124, 199, 201, 205, 209-216, 233, 254, 257, 442, 469n
event horizon, 489
everything, theory of, 472, 476
evolution, 348, 370, 402, 407, 411, 451-452, 493, 494
exchange force, 90
expanding universe. *See* universe, contracting and expanding
experience, 127, 181, 219, 232, 260, 277, 371, 375, 407, 419, 435, 439, 462, 465, 496
experiments, 363-364
expertism, 379, 414-417, 462, 465, 466
eye cones and rods, 332
faces-vase, 11, 16, 71, 73, 76, 271, 496
facts, 362-363, 388-392, 402, 425, 426, 428, 431
false symmetry, 20, 80n, 103, 269-272, 312, 352, 357, 480
faster than light. *See* instantaneous and over-time transmission
female and male. *See* male and female
ferromagnetic material, 96
Feynman diagram, 288n
fictitious force, 138
field, 5
figure-ground, 11, 71, 212, 269, 272, 290, 332, 485

Fitzgerald contraction. *See* Lorentz-Fitzgerald contraction
flat earth. *See* earth, flat and round
focus and background of attention, 496
focus-background, 16
force-free region, 117, 125-128, 141
formative causation, 490n
Foucault pendulum, 144
frame, 5
frame, inertial and non-inertial, 118, 125-128, 136-137, 150, 170-171, 228, 231
frame dragging, 165
frame hopping, 271
free fall, stable and changing, 165-167, 219
free invention, 391, 405. *See also* discovery vs. invention
free neutron, 112
fundamental forces, complementary, 3, 7, 8-10, 12-14, 20-22, 24, 25, 27, 29
fundamental forces, physics, 8, 20, 27, 103, 107, 109, 469-470, 472, 473, 474, 476, 477, 478, 486
fundamental vs. effective speed, 185-186
fusion and fission, 98
Galilean transformation, 124n, 135, 252, 253, 268
gedankenexperimente. *See* thought experiments
generality, 2, 26, 32, 109, 263, 264, 327, 366, 395-396, 427, 432, 452, 503
gestalt psychology, 11
global positioning system, 241
grand unification, 472
gravitation, attraction and repulsion, 159-164
gravitation, matter and medium, 173-174
gravitation, spin and orbit, 164-165
gravitational field, 150
gravitational lensing. *See* light, bending by gravity
gravitational mass. *See* mass, inertial and gravitational
gravitational redshift, 293
gravitation and acceleration. *See* acceleration and gravitation
gravity, 149-180
gravity, artificial, 178
gravity, direct and indirect, 174
gravity, Einstein vs. Newton models, 150-156, 157-159, 174, 320
gravity, hierarchical, 175-180
gravity, negative, 163, 487
gravity, stable and changing, 158-159
gravity, structural and process, 165-167
gravity and clocks, 294, 296
gravity bending light. *See* light, bending by gravity
gravity wave, 173-174
growth and dormancy, 493
gyroscope, 145n
Hafele-Keating experiment, 240
heat death, 347
hierarchical layers of stasis and motion, 132
hierarchy, 4, 297-304
hierarchy problem, 486
Higgs boson, 473
holes, theory of, 90
hologram, 102n
holon, 298n
horizontal and vertical, 7, 26, 62, 140, 167n, 168, 470
house called science, 443-445
Hubble's law, 293, 486
hydrogen, 84, 293n
hypnosis, 97
idolatry, 258, 284, 449
indeterminism, 40, 42, 43, 56, 70, 77. *See also* uncertainty
indirect contact. *See* contact, direct and indirect
indoctrination and propaganda, 412-414. *See also* science education
induction. *See* deduction and induction
inertia and outertia, 136, 140
inertial frames. *See* frames, inertial and non-inertial
inertial mass. *See* mass, inertial and gravitational
inertialness test, 119, 130, 131, 133
inside-outside dimension, 5, 137, 140-145, 158
inside realm. *See* inside-outside dimension
instantaneous and over-time transmission, 151, 325-326, 344
instrumentalism, 52, 359, 362, 422, 428n
intelligent design, 451
interactions between nucleons, 106-108
interactions between nucleuses, 105-106
interval, space-time. *See* space-time interval
intraaction and interaction, 146, 174, 322, 325, 334, 339, 344
introvert and extrovert, 499
invention. *See* discovery vs. invention
inverse square law, 85, 141
isotropy and anisotropy, 192, 196
Jaina physics, 287
jargon. *See* terminology
jiffyland, 54n
lab experiments and quantum theory, 74
laser, 101
law of nature, 121-123, 133-135, 252
left hemisphere and right hemisphere, 498
light, bending by gravity, 152, 219, 294
light, speed of, 181-225, 442
linear motion. *See* motion, uniform and non-uniform
line of demarcation. *See* truth and falseness
little bang, 353

local time and universal time, 350. *See also* simultaneity
Lorentz-Fitzgerald contraction, 199, 253n
Lorentz transformation, 122, 193, 196, 199, 229, 234, 237, 241, 247, 249, 253, 265-267, 268, 285, 384
lottery, 434, 456
magic, 97, 151, 156, 174, 200, 227, 247, 260, 272, 314, 320, 356, 395, 400n, 422, 447-448, 476
magnetic domains, 95
magnetic monopole, 85
male and female, 1, 14, 23, 330
many worlds interpretation, 53-54, 65, 71, 72, 488, 501
map, 92, 376, 407, 427, 496
mass, inertial and gravitational, 117, 155-156, 171-172
mass and energy, 262-263
materialism and idealism, 490n
mathematics, 54, 55, 124n, 201, 284, 365, 367-368, 398-402, 426, 442, 464
matrix mechanics and wave mechanics, 58
matter domain. *See* matter-medium dimension
matter-medium dimension, 5, 310
matter waves, 35, 36
Maxwell's equations, 123, 242
meaning. *See* prediction and understanding
measurability. *See* testability
measurement problem, 67-68
mechanism and vitalism, 490
media, 413-414
meiosis and mitosis, 491
memory, 496
meson, 90, 104, 108
metal and non-metal, 98
metaphysics, 359-362, 363, 381-388, 419, 444, 496
Michelson-Gale experiment, 218n
Michelson-Morley experiment, 125, 204-209, 215-218, 251, 280, 335
micro-macro split, 67, 75, 81, 396, 397, 472, 497
missing matter, 485-486
molecular bonding, 87, 106
molecular chaos, 77
molecule, 85
momentum and position, 40-41, 42, 43, 44, 50, 72, 341, 373
motion, absolute and relative, 120-121, 129-133
motion, high-speed, 232-234
motion, uniform and non-uniform, 116, 140-144, 441
multiple meaning, 24-27, 201n, 333n, 427n
muon, 240
mysticism, 227, 447-453, 454
myths, 463
negentropy. *See* entropy and order
nesting hierarchy, 97, 176, 270, 282, 303, 485
nesting time metrics, 350
neutrino, 91, 469n
neutron, 110. *See also* proton and neutron
Newton's laws of mechanics, 116-118, 123
non-inertial frames. *See* frames, inertial and non-inertial
non-uniform motion. *See* motion, uniform and non-uniform
normal/extraordinary science. *See* revolutions, Kuhnian
no symmetry, 20, 104, 129, 135, 267-269, 291, 312, 347-348, 396, 480, 485
nuclear force, 89-90, 106-108, 469, 476
nucleon, 88, 104, 106-108
nucleus, 85
nuclon, 114n
object vs. function, 137-138, 229
observation, 15, 17, 39, 42, 43, 50, 65, 69, 83, 390n, 495. *See also* wave function
order. *See* entropy and order
oscillating dominance, 15, 303n
outside realm. *See* inside-outside dimension
pair production, 91, 102, 105
paradigm, Kuhnian, 376n, 387n, 391, 404n
paramagnetism, 95
particle collision experiments, 363, 394
particle. *See* wave and particle
parts and whole, 4, 21, 24, 26, 80, 138, 271, 289, 290, 318, 328, 329, 334, 353, 388, 394, 425, 434, 435, 458, 486, 501
Pauli exclusion principle, 373
perception, 11, 17, 19, 69, 70, 84, 270n, 271, 332, 351, 353, 362, 385, 391, 438, 496
periodic table of elements, 97-99, 109
P-Force. *See* fundamental forces, complementary
photoelectric effect, 33, 319n
photon. *See* electron and photon
Planck's constant, 34, 40, 45, 81
planets, 117, 165, 177, 241, 297n, 479, 482
Plato's cave, 274, 333n
playing dice, 42, 311, 356
polarity, 92, 95
polynucleotides, 491
positivism, 359-362, 363, 381, 384, 423
positron emission tomography, 91
positron. *See* electron and positron
preconceptions. *See* metaphysics
prediction and understanding, 64, 152, 258, 264, 272, 360, 361, 420-423, 425, 433, 435, 443, 455-458, 468, vii
principle of complementarity. *See* complementarity
principle of local causes, 335-339

principle of relativity. *See* relativity, principle of
probability, 41, 43, 46, 47, 48, 71, 75, 78, 311, 394, 402
propensity, 69, 72n
proper time vs. relative time, 283n
proton, 85, 86-92, 110, 484
proton and neutron, 88-90, 104, 110, 112-113
pseudoscience, 450
puzzle, conceptual pieces of, vi, 109, 383, 409
P-view. *See* seeing
quantum jump, 49, 57
quantum paradox, 63-64
quantum postulate, 34, 46, 58n, 63
quantum theory vs. relativity theory, 317, 319, 324, 329n, 335, 470-472, 475, 476. *See also* Bohr-Einstein dispute
quark, 90, 98
quasar, 481
real vs. apparent, 233, 247-249
recessive entity. *See* dominant-recessive dimension
redshift and blueshift, 292, 486, 489
reductionism, 153, 157, 251, 263, 264, 265, 308, 356, 388, 396, 470, 506
reflection, 100
refraction, 100, 220, 295, 296, 322
refutability/falsifiability. *See* truth and falseness
reinforcement, positive and negative, 497, 500
relative motion. *See* motion, absolute and relative
relativity, principle of, 118-135, 170, 187, 193, 196, 234, 236, 238, 252, 253, 265, 280, 290
relativity vs. thermodynamics, 353-355
religion, 14, 330, 360, 392, 412, 420, 447-454
reporting science, 413-414
resonance and non-resonance, 94n, 99-100
retarded wave. *See* wave, retarded and advanced
revolutions, Kuhnian, 375-378, 408-412, 444
rods and cones of eyes, 332
rubber sheet, 150, 164, 276, 284
Rubens illustration, 11, 16, 71, 271, 496
Sagnac experiment, 218n
same and different, 21-24, 107, 172, 289
scalar hierarchy, 301
scattering, 99-100
Schrödinger's cat experiment, 66-67, 74, 338n
science awards, 416, 448, 449
science education, 376, 378-379, 412-413, 452
science reporting, 413-414
science terminology. *See* terminology
science textbooks, 378-379, 413
scientific method, 359-380
scientism, 449
scientists as gods, 259, 449
second law of thermodynamics. *See* thermodynamics
seeing, 15-18, 94, 328-329, 427, 427n, 428. *See also* being and seeing
shadow, 153, 274, 289, 290, 310, 333n
simple harmonic motion, 18, 108, 144, 146n, 302, 343n
simplicity, 2, 32, 46, 252, 364-366, 395, 422, 432, 452, 503
simultaneity, 13, 44, 194, 195n, 244, 265, 278-283, 285, 288, 329, 333-334, 339, 343, 345, 356, 497, 499, 500
singularity, 22, 40, 49, 353, 480-482
social construction, 416n
space, absolute and relative, 121, 129-133, 230, 442
space and time, 273-291
spaceship, 229, 231, 234, 248
space-time continuum, 150-153, 157, 273-276, 282, 284, 318, 442
space-time event. *See* space-time interval
space-time interval, space-like and time-like, 284-286, 288, 326, 328, 334, 337, 339
specialists, 379-380, 414-417, 462, 466, 507
special relativity, historical context, 253-258
special relativity, psychosocial factors, 259-260
specification hierarchy, 301
speed of light. *See* light, speed of
spin and orbit, 145
spiritual, 14, 392, 450-454, 490
spooky action at a distance, 74n, 338n
stable-changing dimension, 5, 128, 137, 147-148, 158, 441
stable region. *See* stable-changing dimension
standard model, 472-475
standing and traveling wave. *See* wave, standing and traveling
stasis and motion, 116, 118-119, 148
stasis-motion dimension, 5, 145-146, 270, 299-326
states of matter, 303
static and dynamic relationships, 5
static relationship. *See* stasis-motion dimension
string theory, 475-478
subatomic wave, 104, 309, 310
subatomic zoo, 364
subjectivism, 68
summer and winter, 493
sun and nus, 482-483
superspace, 53, 72
supersymmetry, 473-474, 476
suspended animation, 67, 73
symmetry, 18-20, 355-357, 473
symmetry and antimatter, 102-105
symmetry and relativity, 90, 103, 120, 228, 233, 251, 267-272, 312

symmetry and twin paradox, 234-240
symmetry conviction. *See* false symmetry
synchronizing distant clocks. *See* simultaneity
Taoism, 21, 330
technology, 455-460
temperature, 303, 345, 473, 481, 484
terminology, 10, 19, 24, 56, 97, 104, 157, 166, 223, 259, 311, 379n, 445-447, 448, 491n, 507
testability, 43, 309, 363, 392-394, 452
textbooks, science, 378-379, 413
theory of everything, 472, 476
thermodynamics, 256, 345-355, 373, 481, 488
thesis-antithesis, 4, 13, 22, 161, 289, 434, 435, 458, 501
thought experiments, 154, 170, 234, 244, 260-262, 269-272, 364, 395
threads to weave, 366, 396, 432, 463
time, forward and backward, 111, 288, 345, 346, 353, 481
time asymmetry. *See* arrow of time
tools of the mind, 164, 446
tower of Pisa, 154
train, 120, 124, 157, 231, 234, 244-245, 261, 269-271
transverse Doppler. *See* Doppler effect
traveling wave. *See* wave, standing and traveling
true time, 254n, 257, 281-283
truth and falseness, 362, 368-370, 402-405, 428-435
twin paradox, 234-240
2FT. *See* Two-Force Theory
two forces. *See* fundamental forces, complementary
Two-Force Theory, 1, 3-4, 7, 8, 20, 22
ultra-violet catastrophe, 32
uncertainty, 13, 23, 69, 76, 77, 78, 311, 394, 397, 453, 471, 493
uncertainty principle, 40-44, 55, 76-81, 327, 340, 373
uniform motion. *See* motion, uniform and non-uniform
unitarity, 22, 80, 135, 157, 164, 263, 264, 268, 308, 313, 321, 326, 485, 506
universe, contracting and expanding, 162
universes, complementary, 488-490
vacuum, 219n, 295, 480
Van der Waals' attraction, 108n
wave, gravity, 173-174
wave, phased and unphased, 101
wave, plane and spherical, 101
wave, polarized and unpolarized, 101
wave, retarded and advanced, 343-345, 346-347, 351, 353
wave, standing and traveling, 5, 86n, 94n, 100, 102, 174, 332, 333n, 334, 344, 485
wave, transverse and longitudinal, 86, 100
wave and particle, 34-36, 327
wave dualities, 100-102
wave function, 46-54, 57, 65, 68, 344, 497
wave mechanics and matrix mechanics, 58
wave-particle duality, 34-36, 45, 50, 56, 59, 69, 77, 79, 82
wavicle, 39, 80, 135
Western and Eastern, 24, 29, 60, 278, 423, 499, 500
whole. *See* parts and whole
wisdom, 115, 451, 458, 460, 465
women in science, 330
worm holes, 54n, 482
X-Force. *See* fundamental forces, complementary
X-view. *See* seeing
yes and no, 23, 289
yin-yang, 14, 21, 330, 484

Made in the USA
Middletown, DE
10 May 2021

39401354R10319